Predictive Control

for Linear and Hybrid Systems

Model Predictive Control (MPC), the dominant advanced control approach in industry over the past 25 years, is presented comprehensively in this unique book. With a simple, unified approach, and with attention to real-time implementation, it covers predictive control theory including the stability, feasibility, and robustness of MPC controllers. The theory of explicit MPC, where the nonlinear optimal feedback controller can be calculated efficiently, is presented in the context of linear systems with linear constraints, switched linear systems, and, more generally, linear hybrid systems. Drawing upon years of practical experience and using numerous examples and illustrative applications, the authors discuss:

- The techniques required to design predictive control laws, including algorithms for polyhedral manipulations, mathematical, and multiparametric programming.

- How to validate the theoretical properties and to implement predictive control policies.

The most important algorithms feature in an accompanying free online MATLAB toolbox, which allows easy access to sample solutions. Predictive Control for Linear and Hybrid Systems is an ideal reference for graduate, postgraduate and advanced control practitioners interested in theory and/or implementation aspects of predictive control.

Francesco Borrelli is a chaired professor at the Department of Mechanical Engineering of the University of California, Berkeley. Since 2004 he has served as a consultant for major international corporations in the area of real-time predictive control. He was the founder and CTO of BrightBox Technologies Inc., and is the co-director of the Hyundai Center of Excellence in Integrated Vehicle Safety Systems and Control at UC Berkeley. His research interests include constrained optimal control, model predictive control and its application to advanced automotive control, robotics, and energy-efficient building operation.

Alberto Bemporad is a professor and former director of the IMT School for Advanced Studies Lucca. He has published numerous papers on model predictive control and its application in multiple domains. He has been a consultant for major automotive companies and cofounder of ODYS S.r.l., a company specialized in advanced control and optimization software for industrial production. He is the author or co-author of various MATLAB toolboxes for model predictive control design, including the Model Predictive Control Toolbox and the Hybrid Toolbox.

Manfred Morari was a professor and head of the Department of Information Technology and Electrical Engineering at ETH Zurich. During the last three decades he has shaped many of the developments and applications of model predictive control through his academic research and interactions with companies from a wide range of sectors. The analysis techniques and software developed in his group are used throughout the world. He received numerous awards and was elected to the National Academy of Engineering (US) and is a Fellow of the Royal Academy of Engineering (UK).

Predictive Control
for Linear and Hybrid Systems

Francesco Borrelli

University of California, Berkeley

Alberto Bemporad

IMT School for Advanced Studies, Lucca

Manfred Morari

ETH Zurich

CAMBRIDGE
UNIVERSITY PRESS

University Printing House, Cambridge CB2 8BS, United Kingdom

One Liberty Plaza, 20th Floor, New York, NY 10006, USA

477 Williamstown Road, Port Melbourne, VIC 3207, Australia

314-321, 3rd Floor, Plot 3, Splendor Forum, Jasola District Centre, New Delhi - 110025, India

79 Anson Road, #06-04/06, Singapore 079906

Cambridge University Press is part of the University of Cambridge.

It furthers the University's mission by disseminating knowledge in the pursuit of education, learning and research at the highest international levels of excellence.

www.cambridge.org
Information on this title: www.cambridge.org/9781107652873
10.1017/9781139061759

First published 2017

A catalogue record for this publication is available from the British Library

Library of Congress Cataloging in Publication data
Names: Borrelli, Francesco, author. | Bemporad, Alberto, author. | Morari, Manfred, author.
Title: Predictive control for linear and hybrid systems / Francesco Borrelli, University of California, Berkeley, Alberto Bemporad, IMT Institute for Advanced Studies, Manfred Morari, Swiss Federal Institute of Technology (ETH).
Description: New York : Cambridge University Press, 2017. |
Includes bibliographical references. | Includes bibliographical references and index.
Identifiers: LCCN 2016042160| ISBN 9781107016880 (Hardback) |
ISBN 9781107652873 (Paperback)
Subjects: LCSH: Predictive control.
Classification: LCC TJ217.6 .B67 2017 | DDC 629.8–dc23 LC record available at https://lccn.loc.gov/2016042160

ISBN   978-1-107-01688-0   Hardback
ISBN   978-1-107-65287-3   Paperback

To

Maryan, Federica and Marina

and our families

Contents

V Constrained Optimal Control of Hybrid Systems   347

Preface

Dynamic optimization has become a standard tool for decision making in a wide range of areas. The search for the most fuel-efficient strategy to send a rocket into orbit or the most economical way to start up a chemical production facility can be expressed as dynamic optimization problems that are solved almost routinely nowadays.

The basis for these dynamic optimization problems is a dynamic model, for example,

$$x_{k+1} = g(x_k, u_k), \ x_0 = x(0)$$

that describes the evolution of the state x_k with time, starting from the initial condition $x(0)$, as it is affected by the manipulated input u_k. Here, $g(x, u)$ is some nonlinear function. Throughout the book we are assuming that this discrete-time description, i.e., the model of the underlying system, is available. The goal of the dynamic optimization procedure is to find the vector of manipulated inputs $U_N = [u'_0, ..., u'_{N-1}]'$ such that the objective function is optimized over some time horizon N, typically

$$\min_{U_N} \sum_{k=0}^{N-1} q(x_k, u_k) + p(x_N)$$

The terms $q(x, u)$ and and $p(x)$ are referred to as the *stage cost* and *terminal cost*, respectively. Many practical problems can be put into this form and many algorithms and software packages are available to determine the optimal solution vector U_N^*, the *optimizer*. The various algorithms exploit the structure of the particular problem, e.g., linearity and convexity, so that even large problems described by complex models and involving many degrees of freedom can be solved efficiently and reliably.

One difficulty with this idea is that, in practice, the sequence of $u_0, u_1, ...,$ which is obtained by this procedure cannot be simply applied. The model of the system predicting its evolution is usually inaccurate and the system may be affected by external disturbances that may cause its path to deviate significantly from the one that is predicted. Therefore, it is common practice to measure the state after some time period, say one time step, and to solve the dynamic optimization problem again, starting from the measured state $x(1)$ as the new initial condition. This *feedback* of the measurement information to the optimization endows the whole procedure with a *robustness* typical for closed-loop systems.

What we have described above is usually referred to as Model Predictive Control (MPC), but other names like Open Loop Optimal Feedback and Reactive Scheduling have been used as well. Over the last 25 years MPC has evolved to dominate the process industry, where it has been employed for thousands of problems [241].

The popularity of MPC stems from the fact that the resulting operating strategy respects all the system and problem details, including interactions and constraints, something that would be very hard to accomplish in any other way.

Indeed, often MPC is used for the regulatory control of large multivariable linear systems with constraints, where the objective function is not related to an economical objective, but is simply chosen in a mathematically convenient way, namely quadratic in the states and inputs, to yield a "good" closed-loop response. Again, there is no other controller design method available today for such systems that provides constraint satisfaction and stability guarantees.

One limitation of MPC is that running the optimization algorithm on-line at each time step requires substantial time and computational resources. Today, fast computational platforms together with advances in the field of operations research and optimal control have enlarged in a very significant way the scope of applicability of MPC to fast-sampled applications. One approach is to use tailored optimization routines which exploit both the structure of the MPC problem and the architecture of the embedded computing platform to implement MPC in the order of milliseconds.

The second approach is to have the result of the optimization precomputed and stored for each x in the form of a look-up table or as an algebraic function $u_k = f(x(k))$ which can be easily evaluated. In other words, we want to determine the (generally nonlinear) feedback control law $f(x)$ that generates the optimal $u_k = f(x(k))$ explicitly and not just implicitly as the result of an optimization problem. It requires the solution of the Bellman equation and has been a long-standing problem in optimal control. A clean, simple solution exists only in the case of linear systems with a quadratic objective function, where the optimal controller turns out to be a linear function of the state (Linear Quadratic Regulator, LQR). For all other cases a solution of the Bellman equation was considered prohibitive except for systems of low dimension (2 or 3), where a look-up table can be generated by gridding the state space and solving the optimization problem off-line for each grid point.

The major contribution of this book is to show how the nonlinear optimal feedback controller can be calculated efficiently for some important classes of systems, namely linear systems with constraints and switched linear systems or, more generally, hybrid systems. Traditionally, the design of feedback controllers for linear systems with constraints, for example, antiwindup techniques, was *ad hoc* requiring both much experience and trial and error. Though significant progress has been achieved on antiwindup schemes over the last decade, these techniques deal with input constraints only and cannot be extended easily.

The classes of constrained linear systems and linear hybrid systems treated in this book cover many, if not most, practical problems. The new design techniques hold the promise to lead to better performance and a dramatic reduction in the required engineering effort.

The book is structured in five parts.

- In the first part of the book (Part I) we recall the main concepts and results of convex and discrete optimization. Our intent is to provide only the necessary background for the understanding of the rest of the book. The material of this part follows closely the presentation from the following books and lecture notes: "Convex Optimization" by Boyd and Vandenberghe [65], "Nonlinear Programming Theory and Algorithms" by Bazaraa, Sherali and Shetty [27], "LMIs in Control" by Scherer and Weiland [258] and "Lectures on Polytopes" by Ziegler [296].

  Continuous problems as well as integer and mixed-integer problems are presented in Chapter 1. Chapter 1 also discusses the classical results of Lagrange duality. In Chapter 2, linear and quadratic programs are presented together with their properties and some fundamental results. Chapter 3 introduces algorithms for the solution of unconstrained and constrained optimization problems. We only discuss those that are important for the problems encountered in this book and explain the underlying concepts. Since polyhedra are the fundamental geometric objects used in this book, Part I closes with Chapter 4, where we introduce the main definitions and the algorithms, which describe standard operations on polyhedra.

- The second part of the book (Part II) is a self-contained introduction to multiparametric programming. In our framework, parametric programming is the main technique used to study and compute state feedback optimal control laws. In fact, we formulate the finite time optimal control problems as mathematical programs where the input sequence is the optimization vector. Depending on the dynamical model of the system, the nature of the constraints, and the cost function used, a different mathematical program is obtained. The current state of the dynamical system enters the cost function and the constraints as a parameter that affects the solution of the mathematical program. We study the structure of the solution as this parameter changes and we describe algorithms for solving multiparametric linear, quadratic and mixed integer programs. They constitute the basic tools for computing the state feedback optimal control laws for these more complex systems in the same way as algorithms for solving the Riccati equation are the main tools for computing optimal controllers for linear systems. In Chapter 5, we introduce the concept of multiparametric programming and we recall the main results of nonlinear multiparametric programming. Then, in Chapter 6, we describe three algorithms for solving multiparametric linear programs (mp-LP), multiparametric quadratic programs (mp-QP) and multiparametric mixed-integer linear programs (mp-MILP).

- In the third part of the book (Part III) we introduce the general class of optimal control problems studied in the book. Chapter 7 contains the

basic definitions and essential concepts. Chapter 8 presents standard results
on Linear Quadratic Optimal Control, while in Chapter 9, unconstrained
optimal control problems for linear systems with cost functions based on 1
and ∞ norms are analyzed.

- In the fourth part of the book (Part IV) we focus on linear systems
  with polyhedral constraints on inputs and states. We start with a self-
  contained introduction to controllability, reachability and invariant set
  theory in Chapter 10. The chapter focuses on computational algorithms for
  constrained linear systems and constrained linear systems subject to additive
  and parametric uncertainty.

  In Chapter 11 we study *finite time and infinite time* constrained optimal
  control problems with cost functions based on 2, 1 and ∞ norms. We first
  show how to transform them into LP or QP optimization problems for a
  fixed initial condition. Then we show that the solution to all these optimal
  control problems can be expressed as a *piecewise affine* state feedback law.
  Moreover, the optimal control law is continuous and the value function is
  convex and continuous. The results form a natural extension of the theory
  of the Linear Quadratic Regulator to constrained linear systems.

  Chapter 12 presents the concept of MPC. Classical feasibility and stability
  issues are shown through simple examples and explained by using invariant
  set methods. Finally, we show how they can be addressed with a proper
  choice of the terminal constraints and the cost function.

  The result in Chapter 11 and Chapter 12 have important consequences
  for the implementation of MPC laws. Precomputing off-line the explicit
  piecewise affine feedback policy reduces the on-line computation for the
  receding horizon control law to a function evaluation, therefore avoiding
  the on-line solution of a mathematical program. However, the number
  of polyhedral regions of the explicit optimal control laws could grow
  exponentially with the number of constraints in the optimal control problem.
  Chapter 13 discusses approaches to define approximate explicit control laws
  of desired complexity that provide certificates of recursive feasibility and
  stability.

  Chapter 14 focuses on efficient on-line methods for the computation of
  MPC control laws. If the state-feedback solution is available explicitly, we
  present efficient on-line methods for the evaluation of explicit piecewise
  affine control laws. In particular, we present algorithms to reduce its storage
  demands and computational complexity. If the on-line solution of a quadratic
  or linear program is preferred, we briefly discuss how to improve the efficiency
  of a mathematical programming solver by exploiting the structure of the
  MPC control problem.

  Part IV closes with Chapter 15 where we address the robustness of the
  optimal control laws. We discuss min–max control problems for uncertain
  linear systems with polyhedral constraints on inputs and states and present
  an approach to compute their state feedback solutions. Robustness is
  achieved against additive norm-bounded input disturbances and/or polyhedral parametric uncertainties in the state space matrices.

- In the fifth part of the book (Part V) we focus on linear hybrid systems. We give an introduction to the different formalisms used to model hybrid systems focusing on computation-oriented models (Chapter 16). In Chapter 17, we study finite time optimal control problems with cost functions based on 2, 1 and ∞ norms. The optimal control law is shown to be, in general, piecewise affine over nonconvex and disconnected sets. Along with the analysis of the solution properties, we present algorithms that compute the optimal control law for all the considered cases.

Francesco Borrelli
Alberto Bemporad
Manfred Morari

Acknowledgments

- Large parts of the material presented in this book are extracted from the work of the authors Francesco Borrelli, Alberto Bemporad and Manfred Morari.

- Several sections contain the results of the PhD theses of Miroslav Baric, Mato Baotic, Fabio Torrisi, Domenico Mignone, Pascal Grieder, Eric Kerrigan, Tobias Geyer and Frank J. Christophersen.

- We are extremely grateful to Martin Herceg for his meticulous work on the Matlab examples and all the book figures.

- A special thanks goes to Michal Kvasnica, Stefan Richter, Valerio Turri, Thomas Besselmann and Rick Meyer for their help with the construction of the examples.

- Our gratitude also goes to the colleagues who have carefully read preliminary versions of the book and gave us suggestions on how to improve them. They include Dimitri Bertsekas, Miroslav Fikar, Paul Goulart, Per Olof Gutman, Diethard Klatte, Bill Levine, David Mayne and all the students of our classes on Model Predictive Control.

- The authors of the LaTex book style files and macros are Stephen Boyd and Lieven Vandenberghe. We are most grateful to Stephen and Lieven for sharing them with us.

- Over many years ABB provided generous financial support making possible the fundamental research reported in this book.

- The authors of Chapter 3 are Dr. Alexander Domahidi and Dr. Stefan Richter.

- The author of Chapter 13 is Professor Colin N. Jones.

Symbols and Acronyms

Logic Operators and Functions

$A \Rightarrow B$	A implies B, i.e., if A is true then B must be true
$A \Leftrightarrow B$	A implies B and B implies A, i.e., A is true if and only if (iff) B is true

Sets

$\mathbb{R}$ ($\mathbb{R}_+$)	Set of (nonnegative) real numbers
$\mathbb{N}$ ($\mathbb{N}_+$)	Set of (nonnegative) integers
$\mathbb{R}^n$	Set of real vectors with n elements
$\mathbb{R}^{n \times m}$	Set of real matrices with n rows and m columns

Algebraic Operators and Matrices

A'	Transpose of matrix A
A^{-1}	Inverse of matrix A
$A^{\dagger}$	Generalized Inverse of A, $A^{\dagger} = (A'A)^{-1}A'$
$\det(A)$	Determinant of matrix A
$A \succ (\succeq)0$	A symmetric positive (semi)definite matrix, $x'Ax > (\geq)0, \forall x \neq 0$
$A \prec (\preceq)0$	A symmetric negative (semi)definite matrix, $x'Ax < (\leq)0, \forall x \neq 0$
A_i	i-th row of matrix A
x_i	i-th element of vector x
$x \in \mathbb{R}^n, \ x > 0 \ (x \geq 0)$	True iff $x_i > 0 \ (x_i \geq 0) \ \forall \ i = 1, \ldots, n$
$x \in \mathbb{R}^n, \ x < 0 \ (x \leq 0)$	True iff $x_i < 0 \ (x_i \leq 0) \ \forall \ i = 1, \ldots, n$
$|x|, \ x \in \mathbb{R}$	Absolute value of x
$\|x\|$	Any vector norm of x
$\|x\|_2$	Euclidian norm of vector $x \in \mathbb{R}^n$, $\|x\|_2 = \sqrt{\sum_{i=1}^{n} |x_i|^2}$

$\|x\|_1$	Sum of absolute elements of vector $x \in \mathbb{R}^n$, $\|x\|_1 = \sum_{i=1}^n |x_i|$
$\|x\|_\infty$	Largest absolute value of the vector $x \in \mathbb{R}^n$, $\|x\|_\infty = \max_{i \in \{1,\dots,n\}} |x_i|$
$\|S\|_\infty$	Matrix ∞-norm of $S \in \mathbb{C}^{m \times n}$, i.e., $\|S\|_\infty = \max_{i \in \{1,\dots,m\}} \sum_{j=1}^n |s_{i,j}|$
$\|S\|_1$	Matrix 1-norm of $S \in \mathbb{C}^{m \times n}$, i.e., $\|S\|_1 = \max_{j \in \{1,\dots,n\}} \sum_{i=1}^m |s_{i,j}|$
$\mathbf{I}$	Identity matrix
$\mathbf{1}$	Vector of ones, $\mathbf{1} = [1\ 1\ \dots\ 1]'$
$\mathbf{0}$	Vector of zeros, $\mathbf{0} = [0\ 0\ \dots\ 0]'$

Set Operators and Functions

$\emptyset$	The empty set
$:$	"Such that"
$\partial\mathcal{P}$	The boundary of $\mathcal{P}$
$\text{int}(\mathcal{P})$	The interior of $\mathcal{P}$, i.e., $\text{int}(\mathcal{P}) = \mathcal{P} \setminus \partial\mathcal{P}$
$|\mathcal{P}|$	The cardinality of $\mathcal{P}$, i.e., the number of elements in $\mathcal{P}$
$\mathcal{P} \cap \mathcal{Q}$	Set intersection $\mathcal{P} \cap \mathcal{Q} = \{x : x \in \mathcal{P} \text{ and } x \in \mathcal{Q}\}$
$\mathcal{P} \cup \mathcal{Q}$	Set union $\mathcal{P} \cup \mathcal{Q} = \{x : x \in \mathcal{P} \text{ or } x \in \mathcal{Q}\}$
$\bigcup_{r \in \{1,\dots,R\}} \mathcal{P}_r$	Union of R sets $\mathcal{P}_r$, i.e., $\bigcup_{r \in \{1,\dots,R\}} \mathcal{P}_r = \{x : x \in \mathcal{P}_0$ or $\dots$ or $x \in \mathcal{P}_R\}$
$\mathcal{P}^c$	Complement of the set $\mathcal{P}$, $\mathcal{P}^c = \{x : x \notin \mathcal{P}\}$
$\mathcal{P} \setminus \mathcal{Q}$	Set difference $\mathcal{P} \setminus \mathcal{Q} = \{x : x \in \mathcal{P} \text{ and } x \notin \mathcal{Q}\}$
$\mathcal{P} \subseteq \mathcal{Q}$	The set $\mathcal{P}$ is a subset of $\mathcal{Q}$, $x \in \mathcal{P} \Rightarrow x \in \mathcal{Q}$
$\mathcal{P} \subset \mathcal{Q}$	The set $\mathcal{P}$ is a strict subset of $\mathcal{Q}$, $x \in \mathcal{P} \Rightarrow x \in \mathcal{Q}$ and $\exists x \in (\mathcal{Q} \setminus \mathcal{P})$
$\mathcal{P} \supseteq \mathcal{Q}$	The set $\mathcal{P}$ is a superset of $\mathcal{Q}$
$\mathcal{P} \supset \mathcal{Q}$	The set $\mathcal{P}$ is a strict superset of $\mathcal{Q}$
$\mathcal{P} \ominus \mathcal{Q}$	Pontryagin difference $\mathcal{P} \ominus \mathcal{Q} = \{x : x + q \in \mathcal{P},\ \forall q \in \mathcal{Q}\}$
$\mathcal{P} \oplus \mathcal{Q}$	Minkowski sum $\mathcal{P} \oplus \mathcal{Q} = \{x + q : x \in \mathcal{P},\ q \in \mathcal{Q}\}$
$f(x)$	With abuse of notation denotes the value of the function f at x or the function f, $f : x \to f(x)$. We use the notation $f : \mathbb{R}^n \to \mathbb{R}^s$ to mean that f is a $\mathbb{R}^s$-valued function on some subset of $\mathbb{R}^n$, its domain, which we denote by $\text{dom } f$
$f(x)$ continuous and positive definite	$f : \mathbb{R}^n \to \mathbb{R}$ continuous for all $x \in \mathbb{R}^n$, $f(0) = 0$ and $f(x) > 0\ \forall x \in \mathbb{R}^n \setminus \{0\}$
$f(x) \succ 0$	$f(x)$ continuous and positive definite
$f(x)$ continuous and positive semi-definite	$f : \mathbb{R}^n \to \mathbb{R}$ continuous for all $x \in \mathbb{R}^n$, $f(x) \geq 0\ \forall x \in \mathbb{R}^n$
$f(x) \succeq 0$	$f(x)$ continuous and positive semi-definite

Acronyms

ARE	Algebraic Riccati Equation
CLQR	Constrained Linear Quadratic Regulator
CFTOC	Constrained Finite Time Optimal Control
CITOC	Constrained Infinite Time Optimal Control
DP	Dynamic Program(ming)
LMI	Linear Matrix Inequality
LP	Linear Program(ming)
LQR	Linear Quadratic Regulator
LTI	Linear Time Invariant
MILP	Mixed Integer Linear Program
MIQP	Mixed Integer Quadratic Program
MPC	Model Predictive Control
mp-LP	multiparametric Linear Program
mp-QP	multiparametric Quadratic Program
PWA	Piecewise Affine
PPWA	Piecewise Affine on Polyhedra
PWP	Piecewise Polynomial
PWQ	Piecewise Quadratic
QP	Quadratic Program(ming)
RHC	Receding Horizon Control
rhs	right-hand side
SDP	Semi Definite Program(ming)

Part I

Basics of Optimization

1

Main Concepts

In this chapter, we recall the main concepts and definitions of continuous and discrete optimization. Our intent is to provide only the necessary background for the understanding of the rest of the book. The notions of feasibility, optimality, convexity and active constraints introduced in this chapter will be widely used in this book.

1.1 Optimization Problems

An optimization problem is generally formulated as

$$\begin{aligned} \inf_z \quad & f(z) \\ \text{subj. to} \quad & z \in S \subseteq Z, \end{aligned} \tag{1.1}$$

where the vector z collects the decision variables, Z is the optimization problem *domain*, and $S \subseteq Z$ is the set of *feasible* or *admissible* decisions. The function $f \colon Z \to \mathbb{R}$ assigns to each decision z a *cost* $f(z) \in \mathbb{R}$. We will often use the following shorter form of problem (1.1)

$$\inf_{z \in S \subseteq Z} f(z). \tag{1.2}$$

Solving problem (1.2) means to compute the least possible cost f^*

$$f^* = \inf_{z \in S} f(z).$$

The number f^* is the *optimal value* of problem (1.2), i.e.,

$$f(z) \geq f(z^*) = f^* \; \forall z \in S, \text{ with } z^* \in S,$$

or the greatest lower bound of $f(z)$ over the set S:

$$f(z) > f^* \; \forall z \in S \text{ and } (\forall \varepsilon > 0 \; \exists z \in S \colon \; f(z) \leq f^* + \varepsilon).$$

If $f^* = -\infty$ we say that the problem is *unbounded below*. If the set S is empty then the problem is said to be *infeasible* and we set $f^* = +\infty$ by convention. If $S = Z$ the problem is said to be *unconstrained*.

In general, one is also interested in finding an *optimal solution*, that is in finding a decision whose associated cost equals the optimal value, i.e., $z^* \in S$ with $f(z^*) = f^*$. If such z^* exists, then we rewrite problem (1.2) as

$$f^* = \min_{z \in S} f(z) \tag{1.3}$$

and z^* is called an *optimizer*, *global optimizer* or *optimal solution*. *Minimizer* or *global minimizer* are also used to refer to an optimizer of a minimization problem. The set of all optimal solutions is denoted by

$$\mathrm{argmin}_{z \in S} f(z) = \{z \in S : f(z) = f^*\} .$$

A problem of determining whether the set of feasible decisions is empty and, if not, to find a point which is feasible, is called a *feasibility problem*.

1.1.1 Continuous Problems

In continuous optimization the problem domain Z is a subset of the finite-dimensional Euclidian vector-space $\mathbb{R}^s$ and the subset of admissible vectors is defined through a list of equality and inequality constraints:

$$
\begin{aligned}
\inf_z \quad & f(z) \\
\text{subj. to} \quad & g_i(z) \leq 0 \quad \text{for } i = 1, \ldots, m \\
& h_j(z) = 0 \quad \text{for } j = 1, \ldots, p \\
& z \in Z,
\end{aligned}
\tag{1.4}
$$

where $f, g_1, \ldots, g_m, h_1, \ldots, h_p$ are real-valued functions defined over $\mathbb{R}^s$, i.e., $f : \mathbb{R}^s \to \mathbb{R}$, $g_i : \mathbb{R}^s \to \mathbb{R}$, $h_i : \mathbb{R}^s \to \mathbb{R}$. The domain Z is the intersection of the domains of the cost and constraint functions:

$$Z = \{z \in \mathbb{R}^s \ : \ z \in \mathrm{dom}\, f, \ z \in \mathrm{dom}\, g_i, \ i = 1, \ldots, m, \ z \in \mathrm{dom}\, h_j, \ j = 1, \ldots, p\}. \tag{1.5}$$

In the sequel we will consider the constraint $z \in Z$ implicit in the optimization problem and often omit it. Problem (1.4) is unconstrained if $m = p = 0$.

The inequalities $g_i(z) \leq 0$ are called *inequality constraints* and the equations $h_i(z) = 0$ are called *equality constraints*. A point $\bar{z} \in \mathbb{R}^s$ is *feasible* for problem (1.4) if: (i) it belongs to Z, (ii) it satisfies all inequality and equality constraints, i.e., $g_i(\bar{z}) \leq 0$, $i = 1, \ldots, m$, $h_j(\bar{z}) = 0$, $i = j, \ldots, p$. The set of feasible vectors is

$$S = \{z \in \mathbb{R}^s : z \in Z, \ g_i(z) \leq 0, \ i = 1, \ldots, m, \ h_j(z) = 0, \ j = 1, \ldots, p\}. \tag{1.6}$$

Problem (1.4) is a continuous finite-dimensional optimization problem (since Z is a finite-dimensional Euclidian vector space). We will also refer to (1.4) as a *nonlinear mathematical program* or simply *nonlinear program*. Let f^* be the optimal value of problem (1.4). An optimizer, if it exists, is a feasible vector z^* with $f(z^*) = f^*$.

A feasible point $\bar{z}$ is *locally optimal* for problem (1.4) if there exists an $R > 0$ such that

$$
\begin{aligned}
f(\bar{z}) = \inf_z \quad & f(z) \\
\text{subj. to} \quad & g_i(z) \le 0 \quad \text{for } i = 1, \dots, m \\
& h_i(z) = 0 \quad \text{for } i = 1, \dots, p \\
& \|z - \bar{z}\| \le R \\
& z \in Z.
\end{aligned}
\tag{1.7}
$$

Roughly speaking, this means that $\bar{z}$ is the minimizer of $f(z)$ in a feasible neighborhood of $\bar{z}$ defined by $\|z - \bar{z}\| \le R$. The point $\bar{z}$ is called a *local optimizer* or *local minimizer*.

Active, Inactive and Redundant Constraints

Consider a feasible point $\bar{z}$. We say that the i-th inequality constraint $g_i(z) \le 0$ is *active* at $\bar{z}$ if $g_i(\bar{z}) = 0$. If $g_i(\bar{z}) < 0$ we say that the constraint $g_i(z) \le 0$ is *inactive* at $\bar{z}$. Equality constraints are always active for all feasible points.

We say that a constraint is *redundant* if removing it from the list of constraints does not change the feasible set S. This implies that removing a redundant constraint from problem (1.4) does not change its solution.

Problems in Standard Forms

Optimization problems can be cast in several forms. In this book, we use the form (1.4) where we adopt the convention to minimize the cost function and to have the right-hand side of the inequality and equality constraints equal to zero. Any problem in a different form (e.g., a maximization problem or a problem with "box constraints") can be transformed and arranged into this form. The interested reader is referred to Chapter 4 of [65] for a detailed discussion on transformations of optimization problems into different standard forms.

Eliminating Equality Constraints

Often in this book we will restrict our attention to problems without equality constraints, i.e., $p = 0$

$$
\begin{aligned}
\inf_z \quad & f(z) \\
\text{subj. to} \quad & g_i(z) \le 0 \quad \text{for } i = 1, \dots, m.
\end{aligned}
\tag{1.8}
$$

The simplest way to remove equality constraints is to replace them with two inequalities for each equality, i.e., $h_i(z) = 0$ is replaced by $h_i(z) \le 0$ and $-h_i(z) \le 0$. Such a method, however, can lead to poor numerical conditioning and may ruin the efficiency and accuracy of a numerical solver.

If one can explicitly parameterize the solution of the equality constraint $h_i(z) = 0$, then the equality constraint can be *eliminated* from the problem. This process can be described in a simple way for linear equality constraints. Assume the equality constraints to be linear, $Az - b = 0$, with $A \in \mathbb{R}^{p \times s}$. If $Az = b$ is inconsistent then the problem is infeasible. The general solution of the equation $Az = b$ can be expressed as $z = Fx + z_0$ where F is a matrix of full rank whose spanned space coincides with the null space of the A matrix, i.e., $\mathcal{R}(F) = \mathcal{N}(A)$,

$F \in \mathbb{R}^{s \times k}$, where k is the dimension of the null space of A. The variable $x \in \mathbb{R}^k$ is the new optimization variable and the original problem becomes

$$
\begin{aligned}
\inf_x \quad & f(Fx + z_0) \\
\text{subj. to} \quad & g_i(Fx + z_0) \leq 0 \quad \text{for } i = 1, \ldots, m.
\end{aligned}
\tag{1.9}
$$

We want to point out that in some cases the elimination of equality constraints can make the problem harder to analyze and understand and can make a solver less efficient. In large problems it can destroy useful structural properties of the problem such as sparsity. Some advanced numerical solvers perform elimination automatically.

Problem Description

The functions f, g_i and h_i can be available in analytical form or can be described through an *oracle model* (also called "black box" or "subroutine" model). In an oracle model, f, g_i and h_i are not known explicitly but can be evaluated by querying the oracle. Often the oracle consists of subroutines which, called with the argument z, return $f(z)$, $g_i(z)$ and $h_i(z)$ and their gradients $\nabla f(z), \nabla g_i(z), \nabla h_i(z)$. In the rest of the book we assume that analytical expressions of the cost and the constraints of the optimization problem are available.

1.1.2 Integer and Mixed-Integer Problems

If the decision set Z in the optimization problem (1.2) is finite, then the optimization problem is called *combinatorial* or *discrete*. If $Z \subseteq \{0,1\}^s$, then the problem is said to be *integer*.

If Z is a subset of the Cartesian product of an integer set and a real Euclidian space, i.e., $Z \subseteq \{[z_c, z_b] : z_c \in \mathbb{R}^{s_c}, z_b \in \{0,1\}^{s_b}\}$, then the problem is said to be *mixed-integer*. The standard formulation of a *mixed-integer nonlinear program* is

$$
\begin{aligned}
\inf_{[z_c, z_b]} \quad & f(z_c, z_b) \\
\text{subj. to} \quad & g_i(z_c, z_b) \leq 0 \quad \text{for } i = 1, \ldots, m \\
& h_j(z_c, z_b) = 0 \quad \text{for } j = 1, \ldots, p \\
& z_c \in \mathbb{R}^{s_c}, \ z_b \in \{0,1\}^{s_b}
\end{aligned}
\tag{1.10}
$$

where $f, g_1, \ldots, g_m, h_1, \ldots, h_p$ are real-valued functions defined over Z.

For combinatorial, integer and mixed-integer optimization problems, all definitions introduced in the previous section apply.

1.2 Convexity

A set $S \in \mathbb{R}^s$ is *convex* if

$$
\lambda z_1 + (1 - \lambda) z_2 \in S \text{ for all } z_1 \in S, z_2 \in S \text{ and } \lambda \in [0, 1].
$$

A function $f : S \to \mathbb{R}$ is convex if S is convex and

$$f(\lambda z_1 + (1 - \lambda)z_2) \le \lambda f(z_1) + (1 - \lambda)f(z_2)$$
$$\text{for all } z_1 \in S, z_2 \in S \text{ and } \lambda \in [0, 1].$$

A function $f : S \to \mathbb{R}$ is *strictly convex* if S is convex and

$$f(\lambda z_1 + (1 - \lambda)z_2) < \lambda f(z_1) + (1 - \lambda)f(z_2)$$
$$\text{for all } z_1 \in S, z_2 \in S \text{ and } \lambda \in (0, 1).$$

A twice differentiable function $f : S \to \mathbb{R}$ is *strongly convex* if the Hessian

$$\nabla^2 f(z) \succ 0 \text{ for all } z \in S.$$

A function $f : S \to \mathbb{R}$ is concave if S is convex and $-f$ is convex.

Operations Preserving Convexity

Various operations preserve convexity of functions and sets. A detailed list can be found in Chapter 3.2 of [65]. A few operations used in this book are mentioned below.

1. The intersection of an arbitrary number of convex sets is a convex set:

   if $S_1, S_2, \ldots, S_k$ are convex, then $S_1 \cap S_2 \cap \ldots \cap S_k$ is convex.

   This property extends to the intersection of an infinite number of sets:

   $$\text{if } S_n \text{ is convex } \forall n \in \mathbb{N}_+ \text{ then } \bigcap_{n \in \mathbb{N}_+} S_n \text{ is convex.}$$

   The empty set is convex because it satisfies the definition of convexity.

2. The sublevel sets of a convex function f on S are convex:

   if $f(z)$ is convex then $S_\alpha = \{z \in S : f(z) \le \alpha\}$ is convex $\forall \alpha \in \mathbb{R}$.

3. If $f_1, \ldots, f_N$ are convex functions, then $\sum_{i=1}^{N} \alpha_i f_i$ is a convex function for all $\alpha_i \ge 0$, $i = 1, \ldots, N$.

4. The composition of a convex function $f(z)$ with an affine map $z = Ax + b$ generates a convex function $f(Ax + b)$ of x:

   if $f(z)$ is convex then $f(Ax + b)$ is convex on $\{x : Ax + b \in \text{dom}(f)\}$.

5. Suppose $f(x) = h(g(x)) = h(g_1(x), \ldots, g_k(x))$ with $h : \mathbb{R}^k \to R$, $g_i : \mathbb{R}^s \to R$. Then,

   (a) f is convex if h is convex, h is nondecreasing in each argument, and g_i are convex,

(b) f is convex if h is convex, h is nonincreasing in each argument, and g_i are concave,

(c) f is concave if h is concave, h is nondecreasing in each argument, and g_i are concave.

6. The pointwise maximum of a set of convex functions is a convex function:

$$f_1(z), \ldots, f_k(z) \text{ convex functions} \Rightarrow f(z)$$
$$= \max\{f_1(z), \ldots, f_k(z)\} \text{ is a convex function.}$$

This property holds also when the set is infinite.

Linear and Quadratic Convex Functions

1. A linear function $f(z) = c'z + r$ is both convex and concave.

2. A quadratic function $f(z) = z'Hz + 2q'z + r$ is convex if and only if $H \succeq 0$.

3. A quadratic function $f(z) = z'Hz + 2q'z + r$ is strictly convex if and only if $H \succ 0$. A strictly convex quadratic function is also strongly convex.

Convex Optimization Problems

The standard optimization problem (1.4) is said to be *convex* if the cost function f is convex on Z and S is a convex set. A fundamental property of convex optimization problems is that local optimizers are also global optimizers. This is proven next.

Theorem 1.1 *Consider a convex optimization problem and let $\bar{z}$ be a local optimizer. Then, $\bar{z}$ is a global optimizer.*

Proof: By hypothesis $\bar{z}$ is feasible and there exists R such that

$$f(\bar{z}) = \min\{f(z) \, : \, g_i(z) \leq 0 \; i = 1, \ldots, m, \; h_j(z) = 0, \; j = 1, \ldots, p \; \|z - \bar{z}\| \leq R\}. \tag{1.11}$$

Now suppose that $\bar{z}$ is not globally optimal. Then, there exist a feasible y such that $f(y) < f(\bar{z})$, which implies that $\|y - \bar{z}\| > R$. Now consider the point z given by

$$z = (1 - \theta)\bar{z} + \theta y, \quad \theta = \frac{R}{2\|y - \bar{z}\|}.$$

Then $\|z - \bar{z}\| = R/2 < R$ and by convexity of the feasible set z is feasible. By convexity of the cost function f

$$f(z) \leq (1 - \theta)f(\bar{z}) + \theta f(y) < f(\bar{z}),$$

which contradicts (1.11).   ∎

Theorem 1.1 does not make any statement about the existence of a solution to problem (1.4). It merely states that all local minima of problem (1.4) are also global minima. For this reason, convexity plays a central role in the solution of continuous optimization problems. It suffices to compute a local minimum to problem (1.4) to determine its global minimum. Convexity also plays a major role in most non-convex optimization problems which are solved by iterating between the solutions of convex subproblems.

It is difficult to determine whether the feasible set S of the optimization problem (1.4) is convex or not except in special cases. For example, if the functions $g_1(z), \ldots, g_m(z)$ are convex and all the $h_i(z)$ (if any) are affine in z, then the feasible set S in (1.6) is an intersection of convex sets and is therefore convex. Moreover there are nonconvex problems which can be transformed into convex problems through a change of variables and manipulations of cost and constraints. The discussion of this topic goes beyond the scope of this overview on optimization. The interested reader is referred to [65].

Remark 1.1 With the exception of trivial cases, integer and mixed-integer optimization problems are always nonconvex problems because $\{0, 1\}$ is not a convex set.

1.3 Optimality Conditions

In general, an analytical solution to problem (1.4), restated below, does not exist.

$$
\begin{aligned}
\inf_z \quad & f(z) \\
\text{subj. to} \quad & g_i(z) \leq 0 && \text{for } i = 1, \ldots, m \\
& h_j(z) = 0 && \text{for } j = 1, \ldots, p \\
& z \in Z.
\end{aligned}
\tag{1.12}
$$

Solutions are usually computed by iterative algorithms which start from an initial guess z_0 and at step k generate a point z_k such that the sequence $\{f(z_k)\}_{k=0,1,2,\ldots}$ converges to f^* as k increases. These algorithms iteratively use and/or solve conditions for optimality, i.e., analytical conditions that a point z must satisfy in order to be an optimizer. For instance, for *convex, unconstrained* optimization problems with a *smooth* cost function the most commonly used optimality criterion requires the gradient to vanish at the optimizer, i.e., z is an optimizer if and only if $\nabla f(z) = 0$. In this chapter we summarize necessary and sufficient optimality conditions for unconstrained and constrained optimization problems.

1.3.1 Optimality Conditions for Unconstrained Problems

The proofs of the theorems presented next can be found in Chapter 4 and Section 8.6.1 of [27].

Necessary Conditions

> **Theorem 1.2** *Suppose that $f : \mathbb{R}^s \to \mathbb{R}$ is differentiable at $\bar{z}$. If there exists a vector d such that $\nabla f(\bar{z})'d < 0$, then there exists a $\delta > 0$ such that $f(\bar{z} + \lambda d) < f(\bar{z})$ for all $\lambda \in (0, \delta)$.*

The vector d in the theorem above is called a *descent direction*. At a given point $\bar{z}$ a descent direction d satisfies the condition $\nabla f(\bar{z})'d < 0$. Theorem 1.2 states that if a descent direction exists at a point $\bar{z}$, then it is possible to move from $\bar{z}$ towards a new point $\tilde{z}$ whose associated cost $f(\tilde{z})$ is lower than $f(\bar{z})$. The direction of *steepest descent* d_s at a given point $\bar{z}$ is defined as the normalized direction where $\nabla f(\bar{z})'d_s < 0$ is minimized. The direction d_s of steepest descent is $d_s = -\frac{\nabla f(\bar{z})}{\|\nabla f(\bar{z})\|}$.

Two corollaries of Theorem 1.2 are stated next.

Corollary 1.1   *Suppose that $f : \mathbb{R}^s \to \mathbb{R}$ is differentiable at $\bar{z}$. If $\bar{z}$ is a local minimizer, then $\nabla f(\bar{z}) = 0$.*

Corollary 1.2   *Suppose that $f : \mathbb{R}^s \to \mathbb{R}$ is twice differentiable at $\bar{z}$. If $\bar{z}$ is a local minimizer, then $\nabla f(\bar{z}) = 0$ and the Hessian $\nabla^2 f(\bar{z})$ is positive semidefinite.*

Sufficient Condition

Theorem 1.3   *Suppose that $f : \mathbb{R}^s \to \mathbb{R}$ is twice differentiable at $\bar{z}$. If $\nabla f(\bar{z}) = 0$ and the Hessian of $f(z)$ at $\bar{z}$ is positive definite, then $\bar{z}$ is a local minimizer.*

Necessary and Sufficient Condition

Theorem 1.4   *Suppose that $f : \mathbb{R}^s \to \mathbb{R}$ is differentiable at $\bar{z}$. If f is convex, then $\bar{z}$ is a global minimizer if and only if $\nabla f(\bar{z}) = 0$.*

When the optimization is constrained and the cost function is not sufficiently smooth, the conditions for optimality become more complicated. The intent of this chapter is to give an overview of some important optimality criteria for constrained nonlinear optimization. The optimality conditions derived here will be the main building blocks for the theory developed later in this book.

1.4 Lagrange Duality Theory

Consider the nonlinear program (1.12). Let f^* be the optimal value. Denote by Z the domain of cost and constraints (1.5). Any feasible point $\bar{z}$ provides an upper bound to the optimal value $f(\bar{z}) \geq f^*$. Next, we will show how to generate a lower bound on f^*.

Starting from the standard nonlinear program (1.12) we construct another problem with different variables and constraints. The original problem (1.12) will be called the primal problem while the new one will be called the *dual* problem. First, we augment the objective function with a weighted sum of the constraints. In this way the *Lagrange dual function* (or Lagrangian) L is obtained

$$
\begin{aligned}
L(z, u, v) = f(z) &+ u_1 g_1(z) + \cdots + u_m g_m(z) + \\
&+ v_1 h_1(z) + \cdots + v_p h_p(z),
\end{aligned} \tag{1.13}
$$

where the scalars $u_1, \ldots, u_m, v_1, \ldots, v_p$ are real variables called *dual variables* or *Lagrange multipliers*. We can write Equation (1.13) in the compact form

$$L(z, u, v) = f(z) + u'g(z) + v'h(z), \tag{1.14}$$

where $u = [u_1, \ldots, u_m]'$, $v = [v_1, \ldots, v_p]'$ and $L : \mathbb{R}^s \times \mathbb{R}^m \times \mathbb{R}^p \to \mathbb{R}$. The components u_i and v_i are called dual variables. Note that the i-th dual variable u_i is associated with the i-th inequality constraint of problem (1.12), the i-th dual variable v_i is associated with the i-th equality constraint of problem (1.12).

Let z be a feasible point: for arbitrary vectors $u \geq 0$ and v we trivially obtain a lower bound on $f(z)$

$$L(z, u, v) \leq f(z). \tag{1.15}$$

We minimize both sides of Equation (1.15)

$$\inf_{z \in Z, \ g(z) \leq 0, \ h(z) = 0} L(z, u, v) \leq \inf_{z \in Z, \ g(z) \leq 0, \ h(z) = 0} f(z) \tag{1.16}$$

in order to reconstruct the original problem on the right-hand side of the expression. Since for arbitrary $u \geq 0$ and v

$$\inf_{z \in Z} L(z, u, v) \leq \inf_{z \in Z, \ g(z) \leq 0, \ h(z) = 0} L(z, u, v), \tag{1.17}$$

we obtain

$$\inf_{z \in Z} L(z, u, v) \leq \inf_{z \in Z, \ g(z) \leq 0, \ h(z) = 0} f(z). \tag{1.18}$$

Equation (1.18) implies that for arbitrary $u \geq 0$ and v the solution to

$$\inf_{z \in Z} L(z, u, v) \tag{1.19}$$

provides us with a lower bound to the original problem. The "best" lower bound is obtained by maximizing problem (1.19) over the dual variables

$$\sup_{(u,v), \ u \geq 0} \inf_{z \in Z} L(z, u, v) \leq \inf_{z \in Z, \ g(z) \leq 0, \ h(z) = 0} f(z).$$

Define the dual cost $d(u, v)$ as follows

$$d(u, v) = \inf_{z \in Z} L(z, u, v) \in [-\infty, +\infty]. \tag{1.20}$$

Then the Lagrange dual problem is defined as

$$\sup_{(u,v), \ u \geq 0} d(u, v), \tag{1.21}$$

and its optimal solution, if it exists, is denoted by (u^*, v^*). The dual cost $d(u, v)$ is the optimal value of an *unconstrained optimization problem*. Problem (1.20) is called the *Lagrange dual subproblem*. Only points (u, v) with $d(u, v) > -\infty$ are interesting for the Lagrange dual problem. A point (u, v) will be called *dual feasible* if $u \geq 0$ and $d(u, v) > -\infty$. $d(u, v)$ is *always a concave function* since it is the pointwise infimum of a family of affine functions of (u, v). This implies that the *dual problem* *is a convex optimization problem (max of a concave function over a convex set) even if the original problem is not convex.* Therefore, it is easier in principle to solve the dual problem than the primal (which is in general nonconvex). However, in general, the solution to the dual problem is only a lower bound of the primal problem:

$$\sup_{(u,v),\ u\geq 0} d(u,v) \leq \inf_{z\in Z,\ g(z)\leq 0,\ h(z)=0} f(z).$$

Such a property is called *weak duality*. In a simpler form, let f^* and d^* be the primal and dual optimal value, respectively,

$$f^* = \inf_{z\in Z,\ g(z)\leq 0,\ h(z)=0} f(z), \tag{1.22a}$$

$$d^* = \sup_{(u,v),\ u\geq 0} d(u,v), \tag{1.22b}$$

then, we always have

$$f^* \geq d^* \tag{1.23}$$

and the difference $f^* - d^*$ is called the *optimal duality gap*. The weak duality inequality (1.23) holds also when d^* and f^* are infinite. For example, if the primal problem is unbounded below, so that $f^* = -\infty$, we must have $d^* = -\infty$, i.e., the Lagrange dual problem is infeasible. Conversely, if the dual problem is unbounded above, so that $d^* = +\infty$, we must have $f^* = +\infty$, i.e., the primal problem is infeasible.

Remark 1.2   We have shown that the dual problem is a convex optimization problem even if the original problem is nonconvex. As stated in this section, for nonconvex optimization problems it is "easier" to solve the dual problem than the primal problem. However the evaluation of $d(\bar{u},\bar{v})$ at a point $(\bar{u},\bar{v})$ requires the solution of the nonconvex unconstrained optimization problem (1.20), which, in general, is "not easy."

1.4.1 Strong Duality and Constraint Qualifications

If $d^* = f^*$, then the duality gap is zero and we say that **strong duality** holds:

$$\sup_{(u,v),\ u\geq 0} d(u,v) = \inf_{z\in Z,\ g(z)\leq 0,\ h(z)=0} f(z). \tag{1.24}$$

This means that the best lower bound obtained by solving the dual problem coincides with the optimal cost of the primal problem. In general, strong duality does not hold, even for convex primal problems. Constraint qualifications are conditions on the constraint functions which imply strong duality for convex problems. A detailed discussion on constraint qualifications can be found in Chapter 5 of [27].

A well-known simple constraint qualification is "Slater's condition":

Definition 1.1 (Slater's condition)   *Consider problem (1.12). There exists $\hat{z} \in \mathbb{R}^s$ which belongs to the relative interior of the problem domain Z, which is feasible ($g(\hat{z}) \leq 0$, $h(\hat{z}) = 0$) and for which $g_j(\hat{z}) < 0$ for all j for which g_j is not an affine function.*

Remark 1.3   Note that Slater's condition reduces to feasibility when all inequality constraints are linear and $Z = \mathbb{R}^n$.

Theorem 1.5 (Slater's theorem)   *Consider the primal problem (1.22a) and its dual problem (1.22b). If the primal problem is convex, Slater's condition holds and f^* is bounded then $d^* = f^*$.*

1.4.2   Certificate of Optimality

Consider the (primal) optimization problem (1.12) and its dual (1.21). Any feasible point z gives us information about an upper bound on the cost, i.e., $f^* \leq f(z)$. If we can find a dual feasible point (u, v) then we can establish a lower bound on the optimal value of the primal problem: $d(u, v) \leq f^*$. In summary, without knowing the exact value of f^* we can give a bound on how suboptimal a given feasible point is. In fact, if z is primal feasible and (u, v) is dual feasible then $d(u, v) \leq f^* \leq f(z)$. Therefore z is ε-suboptimal, with ε equal to the primal-dual gap, i.e., $\varepsilon = f(z) - d(u, v)$.

The optimal value of the primal (and dual) problems will lie in the same interval

$$f^* \in [d(u, v), f(z)] \text{ and } d^* \in [d(u, v), f(z)].$$

For this reason (u, v) is also called a *certificate* that proves the (sub)optimality of z. Optimization algorithms make extensive use of such criteria. Primal-dual algorithms iteratively solve primal and dual problems and generate a sequence of primal and dual feasible points z_k, (u_k, v_k), $k > 0$ until a certain ε is reached. The condition

$$f(z_k) - d(u_k, v_k) < \varepsilon,$$

for terminating the algorithm guarantees that when the algorithm terminates, z_k is ε-suboptimal. If strong duality holds the condition can be met for arbitrarily small tolerances ε.

1.5   Complementary Slackness

Consider the (primal) optimization problem (1.12) and its dual (1.21). Assume that strong duality holds. Suppose that z^* and (u^*, v^*) are primal and dual feasible with zero duality gap (hence, they are primal and dual optimal):

$$f(z^*) = d(u^*, v^*).$$

By definition of the dual problem, we have

$$f(z^*) = \inf_{z \in Z} \left(f(z) + u^{*\prime} g(z) + v^{*\prime} h(z) \right).$$

Therefore

$$f(z^*) \leq f(z^*) + u^{*\prime} g(z^*) + v^{*\prime} h(z^*), \tag{1.25}$$

and since $h(z^*) = 0$, $u^* \geq 0$ and $g(z^*) \leq 0$ we have

$$f(z^*) \leq f(z^*) + u^{*\prime}g(z^*) \leq f(z^*). \tag{1.26}$$

From the last equation we can conclude that $u^{*\prime}g(z^*) = \sum_{i=1}^{m} u_i^* g_i(z^*) = 0$ and since $u_i^* \geq 0$ and $g_i(z^*) \leq 0$, we have

$$u_i^* g_i(z^*) = 0, \ i = 1, \ldots, m. \tag{1.27}$$

Conditions (1.27) are called *complementary slackness* conditions. Complementary slackness conditions can be interpreted as follows. If the i-th inequality constraint of the primal problem is inactive at the optimum $(g_i(z^*) < 0)$, then the i-th dual optimizer has to be zero $(u_i^* = 0)$. Vice versa, if the i-th dual optimizer is different from zero $(u_i^* > 0)$, then the i-th constraint is active at the optimum $(g_i(z^*) = 0)$.

Relation (1.27) implies that the inequality in (1.25) holds as equality

$$f(z^*) + \sum_i u_i^* g_i(z^*) + \sum_j v_j^* h_j(z^*) = \min_{z \in Z} \left(f(z) + \sum_i u_i^* g_i(z) + \sum_j v_j^* h_j(z) \right). \tag{1.28}$$

Therefore, complementary slackness conditions implies that z^* is a minimizer of $L(z, u^*, v^*)$.

1.6 Karush-Kuhn-Tucker Conditions

Consider the (primal) optimization problem (1.12) and its dual (1.21). Assume that strong duality holds. Assume that the cost functions and constraint functions f, g_i, h_i are differentiable. Let z^* and (u^*, v^*) be primal and dual optimal points, respectively. Complementary slackness conditions implies that z^* minimizes $L(z, u^*, v^*)$ under no constraints (Equation (1.28)). Since f, g_i, h_i are differentiable, the gradient of $L(z, u^*, v^*)$ must be zero at z^*

$$\nabla f(z^*) + \sum_i u_i^* \nabla g_i(z^*) + \sum_j v_j^* \nabla h_j(z^*) = 0.$$

In summary, the primal and dual optimal pair z^*, (u^*, v^*) of an optimization problem with differentiable cost and constraints and zero duality gap, have to satisfy the following conditions:

$$\nabla f(z^*) + \sum_{i=1}^{m} u_i^* \nabla g_i(z^*) + \sum_{j=1}^{p} v_j^* \nabla h_i(z^*) = 0, \tag{1.29a}$$

$$u_i^* g_i(z^*) = 0, \quad i = 1, \ldots, m \tag{1.29b}$$

$$u_i^* \geq 0, \quad i = 1, \ldots, m \tag{1.29c}$$

$$g_i(z^*) \leq 0, \quad i = 1, \ldots, m \tag{1.29d}$$

$$h_j(z^*) = 0, \quad j = 1, \ldots, p \tag{1.29e}$$

where Equations (1.29d)–(1.29e) are the primal feasibility conditions, Equation (1.29c) is the dual feasibility condition and Equation (1.29b) are the complementary slackness conditions.

Conditions (1.29a)–(1.29e) are called the *Karush-Kuhn-Tucker* (KKT) conditions. We have shown that the KKT conditions are necessary conditions for any primal-dual optimal pair if strong duality holds and the cost and constraints are differentiable, i.e., any primal and dual optimal points z^*, $(u^*,\ v^*)$ must satisfy the KKT conditions (1.29). If the primal problem is also convex then the KKT conditions are sufficient, i.e., a primal dual pair z^*, $(u^*,\ v^*)$ which satisfies conditions (1.29a)–(1.29e) is a primal dual optimal pair with zero duality gap.

There are several theorems which characterize primal and dual optimal points z^* and $(u^*,\ v^*)$ by using KKT conditions. They mainly differ on the type of constraint qualification chosen for characterizing strong duality. Next we report just two examples.

If a convex optimization problem with differentiable objective and constraint functions satisfies Slater's condition, then the KKT conditions provide necessary and sufficient conditions for optimality.

Theorem 1.6   *[27, p. 244] Consider problem (1.12) and let Z be a nonempty set of $\mathbb{R}^s$. Suppose that problem (1.12) is convex and that cost and constraints f, g_i and h_i are differentiable at a feasible z^*. If problem (1.12) satisfies Slater's condition then z^* is optimal if and only if there are $(u^*,\ v^*)$ that, together with z^*, satisfy the KKT conditions (1.29).*

If a convex optimization problem with differentiable objective and constraint functions has a linearly independent set of active constraints gradients, then the KKT conditions provide necessary and sufficient conditions for optimality.

Theorem 1.7 (Section 4.3.7 in [27])   *Consider problem (1.12) and let Z be a nonempty open set of $\mathbb{R}^s$. Let z^* be a feasible solution and $A = \{i\colon g_i(z^*) = 0\}$ be the set of active constraints at z^*. Suppose cost and constraints f, g_i are differentiable at z^* for all i and that h_j are continuously differentiable at z^* for all j. Further, suppose that $\nabla g_i(z^*)$ for $i \in A$ and $\nabla h_j(z^*)$ for $j - 1,\ldots,p$, are linearly independent. If z^*, $(u^*,\ v^*)$ are primal and dual optimal points, then they satisfy the KKT conditions (1.29). In addition, if problem (1.12) is convex, then z^* is optimal if and only if there are $(u^*,\ v^*)$ that, together with z^*, satisfy the KKT conditions (1.29).*

The KKT conditions play an important role in optimization. In a few special cases, it is possible to solve the KKT conditions (and therefore, the optimization problem) analytically. Many algorithms for convex optimization are conceived as, or can be interpreted as, methods for solving the KKT conditions as Boyd and Vandenberghe observe in [65].

The following example [27] shows a convex problem where the KKT conditions are not fulfilled at the optimum. In particular, both the constraint qualifications of Theorem 1.7 and Slater's condition in Theorem 1.6 are violated.

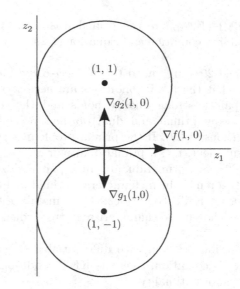

Figure 1.1 Example 1.1. Constraints, feasible set and gradients. The feasible set is the point $(1,0)$ that satisfies neither Slater's condition nor the constraint qualification condition in Theorem 1.7.

Example 1.1   [27, p. 196] Consider the convex optimization problem

$$\min \quad z_1$$

$$\text{subj. to} \quad (z_1 - 1)^2 + (z_2 - 1)^2 \leq 1 \\ (z_1 - 1)^2 + (z_2 + 1)^2 \leq 1. \qquad (1.30)$$

From the graphical interpretation in Figure 1.1 it is immediate that the feasible set is a single point $\bar{z} = [1,0]'$. The optimization problem does not satisfy Slater's conditions and moreover $\bar{z}$ does not satisfy the constraint qualifications in Theorem 1.7. At the optimum $\bar{z}$ Equation (1.29a) cannot be satisfied for any pair of nonnegative real numbers u_1 and u_2.

1.6.1   Geometric Interpretation of KKT Conditions

A geometric interpretation of the KKT conditions is depicted in Figure 1.2 for an optimization problem in two dimensions with inequality constraints and no equality constraints.

Equation (1.29a) and equation (1.29c) can be rewritten as

$$-\nabla f(z) = \sum_{i \in A} u_i \nabla g_i(z), \quad u_i \geq 0, \qquad (1.31)$$

where $A = \{1,2\}$ at z_1 and $A = \{2,3\}$ at z_2. This means that the negative gradient of the cost at the optimum $-\nabla f(z^*)$ (which represents the direction of steepest descent) has to belong to the cone spanned by the gradients of the active constraints

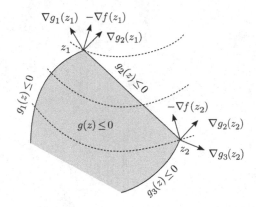

Figure 1.2 Geometric interpretation of KKT conditions [27].

∇g_i (since inactive constraints have the corresponding Lagrange multipliers equal to zero). In Figure 1.2, condition (1.31) is not satisfied at z_2. In fact, one can move within the set of feasible points $g(z) \leq 0$ and decrease f, which implies that z_2 is not optimal. At point z_1, on the other hand, the cost f can only decrease if some constraint is violated. Any movement in a feasible direction increases the cost. Conditions (1.31) are fulfilled and hence z_1 is optimal.

2

Linear and Quadratic Optimization

This chapter focuses on two widely known and used subclasses of convex optimization problems: linear and quadratic programs. They are popular because many important practical problems can be formulated as linear or quadratic programs and because they can be solved efficiently. Also, they are the basic building blocks of many other optimization algorithms. This chapter presents their formulation together with their main properties and some fundamental results.

2.1 Polyhedra and Polytopes

We first introduce a few concepts needed for the geometric interpretation of linear an quadratic optimization. They will be discussed in more detail in Section 4.2.

A *polyhedron* $\mathcal{P}$ in $\mathbb{R}^n$ denotes an intersection of a finite set of closed halfspaces in $\mathbb{R}^n$:

$$\mathcal{P} = \{x \in \mathbb{R}^n : \ Ax \leq b\}, \tag{2.1}$$

where $Ax \leq b$ is the usual shorthand form for a system of inequalities, namely $a_i'x \leq b_i$, $i = 1, \ldots, m$, where $a_1', \ldots, a_m'$ are the rows of A, and $b_1, \ldots, b_m$ are the components of b. A *polytope* is a bounded polyhedron. In Figure 2.1 a two-dimensional polytope is plotted.

A linear inequality $c'z \leq c_0$ is said to be *valid* for $\mathcal{P}$ if it is satisfied for all points $z \in \mathcal{P}$. A *face* of $\mathcal{P}$ is any nonempty set of the form

$$\mathcal{F} = \mathcal{P} \cap \{z \in \mathbb{R}^s : \ c'z = c_0\}, \tag{2.2}$$

where $c'z \leq c_0$ is a *valid* inequality for $\mathcal{P}$. All faces of $\mathcal{P}$ satisfying $\mathcal{F} \subset \mathcal{P}$ are called proper faces and have dimension less than $\dim(\mathcal{P})$. The faces of dimension 0,1, $\dim(\mathcal{P})$-2 and $\dim(\mathcal{P})$-1 are called *vertices*, *edges*, *ridges*, and *facets*, respectively.

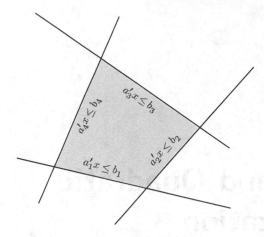

Figure 2.1 Polytope. A polytope is a bounded polyhedron defined by the intersection of closed halfspaces. The planes (here lines) defining the boundary of the halfspaces are $a_i'x - b_i = 0$.

2.2 Linear Programming

When the cost and the constraints of the continuous optimization problem (1.4) are affine, then the problem is called a *linear program* (LP). The most general form of a linear program is

$$\begin{aligned}
\inf_z \quad & c'z \\
\text{subj. to} \quad & Gz \leq w,
\end{aligned} \tag{2.3}$$

where $G \in \mathbb{R}^{m \times s}$, $w \in \mathbb{R}^m$. Linear programs are convex optimization problems.

Two other common forms of linear programs include both equality and inequality constraints:

$$\begin{aligned}
\inf_z \quad & c'z \\
\text{subj. to} \quad & Gz \leq w \\
& Az = b,
\end{aligned} \tag{2.4}$$

where $A \in \mathbb{R}^{p \times s}$, $b \in \mathbb{R}^p$, or only equality constraints and positive variables:

$$\begin{aligned}
\inf_z \quad & c'z \\
\text{subj. to} \quad & Az = b \\
& z \geq 0.
\end{aligned} \tag{2.5}$$

By standard simple manipulations [65, p. 146] it is always possible to convert one of the three forms (2.3), (2.4) and (2.5) into the others.

2.2.1 Geometric Interpretation and Solution Properties

Let $\mathcal{P}$ be the feasible set (1.6) of problem (2.3). As $Z = \mathbb{R}^s$, this implies that $\mathcal{P}$ is a polyhedron defined by the inequality constraints in (2.3). If $\mathcal{P}$ is empty, then the problem is infeasible. We will assume for the following discussion that $\mathcal{P}$ is not empty. Denote by f^* the optimal value and by Z^* the set of optimizers of problem (2.3)

$$Z^* = \operatorname*{argmin}_{z \in \mathcal{P}} \ c'z.$$

Three cases can occur.

Case 1. The LP solution is unbounded, i.e., $f^* = -\infty$.

Case 2. The LP solution is bounded, i.e., $f^* > -\infty$ and the optimizer is unique. $z^* = Z^*$ is a singleton.

Case 3. The LP solution is bounded and there are multiple optima. Z^* is an subset of $\mathbb{R}^s$ which can be bounded or unbounded.

The two-dimensional geometric interpretation of the three cases discussed above is depicted in Figure 2.2.

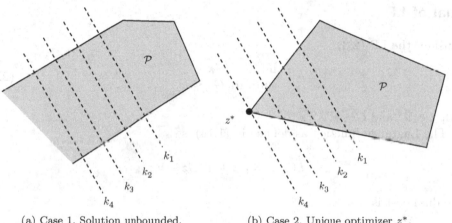

    (a) Case 1. Solution unbounded.   (b) Case 2. Unique optimizer z^*.

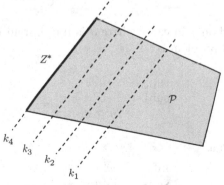

(c) Case 3. Multiple optimizers Z^*.

Figure 2.2 Linear program. Geometric interpretation, level curve parameters $k_i < k_{i-1}$.

The level curves of the cost function $c'z$ are represented by the parallel lines. All points z belonging both to the line $c'z = k_i$ and to the polyhedron $\mathcal{P}$ are feasible points with an associated cost k_i, with $k_i < k_{i-1}$. Solving (2.3) amounts to finding a feasible z which belongs to the level curve with the smallest cost k_i. Since the gradient of the cost is c, the direction of steepest descent is $-c/\|c\|$.

Case 1 is depicted in Figure 2.2(a). The feasible set $\mathcal{P}$ is unbounded. One can move in the direction of steepest descent $-c$ and be always feasible, thus decreasing the cost to $-\infty$. Case 2 is depicted in Figure 2.2(b). The optimizer is unique and it coincides with one of the vertices of the feasible polyhedron. Case 3 is depicted in Figure 2.2(c). The whole facet of the feasible polyhedron $\mathcal{P}$ is optimal, i.e., the cost for any point z belonging to the facet equals the optimal value f^*. In general, the optimal facet will be a facet of the polyhedron $\mathcal{P}$ parallel to the hyperplane $c'z = 0$.

From the analysis above we can conclude that the optimizers of any bounded LP always lie on the boundary of the feasible polyhedron $\mathcal{P}$.

2.2.2 Dual of LP

Consider the LP (2.3)

$$\begin{aligned} \inf_z \quad & c'z \\ \text{subj. to} \quad & Gz \leq w, \end{aligned} \tag{2.6}$$

with $z \in \mathbb{R}^s$ and $G \in \mathbb{R}^{m \times s}$.

The Lagrange function as defined in (1.14) is

$$L(z, u) = c'z + u'(Gz - w).$$

The dual cost is

$$d(u) = \inf_z L(z, u) = \inf_z (c' + u'G)z - u'w = \begin{cases} -u'w & \text{if} - G'u = c \\ -\infty & \text{if} - G'u \neq c. \end{cases}$$

Since we are interested only in cases where d is finite, from the relation above we conclude that the dual problem is

$$\begin{aligned} \sup_u \quad & -u'w \\ \text{subj. to} \quad & -G'u = c \\ & u \geq 0, \end{aligned} \tag{2.7}$$

which can be rewritten as

$$\begin{aligned} \inf_u \quad & w'u \\ \text{subj. to} \quad & G'u = -c \\ & u \geq 0. \end{aligned} \tag{2.8}$$

Note that for LPs, feasibility implies strong duality (Remark 1.3).

2.2.3 KKT condition for LP

The KKT conditions (1.29a)–(1.29e) for the LP (2.3) become

$$G'u + c = 0, \tag{2.9a}$$
$$u_i(G_i z - w_i) = 0, \; i = 1, \ldots, m \tag{2.9b}$$
$$u \geq 0, \tag{2.9c}$$
$$Gz - w \leq 0. \tag{2.9d}$$

They are: stationarity condition (2.9a), complementary slackness conditions (2.9b), dual feasibility (2.9c) and primal feasibility (2.9d). Often dual feasibility in linear programs refers to both (2.9a) and (2.9c).

2.2.4 Active Constraints and Degeneracies

Consider the LP (2.3). Let $I = \{1, \ldots, m\}$ be the set of constraint indices. For any $A \subseteq I$, let G_A and w_A be the submatrices of G and w, respectively, comprising the rows indexed by A and denote with G_j and w_j the j-th row of G and w, respectively. Let z be a feasible point and consider the set of active and inactive constraints at z:

$$A(z) = \{i \in I : G_i z = w_i\}$$
$$NA(z) = \{i \in I : G_i z < w_i\}. \tag{2.10}$$

From (2.10) we have

$$G_{A(z^*)} z^* = w_{A(z^*)}$$
$$G_{NA(z^*)} z^* < w_{NA(z^*)}. \tag{2.11}$$

Definition 2.1 (Linear Independence Constraint Qualification (LICQ))
We say that LICQ holds at z^* if the matrix $G_{A(z^*)}$ has full row rank.

Lemma 2.1   *Assume that the feasible set $\mathcal{P}$ of problem (2.3) is bounded. If the LICQ is violated at $z_1^* \in Z^*$ then there exists $z_2^* \subset Z^*$ such that $|A(z_2^*)| > s$.*

Consider $z_1^* \in Z^*$ on an optimal facet. Lemma 2.1 states the simple fact that if LICQ is violated at z_1^*, then there is a vertex $z_2^* \in Z^*$ on the same facet where $|A(z_2^*)| > s$. Thus, violation of LICQ is equivalent to having more than s constraints active at an optimal vertex.

Definition 2.2   *The LP (2.3) is said to be primal degenerate if there exists a $z^* \in Z^*$ such that the LICQ does not hold at z^*.*

Figure 2.3 depicts a case of primal degeneracy with four constraints active at the optimal vertex, i.e., more than the minimum number two.

Definition 2.3   *The LP (2.3) is said to be dual degenerate if its dual problem is primal degenerate.*

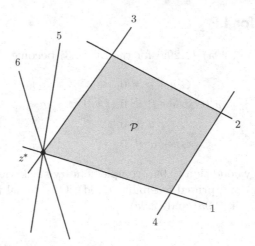

Figure 2.3 Primal degeneracy in a linear program.

The LICQ condition is invoked so that the optimization problem is well behaved in a way we will explain next. Let us look at the equality constraints of the dual problem (2.8) at the optimum $G'u^* = -c$ or equivalently $G'_A u^*_A = -c$ since $u^*_{NA} = 0$. If LICQ is satisfied, then the equality constraints will allow only a unique optimizer u^*. If LICQ is not satisfied, then the dual *may* have multiple optimizers. Thus we have the following Lemma.

Lemma 2.2 *If the primal problem (2.3) is not degenerate then the dual problem has a unique optimizer. If the dual problem (2.8) is not degenerate, then the primal problem has a unique optimizer.*

Multiple dual optimizers imply primal degeneracy and multiple primal optimizers imply dual degeneracy but the reverse is not true as we will illustrate next. In other words, LICQ is only a sufficient condition for uniqueness of the dual.

Example 2.1 Primal and dual degeneracies
Consider the following pair of primal and dual LPs

$$\text{Primal}$$
$$\inf \; [-1 \;\; -1]x$$
$$\begin{bmatrix} 0 & 1 \\ 1 & 1 \\ -1 & 0 \\ 0 & -1 \end{bmatrix} x \le \begin{bmatrix} 1 \\ 1 \\ 0 \\ 0 \end{bmatrix} \tag{2.12}$$

$$\text{Dual}$$
$$\inf \; [1\;1\;0\;0]u$$
$$\begin{bmatrix} 0 & 1 & -1 & 0 \\ 1 & 1 & 0 & -1 \end{bmatrix} u = \begin{bmatrix} 1 \\ 1 \end{bmatrix} \tag{2.13}$$
$$u \ge 0$$

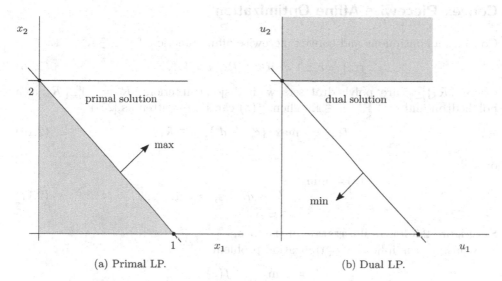

(a) Primal LP.   (b) Dual LP.

Figure 2.4 Example 2.1. LP with primal and dual degeneracy. The vectors max and min point in the direction for which the objective improves. The feasible sets are shaded.

Substituting for u_3 and u_4 from the equality constraints in (2.13) we can rewrite the dual as

$$
\begin{array}{l}
\text{Dual} \\[4pt]
\inf \ [1 \ \ 1] \begin{bmatrix} u_1 \\ u_2 \end{bmatrix} \\[12pt]
\begin{bmatrix} 1 & 0 \\ 0 & 1 \\ 1 & 1 \\ 0 & 1 \end{bmatrix} \begin{bmatrix} u_1 \\ u_2 \end{bmatrix} \geq \begin{bmatrix} 0 \\ 1 \\ 1 \\ 0 \end{bmatrix}
\end{array}
\tag{2.14}
$$

The situation is portrayed in Figure 2.4.

Consider the two solutions for the primal LP denoted with 1 and 2 in Figure 2.4(a) and referred to as "basic" solutions. Basic solution 1 is primal nondegenerate, since it is defined by exactly as many active constraints as there are variables. Basic solution 2 is primal degenerate, since it is defined by three active constraints, i.e., more than two. Any convex combination of optimal solutions 1 and 2 is also optimal. This continuum of optimal solutions in the primal problem corresponds to a degenerate solution in the dual space, that is, the dual problem is primal-degenerate. Hence the primal problem is dual-degenerate. In conclusion, Figures 2.4(a) and 2.4(b) show an example of a primal problem with multiple optima and the corresponding dual problem being primal degenerate.

Next we want to show that the statement "if the dual problem is primal degenerate then the primal problem has multiple optima" is, in general, not true. Switch dual and primal problems, i.e., call the "dual problem" primal problem and the "primal problem" dual problem (this can be done since the dual of the dual problem is the primal problem). Then, we have a dual problem which is primal degenerate in solution 2 while the primal problem does not present multiple optimizers.

2.2.5   Convex Piecewise Affine Optimization

Consider a continuous and convex piecewise affine function $f : \mathcal{R} \subseteq \mathbb{R}^s \to \mathbb{R}$:

$$f(z) = c_i'z + d_i \text{ if } z \in \mathcal{R}_i, \ i = 1, \dots, p, \tag{2.15}$$

where $\{\mathcal{R}_i\}_{i=1}^p$ are polyhedral sets with disjoint interiors, $\mathcal{R} = \bigcup_{i=1}^p \mathcal{R}_i$ is a polyhedron and $c_i \in \mathbb{R}^s$, $d_i \in \mathbb{R}$. Then, $f(z)$ can be rewritten as [257]

$$f(z) = \max_{i=1,\dots,k} \{c_i'z + d_i\}, \ z \in \mathcal{R}, \tag{2.16}$$

or [65]:

$$
\begin{aligned}
f(z) = \min_\varepsilon \quad & \varepsilon \\
& c_i'z + d_i \leq \varepsilon, \quad i = 1, \dots, k \\
& z \in \mathcal{R}.
\end{aligned}
\tag{2.17}
$$

See Figure 2.5 for an illustration of the idea.

Consider the following optimization problem

$$
\begin{aligned}
f^* = \min_z \quad & f(z) \\
\text{subj. to} \quad & Gz \leq w \\
& z \in \mathcal{R},
\end{aligned}
\tag{2.18}
$$

where the cost function has the form (2.15). Substituting (2.17) this becomes

$$
\begin{aligned}
f^* = \min_{z,\varepsilon} \quad & \varepsilon \\
\text{subj. to} \quad & Gz \leq w \\
& c_i'z + d_i \leq \varepsilon, \quad i = 1, \dots, k \\
& z \in \mathcal{R}.
\end{aligned}
\tag{2.19}
$$

The previous result can be extended to the sum of continuous and convex piecewise affine functions. Let $f : \mathcal{R} \subseteq \mathbb{R}^s \to \mathbb{R}$ be defined as:

$$f(z) = \sum_{j=1}^r f^j(z), \tag{2.20}$$

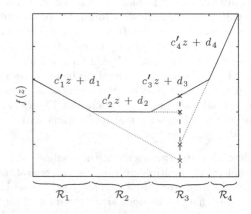

Figure 2.5 Convex piecewise affine (PWA) function described as the max of affine functions.

with

$$f^j(z) = \max_{i=1,\dots,k^j} \{c_i^{j'}z + d_i^j\}, \; z \in \mathcal{R}. \tag{2.21}$$

Then the following optimization problem

$$\begin{aligned} f^* = \min_z \quad & f(z) \\ \text{subj. to} \quad & Gz \leq w \\ & z \in \mathcal{R}, \end{aligned} \tag{2.22}$$

where the cost function has the form (2.20), can be solved by the following linear program:

$$\begin{aligned} f^* = \min_{z,\varepsilon^1,\dots,\varepsilon^r} \quad & \varepsilon^1 + \cdots + \varepsilon^r \\ \text{subj. to} \quad & Gz \leq w \\ & c_i^{1'}z + d_i^1 \leq \varepsilon^1, \quad i = 1,\dots,k^1 \\ & c_i^{2'}z + d_i^2 \leq \varepsilon^2, \quad i = 1,\dots,k^2 \\ & \quad \vdots \\ & c_i^{r'}z + d_i^r \leq \varepsilon^r, \quad i = 1,\dots,k^r \\ & z \in \mathcal{R}. \end{aligned} \tag{2.23}$$

Remark 2.1 Note that the results of this section can be immediately applied to the minimization of one or infinity norms. For any $y \subset \mathbb{R}$, $|y| = \max\{y, -y\}$. Therefore for any $Q \in \mathbb{R}^{k \times s}$ and $p \in \mathbb{R}^k$:

$$\|Qz - p\|_\infty = \max\{Q_1'z + p_1, -Q_1'z - p_1, \dots, Q_k'z + p_k, -Q_k'z - p_k\},$$

and

$$\|Qz - p\|_1 = \sum_{i=1}^k |Q_i'z + p_i| = \sum_{i=1}^k \max\{Q_i'z + p_i, -Q_i'z - p_i\}.$$

2.3 Quadratic Programming

The continuous optimization problem (1.4) is called a *quadratic program* (QP) if the constraint functions are affine and the cost function is a convex quadratic function. In this book we will use the form:

$$\begin{aligned} \min_z \quad & \tfrac{1}{2}z'Hz + q'z + r \\ \text{subj. to} \quad & Gz \leq w, \end{aligned} \tag{2.24}$$

where $z \in \mathbb{R}^s$, $H = H' \succ 0 \in \mathbb{R}^{s \times s}$, $q \in \mathbb{R}^s$, $G \in \mathbb{R}^{m \times s}$. In (2.24) the constant term can be omitted if one is only interested in the optimizer.

Other QP forms often include equality and inequality constraints:

$$\begin{aligned} \min_z \quad & \tfrac{1}{2}z'Hz + q'z + r \\ \text{subj. to} \quad & Gz \leq w \\ & Az = b. \end{aligned} \tag{2.25}$$

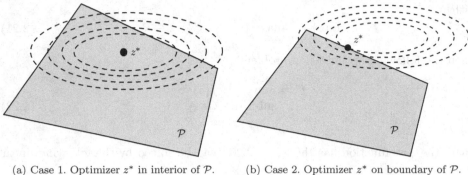

(a) Case 1. Optimizer z^* in interior of $\mathcal{P}$.   (b) Case 2. Optimizer z^* on boundary of $\mathcal{P}$.

Figure 2.6 Geometric interpretation of the quadratic program solution.

2.3.1 Geometric Interpretation and Solution Properties

Let $\mathcal{P}$ be the feasible set (1.6) of problem (2.24). As $Z = \mathbb{R}^s$, this implies that $\mathcal{P}$ is a polyhedron defined by the inequality constraints in (2.24). The two dimensional geometric interpretation is depicted in Figure 2.6. The level curves of the cost function $\frac{1}{2}z'Hz+q'z+r$ are represented by the ellipsoids. All the points z belonging both to the ellipsoid $\frac{1}{2}z'Hz + q'z + r = k_i$ and to the polyhedron $\mathcal{P}$ are feasible points with an associated cost k_i. The smaller the ellipsoid, the smaller is its cost k_i. Solving (2.24) amounts to finding a feasible z which belongs to the level curve with the smallest cost k_i. Since H is strictly positive definite, the QP (2.24) cannot have multiple optima nor unbounded solutions. If $\mathcal{P}$ is not empty the optimizer is unique. Two cases can occur if $\mathcal{P}$ is not empty:

Case 1. The optimizer lies strictly inside the feasible polyhedron (Figure 2.6(a)).

Case 2. The optimizer lies on the boundary of the feasible polyhedron (Figure 2.6(b)).

In Case 1 the QP (2.24) is unconstrained and we can find the minimizer by setting the gradient equal to zero

$$Hz^* + q = 0. \tag{2.26}$$

Since $H \succ 0$ we obtain $z^* = -H^{-1}q$.

2.3.2 Dual of QP

Consider the QP (2.24)

$$\min_z \quad \tfrac{1}{2}z'Hz + q'z$$
$$\text{subj. to} \quad Gz \le w.$$

The Lagrange function as defined in (1.14) is

$$L(z, u) = \frac{1}{2}z'Hz + q'z + u'(Gz - w).$$

The dual cost is

$$d(u) = \min_z \frac{1}{2} z'Hz + q'z + u'(Gz - w) \qquad (2.27)$$

and the dual problem is

$$\max_{u \geq 0} \min_z \frac{1}{2} z'Hz + q'z + u'(Gz - w). \qquad (2.28)$$

For a given u the Lagrange function $\frac{1}{2}z'Hz + q'z + u'(Gz - w)$ is convex. Therefore it is necessary and sufficient for optimality that the gradient is zero

$$Hz + q + G'u = 0.$$

From the equation above we can derive $z = -H^{-1}(q + G'u)$ and, substituting this in Equation (2.27), we obtain:

$$d(u) = -\frac{1}{2}u'(GH^{-1}G')u - u'(w + GH^{-1}q) - \frac{1}{2}q'H^{-1}q. \qquad (2.29)$$

By using (2.29) the dual problem (2.28) can be rewritten as:

$$\begin{array}{ll} \min_u & \frac{1}{2}u'(GH^{-1}G')u + u'(w + GH^{-1}q) + \frac{1}{2}q'H^{-1}q \\ \text{subj. to} & u \geq 0. \end{array} \qquad (2.30)$$

Note that for convex QPs feasibility implies strong duality (Remark 1.3).

2.3.3 KKT conditions for QP

Consider the QP (2.24). Then, $\nabla f(z) = Hz + q$, $g_i(z) = G_iz - w_i$ (where G_i is the i-th row of G), $\nabla g_i(z) = G_i'$. The KKT conditions become

$$Hz + q + G'u = 0 \qquad (2.31a)$$
$$u_i(G_iz - w_i) = 0, \; i = 1, \ldots, m \qquad (2.31b)$$
$$u \geq 0 \qquad (2.31c)$$
$$Gz - w \leq 0. \qquad (2.31d)$$

2.3.4 Active Constraints and Degeneracies

Consider the definition of active set $A(z)$ in (2.10). Note that $A(z^*)$ may be empty in the case of a QP. We define primal and dual degeneracy as in the LP case.

Definition 2.4 *The QP (2.24) is said to be primal degenerate if there exists a $z^* \in Z^*$ such that the LICQ does not hold at z^*.*

Note that if the QP (2.24) is not primal degenerate, then the dual QP (2.30) has a unique solution since $u_{NA}^* = 0$ and $(G_A H^{-1} G_A')$ is invertible.

Definition 2.5 *The QP (2.24) is said to be dual degenerate if its dual problem is primal degenerate.*

We note from (2.30) that all the constraints are independent. Therefore LICQ always holds for dual QPs and dual degeneracy can never occur for QPs with $H \succ 0$.

2.3.5 Constrained Least-Squares Problems

The problem of minimizing the convex quadratic function

$$\|Az - b\|_2^2 = z'A'Az - 2b'Az + b'b \tag{2.32}$$

is an (unconstrained) QP. It arises in many fields and has many names, e.g., linear regression or least-squares approximation. From (2.26) we find the minimizer

$$z^* = (A'A)^{-1}A'b = A^\dagger b,$$

where $A^\dagger$ is the *generalized inverse* of A. When linear inequality constraints are added, the problem is called constrained linear regression or *constrained least-squares*, and there is no longer a simple analytical solution. As an example we can consider regression with lower and upper bounds on the variables, i.e.,

$$\begin{aligned}
\min_z \quad & \|Az - b\|_2^2 \\
\text{subj. to} \quad & l_i \le z_i \le u_i, \quad i = 1, \dots, n,
\end{aligned} \tag{2.33}$$

which is a QP. In Chapter 6.3.1 we will show how to compute an analytical solution to the constrained least-squares problem. In particular we will show how to compute the solution z^* as a function of b, u_i and l_i.

2.4 Mixed-Integer Optimization

As discussed in Section 1.1.2, if the decision set Z in the optimization problem (1.2) is the Cartesian product of a binary set and a real Euclidian space, i.e., $Z \subseteq \{[z_c, z_b] : z_c \in \mathbb{R}^{s_c}, z_b \in \{0,1\}^{s_b}\}$, then the optimization problem is said to be *mixed-integer*. In this section Mixed Integer Linear Programming (MILP) and Mixed Integer Quadratic Programming (MIQP) are introduced.

When the cost of the optimization problem (1.10) is quadratic and the constraints are affine, then the problem is called a *mixed integer quadratic program* (MIQP). The most general form of an MIQP is

$$\begin{aligned}
\inf_{[z_c, z_b]} \quad & \tfrac{1}{2}z'Hz + q'z + r \\
\text{subj. to} \quad & G_c z_c + G_b z_b \le w \\
& A_c z_c + A_b z_b = b \\
& z_c \in \mathbb{R}^{s_c}, \ z_b \in \{0,1\}^{s_b} \\
& z = [z_c, z_b],
\end{aligned} \tag{2.34}$$

where $H \succeq 0 \in \mathbb{R}^{s \times s}$, $G_c \in \mathbb{R}^{m \times s_c}$, $G_b \in \mathbb{R}^{m \times s_b}$, $w \in \mathbb{R}^m$, $A_c \in \mathbb{R}^{p \times s_c}$, $A_b \in \mathbb{R}^{p \times s_b}$, $b \in \mathbb{R}^p$ and $s = s_c + s_d$. Mixed integer quadratic programs are nonconvex optimization problems, in general. When $H = 0$ the problem is called a *mixed integer linear program* (MILP). Often the term r is omitted from the cost since it does not affect the optimizer, but r has to be considered when computing the optimal value. In this book, we will often use the form of MIQP with inequality constraints only

$$
\begin{aligned}
\inf_{[z_c, z_b]} \quad & \tfrac{1}{2} z' H z + q' z + r \\
\text{subj. to} \quad & G_c z_c + G_b z_b \leq w \\
& z_c \in \mathbb{R}^{s_c}, \; z_b \in \{0,1\}^{s_b} \\
& z = [z_c, z_b].
\end{aligned}
\tag{2.35}
$$

The general form can always be translated into the form with inequality constraints only by standard simple manipulations.

For a fixed integer value $\bar{z}_b$ of z_b, the MIQP (2.34) becomes a quadratic program:

$$
\begin{aligned}
\inf_{z_c} \quad & \tfrac{1}{2} z_c' H_c z_c + q_c(z_b)' z_c + k(z_b) \\
\text{subj. to} \quad & G_c z_c \leq w - G_b \bar{z}_b \\
& A_c z_c = b_{eq} - A_b \bar{z}_b \\
& z_c \in \mathbb{R}^{s_c}.
\end{aligned}
\tag{2.36}
$$

Therefore the most obvious way to interpret and solve an MIQP is to enumerate all the 2^{s_b} integer values of the variable z_b and solve the corresponding QPs. By comparing the 2^{s_b} optimal costs one can derive the optimizer and the optimal cost of the MIQP (2.34). Although this approach is not used in practice, it gives a simple way for proving what is stated next. Let $\mathcal{P}_{\bar{z}_b}$ be the feasible set (1.6) of problem (2.35) for a fixed $z_b = \bar{z}_b$. The cost is a quadratic function defined over $\mathbb{R}^{s_c}$ and $\mathcal{P}_{\bar{z}_b}$ is a polyhedron defined by the inequality constraints

$$
G_c z_c \leq w - G_b \bar{z}_b.
\tag{2.37}
$$

Denote by f^* the optimal value and by Z^* the set of optimizers of problem (2.34) If $\mathcal{P}_{\bar{z}_b}$ is empty for all $\bar{z}_b$, then the problem (2.35) is infeasible. Five cases can occur if $\mathcal{P}_{\bar{z}_b}$ is not empty for at last one $\bar{z}_b \in \{0,1\}^{s_b}$:

Case 1. The MIQP solution is unbounded, i.e., $f^* = -\infty$. This cannot happen if $H_c \succ 0$.

Case 2. The MIQP solution is bounded, i.e., $f^* > -\infty$ and the optimizer is unique. Z^* is a singleton.

Case 3. The MIQP solution is bounded and there are infinitely many optimizers corresponding to the same integer value. Z^* is the Cartesian product of an infinite dimensional subset of $\mathbb{R}^s$ and an integer number z_b^*. This cannot happen if $H_c \succ 0$.

Case 4. The MIQP solution is bounded and there are finitely many optimizers corresponding to different integer values. Z^* is a finite set of optimizers $\{(z_{1,c}^*, z_{1,b}^*), \ldots, (z_{N,c}^*, z_{N,b}^*)\}$.

Case 5. The union of Case 3 and Case 4.

3

Numerical Methods for Optimization

Contributed by Dr. Alexander Domahidi and Dr. Stefan Richter
ETH Zurich
alex.domahidi@gmail.com, stefan.richter@alumni.ethz.ch

There is a great variety of algorithms for the solution of unconstrained and constrained optimization problems. In this chapter we introduce those algorithms, which are important for the problems encountered in this book and explain the underlying concepts. In particular, we present numerical methods for finding a *minimizer* z^* to the optimization problem

$$\min_z \quad f(z)$$
$$\text{subj. to} \quad z \in S, \tag{3.1}$$

where both the objective function $f : \mathbb{R}^s \to \mathbb{R}$ and the feasible set S are convex, hence (3.1) is a convex program. We will assume that if the problem is feasible, an optimizer $z^* \in S$ exists, i.e., $f^* = f(z^*)$.

3.1 Convergence

Many model predictive control (MPC) problems for linear systems have the form of problem (3.1) with a linear or convex quadratic objective function and a polyhedral feasible set S defined by linear equalities and inequalities. In all but the simplest cases, analytically finding a minimizer z^* is impossible. Hence, one has to resort to numerical methods, which compute an approximate minimizer that is "good enough" (we give a precise definition of this term below). Any such method proceeds in an *iterative* manner, i.e., starting from an initial guess z^0, it computes a sequence $\{z^k\}_{k=1}^{k=k_{\max}}$, where

$$z^{k+1} = \Psi\left(z^k, f, S\right),$$

with Ψ being an update rule depending on the method. For a method to be useful, it should terminate after $k_{\max}$ iterations and return an approximate minimizer $z^{k_{\max}}$ that satisfies

$$|f\left(z^{k_{\max}}\right) - f\left(z^*\right)| \leq \epsilon \quad \text{and} \quad \text{dist}(z^{k_{\max}}, S) \leq \delta,$$

where

$$\text{dist}(z, S) = \min_{y \in S} \|y - z\|$$

is the shortest distance between a point and a set in $\mathbb{R}^s$ measured by the norm $\|\cdot\|$. The parameters $\epsilon, \delta > 0$ define the required accuracy of the approximate minimizer.

There exist mainly two classes of optimization methods for solving (3.1): first-order methods that make use of first-order information of the objective function (subgradients or gradients) and second-order methods using in addition second-order information (Hessians). Note that either method may also use zero-order information, i.e., the objective value itself, for instance, to compute step sizes.

In the following list, key aspects that are important for *any* optimization method are summarized. Some of these aspects are discussed in detail for the corresponding optimization methods later in this chapter.

- *Convergence:* Is $k_{\max}$ finite for all accuracies $\epsilon, \delta > 0$?

- *Convergence rate:* How do errors $|f\left(z^k\right) - f\left(z^*\right)|$ and $\text{dist}(z^k, S)$ depend on the iteration counter k?

- *Feasibility:* Does the method produce feasible iterates, i.e., $\text{dist}(z^k, S) = 0$ for all iterations k?

- *Numerical robustness:* Do the properties above change in presence of finite precision arithmetics when using fixed or floating point number representations?

- *Warm-starting:* Can the method take advantage of z^0 being close to a minimizer z^*?

- *Preconditioning:* Is it possible to transform (3.1) into an equivalent problem that can be solved in fewer iterations?

- *Computational complexity:* How many arithmetic operations are needed for computing the approximate solution $z^{k_{\max}}$?

The speed or rate of convergence relates to the second point in the list above and is an important distinctive feature between various methods. We present a definition of convergence rates next.

Definition 3.1 (Convergence Rates (cf. [223, Section 2.2])) *Let $\{e^k\}$ be a sequence of positive reals that converges to 0. The convergence rate of this sequence is called* linear *if there exists a constant $q \in (0, 1)$ such that*

$$\limsup_{k \to \infty} \frac{e^{k+1}}{e^k} \leq q.$$

If $q = 1$, the convergence rate is said to be sublinear, *whereas if $q = 0$, the sequence has a* superlinear *convergence rate. A superlinearly converging sequence converges with* order p *if*

$$\limsup_{k \to \infty} \frac{e^{k+1}}{e^{k^p}} \leq C,$$

where C is a positive constant, not necessarily less than 1. In particular, superlinear convergence with order $p = 2$ is called quadratic convergence.

For an optimization method, the sequence $\{e^k\}$ in the definition above can be any sequence of errors associated with the particular method, e.g., e^k can be defined as the absolute error in the objective value

$$e^k = |f(z^k) - f(z^*)|$$

or any other nonnegative error measure that vanishes only at a minimizer z^*.

In the previous classification of convergence rates, sublinear convergence is eventually slower than linear convergence which itself is eventually slower than superlinear convergence. Within superlinear convergence rates, the ones with higher order converge faster eventually. However, for a fixed number of iterations, the actual ordering of convergence rates also depends on constants q and C, for instance, a linearly converging sequence with a small value of q might have a smaller error after 10 steps than a quadratically converging sequence.

In the remainder of this chapter we first introduce methods that solve the unconstrained problem, i.e., $S = \mathbb{R}^s$, in Section 3.2. They are the prerequisite for understanding constrained optimization methods, i.e., $S \subset \mathbb{R}^s$, which are then presented in Section 3.3.

3.2 Unconstrained Optimization

In unconstrained *smooth* optimization we are interested in solving the problem

$$\min_z \ f(z), \tag{3.2}$$

where $f : \mathbb{R}^s \to \mathbb{R}$ is convex and *continuously differentiable*.

For the purpose of this section, we focus on so-called *descent methods* for unconstrained optimization. These methods obtain the next iterate z^{k+1} from the current iterate z^k by taking a step of size $h^k > 0$ along a certain direction d^k:

$$z^{k+1} = z^k + h^k d^k \tag{3.3}$$

Particular methods differ in the way d^k and h^k are chosen. In general, d^k has to be a descent direction, i.e., for all $z^k \neq z^*$ it must satisfy

$$\nabla f(z^k)' d^k < 0. \tag{3.4}$$

The step sizes h^k have to be chosen carefully in order to ensure a sufficiently large decrease in the function value, which is crucial for convergence of descent methods. Since solving for the optimal step size,

$$h^{k,*} \in \arg\min_{h \geq 0} f(z^k + hd^k),$$

is generally too expensive, *inexact line search* methods are used instead. They compute h^k cheaply and achieve a reasonably good decrement in the function value ensuring convergence of the algorithm. We discuss line search methods in Section 3.2.3.

In the following, we present two of the most important instances of descent methods: the classic gradient method in Section 3.2.1 and Newton's method in Section 3.2.2. We include Nesterov's fast gradient method in the discussion along with gradient methods, although it is not a descent method. We shall see that gradient methods, under certain assumptions on the gradient of f, allow for a constant step size, which results in an extremely simple and efficient algorithm in practice. Newton's method, on the contrary, requires a line search to ensure global convergence, but it has a better (local) convergence rate than gradient methods.

3.2.1 Gradient Methods

Classic Gradient Method

One property of the gradient of f evaluated at z^k, $\nabla f(z^k)$, is that it points into the direction of steepest local ascent. The main idea of gradient methods for minimization is therefore to use the anti-gradient direction as a descent direction,

$$d^k = -\nabla f(z^k),$$

trivially satisfying (3.4). Before turning to the question of appropriate step sizes h^k, we introduce an additional assumption on the gradient of f. For the remainder of this section, we assume that f has a Lipschitz continuous gradient, i.e., that f is so-called L-smooth. All of the upcoming results regarding Lipschitz continuity of the gradient are taken from [220, Section 1.2.2].

Definition 3.2 (L-Smoothness of a Function)   *Let $f : \mathbb{R}^s \to \mathbb{R}$ be once continuously differentiable on $\mathbb{R}^s$. The gradient ∇f is Lipschitz continuous on $\mathbb{R}^s$ with Lipschitz constant $L > 0$ if for all pairs $(z, y) \in \mathbb{R}^s \times \mathbb{R}^s$*

$$\|\nabla f(z) - \nabla f(y)\| \leq L \|z - y\| .$$

If function f is twice continuously differentiable, L-smoothness can be characterized using the largest eigenvalue of the Hessian of f. This alternative characterization can be useful in order to compute the actual value of the Lipschitz constant.

Lemma 3.1 (L-Smoothness: Second-Order Characterization)   *Let function $f : \mathbb{R}^s \to \mathbb{R}$ be twice continuously differentiable on $\mathbb{R}^s$. Then the gradient ∇f*

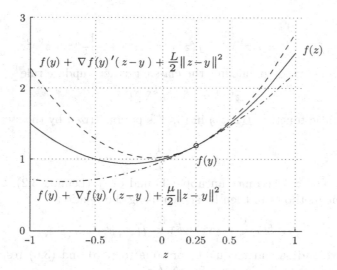

Figure 3.1 Upper and lower bounds on functions: The quadratic upper bound stems from L-smoothness of f (Definition 3.2), while the quadratic lower bound is obtained from strong convexity of f (Theorem 3.1). In this example, $f(z) = z^2 + \frac{1}{2}z + 1$ and we have chosen $L = 3$ and $\mu = 1$ for illustration.

is Lipschitz continuous on $\mathbb{R}^s$ with Lipschitz constant $L > 0$ if and only if for all $z \in \mathbb{R}^s$

$$\left\| \nabla^2 f(z) \right\| \leq L.$$

The most important consequence of Lipschitz continuity of the gradient is that f can be globally upper-bounded by a quadratic. This is expressed in the next lemma and illustrated in Figure 3.1.

Algorithm 3.1   *Gradient method for smooth convex optimization*

Input: Initial iterate $z^0 \in \mathbb{R}^s$, Lipschitz constant L of ∇f
Output: Point close to z^*

> **Repeat**
>
> $$z^{k+1} \leftarrow z^k - \frac{1}{L} \nabla f(z^k)$$
>
> **Until** stopping criterion is satisfied

Lemma 3.2 (Descent Lemma)   *Let f be L-smooth on $\mathbb{R}^s$. Then for any pair $(z, y) \in \mathbb{R}^s \times \mathbb{R}^s$ we have*

$$f(z) \leq f(y) + \nabla f(y)'(z - y) + \frac{L}{2} \left\| z - y \right\|^2. \qquad (3.5)$$

We now return to the choice of step sizes h^k in (3.3). Notice that interpreting the next iterate z^{k+1} as the result of minimizing a quadratic at the previous iterate z^k,

$$z^{k+1} = \arg\min_z \bar{f}(z, z^k), \qquad (3.6)$$

where we define

$$\bar{f}(z, z^k) = f(z^k) + \nabla f(z^k)'(z - z^k) + \frac{1}{2h^k}\|z - z^k\|^2, \tag{3.7}$$

yields after some standard calculus the classic gradient update rule

$$z^{k+1} = z^k - h^k \nabla f(z^k).$$

The quadratic function $\bar{f}(z, z^k)$ in (3.7) is parametrized by the step length h^k. Choosing it as

$$h^k = \frac{1}{L},$$

the quadratic $\bar{f}(z, z^k)$ becomes an upper bound on f (Lemma 3.2). Convergence of the gradient method can then be shown using the relation

$$f(z^{k+1}) \overset{(1)}{\leq} \bar{f}(z^{k+1}, z^k) \overset{(2)}{\leq} \bar{f}(z^k, z^k) = f(z^k), \tag{3.8}$$

where the first and second inequality are due to (3.5) and (3.6) respectively. In fact, the second inequality is strict for $z^k \neq z^*$.

The gradient method with constant step size summarized in Algorithm 3.1 can be shown to converge sublinearly with the number of iterations growing proportionally to $L/\epsilon \cdot \|z^0 - z^*\|^2$ [220, Corollary 2.1.2]. The required level of suboptimality $\epsilon > 0$ is hereby defined in terms of the error in the objective value $f(z^k) - f^*$. Note that in case of a constant step size, the gradient method is an extremely simple, division-free algorithm (given that the gradient ∇f can be computed without divisions).

Remark 3.1   If the Lipschitz constant L is unknown, a simple strategy is to start with an initial guess $\tilde{L} > 0$, solve for z^{k+1} according to (3.6) with $h^k = 1/\tilde{L}$ and then check whether the first inequality in (3.8) holds true. If so, the quadratic objective in (3.6) was a (local) upper bound of f and minimization consequently leads to a decrease in the function value, i.e., $f(z^{k+1}) < f(z^k)$. Otherwise, the current guess for the Lipschitz constant needs to be increased, e.g., $\tilde{L} = 2\tilde{L}$, and the procedure repeated. Termination is guaranteed because f was assumed L-smooth.

Algorithm 3.2   *Fast gradient method for smooth convex optimization*

Input: Initial iterates $z^0 \in \mathbb{R}^s$, $y^0 = z^0$; $\alpha^0 = \frac{1}{2}(\sqrt{5} - 1)$, Lipschitz constant L of ∇f
Output: Point close to z^*

    Repeat

$$z^{k+1} \leftarrow y^k - \frac{1}{L}\nabla f(y^k)$$

$$\alpha^{k+1} \leftarrow \frac{\alpha^k}{2}\left(\sqrt{\alpha^{k^2} + 4} - \alpha^k\right)$$

$$\beta^k \leftarrow \frac{\alpha^k(1 - \alpha^k)}{\alpha^{k^2} + \alpha^{k+1}}$$

$$y^{k+1} \leftarrow z^{k+1} + \beta^k(z^{k+1} - z^k)$$

    Until stopping criterion is satisfied

Fast Gradient Method

An accelerated version of the gradient method, Nesterov's fast gradient method [220, Section 2.2], is given in Algorithm 3.2. The basic idea behind this method is to take the gradient at an affine combination of the previous two iterates, i.e., the gradient is evaluated at

$$y^k = z^k + \beta^{k-1}(z^k - z^{k-1})$$

for some carefully chosen $\beta^{k-1} \in (0,1)$. This two-step method can be shown to converge sublinearly with the number of iterations growing proportionally to $\sqrt{L/\epsilon}$· $\|z^0 - z^*\|$ [220, Theorem 2.2.2]. This is the best convergence rate for L-smooth problems using only first-order information (see [220, Section 2.1] for details). The mathematical reasoning behind the improved convergence rate and optimality of the fast gradient method amongst all first-order methods is quite involved and beyond the scope of this book.

Example 3.1 Gradient method vs. fast gradient method for quadratic functions
Consider the optimization problem given by

$$\min \frac{1}{2} z' H z + q' z,$$

where the Hessian and the linear cost vector are given by

$$H = \begin{bmatrix} 0.2 & 0.2 \\ 0.2 & 2 \end{bmatrix} \quad \text{and} \quad q = \begin{bmatrix} -0.1 \\ 0.1 \end{bmatrix}.$$

For quadratic functions, a tight Lipschitz constant L of the gradient can be determined by computing the largest eigenvalue of the Hessian (Lemma 3.1), i.e.,

$$L - \lambda_{\max}(H), \tag{3.9}$$

here $L = 2.022$. The upper left plot in Figure 3.2 shows the first 10 iterations of the gradient method with constant step size $1/L$ (Algorithm 3.1). The upper right and lower left plot show the first 10 iterations using larger (but constant) step sizes, which correspond to underestimating the tight Lipschitz constant L. It can be shown that if L is underestimated by a factor of 2 or more, convergence is lost (cf. lower left plot). The lower right plot shows the iterates of the fast gradient method using the tight Lipschitz constant (Algorithm 3.2). Contrary to the gradient method, the fast gradient method generates a nonmonotonically decreasing error sequence.

Strongly Convex Problems

In the case where the objective function f in (3.2) is also strongly convex, *linear* convergence rates can be established for both the gradient method and the fast gradient method. The following theorems from [151, Section B] characterize strong convexity whenever the function is once or twice continuously differentiable. For an illustration of this property see Figure 3.1.

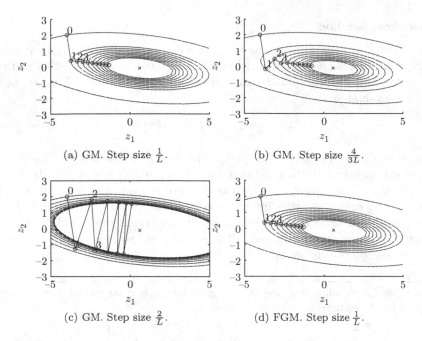

(a) GM. Step size $\frac{1}{L}$.   (b) GM. Step size $\frac{4}{3L}$.

(c) GM. Step size $\frac{2}{L}$.   (d) FGM. Step size $\frac{1}{L}$.

Figure 3.2 Example 3.1. First 10 iterations of the gradient method (GM) with constant step size $1/L$ using the tight, a slightly underestimated and a strongly underestimated Lipschitz constant L. Also shown are the iterates of the fast gradient method (FGM) using the tight Lipschitz constant. The contour lines indicate the function value of each iterate. Note that the gradients are orthogonal to the contour lines.

Theorem 3.1 (Strong Convexity: First-Order Characterization) *Let function $f : \mathbb{R}^s \to \mathbb{R}$ be once continuously differentiable on $\mathbb{R}^s$. Then f is strongly convex on $\mathbb{R}^s$ with convexity parameter $\mu > 0$ if and only if for all pairs $(z, y) \in \mathbb{R}^s \times \mathbb{R}^s$*

$$f(z) \geq f(y) + \nabla f(y)'(z - y) + \frac{\mu}{2} \|z - y\|^2 .$$

Theorem 3.2 (Strong Convexity: Second-Order Characterization) *Let $f : \mathbb{R}^s \to \mathbb{R}$ be twice continuously differentiable on $\mathbb{R}^s$. Then f is strongly convex on $\mathbb{R}^s$ with parameter $\mu > 0$ if and only if for all $z \in \mathbb{R}^s$*

$$\nabla^2 f(z) \succeq \mu I_s .$$

Algorithm 3.3 *Fast gradient method for smooth strongly convex optimization*

Input: Initial iterates $z^0 \in \mathbb{R}^s$, $y^0 = z^0$; $0 < \sqrt{\mu/L} \leq \alpha^0 < 1$, Lipschitz constant L of ∇f, strong convexity parameter μ of f

Output: Point close to z^*

    Repeat

$$z^{k+1} \leftarrow y^k - \frac{1}{L} \nabla f(y^k)$$

$$\text{Compute } \alpha^{k+1} \in (0,1): \alpha^{k+1^2} = (1 - \alpha^{k+1})\alpha^{k^2} + \frac{\mu\alpha^{k+1}}{L}$$

$$\beta^k \leftarrow \frac{\alpha^k(1 - \alpha^k)}{\alpha^{k^2} + \alpha^{k+1}}$$

$$y^{k+1} \leftarrow z^{k+1} + \beta^k(z^{k+1} - z^k)$$

Until stopping criterion is satisfied

From Theorem 3.1 it follows that a differentiable strongly convex function can be lower-bounded by a quadratic. Relating this to the quadratic upper bound in Lemma 3.2, the inequality

$$\kappa_f = \frac{L}{\mu} \geq 1$$

follows intuitively. The constant κ_f is the so-called *condition number* of the L-smooth and strongly convex function f.

With the additional assumption of strong convexity, it can be shown that the gradient method in Algorithm 3.1 converges linearly with the number of iterations growing proportionally to $\kappa_f \cdot \ln(1/\epsilon)$ [220, Theorem 2.1.15] with $\epsilon > 0$ being the required level of suboptimality in terms of the error in the objective. In order for the fast gradient method in Algorithm 3.2 to achieve linear convergence rate, it either needs to be modified or restarted after a number of steps that depends on the condition number κ_f. In both cases it can be shown that the number of iterations grows proportionally to $\sqrt{\kappa_f} \cdot \ln(1/\epsilon)$ [220, Theorem 2.2.2]. This is the best convergence rate for L-smooth and strongly convex problems using only first-order information (see [220, Section 2.1] for details). A modified fast gradient method that takes strong convexity into account explicitly and thus does not need restarting is given in Algorithm 3.3.

Example 3.2 Strong convexity parameter of a quadratic function
Let $f : \mathbb{R}^s \to \mathbb{R}$ be a quadratic function defined as

$$f(z) = \frac{1}{2}z'Hz + q'z$$

with $H \succ 0$. Then, by Theorem 3.2, the (tight) parameter for strong convexity μ is the smallest eigenvalue of the Hessian H, i.e.,

$$\mu = \lambda_{\min}(H).$$

For the Hessian given in Example 3.1 we therefore have $\mu = 0.178$, and the condition number is $\kappa_f = 11.36$. From the convergence results for the gradient method and the fast gradient method for the strongly convex case it follows that condition numbers close to 1 lead to fast convergence. So, for this example, fast convergence can be guaranteed for both methods.

Preconditioning

In the previous section we have seen that for gradient methods the number of iterations to find an approximate minimizer depends on the condition number κ_f.

In practice, a high condition number usually indicates that many iterations of Algorithm 3.1 and Algorithm 3.3 are needed to achieve the specified accuracy. To improve the conditioning of the problem and thus to lower the amount of computation needed to obtain an approximate minimizer, one can transform problem (3.2) into an equivalent problem by a variable transformation

$$y = Pz$$

with P invertible. The aim is to choose the preconditioner P such that the new objective function $h(y) = f(P^{-1}y)$ has a smaller condition number, i.e., $\kappa_h < \kappa_f$. Intuitively, the best choice for P is one that achieves a circular shape of the level sets of the objective function or at least comes close to it. Mathematically speaking, the preconditioner matrix P should be such that $\kappa_h \approx 1$. Finding optimal preconditioners for which $\kappa_h = 1$ is impossible for all but the simplest cases. We give one such example in the following.

Example 3.3 Optimal preconditioner for a quadratic function
Let $f(z) = \frac{1}{2}z'Hz + q'z$ with Hessian $H \succ 0$. It is easily verified that $P = H^{1/2}$ is the optimal preconditioner for which $\kappa_h = 1$, since $h(y) = \frac{1}{2}y'y + q'H^{-\frac{1}{2}}y$. However, computing $H^{1/2}$ and its inverse is more expensive than minimizing f, which only requires the inverse H^{-1}. So nothing has been gained from preconditioning in this example.

Stopping Criteria

Stopping criteria test whether the current iterate z^k is optimal or, more practically, has attained an acceptable level of suboptimality. In optimization, the most natural stopping criterion is

$$f(z^k) - f^* \leq \epsilon, \tag{3.10}$$

where $\epsilon > 0$ is an accuracy measure specified by the user. However, since the optimal value f^* is rarely known, we look for a computable *lower bound* $\underline{f}^{k,*}$ ($\leq f^*$), so that we arrive at the evaluable sufficient condition for (3.10)

$$f(z^k) - \underline{f}^{k,*} \leq \epsilon.$$

The latter stopping criterion works for any $\epsilon > 0$ if the lower bounds $\underline{f}^{k,*}$ converge to the optimal value f^* for an increasing iteration counter k.

For differentiable convex functions such lower bounds can be constructed right from the definition of convexity, i.e.,

$$f^* \geq f(z^k) + \nabla f(z^k)'(z^* - z^k)$$
$$\geq f(z^k) - \|\nabla f(z^k)\|\|z^* - z^k\|,$$

where the last inequality is due to Cauchy-Schwarz. Observe that for any converging method, this lower bound converges to f^* from below since the gradient at the minimizer z^* vanishes in unconstrained optimization.

For many problems, the distance between an iterate and the minimizer can be upper-bounded, i.e., $\|z^* - z^k\| \leq R$ for all iterations k. Consequently, we end up with the evaluable stopping criterion

$$\|\nabla f(z^k)\|\, R \leq \epsilon,$$

which is sufficient for (3.10).

More sophisticated stopping criteria can be derived if properties such as strong convexity of function f are exploited. For more details, we refer the interested reader to [250, Section 6.3].

Summary

In this section, we have discussed the concept of descent methods and presented one of the most important instances, the classic gradient method for smooth convex optimization. Although not a descent method in its basic form, Nesterov's fast gradient method was included due to its simplicity and improved speed of convergence. Both methods are particularly useful in the strongly convex case, where they converge linearly, with the fast gradient method attaining the best theoretical convergence rate possible for first-order methods. Despite a slower theoretical (and most of the time also practical) convergence, the classic gradient method is well-suited for processors with fixed-point arithmetic, for instance field-programmable gate arrays (FPGAs), because of its numerical robustness against inaccurate computations. For the fast gradient method, recent research, e.g., [97], indicates that care has to be taken if inexact gradients are used. When using gradient methods in practice, preconditioning should be applied to decrease the condition number of the problem in order to decrease the number of iterations needed. The latter can also be achieved by warm-starting, i.e., starting the methods from a point z^0 close to a minimizer z^*. This follows immediately from the dependence of the number of iterations on the distance $\|z^0 - z^*\|$. More information and links to the literature can be found in [250].

3.2.2 Newton's Method

The (fast) gradient method uses only first-order information of the function f. A much more quickly converging algorithm, at least locally, can be obtained when using second-order information, i.e., the Hessian $\nabla^2 f$. In the following, we assume f to be μ-strongly convex; hence, the Hessian $\nabla^2 f$ is positive definite on $\mathbb{R}^s$. Furthermore we assume that the Hessian $\nabla^2 f$ is Lipschitz continuous on $\mathbb{R}^s$ with Lipschitz constant M, i.e.,

$$\|\nabla^2 f(z) - \nabla^2 f(y)\| \leq M\|z - y\| \;\forall z, y \in \mathbb{R}^s. \tag{3.11}$$

Intuitively, the constant M measures how well f can be approximated by a quadratic function ($M = 0$ for a quadratic function f).

Algorithm 3.4   *Newton's method for (3.2) (strongly convex case)*

Input: Initial iterate $z^0 \in \mathbb{R}^s$
Output: Point close to z^*

   Repeat

      Newton direction: $d^k \leftarrow - \left[\nabla^2 f(z^k)\right]^{-1} \nabla f(z^k)$

      Line search: find $h^k > 0$ such that $f(z^k + h^k d^k) < f(z^k)$
      $z^{k+1} \leftarrow z^k + h^k d^k$

   Until stopping criterion is satisfied

The main idea of Newton's method is to approximate the function f by a local quadratic model $\theta(d)$ based on a second-order Taylor expansion,

$$f(z^k + d) \approx f(z^k) + \nabla f(z^k)'d + \frac{1}{2}d'\nabla^2 f(z^k)d \; = \theta(d), \qquad (3.12)$$

where $\nabla^2 f(z^k)$ is the Hessian of f evaluated at the current iterate z^k and d is a search direction. The latter is chosen to minimize the quadratic model $\theta(d)$. From the optimality condition $\nabla_d \theta(d) = 0$, the so-called *Newton direction*

$$d_N(z^k) = - \left[\nabla^2 f(z^k)\right]^{-1} \nabla f(z^k) \qquad (3.13)$$

is obtained. The Newton direction is a descent direction, as it satisfies (3.4):

$$\nabla f(z^k)'d_N(z^k) = -\nabla f(z^k)'[\nabla^2 f(z^k)]^{-1}\nabla f(z^k) < 0 \qquad (3.14)$$

for all $z^k \neq z^*$ since, by strong convexity, the inverse of the Hessian $\left[\nabla^2 f(z)\right]^{-1}$ exists for all $z \in \mathbb{R}^s$ and is positive definite. If the Lipschitz constant of the Hessian, M, is small, this method works very well, since the quadratic model remains a good approximation even if a full Newton step is taken. However, the quadratic model $\theta(d)$ is in general neither an upper nor a lower bound on f, and consequently a line search (Section 3.2.3) must be used to ensure a sufficient decrease in the function value, which results in global convergence of the algorithm. Newton's method is summarized in Algorithm 3.4. It can be shown to converge globally in two phases (for technical details see [65, Section 9.5] and [220, Section 1.2.4]):

1. *Damped Newton phase:* When the iterates are "far away" from the optimal point, the function value is decreased *sublinearly*, i.e., it can be shown that there exist constants η, γ (with $0 < \eta < \mu^2/M$ and $\gamma > 0$) such that $f(z^{k+1}) - f(z^k) \leq -\gamma$ for all z^k for which $\|\nabla f(z^k)\|_2 \geq \eta$. The term *damped* is used in this context since during this phase, the line search often returns step sizes $h^k < 1$, i.e., no full Newton steps are taken.

2. *Quadratic convergence phase:* If an iterate is "sufficiently close" to the optimum, i.e., $\|\nabla f(z^k)\|_2 < \eta$, a full Newton step is taken, i.e., the line search returns $h^k = 1$. The set $\{z \mid \|\nabla f(z)\|_2 < \eta\}$ is called the *quadratic convergence zone*, and once an iterate z^k is in this zone, all subsequent iterates remain in it. The function value $f(z^k)$ converges *quadratically* (i.e., with the iteration number growing proportionally to $k_{\max} \sim \mathcal{O}(\ln 1/\epsilon)$) to f^*.

The quadratic convergence property makes Newton's method one of the most powerful algorithms for unconstrained optimization of twice continuously differentiable functions. (The method is also widely employed for solving nonlinear equations.) As a result, it is used in interior point methods for constrained optimization to solve the (unconstrained or equality constrained) subproblems in barrier methods or to generate primal-dual search directions. We discuss these in detail in Section 3.3.2.

A disadvantage of Newton's method is the need to form the Hessian $\nabla^2 f(z^k)$, which can be numerically ill-conditioned, and to solve the linear system

$$\nabla^2 f(z^k) d_N = -\nabla f(z^k) \tag{3.15}$$

for the Newton direction d_N, which costs $\mathcal{O}(s^3)$ floating point operations in general. Each iteration is therefore significantly more expensive than one iteration of any of the gradient methods.

Preconditioning and Affine Invariance

Newton's method is *affine invariant*, i.e., a linear (or affine) transformation of variables as discussed in Section 3.2.1 will lead to the same iterates subject to transformation, i.e.,

$$z^k = P y^k \quad \forall k = 0, \dots, k_{\max}. \tag{3.16}$$

Therefore, the number of steps to reach a desired accuracy of the solution does not change. Consequently, preconditioning has no effect when using Newton's method.

Newton's Method with Linear Equality Constraints

Equality constraints are naturally handled in Newton's method as follows. Assume we want to solve the problem

$$\begin{aligned}
\text{minimize} \quad & f(z) \\
\text{subject to} \quad & Az = b
\end{aligned} \tag{3.17}$$

with $A \in \mathbb{R}^{p \times s}$ (assuming full rank p) and $b \in \mathbb{R}^p$. As discussed above, the Newton direction is given by the minimization of the quadratic model $\theta(d)$ at the current feasible iterate z^k (i.e., $Az^k = b$) with an additional equality constraint:

$$\begin{aligned}
d_N = \arg\min_d \quad & \frac{1}{2} d' \nabla^2 f(z^k) d + \nabla f(z^k)' d + f(z^k) \\
\text{subj. to} \quad & Ad = 0
\end{aligned} \tag{3.18}$$

Since d is restricted to lie in the nullspace of A, we ensure that any iterate generated by Newton's method will satisfy $Az^{k+1} = A(z^k + h^k d_N(z^k)) = b$. From the optimality conditions of (3.18), it follows that d_N is obtained from the solution to the linear system of dimension $s + p$:

$$\begin{bmatrix} \nabla^2 f(z) & A' \\ A & 0 \end{bmatrix} \begin{bmatrix} d \\ y \end{bmatrix} = \begin{bmatrix} -\nabla f(z) \\ 0 \end{bmatrix}, \tag{3.19}$$

where $y \in \mathbb{R}^p$ are the Lagrange multipliers associated with the equality constraint $Ad = 0$.

3.2.3   Line Search Methods

A line search determines the step length h^k that is taken from the current iterate $z^k \in \mathbb{R}^s$ along a search direction $\Delta z^k \in \mathbb{R}^s$ to the next iterate,

$$z^{k+1} = z^k + h^k \Delta z^k. \tag{3.20}$$

In the most general setting, a merit function $\varphi : \mathbb{R}^s \to \mathbb{R}$ is employed to measure the quality of the step taken to determine a value of the step size h^k that ensures convergence of the optimization algorithm. Merit functions can, for example, be the objective function f, or the duality measure μ (3.44) in primal-dual interior point methods.

The general line search problem can be written as a one dimensional optimization problem over the step size h,

$$h^* \in \arg\min_{h \geq 0} \varphi(z^k + h\Delta z^k), \tag{3.21}$$

minimizing the merit function along the ray $\{z^k + h\Delta z^k\}$ for $h \geq 0$. It is often too expensive to solve this problem exactly, and therefore usually *inexact* line search methods are employed in practice that return a step size h^k that approximates the optimal step size h^* from (3.21) sufficiently well. In the following, we give a brief overview on practical inexact line search methods; see [224] for a thorough treatment of the subject.

Wolfe Conditions

In order to ensure convergence of descent methods for unconstrained optimization, for example Newton's method presented in Section 3.2, certain conditions on the selected step size h^k must be fulfilled. Popular conditions for inexact line searches are the so-called *Wolfe conditions* [223]

$$f(z^k + h^k \Delta z^k) \leq f(z^k) + c_1 h^k \nabla f(z^k)' \Delta z^k \tag{3.22a}$$

$$\nabla f(z^k + h^k \Delta z^k)' \Delta z^k \geq c_2 \nabla f(z^k)' \Delta z^k \tag{3.22b}$$

with $0 < c_1 < c_2 < 1$. The first condition (3.22a) is called *Armijo's condition* [9], and ensures that a sufficient decrease in the objective is achieved. The second condition, the *curvature condition* (3.22b), ensures that the step length is not too small [223].

One of the most popular inexact line search methods that fulfills the Wolfe conditions is the backtracking line search as discussed in the following.

Backtracking Line Search for Newton's Method

The backtracking line search for Newton's method is given in Algorithm 3.5. It starts with a unit step size, which is iteratively reduced by a factor $t \in (0, 1)$ until Armijo's condition (3.22a) is satisfied. It can be shown that it is not necessary to check the curvature condition because it is implied by the backtracking rule [223, Algorithm 3.1].

Algorithm 3.5   *Backtracking Line Search for Newton's Method (Algorithm 3.4) [223, Algorithm 3.1]*

Input: Current iterate z^k, descent direction Δz^k, constants $c \in (0,1)$ and $t \in (0,1)$
Output: Step size h^k ensuring Wolfe conditions (3.22)

   $h^k \leftarrow 1$
   Repeat
      $h^k \leftarrow th^k$
   Until Armijo's condition (3.22a) holds, i.e., $f(z^k + h^k\Delta z^k) \leq f(z^k) + ch^k\nabla f(z^k)'\Delta z^k$

Backtracking Line Search for Constrained Optimization

In order to obtain a line search method that ensures in addition feasibility of the next iterate z^{k+1}, the backtracking procedure is preceeded by finding $0 < \bar{h} \leq 1$ such that $z^k + \bar{h}\Delta z^k$ is feasible [65, pp. 465]. The backtracking then starts with $h^k = \bar{h}$ in Algorithm 3.5.

3.3   Constrained Optimization

In this section we present three methods to solve the constrained optimization problem (1.1): gradient projection, interior point and active set methods.

3.3.1   Gradient Projection Methods

Gradient methods presented in Section 3.2.1 for unconstrained L-smooth convex optimization (or strongly convex optimization with parameter μ) have a natural extension to the constrained problem

$$\min_z \quad f(z)$$
$$\text{subj. to} \quad z \in S \tag{3.23}$$

where S is a convex subset of $\mathbb{R}^s$. The treatment of constraints within first-order methods is best explained in terms of the classic gradient method in Algorithm 3.1; the accelerated versions (Algorithm 3.2 and Algorithm 3.3) can be adapted according to the same principle.

As in the unconstrained case, we begin by rewriting the gradient update rule as a minimization of a quadratic upper bound of f defined in (3.7) with $h^k = 1/L$, but restrict the minimization to the set S:

$$z^{k+1} = \arg\min_{z \in S} f(z^k) + \nabla f(z^k)'(z - z^k) + \frac{L}{2}\|z - z^k\|^2. \tag{3.24}$$

The solution to this constrained minimization problem can be recast as

$$z^{k+1} = \pi_S\left(z^k - \frac{1}{L}\nabla f(z^k)\right), \tag{3.25}$$

where π_S denotes the *projection operator* for set S defined as

$$\pi_S(z) = \arg\min_{y \in S} \frac{1}{2}\|y - z\|^2. \tag{3.26}$$

Algorithm 3.6   *Gradient method for constrained smooth convex optimization*

Input: Initial iterate $z^0 \in S$, Lipschitz constant L of ∇f
Output: Point close to z^*

  Repeat

$$z^{k+1} \leftarrow \pi_S\left(z^k - \frac{1}{L}\nabla f(z^k)\right)$$

  Until stopping criterion is satisfied

Algorithm 3.7   *Fast gradient method for constrained smooth strongly convex optimization*

Input: Initial iterates $z^0 \in S$, $y^0 = z^0$; $0 < \sqrt{\mu/L} \leq \alpha^0 < 1$, Lipschitz constant L of ∇f, strong convexity parameter μ of f
Output: Point close to z^*

  Repeat

$$z^{k+1} \leftarrow \pi_S\left(y^k - \frac{1}{L}\nabla f(y^k)\right)$$

Compute $\alpha^{k+1} \in (0,1)$: $\alpha^{k+1^2} = (1 - \alpha^{k+1})\alpha^{k^2} + \frac{\mu\alpha^{k+1}}{L}$

$$\beta^k \leftarrow \frac{\alpha^k(1 - \alpha^k)}{\alpha^{k^2} + \alpha^{k+1}}$$

$$y^{k+1} \leftarrow z^{k+1} + \beta^k(z^{k+1} - z^k)$$

  Until stopping criterion is satisfied

The expressions (3.24) and (3.25) for z^{k+1} can be shown to be equivalent by deriving the optimality conditions and noting that they coincide. Thus, for constrained optimization, Algorithms 3.1, 3.2 and 3.3 remain unchanged, except that the right hand side of the assignment in the first line of these algorithms is projected onto set S. The resulting algorithmic schemes are stated in Algorithm 3.6 (gradient method) and Algorithm 3.7 (fast gradient method for the strongly convex case) and illustrated in Figure 3.3. It can be shown that for both Algorithm 3.6 and Algorithm 3.7 the same convergence results hold as in the unconstrained case [220, Section 2.2.4].

Whenever the projection operator for set S can be evaluated efficiently, as for the convex sets listed in Table 3.1, gradient projection methods have been shown to work very well in practice. Such sets, which are also referred to as 'simple' sets in the literature, can be translated, uniformly scaled and/or rotated without complicating the projection as stated in the next lemma.

Table 3.1 Euclidean norm projection operators of selected convex sets in $\mathbb{R}^s$.

Set Definition	Projection Operator
affine set $S = \{z \in \mathbb{R}^s \mid Az = b\}$ with $(A, b) \in \mathbb{R}^{p \times s} \times \mathbb{R}^p$	$\pi_S(z) = \begin{cases} z + A'(AA')^{-1}(b - Az) & \text{if rank } A = p, \\ z + A'A'^{\dagger}(A^{\dagger}b - z) & \text{otherwise} \end{cases}$
nonnegative orthant $S = \{z \in \mathbb{R}^s \mid z \geq 0\}$	$(\pi_S(z))_i = \begin{cases} z_i & \text{if } z_i \geq 0, \\ 0 & \text{otherwise} \end{cases} \quad i = 1, \ldots, s$
rectangle $S = \{z \in \mathbb{R}^s \mid l \leq z \leq u\}$ with $(l, u) \in \mathbb{R}^s \times \mathbb{R}^s$	$(\pi_S(z))_i = \begin{cases} l_i & \text{if } z_i < l_i, \\ z_i & \text{if } l_i \leq z_i \leq u_i, \quad i = 1, \ldots, s \\ u_i & \text{if } z_i > u_i \end{cases}$
2-norm ball $S = \{z \in \mathbb{R}^s \mid \|z\| \leq r\}, \, r \geq 0$	$\pi_S(z) = \begin{cases} r\frac{z}{\|z\|} & \text{if } \|z\| > r, \\ z & \text{otherwise} \end{cases}$

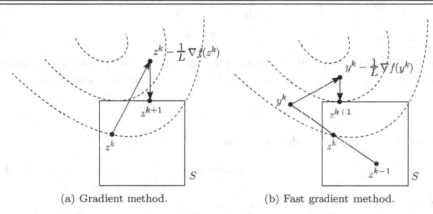

(a) Gradient method.   (b) Fast gradient method.

Figure 3.3 Classic gradient method (left) and fast gradient method (right) for the constrained case. After computing the standard update rule from the unconstrained case, a projection onto the feasible set is applied.

Lemma 3.3 (Projection on Translated, Scaled and Rotated Convex Set)

Let $S \subseteq \mathbb{R}^s$ be a closed convex set and π_S its associated projection operator. Projection of point $\bar{z} \in \mathbb{R}^s$ on the set

$$\hat{S} = \{z \in \mathbb{R}^s \mid z = \gamma W y + c, \, y \in S\}$$

that is obtained from set S by translation with offset $c \in \mathbb{R}^s$, positive scaling by $\gamma > 0$ and a linear transformation with an orthonormal matrix $W \in \mathbb{R}^{s \times s}$, can be accomplished by

$$\pi_{\hat{S}}(\bar{z}) = \gamma W \pi_S\left(\gamma^{-1}W'(\bar{z} - c)\right) + c.$$

If the feasible set S is polyhedral and does not have a computationally efficient projection operator, it can alternatively be precomputed by means of multiparametric programming (see Chapter 6) with the point to project as the parameter. Another possibility is to solve the original problem in the dual domain instead as discussed next.

Solution in Dual Domain

If set S is such that the projection (3.26) cannot be carried out efficiently, one can still use first-order methods to solve the constrained problem (3.23) by adopting a dual approach. In the following, we assume that set S can be written as

$$S = \{z \in \mathbb{R}^s \mid Az = b,\, z \in \mathbb{K}\}, \tag{3.27}$$

where $\mathbb{K}$ is a convex subset of $\mathbb{R}^s$ that allows for an efficient projection and the pair (A, b) is from $\mathbb{R}^{p \times s} \times \mathbb{R}^p$. Many practically relevant sets can be represented in terms of (3.27), i.e., as the intersection of a "simple" convex set and an affine set, e.g., polyhedra. We can now rewrite the original problem (3.23) as

$$\min_{z} \ f(z)$$
$$\text{subj. to } Az = b$$
$$z \in \mathbb{K}.$$

Using a technique called partial Lagrange relaxation [52, Section 4.2.2] we eliminate the complicating equality constraints and define the (concave) *dual function* as

$$d(v) = \min_{z \in \mathbb{K}} f(z) + v'(Az - b), \tag{3.28}$$

where $v \in \mathbb{R}^p$. The concave dual problem is then given by

$$d^* = \max_{v} d(v), \tag{3.29}$$

which we also call the *outer problem*.

In order to apply any of the gradient methods for unconstrained optimization in Section 3.2 for solving the outer problem, the dual function $d(v)$ needs to fulfill two requirements: (i) differentiability, i.e., the gradient $\nabla d(v)$ must exist for all v, and (ii) Lipschitz continuity of the gradient. Next, we will show that if f is strongly convex, both requirements are indeed fulfilled.

In the following it is assumed that the objective function f is strongly convex. Strong convexity ensures uniqueness of the minimizer $z^*(v)$ in the definition of the dual function in (3.28). It is this uniqueness of $z^*(v)$ that leads to a differentiable dual function as shown in [52, Danskin's Theorem, Proposition B.25]. Specifically, it can be proved that

$$\nabla d(v) = Az^*(v) - b,$$

i.e., every evaluation of the dual gradient requires the solution of the so-called *inner problem* (3.28).

The Lipschitz constant L_d of the dual gradient can be upper bounded by

$$L_d \leq \frac{\|A\|^2}{\mu}$$

as stated in [221, Theorem 1]. This upper bound is sufficient for the step size computation for any gradient method (note that the step size is given by $1/L_d$). However, the less conservative the Lipschitz constant is, the larger the step size and the smaller the number of iterations. In [251, Theorem 7] it is shown that for the case of quadratic objective functions $f(z) = \frac{1}{2}z'Hz + q'z$, $H \succ 0$, and under mild assumptions, the *tight* Lipschitz constant of the dual gradient is

$$L_d^* = \|AH^{-\frac{1}{2}}\|^2.$$

So, strong convexity of the objective function f is key for using the gradient methods in Section 3.2.1 for the solution of the dual problem (3.29). If f is not strongly convex, there are different approaches available in the literature, e.g., a smoothing approach for the dual function [221].

Note that the (unique) primal minimizer z^* can be recovered from the maximizer of the dual problem v^*, assuming it exists, by

$$z^* = z^*(v^*).$$

It is also important to notice that, in general, the dual function $d(v)$ lacks strong concavity, so only sublinear convergence rates (instead of linear rates for the primal problem) can be guaranteed for the solution of the dual problem. This and the fact that each evaluation of the dual gradient requires solving the inner problem can make the dual approach, as presented in this section, slow in practice. More advanced first-order methods can significantly speed up convergence of the dual approach, e.g., the alternating direction method of multipliers (ADMM). For more details on ADMM, the interested reader is referred to the survey paper [64].

Summary

This section has introduced the generalization of gradient methods to the constrained case, the so-called *gradient projection methods*. As the name suggests, the generalization comes from having to compute the projection of a gradient step onto the feasible set. Apart from this additional operation, these methods are identical to the ones introduced for the unconstrained case.

There exist many important convex sets for which the projection operator can be evaluated in an efficient way, some of which are stated in Table 3.1. More sets can be found, e.g., in [250, Section 5.4], which also contains further literature links. One important consequence of having a set constraint is that a full preconditioner matrix P (Section 3.2.1 and Example 3.3) is not admissible anymore since in the new basis the set can lose its favorable projection properties. For instance, consider a box constraint with its projection operator given in Table 3.1. Only for a diagonal preconditioner the set remains a box in the new basis and thus easy to project. For quadratic objective functions, finding the best preconditioner under set-related structure constraints turns out to be a convex semi-definite program and hence can be solved efficiently [250, Section 8.4].

Finally, we point the reader to the recent *proximity gradient methods*, which can be understood as a natural generalization of gradient projection methods. An introduction to these methods and literature links can be found in [250, Chapter 5] and [228].

3.3.2 Interior Point Methods

Today interior point methods are among the most widely used numerical methods for solving convex optimization problems. This is the result of 25 years of intensive research that has been initiated by Karmarkar's seminal paper in 1984 on solving LPs [166]. In 1994, Nesterov and Nemirovskii generalized interior point methods to nonlinear convex problems, including second-order cone and semi-definite programming [222]. Issues like numerical ill-conditioning, efficient and stable solution of linear systems, etc., are now well understood, and numerous free and commercial codes are available that can solve linear and quadratic programs with high reliability.

We first discuss barrier methods, which were the first polynomial-time algorithms for linear programming of practical relevance. Then we present modern primal-dual method. This powerful class of interior point methods forms the basis of almost all implementations today.

Throughout this section, we consider the primal P-IPM problem

$$\begin{aligned} \min_z \quad & f(z) \\ \text{subj. to} \quad & Az = b \\ & g(z) \leq 0 \end{aligned} \tag{3.30}$$

where $z \in \mathbb{R}^s$ are the primal variables, the functions $f : \mathbb{R}^s \to \mathbb{R}$ and $g : \mathbb{R}^s \to \mathbb{R}^m$ are convex and twice continuously differentiable (for g, this holds component-wise for each function g_i, $i = 1, \ldots, m$). We assume that there exists a strictly feasible point z^0 with respect to $g(z) \leq 0$, i.e., the set

$$S = \{z \in \mathbb{R}^s \mid g_i(z) \leq 0, i = 1, \ldots, m\}, \tag{3.31}$$

has a nonempty interior. We furthermore assume that a minimizer z^* exists and that it is attained.

Primal Barrier Methods

The main idea of the barrier method is to convert the constrained optimization P-IPM problem (3.30) into an unconstrained problem (with respect to inequalities) by means of a *barrier function* $\Phi_g : \mathbb{R}^s \to \mathbb{R}$. We denote by $P(\mu)$ the problem

$$\begin{aligned} z^*(\mu) \in \arg\min \quad & f(z) + \mu \Phi_g(z) \\ \text{subj. to} \quad & Az = b \end{aligned} \tag{3.32}$$

where $\mu > 0$ is called *barrier parameter*. The purpose of the barrier function Φ_g is to "trap" an optimal solution of problem $P(\mu)$ (3.32), which we denote by $z^*(\mu)$, in the set S. Thus, $\Phi_g(z)$ must take on the value $+\infty$ whenever $g_i(z) > 0$ for some i, and a finite value otherwise. As a result, $z^*(\mu)$ is feasible with respect to S, but it differs from z^* since we have perturbed the objective function by the barrier term.

With the above idea in mind, we are now ready to outline the (primal) barrier method summarized in Algorithm 3.8. Starting from (a possibly large) μ^0, a solution $z^*(\mu^0)$ to problem $P(\mu^0)$ is computed, for example by any of the methods for unconstrained optimization presented in Section 3.2. For this reason, Φ_g should be twice continuously differentiable, allowing one to apply Newton's method. After $z^*(\mu^0)$ has been computed, the barrier parameter is decreased by a constant factor, $\mu^1 = \mu^0/\kappa$ (with $\kappa > 1$), and $z^*(\mu^1)$ is computed. This procedure is repeated until μ has been sufficiently decreased. It can be shown under mild conditions that $z^*(\mu) \to z^*$ from the interior of S as $\mu \to 0$. The points $z^*(\mu)$ form the so-called *central path*, which is a continuously differentiable curve in S that ends in the solution set (see Example 3.4). Solving a subproblem $P(\mu)$ is therefore called *centering*.

If the decrease factor κ is not too large, $z^*(\mu^k)$ will be a good starting point for solving $P(\mu^{k+1})$, and therefore Newton's method can be applied to $P(\mu^{k+1})$ to exploit the local quadratic convergence rate. Note that the equality constraints in (3.30) have been preserved in problem $P(\mu)$ (3.32), since they do not cause substantial difficulties for unconstrained methods as shown in Section 3.2. In particular, we refer the reader to Section 3.2.2 for the equality constrained Newton method.

Logarithmic Barrier

So far, we have not specified which function to use as a barrier. For example, the indicator function

$$I_g(z) = \begin{cases} 0 & \text{if } g(z) \le 0 \\ +\infty & \text{otherwise} \end{cases}, \tag{3.33}$$

trivially achieves the purpose of a barrier, but it is not useful since methods for smooth convex optimization as discussed in Section 3.2 require the barrier function to be convex and continuously differentiable. A twice continuously differentiable convex barrier function that approximates I_g well is the *logarithmic barrier function*

$$\Phi_g(z) = -\sum_{i=1}^{m} \ln\left(-g_i(z)\right), \tag{3.34}$$

with domain $\{z \in \mathbb{R}^s \mid g_i(z) < 0 \ \forall \ i = 1, \ldots, m\}$ (the logarithmic barrier confines z to the *interior* of the set S), gradient

$$\nabla \Phi_g(z) = \sum_{i=1}^{m} \frac{1}{-g_i(z)} \nabla g_i(z) \tag{3.35}$$

and continuous Hessian

$$\nabla^2 \Phi_g(z) = \sum_{i=1}^{m} \frac{1}{g_i(z)^2} \nabla g_i(z) \nabla g_i(z)' + \frac{1}{-g_i(z)} \nabla^2 g_i(z), \tag{3.36}$$

which makes it possible to use Newton's method for solving the barrier subproblems. The logarithmic barrier function is used in almost all implementations of barrier methods. In fact, the existence of polynomial-time algorithms for convex optimization is closely related to the existence of barrier functions for the underlying feasible sets [222].

Algorithm 3.8   *Barrier interior point method for (3.30)*

Input: Strictly feasible initial iterate z^0 w.r.t. $g(z) \le 0$, μ^0, $\kappa > 1$, tolerance $\epsilon > 0$
Output: Point close to z^*

> **Repeat**
>> Compute $z^*(\mu^k)$ by minimizing $f(z) + \mu^k \Phi_g(z)$ subject to $Az = b$ starting from the previous solution z^{k-1} (usually by Newton's method, Algorithm 3.4). This is called "centering step."
>> Update: $z^k \leftarrow z^*(\mu^k)$
>> Stopping criterion: Stop if $m\mu^k < \epsilon$
>> Decrease barrier parameter: $\mu^{k+1} \leftarrow \mu^k / \kappa$
>
> **Until** stopping criterion is satisfied

Example 3.4   Central Path of a QP
Consider the QP

$$\min \quad \tfrac{1}{2} z' \left[\begin{smallmatrix} 2 & 0 \\ 0 & 1 \end{smallmatrix}\right] z + \left[\, 4 \ \tfrac{1}{2} \,\right] z$$

$$\text{subj. to} \quad \begin{bmatrix} -1 & 2.0588 \\ -1 & -1.7527 \\ -3.9669 & 1 \\ 1.1997 & -1.3622 \\ 1.1673 & 1.3111 \end{bmatrix} z \le \begin{bmatrix} 1.0890 \\ 1.0623 \\ 2.0187 \\ 1 \\ 1 \end{bmatrix} \tag{3.37}$$

The solutions of $P(\mu)$ (3.32) for this Quadratic Program for different values of μ are depicted in Figure 3.4. For large μ, the level sets are almost identical to the level sets of $\Phi_g(z)$, i.e., $z^*(5)$ is very close to the *analytic center* $z^a = \arg\min \Phi_g(z)$ of the feasible set. As μ decreases, $z^*(\mu)$ approaches z^*, by following the central path.

Example 3.5   Barrier method for QPs
In the following, we derive the barrier method for quadratic programs of the form

$$\begin{aligned} \text{minimize} \quad & \frac{1}{2} z' H z + q' z \\ \text{subject to} \quad & Az = b, \\ & Gz \le w \end{aligned} \tag{3.38}$$

with $H \succ 0$, $A \in \mathbb{R}^{p \times s}$, $b \in \mathbb{R}^p$, $G \in \mathbb{R}^{m \times s}$ and $w \in \mathbb{R}^m$ from the general method described above. The logarithmic barrier function for linear inequalities takes the form

$$\Phi_g(z) = -\sum_{i=1}^m \ln(w_i - G_i z), \ \nabla\Phi_g(z) = \sum_{i=1}^m \frac{1}{w_i - G_i z} G_i',$$

$$\nabla^2 \Phi_g(z) = \sum_{i=1}^m \frac{1}{(w_i - G_i z)^2} G_i' G_i \tag{3.39}$$

where G_i denotes the i-th row of G and w_i the i-th element of w. Hence problem $P(\mu)$ (3.32) takes the form

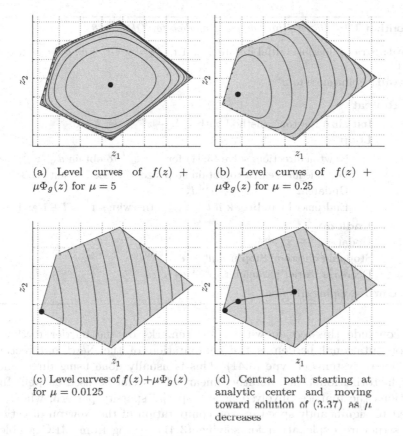

(a) Level curves of $f(z) +$ $\mu\Phi_g(z)$ for $\mu = 5$

(b) Level curves of $f(z) +$ $\mu\Phi_g(z)$ for $\mu = 0.25$

(c) Level curves of $f(z)+\mu\Phi_g(z)$ for $\mu = 0.0125$

(d) Central path starting at analytic center and moving toward solution of (3.37) as μ decreases

Figure 3.4 Example 3.4. Illustration of interior point method for example problem (3.37). Black dots are the solutions $z^*(\mu)$.

$$\min_z \frac{1}{2}z'Hz + q'z - \mu\sum_{i=1}^{m}\ln(w_i - G_i z) \tag{3.40}$$

$$\text{subj. to } Az = b,$$

which we solve by Newton's method. Using the formula for the Newton direction d_N with equality constraints (3.19), we arrive at the following linear system of size $s + p$, which has to be solved in each Newton step at the current iterate z:

$$\begin{bmatrix} H + \mu\sum_{i=1}^{m}\frac{1}{(w_i - G_i z)^2}G_i'G_i & A' \\ A & 0 \end{bmatrix}\begin{bmatrix} d_N \\ y \end{bmatrix} = -\begin{bmatrix} Hz + q + \mu\sum_{i=1}^{m}\frac{1}{w_i - G_i z}G_i' \\ 0 \end{bmatrix}. \tag{3.41}$$

The resulting barrier method for solving (3.38) is summarized in Algorithm 3.9. The inner loop resembles the Newton iterations of the algorithm, computing $z^*(\mu)$. Note that a primal feasible line search must be employed to ensure that the iterates remain in the interior of S, see Section 3.2.3 for details.

Algorithm 3.9   *Barrier interior point method for QPs (3.38)*

Input: Strictly feasible initial iterate z^0 w.r.t. $Az = b$, $Gz \leq w$, μ^0, $\kappa > 1$, tolerance $\epsilon > 0$
Output: Point close to z^*

    Repeat
        Initialize Newton's method with $z^{k,0} \leftarrow z^{k-1}$. Set $l = 0$.
        Loop
            Newton direction: solve (3.41) for $z = z^{k,l}$ to obtain $d_N^{k,l}(z^{k,l})$.
            Feasible line search to obtain step size $h^{k,l}$ (see Section 3.2.3).
            Update: $z^{k,l+1} \leftarrow z^{k,l} + h^{k,l} d_N^{k,l}$
            End inner loop: **break** if $l = l_{\max}$, otherwise set $l = l + 1$ and continue.
        Endloop
        Update: $z^k \leftarrow z^{k,l_{\max}}$
        Stopping criterion: Stop if $m\mu^k < \epsilon$
        Decrease barrier parameter: $\mu^{k+1} \leftarrow \mu^k / \kappa$
    Until stopping criterion

We conclude this section with a few remarks on the barrier method. The main computational burden is the computation of the Newton directions by solving linear systems of type (3.41). This is usually done using direct methods (matrix factorizations), but iterative linear solvers can be used as well. In most applications the matrices involved have a special sparsity pattern, which can be exploited to significantly speed up the computation of the Newton direction. We discuss structure exploitation for solving (3.41) arising from MPC problems in Section 14.3.

If the barrier parameter μ is decreased too quickly, many inner iterations are needed, but only a few outer iterations. The situation is reversed if μ is decreased only slowly, which leads to a quick convergence of the inner problems (usually in one or two Newton steps), but more outer iterations are needed. This tradeoff is not much of an issue in practice. Usually, the total number of Newton steps is in the range of 20−40, essentially independent of the conditioning of the problem. The barrier method can be shown to converge linearly; more details can be found in [65, Section 11.3].

Finally, we would like to point out that the barrier method is a *feasible* method, which means that it starts with a feasible point and all iterates remain feasible. This complicates the implementation in practice, as a line search must be employed that maintains feasibility, and a strictly feasible initial iterate z^0 has to be supplied by the user or computed first by what is called a *Phase I* method. See [65] for more details.

Computational experience has shown that modern primal-dual methods are significantly more effective with only little additional computational cost when compared to the barrier method. Therefore, primal-dual methods are now predominant in commercial codes. They also allow infeasible iterates, that is, the equality and inequality constraints on the primal variables are satisfied only at convergence, which alleviates the necessity for finding a feasible initial point. We discuss this class of methods next.

Primal-Dual Methods

The general idea of primal-dual interior point methods is to solve the Karush-Kuhn-Tucker (KKT) conditions by a modified version of Newton's method. The nonlinear KKT equations represent necessary and sufficient conditions for optimality for convex problems under the assumptions discussed in Section 1.6. For the P-IPM problem (3.30), the KKT conditions are

$$\nabla f(z) + A'v + G(z)'u = 0 \tag{3.42a}$$

$$Az - b = 0 \tag{3.42b}$$

$$g(z) + s = 0 \tag{3.42c}$$

$$s_i u_i = 0, \ i = 1, \ldots, m \tag{3.42d}$$

$$(s, u) \geq 0 \tag{3.42e}$$

where the vector $s \in \mathbb{R}_+^m$ denotes slack variables for the inequality constraints $g(z) \leq 0$, and $u \in \mathbb{R}_+^m$ is the vector of the associated Lagrange multipliers. The vector $v \in \mathbb{R}^p$ denotes the Lagrange multipliers for the equality constraints $Az = b$, and the matrix $G(z) \in \mathbb{R}^{m \times s}$ is the Jacobian of g evaluated at z. The variables z and s are from the primal, the variables v and u from the dual space.

If a primal-dual pair of variables (z^*, v^*, u^*, s^*) is found that satisfies these conditions, the corresponding primal variables (z^*, s^*) are optimal for (3.30) (and the dual variables (v^*, u^*) are optimal for the dual of (3.30)). Since Newton's method is a powerful tool for solving nonlinear equations, the idea of applying it to the system of nonlinear KKT equations is apparent. However, certain modifications to a pure Newton method are necessary to obtain a useful method for constrained optimization. The name *primal-dual* indicates that the algorithm operates in both the primal and dual space.

There are many variants of primal-dual interior point methods. For a thorough treatment of the theory, we refer the reader to the book by Wright [290], which gives a comprehensive overview of the different approaches. In this section, after introducing the general framework of primal-dual methods, we restrict ourselves to the presentation of a variant of Mehrotra's predictor-corrector method [206], because it forms the basis of most existing implementations of primal-dual interior point methods for convex optimization and has proven particularly effective in practice. The method we present here allows infeasible starting points.

Central Path

We start by outlining the basic idea. Most methods track the central path, which we have encountered already in barrier methods, to the solution set. These methods are called path-following methods. In the primal-dual space, the central path is defined as the set of points (z, v, u, s), for which the following *relaxed* KKT conditions hold:

$$\nabla f(z) + A'v + G(z)'u = 0 \tag{3.43a}$$

$$Az - b = 0 \tag{3.43b}$$

$$g(z) + s = 0 \tag{3.43c}$$

$$s_i u_i = \tau, \ i = 1, \ldots, m, \ \tau > 0 \tag{3.43d}$$

$$(s, u) > 0 \tag{3.43e}$$

In (3.43) as compared to (3.42), the complementarity condition (3.42d) is relaxed by a scalar $\tau > 0$, the path parameter, and slacks s and multipliers u are required to be positive instead of nonnegative, i.e., s and u lie in the interior of the positive orthant. The main idea of primal-dual methods is to solve (3.43) for successively decreasing values of τ, and thereby to generate iterates (z^k, v^k, u^k, s^k) that approach (z^*, v^*, u^*, s^*) as $\tau \to 0$. The similarity with the barrier method in this regard is evident. However, the Newton step is modified in primal-dual interior point methods, as we shall see below.

Measure of Progress

Similar to primal barrier methods, the subproblem (3.43) is usually solved only approximately, here by applying only one Newton step. As a result, the generated iterates are not exactly on the central path, i.e., $s_i u_i \neq \tau$ for some i. Hence a more useful measure than the path parameter τ for the progress of the algorithm is the *average* value of the complementarity condition (3.43d),

$$\mu = (s'u)/m \tag{3.44}$$

evidently, $\mu \to 0$ implies $\tau \to 0$. In interior point nomenclature, μ is called the *duality measure*, as it corresponds to the duality gap scaled by $1/m$ if all other equalities in (3.43) are satisfied. To be able to reduce μ substantially (at least by a constant factor) at each iteration of the method, the pure Newton step is enhanced with a centering component, as discussed next.

Newton Directions and Centering

A search direction is obtained by linearizing (3.43) at the current iterate (z, v, u, s) and solving

$$\begin{bmatrix} H(z,u) & A' & G(z)' & 0 \\ A & 0 & 0 & 0 \\ G(z) & 0 & 0 & I \\ 0 & 0 & S & Z \end{bmatrix} \begin{bmatrix} \Delta z \\ \Delta v \\ \Delta u \\ \Delta s \end{bmatrix} = - \begin{bmatrix} \nabla f(z) + A'v + G(z)'u \\ Az - b \\ g(z) + s \\ Su - \nu \end{bmatrix} \tag{3.45}$$

where $S = \mathrm{diag}(s_1, \ldots, s_m)$ and $Z = \mathrm{diag}(u_1, \ldots, u_m)$, the (1,1) block in the coefficient matrix is

$$H(z,u) = \nabla^2 f(z) + \sum_{i=1}^{m} u_i \nabla^2 g_i(z) \tag{3.46}$$

and the vector $\nu \in \mathbb{R}^m$ allows a modification of the right-hand side, thereby generating different search directions that depend on the particular choice of ν. For brevity, we denote the solution to (3.45) by $\Delta[z, v, u, s](\nu)$.

The vector ν can take a value between the following two extremes:

- If $\nu = 0$, the right hand side in (3.45) corresponds to the residual of the KKT conditions (3.42), i.e., $\Delta[z, v, u, s](0)$ is the pure Newton direction

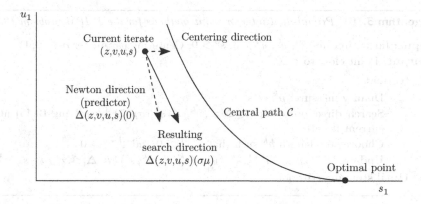

Figure 3.5 Affine-scaling (pure Newton) and centering search directions of classic primal-dual methods. The Newton direction makes only progress towards the solution when it is computed from a point close to the central path, hence centering must be added to the pure Newton step to ensure that the iterates stay sufficiently close to the central path.

that aims at satisfying the KKT conditions (3.42) in one step based on a linearization. This direction is called *affine-scaling* direction. However, pure Newton steps usually decrease some products $s_i u_i$ prematurely towards zero, i.e., the generated iterate is close to the boundary of the feasible set (if $s_i u_i \approx 0$, either s_i or u_i must be small). This is disadvantageous since from such points only very small steps can be taken, and convergence is prohibitively slow. Hence the iterates must be *centered* by bringing them closer to the central path.

- If $\nu = \mu \mathbf{1}$, the complementarity condition (3.43d) is modified such that the resulting search direction aims at making the individual products $s_i u_i$ equal to their average value μ. This is called *centering*. A centering step does not decrease the average value, so no progress towards the solution is made.

A tradeoff between the two goals of progressing towards the solution and centering is achieved by computing search directions $\Delta[z, v, u, s](\nu)$ by solving (3.45) for

$$\nu = \sigma \mu \mathbf{1} \tag{3.47}$$

where $\sigma \in (0, 1)$ is called the *centering parameter*. This is the key difference from pure Newton steps, and the main ingredient (together with tracking the central path) in making Newton's method work efficiently for solving the nonlinear KKT system (3.42). Figure 3.5 depicts this combination of search directions schematically, and a principle framework for primal-dual methods is given in Algorithm (3.10). We will instantiate it with Mehrotra's method below.

In the discussion above, we have excluded almost all theoretical aspects concerned with the convergence of primal-dual methods. For example, the terms "sufficiently close to the central path" and "sufficient reduction in μ" have precise mathematical counterparts that allow one to analyze the convergence properties. The book by Wright [290] is a good starting point for the interested reader.

Algorithm 3.10　*Primal-dual interior point methods for the P-IPM problem (3.30)*

Input: Initial iterates z^0, v^0, $u^0 > 0$, $s^0 > 0$, Centering parameter $\sigma \in (0, 1)$
Output: Point close to z^*

    Repeat

        Duality measure: $\mu^k \leftarrow s^{k\prime} u^k / m$
        Search direction: compute $\Delta[z^k, v^k, u^k, s^k](\sigma\mu^k \mathbf{1})$ by solving (3.45) at the
        current iterate
        Choose step length h^k such that $s^{k+1} > 0$ and $u^{k+1} > 0$
        Update: $(z^{k+1}, v^{k+1}, u^{k+1}, s^{k+1}) \leftarrow (z^k, v^k, u^k, s^k) + h^k \Delta[z^k, v^k, u^k, s^k](\sigma^k\mu^k \mathbf{1})$

    Until stopping criterion

Predictor-Corrector Methods

Modern methods use the full Newton step $\Delta[z, v, u, s](0)$ merely as a *predictor* to estimate the error made by using a linear model of the central path. If a full step along the affine-scaling direction were taken, the complementarity condition would evaluate to

$$(S + \Delta S^{\mathrm{aff}})(u + \Delta u^{\mathrm{aff}}) = \underbrace{Su + \Delta S^{\mathrm{aff}} u + S \Delta u^{\mathrm{aff}}}_{=0} + \Delta S^{\mathrm{aff}} \Delta u^{\mathrm{aff}} \qquad (3.48)$$

where capital letters denote diagonal matrices constructed from the corresponding vectors. The first three terms sum up to zero by the last equality of (3.45). Hence the full Newton step produces an error of $\Delta S^{\mathrm{aff}} \Delta u^{\mathrm{aff}}$ in the complementarity condition. In order to compensate for this error, a *corrector* direction $\Delta[z, v, u, s](-\Delta S^{\mathrm{aff}} \Delta u^{\mathrm{aff}})$ can be added to the Newton direction. This compensation is not perfect as the predictor direction is only an approximation; nevertheless, predictor-corrector methods work usually better than single-direction methods. The principle of predictor-corrector methods is depicted in Figure 3.6.

An important contribution of Mehrotra [206] was to define an adaptive rule for choosing the centering parameter σ based on the information from the predictor step. His rule

$$\sigma = (\mu^{\mathrm{aff}}/\mu)^3 \qquad (3.49)$$

increases the centering if the progress would be poor, that is, if the predicted value of the duality measure, μ^{aff}, is not significantly smaller than μ. If good progress can be made, then $\sigma \ll 1$, and the direction along which the iterate is moved is closer to the pure Newton direction. The corrector and centering direction can be calculated in one step because $\Delta[z, v, u, s](\nu)$ is linear in ν; the final search direction is therefore given by $\Delta[z, v, u, s](\sigma\mu\mathbf{1} - \Delta S^{\mathrm{aff}} \Delta u^{\mathrm{aff}})$.

A variant of Mehrotra's predictor-corrector method that works well in practice is given in Algorithm 3.11. The parameter γ is a safeguard against numerical errors, keeping the iterates away from the boundary. Because of the predictor-corrector scheme, two linear systems have to be solved in each iteration. Since the coefficient matrix is the same for both solves, direct methods are preferred to solve the linear systems in Steps 3 and 7 of the algorithm, as the computational bottleneck of the

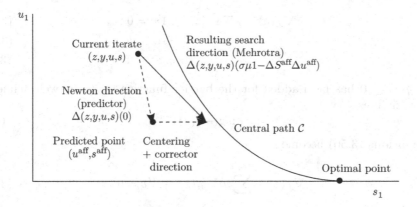

Figure 3.6 Search direction generation in predictor-corrector methods. The affine search direction can be used to correct for linearization errors, which results in better performance in practice due to closer tracking of the central path.

method is the factorization of the coefficient matrix. Once it has been factored in Step 3, the computational cost of the additional solve in Step 7 is negligible, and is more than compensated by the savings in the number of iterations.

Algorithm 3.11 *Mehrotra primal-dual interior point method for (3.30)*

Input: Initial iterates z^0, v^0, $u^0 > 0$, $s^0 > 0$, $\gamma < 1$ (typically $\gamma = 0.99$)
Output: Point close to z^*

> **Repeat**
>> Duality measure: $\mu^k \leftarrow s^{k\prime} u^k / m$
>> Newton direction (predictor): compute $\Delta[z^{k,\mathrm{aff}}, v^{k,\mathrm{aff}}, u^{k,\mathrm{aff}}, s^{k,\mathrm{aff}}](0)$ by solving (3.45) at the current iterate
>> Line search: $h^{k,\mathrm{aff}} \leftarrow \max\{h \in [0,1] \mid s^k + h\Delta s^{k,\mathrm{aff}} \geq 0, \; u^k + h\Delta u^{k,\mathrm{aff}} \geq 0\}$
>> Predicted (affine) duality measure: $\mu^{k,\mathrm{aff}} \leftarrow (s^k + h^{k,\mathrm{aff}}\Delta s^{k,\mathrm{aff}})'(u^k + h^{k,\mathrm{aff}}\Delta u^{k,\mathrm{aff}})/m$
>> Centering parameter: $\sigma^k \leftarrow (\mu^{k,\mathrm{aff}}/\mu^k)^3$
>> Final search direction: compute $\Delta[z^k, v^k, u^k, s^k](\sigma^k\mu^k\mathbf{1} - \Delta S^{k,\mathrm{aff}}\Delta u^{k,\mathrm{aff}})$ by solving (3.45) at current iterate
>> Line search: $h^k \leftarrow \max\{h \in [0,1] \mid s^k + h\Delta s^k \geq 0, \; u^k + h\Delta u^k \geq 0\}$
>> Update: $(z^{k+1}, v^{k+1}, u^{k+1}, s^{k+1}) \leftarrow (z^k, v^k, u^k, s^k) + \gamma h^k \Delta[z^k, v^k, u^k, s^k](\sigma^k\mu^k\mathbf{1} - \Delta S^{k,\mathrm{aff}}\Delta u^{k,\mathrm{aff}})$
> **Until** stopping criterion is satisfied

Relation of Primal-Dual to Primal Barrier Methods

An insightful relation between primal barrier and primal-dual methods can be established as follows. Writing the optimality conditions for problem $P(\mu)$ (3.32) gives

$$\nabla f(z) + \mu \nabla \Phi_g(z) + A'v = 0 \tag{3.50a}$$

$$Az = b \tag{3.50b}$$

$$g(z) \le 0 \tag{3.50c}$$

where $g(z) \le 0$ has been added for the barrier function Φ_g to be well defined. If one now defines

$$u = \mu \operatorname{diag}(-g(z))^{-1} \mathbf{1} \,, \tag{3.51}$$

the conditions (3.50) become

$$\nabla f(z) + \sum_{i=1}^{m} u_i \nabla g_i(z) + A'v = 0 \tag{3.52a}$$

$$Az = b \tag{3.52b}$$

$$g(z) + s = 0 \tag{3.52c}$$

$$Su = \mu \mathbf{1} \tag{3.52d}$$

$$(s, u) \ge 0 \tag{3.52e}$$

where we have introduced slack variables s and used the expression for the gradient of the barrier function Φ_g and the relations $s = -g(z)$, $S = \operatorname{diag}(s)$. Using

$$\sum_{i=1}^{m} u_i \nabla g_i(z) = G(z)'u \,, \tag{3.53}$$

where $G(z)$ is the Jacobian of g evaluated at the point z, yields the same relaxed optimality conditions as (3.43).

Hence if subproblems $P(\mu)$ (3.32) and (3.43) were solved exactly assuming (3.51), the primal iterates of the primal-barrier method and the primal-dual method would coincide (provided that the path parameters coincide, i.e., $\tau = \mu$). However, primal-dual methods are in practice more efficient than primal barrier methods, since they generate search directions using information from the dual space, and therefore the iterates generated by the two algorithms do not coincide in general.

3.3.3   Active Set Methods

In this section we will discuss methods for solving the problem

$$\begin{aligned} \min_z \quad & f(z) \\ \text{subj. to} \quad & z \in S, \end{aligned}$$

where the feasible set $S \subset \mathbb{R}^s$ is a polyhedron, i.e., a set defined by linear equalities and inequalities, and the objective f is a linear function (*linear programming* (LP)) or a convex quadratic function (*quadratic programming* (QP)). As the name indicates, active set methods aim to identify the set of active constraints at the solution. Once this set is known, a solution to the problem can be easily compted. Since the number of potentially visited active sets depends combinatorially on the number of decision variables and constraints, these methods have a worst case complexity that is exponential in the problem size (as opposed to first-order and interior point methods presented in Sections 3.3.1 and 3.3.2, which both have polynomial complexity). However, active set methods work quite well in practice showing the worst case number of iterations only on pathological problem instances. Also, their underlying concepts are important ingredients in multiparametric programming discussed in Chapter 6.

Active Set Method for LP (Simplex Method)

In the following we describe an active set method for the LP

$$\min_z \quad c'z$$
$$\text{subj. to} \quad Gz \leq w \tag{3.54}$$

where for the sake of simplicity we assume that this LP has a solution, i.e., it is neither unbounded below nor infeasible. Although we consider the case where the feasible set is described by linear inequalities only, the same method can be used to solve problems of type

$$\max_u \quad -w'u$$
$$\text{subj. to} \quad G'u = -c$$
$$u \geq 0, \tag{3.55}$$

which is another LP with linear equality *and* inequality constraints. In particular, (3.55) is the dual problem associated with (3.54) and plays a central role in the active set method for the solution of the primal problem. In fact, the dual will be solved simultaneously with the primal problem as will be shown below.

Central to active set methods for LP is the observation that a solution is always attained at a vertex of the polyhedral feasible set. This fact is often referred to as the *fundamental theorem of linear programming* and gives reason to the methods' alternative naming *simplex methods*. A simple strategy would be to enumerate all vertices of the polyhedron and declare the vertex with the smallest cost as solution. However, one can do better by only visiting those vertices that improve the cost over the previous ones. This is the main idea behind active set methods for LP. Interestingly, it was observed that the simplex method finds a solution in about $2m$ to $3m$ iterations for most practical problems where $m \geq s$ denotes the number of inequalities in (3.54). Nevertheless, pathological polyhedral sets exist which require the method to visit all vertices for certain cost vectors, e.g., the Klee-Minty cube which is a polyhedral set in $\mathbb{R}^s$ with 2^s vertices [26].

Before describing the algorithmic scheme of the simplex method for the solution of (3.54), we state the associated necessary and sufficient KKT conditions

$$G'u + c = 0, \tag{3.56a}$$
$$u_i(G_i z - w_i) = 0, \ i = 1, \ldots, m, \tag{3.56b}$$
$$u \geq 0, \tag{3.56c}$$
$$Gz - w \leq 0. \tag{3.56d}$$

These conditions are composed of primal feasibility (3.56d), dual feasibility (3.56a), (3.56c) and complementarity conditions (3.56b). In the simplex method, primal feasibility and the complementarity conditions are maintained throughout the iterations whereas dual feasibility is sacrificed and will only be satisfied whenever a solution is found. Consequently, the method not only returns a primal minimizer z^* but also a certificate for optimality u^* which at the same time is a solution to the dual problem (3.55).

Initialization

Assume that a vertex z of the feasible set is given. At every vertex, at least s inequalities are active. For the initial active set A we select *exactly* s indices of active inequalities such that the associated submatrix G_A is invertible. The remaining ones are put into the set NA, i.e., $A \cup NA = \{1, 2, \ldots, m\}$ and $A \cap NA = \emptyset$. Hence, the vertex z is characterized by

$$G_A z = w_A, \tag{3.57a}$$
$$G_{NA} z \leq w_{NA}, \tag{3.57b}$$

with w_A and w_{NA} being defined accordingly. Similarly, the simplex method initializes a dual multiplier u such that

$$G'_A u_A = -c,$$
$$u_{NA} = 0.$$

Thus, the vertex z and the dual multiplier u fulfill the primal feasibility and complementarity conditions of the KKT conditions (3.56), but not necessarily dual feasibility as $u_A = -G'^{-1}_A c$ is nonnegative only when a solution is found.

Main Loop

If $u_A \geq 0$ *(Optimal solution found)*
    The pair (z, u) satisfies the KKT conditions (3.56) and thus is an optimal primal/dual pair.

Else   *(Pivoting)*
    A *leaving index* $l \in A$ is selected such that the related dual variable is negative, i.e., $u_l < 0$. It can be shown that in this case an *entering index* $e \in NA$ can be found such that after interchanging these indices, i.e.,

$$A = A \setminus \{l\} \cup \{e\},$$
$$NA = NA \setminus \{e\} \cup \{l\},$$

the next vertex has a cost that is never worse than the cost at the previous vertex. In fact, if the previous vertex z is nondegenerate, i.e., all inequalities in (3.57b) are strict, then the cost improves strictly. The systematic interchange of indices between sets A and NA is called *pivoting*. Pivoting maintains primal feasibility and the complementarity conditions.

The simplex method described above terminates in a finite number of iterations if all visited vertices are nondegenerate. This follows from strict improvement of the cost in every pivoting step, the finite number of vertices and our assumption that there exists a solution. In the remaining part of this section, we shortly summarize important aspects for every practical implementation of the simplex method that were not discussed before in detail and at the end provide links to the literature.

- *Initialization and detection of infeasibility* For initialization, the simplex method requires a vertex and an associated set of linearly independent constraint vectors (we call such an active set *admissible* from here on). In general, determining both is as hard as solving the original LP. For this reason, simplex implementations often resort to a two-phase approach: In *Phase 1*, an auxiliary LP is solved for which an initial vertex and admissible active set are easy to spot and which has an optimal solution and associated active set that can be used to initialize an LP in *Phase 2*. From the solution of the latter, an optimizer for the original LP can then be reconstructed.

  For the LP in (3.54), a possible auxiliary *Phase 1* LP is

$$\min_{y,z_+,z_-} \quad y$$
$$\text{subj. to} \quad Gz_+ - Gz_- - 1y \leq w, \qquad (3.58)$$
$$z_+, z_-, y \geq 0.$$

  It can be easily verified that the vector (z_+, z_-, y) with $z_+ = z_- = 0$ and $y = \min\{y \geq 0 \,|\, y \geq -w_i, i = 1, \ldots, m\}$ is a vertex of the feasible set in (3.58). In order to find an admissible active set, we distinguish two cases: If $w \geq 0$, then $y = 0$ and the admissible active set is composed of all indices referring to the nonnegativity inequalities in (3.58). Otherwise, there exists at least one index i such that $w_i < 0$ which implies $y > 0$. In this case, the admissible active set is determined by all indices referring to nonnegativity inequalities for variables z_+ and z_- and a single index referring to one of the active constraints in $Gz_+ - Gz_- - 1y \leq w$ at the initial vertex. This active set can be shown to lead to a constraint submatrix that is always invertible, so, the auxiliary LP can be solved by the simplex method next.

  Let us denote the optimal solution to (3.58) as (y^*, z_+^*, z_-^*). If y^* is positive, then the original LP (3.54) is infeasible. Only if $y^* = 0$, a feasible solution exists and is given by $z_+^* - z_-^*$. Since the solution to the *Phase 1* problem is a vertex and comes with an admissible active set, it can be used right away for the initialization of the simplex method for solving the *Phase 2* problem

$$\min_{y,z_+,z_-} \quad c'z_+ - c'z_-$$
$$\text{subj. to} \quad Gz_+ - Gz_- - 1y \leq w,$$
$$z_+, z_-, y \geq 0 \qquad (3.59)$$
$$y \leq 0,$$

  which is equivalent to the original LP (3.54): Every solution $(0, z_+^{**}, z_-^{**})$ to (3.59) implies a solution $z_+^{**} - z_-^{**}$ to the original problem and vice versa.

- *Detection of unboundedness* An LP is unbounded whenever there exists a feasible sequence $\{z^k\}$, $Gz^k \leq w$, such that $c'z^k \to -\infty$. The simplex method can detect unboundedness quite easily by identifying a descent direction along which the objective decreases without bound.

- *Convergence in case of degeneracy* A vertex is degenerate whenever more than s inequalities are active at this point. Degeneracy does not prevent active set updates but does prevent improving the cost from iterate to iterate by revisiting the same vertex. In the worst case, this can lead to *cycling*, i.e., the simplex method visits the same vertex infinitely often and thus fails to find an optimal solution. The main methodology to overcome cycling is to slightly perturb the problem data such that all visited vertices are guaranteed to be nondegenerate and to recover the original (nonperturbed) solution in a final step.

- *Pivoting rules* In the simplex method presented above, a leaving index is selected from the active set such that the associated dual variable is negative. In fact, there can be more than one such index and different selection rules lead to different next vertices and thus progress in the cost. Many selection or *pivoting* rules for the leaving index exist, e.g., choosing the index that corresponds to the most negative dual variable as proposed in the original simplex method by Dantzig [92], and consequently lead to a whole family of different simplex methods.

- *Linear algebra* Efficiency of the simplex method depends not only on the number of visited vertices, which is unknown in advance and can be exponential in the problem size in the worst case, but to a great extent on efficient linear algebra routines for the computation of new vertices when doing pivoting steps. Since in the simplex method only a single index is removed and added to the active set A in every iteration, specific linear algebra routines can be used to update the factorization of the constraint submatrix G_A.

There is a great amount of literature available on linear programming and the simplex method. One of the earliest books on this topic is by G.B. Dantzig [92] who developed the simplex method in 1947 and is generally known as the *father of linear programming*. Other references in this field are [217], [280] and [223, Chapter 13]. The material for the *Phase 1/Phase 2* part of this section is based on the latter two references. Finally, it should be noted that the simplex method consistently ranks among the ten most important algorithms of the 20th century.

Active Set Method for QP

In this section, we consider an active set method for the solution of the convex QP

$$\min_z \quad \frac{1}{2}z'Hz + q'z \tag{3.60}$$
$$\text{subj. to} \quad Gz \leq w.$$

We restrict the discussion to the case when the Hessian H is positive definite and the feasible set determined by the linear inequality constraints is nonempty.

These assumptions imply that both a unique solution z^* to (3.60) and unique solutions to all the subproblems encountered in the active set method exist (this will become clear later in this section). Another consequence of positive definiteness of the Hessian is that the presented active set method can be shown to never cycle, i.e., it can be guaranteed that the solution is found in a finite number of iterations without any safeguard against cycling, in contrast to the LP case. Note that we do not consider explicit linear equality constraints in order to facilitate a streamlined presentation of the main concepts.

Active set methods for QP share many features of active set methods for LP. They too identify the set of active inequality constraints at the optimal solution and return a certificate of optimality u^* which is a solution to the dual problem

$$\max_u \quad -\frac{1}{2}(q + G'u)'H^{-1}(q + G'u) - w'u$$

$$\text{subj. to} \quad u \geq 0.$$

The dual problem is also a QP. If, however, the constraint matrix G has more rows than columns (this is the usual case) then the Hessian of the concave objective is negative *semi*-definite and thus the dual solution might not be unique.

The principal feature that distinguishes the QP from the LP is that the search for a solution cannot be restricted to the vertices of the polyhedral feasible set. The solution can also be attained on an edge, a face or even in the interior. These possibilities, in the nondegenerate case, correspond to differently sized active sets, i.e., the size is s (solution on a vertex), $s - 1$ (solution on an edge), $s - j$ where $j \in \{2, 3, \ldots, s - 1\}$ (solution on a face) or the active set might even be empty (solution in the interior). So, there is a combinatorial number of potential active sets A, each corresponding to a QP with equality constraints

$$\min_z \quad \frac{1}{2}z'Hz + q'z \tag{3.61}$$

$$\text{subj. to} \quad G_A z = w_A,$$

where G_A is the submatrix of constraint matrix G according to the active set A. The r.h.s. vector w_A is defined accordingly.

The main idea behind active set methods for QP is to select active sets in a way such that the corresponding iterates stay primal feasible and lead to a cost that decreases monotonically. By doing so, one usually circumvents the brute force approach of checking all possible active sets although there exists no active set method that excludes this possibility by mathematical analysis, i.e., no active set method for QP with polynomial complexity is known.

A nonconstant active set size has two important consequences: First of all, there is no interchange of indices between set A and the complementary set NA as in the LP case, i.e., in every iteration of the active set method for QP an index corresponding to a row of constraint matrix G is either added *or* removed from the current active set or no action is taken at all. Second, a solution to the equality-constrained QP (3.61) associated with every active set is not necessarily a feasible iterate with respect to the original problem (3.60). So, additional measures must be taken in order to maintain primal feasibility of the iterates.

For the understanding of the active set method presented next, it is instructive to consider the necessary and sufficient KKT conditions for problem (3.60)

$$Hz + q + G'u = 0, \tag{3.62a}$$

$$u_i(G_i z - w_i) = 0, \ i = 1, \dots, m, \tag{3.62b}$$

$$u \geq 0, \tag{3.62c}$$

$$Gz - w \leq 0. \tag{3.62d}$$

The active set method maintains primal feasibility (3.62d) as well as stationarity (3.62a) and the complementarity conditions (3.62b). The dual variable u is feasible (3.62c) only when the active set at the optimal solution is identified. Note, however, that no dual *iterates* are maintained, in contrast to the LP case, since they are not required in every iteration. Only at times when the termination criterion needs to be verified or an index removed from the current active set, dual variables are computed.

Initialization

Assume that a primal feasible point z is given. Define the initial active set A as a nonempty *subset* of all indices of active inequalities at z such that the associated submatrix G_A has full row rank. In fact, it suffices to choose only a single index corresponding to any active inequality. The remaining indices are put into set NA, i.e., $NA = \{1, 2, \dots, m\} \backslash A$.

Main Loop

Minimize the quadratic cost over the affine set determined by the active set A, i.e., solve (3.61). For the sake of convenience, we substitute the decision variable in (3.61) by $z + \delta$, i.e., the sum of the current iterate and an offset, and obtain the optimal offset δ^* from

$$\begin{array}{ll} \min_\delta & \frac{1}{2} \delta' H \delta + (q + Hz)' \delta \\ \text{subj. to} & G_A \delta = 0. \end{array} \tag{3.63}$$

If $\delta^* = 0$

Compute the Lagrange multipliers u_A for the equality constraints in (3.63).

If $u_A \geq 0$ *(Optimal solution found)*

The pair (z, u), where u is partitioned into u_A and $u_{NA} = 0$, satisfies the KKT conditions (3.62) and thus is an optimal primal/dual pair. This follows from the fact that the pair (z, u_A) is an optimal primal/dual solution for (3.63) and so satisfies the stationarity condition

$$Hz + q + G'_A u_A = 0.$$

By definition of u_{NA} as the zero vector, this implies that the stationarity condition for the original QP (3.62a) is satisfied as well.

Else *(Remove index from active set)*

An index $l \in A$ is removed from the active set such that the related dual variable is negative, i.e., $u_l < 0$, and the next iterate is defined as the previous one, i.e.,

$$A = A \backslash \{l\},$$
$$NA = NA \cup \{l\},$$
$$z = z.$$

It can be shown that by choosing the leaving index this way, an offset δ^* is obtained in the next iteration from (3.63) that is a *descent direction* with respect to the quadratic cost. Also, δ^* turns out to be a feasible direction with respect to the removed inequality constraint, i.e., the leaving constraint will not be added again in the next iteration.

Else *($\delta^* \neq 0$)*

The minimizer $z + \delta^*$ of the quadratic cost over the affine set might be infeasible with respect to the original problem (3.60). So, we need to find the largest step size $\alpha^* \in [0,1]$ such that $z + \alpha^* \delta^*$ remains feasible. Note that this step size can be computed explicitly as

$$\alpha^* = \min \left\{ 1, \min_{i \in NA:\, G_i' \delta^* > 0} \frac{w_i - G_i' z}{G_i' \delta^*} \right\}. \tag{3.64}$$

So, a new primal iterate is obtained from

$$z = z + \alpha^* \delta^*.$$

If $\alpha^* = 1$ *(No active set change)*

$$A = A$$
$$NA = NA$$

Else *(Add blocking index)*

Assume that the inequality corresponding to index $e \in NA$ prevents us from taking a full step, i.e., $\alpha^* = \frac{w_e - G_e' z}{G_e' \delta^*} < 1$. Then we update the active set by adding this index, i.e.,

$$A = A \cup \{e\},$$
$$NA = NA \setminus \{e\}.$$

In the following, we shortly describe how to initialize the active set method and how to obtain a solution to the equality-constrained QP (3.63). Note that this part, same as the described active set method, is based on [223, Chapter 16].

- *Initialization and detection of infeasibility* Initialization of the active set method for QP is less restrictive than for LP since only a feasible point of the polyhedral feasible set, as opposed to a vertex, is required. Such a point can be obtained either from practical insight or from solving the *Phase 1* LP in (3.58). An alternative approach is to solve

$$\min_{z,\epsilon} \quad \frac{1}{2} z' H z + q' z + \rho_1 \epsilon' \epsilon + \rho_2 \mathbf{1}' \epsilon$$
$$\text{subj. to} \quad Gz \leq w + \epsilon$$
$$\epsilon \geq 0,$$

for which a feasible point can be spotted easily and which can be shown to have the same solution as (3.60) if $\rho_1 \geq 0$ and $\rho_2 > \|u^*\|_\infty$, where u^* is a

Lagrange multiplier for the inequality constraint in (3.60). This follows from the theory of exact penalty functions (Section 12.6, page 264).

Since the Lagrange multiplier is usually unknown, one starts with a guess for the weight ρ_2 and solves the problem. The value was chosen big enough if at the solution $\epsilon^* = 0$, otherwise, the weight needs to be increased and the problem solved again. If no big enough weight can be found such that $\epsilon^* = 0$, the problem is infeasible.

- *Linear algebra*   Key to good performance of an active set method is an efficient solution of the equality-constrained QP in (3.63). Notice that a solution to (3.63) satisfies the KKT conditions

$$
\begin{bmatrix} H & G'_A \\ G_A & 0 \end{bmatrix} \begin{bmatrix} \delta^* \\ u_A \end{bmatrix} = \begin{bmatrix} -(q + Hz) \\ 0 \end{bmatrix}.
\tag{3.65}
$$

It turns out that the so-called *KKT matrix* in (3.65) is invertible if the Hessian H is positive definite and the submatrix G_A has full row rank. Both conditions are fulfilled in our case. In particular, whenever the method is initialized with an active set that leads to full row rank of G_A, this property can be shown to be maintained throughout all iterations without requiring extra checks. In this case, a good way for solving (3.65) is the *Schur complement method*. Since H can be inverted, we can express the primal solution as

$$
\delta^* = -H^{-1}\big(G'_A u_A + q + Hz\big),
$$

which leads to the equation

$$
G_A H^{-1} G'_A u_A = -G_A H^{-1}(q + Hz)
$$

for the Lagrange multiplier u_A. Since $G_A H^{-1} G'_A$ is positive definite, we can solve for u_A using a Cholesky factorization and compute δ^* afterwards. Note that the factorization needs to be updated every time the active set changes. Since at most one index is added or removed per iteration, efficient update schemes can be used that circumvent a full factorization.

4

Polyhedra and P-Collections

Polyhedra are the fundamental geometric objects used in this book. There is a vast body of literature related to polyhedra because they are important for a wide range of applications. In this chapter we introduce the main definitions and the algorithms which describe some operations on polyhedra. Our intent is to provide only the necessary elements for the readers interested in reproducing the control algorithms reported in this book. Most definitions given here are standard. For additional details the reader is referred to [296, 135, 112].

4.1 General Set Definitions and Operations

An n-**dimensional ball** $\mathcal{B}(x_c, \rho)$ is the set $\mathcal{B}(x_c, \rho) = \{x \in \mathbb{R}^n : \|x - x_c\|_2 \leq \rho\}$. The vector x_c is the center of the ball and ρ is the radius.

Affine sets are sets described by the solutions of a system of linear equations:

$$\mathcal{F} = \{x \in \mathbb{R}^n : Ax = b, \text{ with } A \in \mathbb{R}^{m \times n}, b \in \mathbb{R}^m\}. \tag{4.1}$$

If $\mathcal{F}$ is an affine set and $\bar{x} \in \mathcal{F}$, then the translated set $\mathcal{V} = \{x - \bar{x} : x \in \mathcal{F}\}$ is a subspace.

The **affine combination** of a finite set of points $x^1, \ldots, x^k$ belonging to $\mathbb{R}^n$ is defined as the point $\lambda^1 x^1 + \cdots + \lambda^k x^k$ where $\sum_{i=1}^{k} \lambda^i = 1$.

The **affine hull** of $\mathcal{K} \subseteq \mathbb{R}^n$ is the set of all affine combinations of points in $\mathcal{K}$ and it is denoted as aff($\mathcal{K}$):

$$\text{aff}(\mathcal{K}) = \left\{\lambda^1 x^1 + \cdots + \lambda^k x^k : x^i \in \mathcal{K}, i = 1, \ldots, k, \sum_{i=1}^{k} \lambda^i = 1\right\}. \tag{4.2}$$

The affine hull of $\mathcal{K}$ is the smallest affine set that contains $\mathcal{K}$ in the following sense: if $\mathcal{S}$ is any affine set with $\mathcal{K} \subseteq \mathcal{S}$, then aff($\mathcal{K}$)$\subseteq \mathcal{S}$.

The **dimension** of an affine set, affine combination or affine hull is the dimension of the largest ball of radius $\rho > 0$ included in the set.

Example 4.1 The set
$$\mathcal{F} = \{x \in \mathbb{R}^2 \; : \; x_1 + x_2 = 1\}$$
is an affine set in $\mathbb{R}^2$ of dimension one. The points $x^1 = [0, 1]'$ and $x^2 = [1, 0]'$ belong to the set $\mathcal{F}$. The point $\bar{x} = -0.2x^1 + 1.2x^2 = [1.2, -0.2]'$ is an affine combination of points x^1 and x^2. The affine hull of x^1 and x^2, $\mathrm{aff}(\{x^1, x^2\})$, is the set $\mathcal{F}$.

Convex sets have been defined in Section 1.2.

The **convex combination** of a finite set of points $x^1, \ldots, x^k$ belonging to $\mathbb{R}^n$ is defined as the point $\lambda^1 x^1 + \cdots + \lambda^k x^k$ where $\sum_{i=1}^{k} \lambda^i = 1$ and $\lambda^i \geq 0$, $i = 1, \ldots, k$.

The **convex hull** of a set $\mathcal{K} \subseteq \mathbb{R}^n$ is the set of all convex combinations of points in $\mathcal{K}$ and it is denoted as $\mathrm{conv}(\mathcal{K})$:

$$\mathrm{conv}(\mathcal{K}) = \left\{ \lambda^1 x^1 + \cdots + \lambda^k x^k \; : \; x_i \in \mathcal{K}, \; \lambda^i \geq 0, \; i = 1, \ldots, k, \; \sum_{i=1}^{k} \lambda^i = 1 \right\}. \tag{4.3}$$

The convex hull of $\mathcal{K}$ is the smallest convex set that contains $\mathcal{K}$ in the following sense: if $\mathcal{S}$ is any convex set with $\mathcal{K} \subseteq \mathcal{S}$, then $\mathrm{conv}(\mathcal{K}) \subseteq \mathcal{S}$.

Example 4.2 Consider three points $x^1 = [1, 1]'$, $x^2 = [1, 0]'$, $x^3 = [0, 1]'$ in $\mathbb{R}^2$. The point $\bar{x} = \lambda^1 x^1 + \lambda^2 x^2 + \lambda^3 x^3$ with $\lambda^1 = 0.2$, $\lambda^2 = 0.2$, $\lambda^3 = 0.6$ is $\bar{x} = [0.4, 0.8]'$ and it is a convex combination of the points $\{x^1, \; x^2, \; x^3\}$. The convex hull of $\{x^1, \; x^2, \; x^3\}$ is the triangle plotted in Figure 4.1. Note that any set in $\mathbb{R}^2$ strictly contained in the triangle and containing $\{x^1, \; x^2, \; x^3\}$ is nonconvex. This illustrates that the convex hull is the smallest set that contains these three points.

A **cone** spanned by a finite set of points $\mathcal{K} = \{x^1, \ldots, x^k\}$ is defined as

$$\mathrm{cone}(\mathcal{K}) = \left\{ \sum_{i=1}^{k} \lambda^i x^i, \; \lambda^i \geq 0, i = 1, \ldots, k \right\}. \tag{4.4}$$

We define $\mathrm{cone}(\mathcal{K}) = \{0\}$ if $\mathcal{K}$ is the empty set.

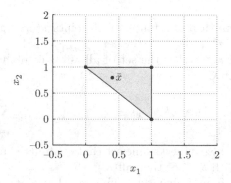

Figure 4.1 Example 4.2. Illustration of the convex hull of three points $x^1 = [1, 1]'$, $x^1 = [1, 0]'$, $x^3 = [0, 1]'$.

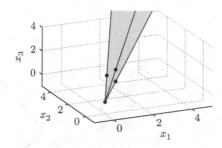

Figure 4.2 Example 4.3. Illustration of a cone spanned by the three points $x^1 = [1, 1, 1]'$, $x^2 = [1, 2, 1]'$, $x^3 = [1, 1, 2]'$.

Example 4.3 Consider three points $x^1 = [1, 1, 1]'$, $x^2 = [1, 2, 1]'$, $x^3 = [1, 1, 2]'$ in $\mathbb{R}^3$. The cone spanned by $\{x^1,\ x^2,\ x^3\}$ is an unbounded set. It is depicted in Figure 4.2.

The **Minkowski sum** of two sets $\mathcal{P}, \mathcal{Q} \subseteq \mathbb{R}^n$ is defined as

$$\mathcal{P} \oplus \mathcal{Q} = \{x + y\ :\ x \in \mathcal{P},\ y \in \mathcal{Q}\}. \tag{4.5}$$

By definition, any point in $\mathcal{P} \oplus \mathcal{Q}$ can be written as the sum of two points, one in $\mathcal{P}$ and one in $\mathcal{Q}$. For instance, the Minkowski sum of two balls ($\mathcal{P}$ and $\mathcal{Q}$) centered at the origin and with radius 1, is a ball ($\mathcal{P} \oplus \mathcal{Q}$) centered at the origin and with radius 2.

4.2 Polyhedra and Representations

Some of the concepts here have been briefly touched in Section 2.1. In the following we give two definitions of a polyhedron. They are mathematically equivalent but they lead to two different polyhedron representations. The proof of equivalence is not trivial and can be found in [296].

An $\mathcal{H}$-*polyhedron* $\mathcal{P}$ in $\mathbb{R}^n$ denotes an intersection of a finite set of closed halfspaces in $\mathbb{R}^n$:

$$\mathcal{P} = \{x \in \mathbb{R}^n :\ Ax \leq b\}, \tag{4.6}$$

where $Ax \leq b$ is the usual shorthand form for a system of inequalities, namely $a_i' x \leq b_i$, $i = 1, \ldots, m$, where $a_1, \ldots, a_m$ are the rows of A, and $b_1, \ldots, b_m$ are the components of b. In Figure 4.3 a two-dimensional $\mathcal{H}$-polyhedron is plotted.

A $\mathcal{V}$-*polyhedron* $\mathcal{P}$ in $\mathbb{R}^n$ denotes the Minkowski sum (defined in (4.5)) of the convex hull of a finite set of points $\{V_1, \ldots, V_k\}$ of $\mathbb{R}^n$ and the cone generated by a finite set of vectors $\{y_1, \ldots, y_{k'}\}$ of $\mathbb{R}^n$:

$$\mathcal{P} = \text{conv}(V) \oplus \text{cone}(Y), \tag{4.7}$$

for some $V = [V_1, \ldots, V_k] \in \mathbb{R}^{n \times k}$, $Y = [y_1, \ldots, y_{k'}] \in \mathbb{R}^{n \times k'}$. The main theorem for polyhedra states that any $\mathcal{H}$-polyhedron can be converted into a $\mathcal{V}$-polyhedron and vice-versa [296, p. 30].

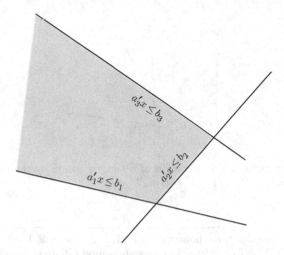

Figure 4.3 $\mathcal{H}$-polyhedron. The planes (here lines) defining the boundary of the halfspaces are $a_i' x - b_i = 0$.

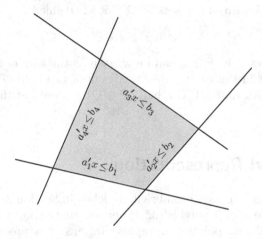

Figure 4.4 $\mathcal{H}$-polytope. The planes (here lines) defining the halfspaces are $a_i' x - b_i = 0$.

An $\mathcal{H}$-*polytope* is a bounded $\mathcal{H}$-polyhedron (in the sense that it does not contain any ray $\{x + ty \;:\; t \geq 0\}$). In Figure 4.4 a two-dimensional $\mathcal{H}$-polytope is plotted.

A $\mathcal{V}$-*polytope* is a bounded $\mathcal{V}$-polyhedron

$$\mathcal{P} = \operatorname{conv}(V) = \{V\lambda \mid \lambda \geq 0, \; \mathbf{1}'\lambda = 1\}. \tag{4.8}$$

In Section 4.4.4 we show how to convert any $\mathcal{H}$-polytope into a $\mathcal{V}$-polytope and vice versa.

The dimension of a polytope (polyhedron) $\mathcal{P}$ is the dimension of its affine hull and is denoted by $\dim(\mathcal{P})$. We say that a polytope $\mathcal{P} \subset \mathbb{R}^n$, $\mathcal{P} = \{x \in \mathbb{R}^n : P^x x \leq P^c\}$, is *full-dimensional* if $\dim(\mathcal{P}) = n$ or, equivalently, if it is possible to fit a nonempty n-dimensional ball in $\mathcal{P}$,

$$\exists x \in \mathbb{R}^n, \epsilon > 0 \;:\; \mathcal{B}(x, \epsilon) \subset \mathcal{P}, \tag{4.9}$$

or, equivalently,

$$\exists x \in \mathbb{R}^n, \ \epsilon > 0 \ : \|\delta\|_2 \le \epsilon \Rightarrow P^x(x + \delta) \le P^c. \tag{4.10}$$

Otherwise, we say that polytope $\mathcal{P}$ is *lower-dimensional*. A polytope is referred to as *empty* if

$$\nexists x \in \mathbb{R}^n : P^x x \le P^c. \tag{4.11}$$

Furthermore, if $\|P_i^x\|_2 = 1$, where P_i^x denotes the i-th row of a matrix P^x, we say that the polytope $\mathcal{P}$ is *normalized*.

Let $\mathcal{P}$ be a polyhedron. A linear inequality $c'z \le c_0$ is said to be *valid* for $\mathcal{P}$ if it is satisfied for all points $z \in \mathcal{P}$. A *face* of $\mathcal{P}$ is any nonempty set of the form

$$\mathcal{F} = \mathcal{P} \cap \{z \in \mathbb{R}^s : \ c'z = c_0\}, \tag{4.12}$$

where $c'z \le c_0$ is a *valid* inequality for $\mathcal{P}$. The *dimension* of a face is the dimension of its affine hull. For the valid inequality $0z \le 0$ we get that $\mathcal{P}$ is a face of $\mathcal{P}$. All faces of $\mathcal{P}$ satisfying $\mathcal{F} \subset \mathcal{P}$ are called proper faces and have dimension less than $\dim(\mathcal{P})$. The faces of dimension 0,1, $\dim(\mathcal{P})$-2 and $\dim(\mathcal{P})$-1 are called *vertices*, *edges*, *ridges*, and *facets*, respectively. The set V of all the vertices of a polytope $\mathcal{P}$ will be denoted as

$$V = \text{extreme}(\mathcal{P}).$$

The next theorem summarizes basic facts about faces.

Theorem 4.1 *Let $\mathcal{P}$ be a polytope, V the set of all its vertices and $\mathcal{F}$ a face.*

1. *$\mathcal{P}$ is the convex hull of its vertices: $\mathcal{P}$=conv(V).*

2. *$\mathcal{F}$ is a polytope.*

3. *Every intersection of faces of $\mathcal{P}$ is a face of $\mathcal{P}$.*

4. *The faces of $\mathcal{F}$ are exactly the faces of $\mathcal{P}$ that are contained in $\mathcal{F}$.*

5. *The vertices of $\mathcal{F}$ are exactly the vertices of $\mathcal{P}$ that are contained in $\mathcal{F}$.*

A *d-simplex* is a polytope of $\mathbb{R}^d$ with $d + 1$ vertices.

In this book we will work mostly with $\mathcal{H}$-polyhedra and $\mathcal{H}$-polytopes. This choice has important consequences for algorithm implementations and their complexity. As a simple example, a unit cube in $\mathbb{R}^d$ can be described through $2d$ equalities as an $\mathcal{H}$-polytope, but requires 2^d points in order to be described as a $\mathcal{V}$-polytope. We want to mention here that many efficient algorithms that work on polyhedra require both $\mathcal{H}$ and $\mathcal{V}$ representations. In Figure 4.5 the $\mathcal{H}$-representation and the $\mathcal{V}$-representation of the same polytope in two dimensions are plotted.

Consider Figure 4.5(b) and notice that no inequality can be removed without changing the polyhedron. Inequalities which can be removed without changing the polyhedron described by the original set are called *redundant*. The representation of an $\mathcal{H}$-polyhedron is *minimal* if it does not contain redundant inequalities. Detecting

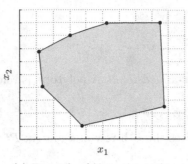

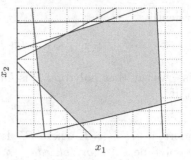

(a) Example of $\mathcal{V}$-representation. The vertices $V_1^P, \ldots, V_7^P$ are depicted as dots.

(b) Example of $\mathcal{H}$-representation. The hyperplanes $P_i^x x = P_i^c$, $i = 1, \ldots, 7$ are depicted as lines.

Figure 4.5 Illustration of a polytope in $\mathcal{H}$- and $\mathcal{V}$-representation.

whether an inequality is redundant for an $\mathcal{H}$-polyhedron requires solving a linear program, as described in Section 4.4.1.

By definition a polyhedron is a closed set. In this book we will also work with sets which are not closed but whose closure is a polyhedron. For this reason the two following nonstandard definitions will be useful.

Definition 4.1 (Open Polyhedron (Polytope)) *A set $\mathcal{C} \subseteq \mathbb{R}^n$ is called an* open polyhedron (polytope) *if it is open and its closure is a polyhedron (polytope).*

Definition 4.2 (Neither Open nor Closed Polyhedron (Polytope)) *A set $\mathcal{C} \subseteq \mathbb{R}^n$ is called* neither an open nor a closed polyhedron (polytope) *if it is neither open nor closed and its closure is a polyhedron (polytope).*

4.3 Polytopal Complexes

According to our definition every polytope represents a convex, compact (i.e., bounded and closed) set. In this book we will also encounter sets that are disjoint or nonconvex but can be represented by the union of a finite number of polytopes. Therefore, it is useful to define the following mathematical concepts.

Definition 4.3 (P-collection) *A set $\mathcal{C} \subseteq \mathbb{R}^n$ is called a* P-collection *if it is a collection of a finite number of n-dimensional polyhedra, i.e.,*

$$\mathcal{C} = \{\mathcal{C}_i\}_{i=1}^{N_C}, \tag{4.13}$$

where $\mathcal{C}_i = \{x \in \mathbb{R}^n \ : \ C_i^x x \leq C_i^c\}$, $\dim(\mathcal{C}_i) = n$, $i = 1, \ldots, N_C$, with $N_C < \infty$.

Definition 4.4 (Underlying Set) *The* underlying set *of a P-collection $\mathcal{C} = \{\mathcal{C}_i\}_{i=1}^{N_C}$ is the*

$$\underline{\mathcal{C}} = \bigcup_{\mathcal{P} \in \mathcal{C}} \mathcal{P} = \bigcup_{i=1}^{N_C} \mathcal{C}_i. \tag{4.14}$$

Example 4.4 A collection $\mathcal{C} = \{[-2, -1], [0, 2], [2, 4]\}$ is a P-collection in $\mathbb{R}^1$ with the underlying set $\underline{\mathcal{C}} = [-2, -1] \cup [0, 4]$. As another example, $\mathcal{C} = \{[-2, 0], [-1, 1], [0, 2]\}$ is a P-collection in $\mathbb{R}^1$ with underlying set $\underline{\mathcal{C}} = [-2, 2]$. Clearly, polytopes that define a P-collection can overlap, the underlying set can be disconnected and nonconvex.

Usually it is clear from the context if we are talking about the P-collection or referring to the underlying set of a P-collection. Therefore, for simplicity, we use the same notation for both.

Definition 4.5 (Strict Polyhedral Partition) *A collection of sets* $\{\mathcal{C}_i\}_{i=1}^{N_C}$ *is a* strict partition *of a set* $\mathcal{C}$ *if* (i) $\bigcup_{i=1}^{N_C} \mathcal{C}_i = \mathcal{C}$ *and* (ii) $\mathcal{C}_i \cap \mathcal{C}_j = \emptyset$, $\forall i \neq j$. *Moreover* $\{\mathcal{C}_i\}_{i=1}^{N_C}$ *is a* strict polyhedral partition *of a polyhedral set* $\mathcal{C}$ *if* $\{\mathcal{C}_i\}_{i=1}^{N_C}$ *is a strict partition of* $\mathcal{C}$ *and* $\bar{\mathcal{C}}_i$ *is a polyhedron for all* i, *where* $\bar{\mathcal{C}}_i$ *denotes the closure of the set* $\mathcal{C}_i$.

Definition 4.6 (Polyhedral Partition) *A collection of sets* $\{\mathcal{C}_i\}_{i=1}^{N_C}$ *is a* partition *of a set* $\mathcal{C}$ *if* (i) $\bigcup_{i=1}^{N_C} \mathcal{C}_i = \mathcal{C}$ *and* (ii) $(\mathcal{C}_i \backslash \partial \mathcal{C}_i) \cap (\mathcal{C}_j \backslash \partial \mathcal{C}_j) = \emptyset$, $\forall i \neq j$. *Moreover* $\{\mathcal{C}_i\}_{i=1}^{N_C}$ *is a* polyhedral partition *of a polyhedral set* $\mathcal{C}$ *if* $\{\mathcal{C}_i\}_{i=1}^{N_C}$ *is a partition of* $\mathcal{C}$ *and* $\mathcal{C}_i$ *is a polyhedron for all* i. *The set* $\partial \mathcal{C}_j$ *is the boundary of the set* $\mathcal{C}_j$.

Note that in a *strict polyhedral partition* of a polyhedron some of the sets $\mathcal{C}_i$ must be open or neither open nor closed. In a *partition* all the sets may be closed and points on the closure of a particular set may also belong to one or several other sets. Also, note that a polyhedral partition is a special class of a P-collection.

4.3.1 Functions on Polytopal Complexes

Definition 4.7 *A function* $h(\theta) : \Theta \rightarrow \mathbb{R}$, *where* $\Theta \subseteq \mathbb{R}^s$, *is* piecewise affine (PWA) *if there exists a strict partition* $R_1, \ldots, R_N$ *of* Θ *and* $h(\theta) = H^i \theta + k^i$, $\forall \theta \in R_i$, $i = 1, \ldots, N$.

Definition 4.8 *A function* $h(\theta) : \Theta \rightarrow \mathbb{R}$, *where* $\Theta \subseteq \mathbb{R}^s$, *is* piecewise affine on polyhedra (PPWA) *if there exists a strict polyhedral partition* $R_1, \ldots, R_N$ *of* Θ *and* $h(\theta) = H^i \theta + k^i$, $\forall \theta \in R_i$, $i = 1, \ldots, N$.

Definition 4.9 *A function* $h(\theta) : \Theta \rightarrow \mathbb{R}$, *where* $\Theta \subseteq \mathbb{R}^s$, *is* piecewise quadratic (PWQ) *if there exists a strict partition* $R_1, \ldots, R_N$ *of* Θ *and* $h(\theta) = \theta' H^i \theta + k^{i'} \theta + l^i$, $\forall \theta \in R_i$, $i = 1, \ldots, N$.

Definition 4.10 *A function* $h(\theta) : \Theta \rightarrow \mathbb{R}$, *where* $\Theta \subseteq \mathbb{R}^s$, *is* piecewise quadratic on polyhedra (PPWQ) *if there exists a strict polyhedral partition* $R_1, \ldots, R_N$ *of* Θ *and* $h(\theta) = \theta' H^i \theta + k^i \theta + l^i$, $\forall \theta \in R_i$, $i = 1, \ldots, N$.

As long as the function $h(\theta)$ we are defining is continuous it is not important if the partition constituting the domain of the function is strict or not. If the function is discontinuous at points on the closure of a set, then this function can only be defined if the partition is strict. Otherwise we may obtain several conflicting definitions of function values (or set-valued functions). Therefore for the statement of theoretical results involving discontinuous functions we will always assume that the partition is strict. For notational convenience, however, when working with continuous functions we will make use of partitions rather than strict partitions.

4.4   Basic Operations on Polytopes

We will now define some basic operations and functions on polytopes. Note that although we focus on polytopes and polytopic objects most of the operations described here are directly (or with minor modifications) applicable to polyhedral objects. Additional details on polytope computation can be found in [296, 135, 112]. All operations and functions described in this chapter are contained in the MultiParametric Toolbox (MPT) [149].

4.4.1   Minimal Representation

We say that a polytope $\mathcal{P} \subset \mathbb{R}^n$, $\mathcal{P} = \{x \in \mathbb{R}^n \ : \ P^x x \leq P^c\}$ is in a *minimal representation* if the removal of any row in $P^x x \leq P^c$ would change it (i.e., if there are no redundant constraints). The computation of the minimal representation (henceforth referred to as *polytope reduction*) of polytopes is discussed in [112]. The redundancy of each constraint is checked, which generally requires the solution of one LP for each half-space defining the nonminimal representation of $\mathcal{P}$. We summarize this simple implementation of the polytope reduction in Algorithm 4.1. An improved algorithm for polytope reduction is discussed in [269] where the authors combine the procedure outlined in Algorithm 4.1 with heuristic methods, such as bounding-boxes and ray-shooting, to discard redundant constraints more efficiently.

It is straightforward to see that a normalized, full-dimensional polytope $\mathcal{P}$ has a *unique* minimal representation. Note that "unique" here means that for $\mathcal{P} = \{x \in \mathbb{R}^n : P^x x \leq P^c\}$ the matrix $[P^x P^c]$ consists of the unique set of row vectors, the rows' order is irrelevant. This fact is very useful in practice. Normalized, full-dimensional polytopes in a minimal representation allow us to avoid any ambiguity when comparing them and very often speed-up other polytope manipulations.

Algorithm 4.1   *Polytope in minimal representation*

Input $\mathcal{P} = \{x \ : \ P^x x \leq P^c\}$, with $P^x \in \mathbb{R}^{n_P \times n}$, $P^c \in \mathbb{R}^{n_P}$

Output $\mathcal{Q} = \{x \ : \ Q^x x \leq Q^c\} = \mathrm{minrep}(\mathcal{P})$

    $\mathcal{I} \leftarrow \{1, \ldots, n_P\}$

    For $i = 1$ **to** n_P

        $\mathcal{I} \leftarrow \mathcal{I} \setminus \{i\}$

        $f^* \leftarrow \max_x P_i^x x, \quad \text{subj. to} \quad P_{(\mathcal{I})}^x x \leq P_{(\mathcal{I})}^c, \ P_i^x x \leq P_i^c + 1$

        If $f^* > P_i^c$ **Then** $\mathcal{I} \leftarrow \mathcal{I} \cup \{i\}$

    End

    $Q^x = P_{\mathcal{I}}^x, \ Q^c = P_{\mathcal{I}}^c$

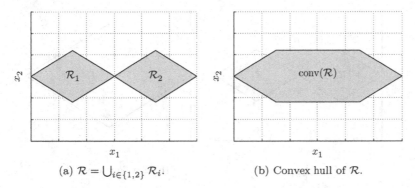

(a) $\mathcal{R} = \bigcup_{i \in \{1,2\}} \mathcal{R}_i$.   (b) Convex hull of $\mathcal{R}$.

Figure 4.6 Illustration of the convex hull operation.

4.4.2   Convex Hull

The convex hull of a set of points $V = \{V^i\}_{i=1}^{N_V}$, with $V^i \in \mathbb{R}^n$, is a polytope defined as

$$\text{conv}(V) = \left\{ x \in \mathbb{R}^n \; : \; x = \sum_{i=1}^{N_V} \alpha^i V^i, \; \alpha^i \geq 0, \; \sum_{i=1}^{N_V} \alpha^i = 1 \right\}. \tag{4.15}$$

The convex hull of a union of polytopes $\mathcal{R}_i \subset \mathbb{R}^n$, $i = 1, \ldots, N_R$, is a polytope

$$\text{conv}\left(\bigcup_{i=1}^{N_R} \mathcal{R}_i \right) = \left\{ x \in \mathbb{R}^n \; : \; x = \sum_{i=1}^{N_R} \omega^i w^i, \; w^i \in \mathcal{R}_i, \; \omega^i \geq 0, \; \sum_{i=1}^{N_R} \alpha^i = 1 \right\}. \tag{4.16}$$

An illustration of the convex hull operation is given in Figure 4.6. Construction of the convex hull of a set of polytopes is an expensive operation which is exponential in the number of facets of the original polytopes. An efficient software implementation is available in [111] and used in the MPT toolbox [149].

4.4.3   Envelope

The envelope of two $\mathcal{H}$-polyhedra $\mathcal{P} = \{x \in \mathbb{R}^n \; : \; P^x x \leq P^c\}$ and $\mathcal{Q} = \{x \in \mathbb{R}^n \; : \; Q^x x \leq Q^c\}$ is an $\mathcal{H}$-polyhedron

$$\text{env}(\mathcal{P}, \mathcal{Q}) = \{x \in \mathbb{R}^n \; : \; \bar{P}^x x \leq \bar{P}^c, \; \bar{Q}^x x \leq \bar{Q}^c\}, \tag{4.17}$$

where $\bar{P}^x x \leq \bar{P}^c$ is the subsystem of $P^x x \leq P^c$ obtained by removing all the inequalities not valid for the polyhedron $\mathcal{Q}$, and $\bar{Q}^x x \leq \bar{Q}^c$ is defined in a similar way with respect to $Q^x x \leq Q^c$ and $\mathcal{P}$ [38]. In a similar fashion, the definition can be extended to the case of the envelope of a P-collection. An illustration of the envelope operation is depicted in Figure 4.7.

The computation of the envelope is relatively cheap since it only requires the solution of one LP for each facet of $\mathcal{P}$ and $\mathcal{Q}$. In particular, if a facet of $\mathcal{Q}$

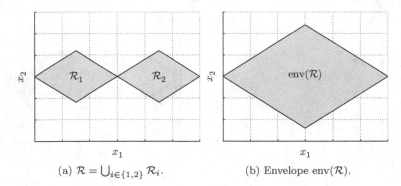

(a) $\mathcal{R} = \bigcup_{i \in \{1,2\}} \mathcal{R}_i$.   (b) Envelope $\mathrm{env}(\mathcal{R})$.

Figure 4.7 Illustration of the envelope operation.

(and $\mathcal{P}$) is found to be "nonredundant" for the polytope $\mathcal{P}$ (for the polytope $\mathcal{Q}$) then it is not part of the envelope. A version of Algorithm 4.1 can be used. Note that the envelope of two (or more) polytopes is not necessarily a bounded set (e.g., when $\mathcal{P} \cup \mathcal{Q}$ is shaped like a star).

4.4.4 Vertex Enumeration

The operation of extracting the vertices $V = \{V^i\}_{i=1}^{N_V}$ of a polytope $\mathcal{P}$ given in $\mathcal{H}$-representation is referred to as vertex enumeration and denoted as $V = \mathrm{extreme}(\mathcal{P})$. The necessary computational effort is exponential in the number of facets. The work in [12] classifies algorithms for vertex enumeration in two groups: graph traversal and incremental. Graph traversal algorithms construct the graph of vertices and edges of the polyhedron starting from a vertex and finding the other vertices by traversing the graph. Going from one vertex to another is equivalent to moving from one base to another base through pivoting in the simplex algorithm. The reverse search [11] approach belongs to this class. Incremental algorithms start from a subset of the polyhedron halfspaces and its associated vertices and iteratively intersect the remaining halfspaces to compute a new set of vertices. The double description method [114] belongs to this class. An efficient implementation of the double description method is available in [111] and is used in the MPT toolbox [149].

Converting a $\mathcal{V}$-polytope into an $\mathcal{H}$-polytope corresponds to the convex hull computation. It can also be done via vertex enumeration. This procedure is based upon the *polar dual* [296, p. 61].

Alternatively, using the polar dual, an $\mathcal{H}$-polytope can be converted into a $\mathcal{V}$-polytope by a convex hull computation. Because of this "symmetry" we say that vertex enumeration and convex hull computation are dual to each other.

4.4.5 Chebyshev Ball

The Chebyshev Ball of a polytope $\mathcal{P} = \{x \in \mathbb{R}^n : P^x x \leq P^c\}$, with $P^x \in \mathbb{R}^{n_P \times n}$, $P^c \in \mathbb{R}^{n_P}$, corresponds to the largest radius ball $\mathcal{B}(x_c, R)$ with center x_c, such that $\mathcal{B}(x_c, R) \subset \mathcal{P}$. The center and radius of the Chebyshev Ball can be easily found by solving the following linear optimization problem

$$\max_{x_c, R} R \tag{4.18a}$$

$$\text{subj. to } P_i^x x_c + R\|P_i^x\|_2 \leq P_i^c, \quad i = 1, \ldots, n_P, \tag{4.18b}$$

where P_i^x denotes the i-th row of P^x. This can be proven as follows. Any point x of the ball can be written as $x = x_c + v$ where v is a vector of length less or equal to R. Therefore the center and radius of the Chebyshev Ball can be found by solving the following optimization problem

$$\max_{x_c, R} R \tag{4.19a}$$

$$\text{subj. to } P_i^x(x_c + v) \leq P_i^c, \quad \forall v \text{ such that } \|v\|_2 \leq R, \; i = 1, \ldots, n_P. \tag{4.19b}$$

Consider the i-th constraint

$$P_i^x(x_c + v) \leq P_i^c, \quad \forall v \text{ such that } \|v\|_2 \leq R.$$

This can be written as

$$P_i^x x_c \leq P_i^c - P_i^x v, \quad \forall v \text{ such that } \|v\|_2 \leq R. \tag{4.20}$$

Constraint (4.20) is satisfied $\forall v$ such that $\|v\|_2 \leq R$ if and only if it is satisfied at $v = \frac{P_i^{x\prime}}{\|P_i^x\|_2}R$. Therefore we can rewrite the optimization problem (4.19) as the linear program (4.18).

If the radius obtained by solving (4.18) is $R = 0$, then the polytope is lower-dimensional. If $R < 0$, then the polytope is empty. Therefore, an answer to the question "is polytope $\mathcal{P}$ full-dimensional/empty?" is obtained at the *cost* of only one linear program. Furthermore, for a full-dimensional polytope we also get a point x_c that is in the strict interior of $\mathcal{P}$. However, the center of a Chebyshev Ball x_c in (4.18) is not necessarily a unique point (e.g., when $\mathcal{P}$ is a rectangle). There are other types of unique interior points one could compute for a full-dimensional polytope, e.g., center of the largest volume ellipsoid, analytic center, etc., but those computations involve the solution of Semi-Definite Programs (SDPs) and therefore may be more expensive than the Chebyshev Ball computation [65]. An illustration of the Chebyshev Ball is given in Figure 4.8.

4.4.6 Projection

Given a polytope $\mathcal{P} = \{[x'y']' \in \mathbb{R}^{n+m} : P^x x + P^y y \leq P^c\} \subset \mathbb{R}^{n+m}$ the projection onto the x-space $\mathbb{R}^n$ is defined as

$$\text{proj}_x(\mathcal{P}) = \{x \in \mathbb{R}^n : \exists y \in \mathbb{R}^m : P^x x + P^y y \leq P^c\}. \tag{4.21}$$

An illustration of a projection operation is given in Figure 4.9. Current projection methods that can operate in general dimensions can be grouped into four classes: Fourier elimination [82, 169], block elimination [15], vertex based approaches [113]

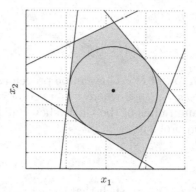

Figure 4.8 Illustration of the Chebyshev Ball contained in a polytope $\mathcal{P}$.

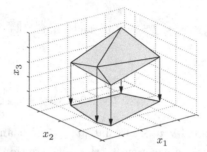

Figure 4.9 Illustration of the projection of a three-dimensional polytope $\mathcal{P}$ onto the plane $x_3 = 0$.

and wrapping-based techniques [161]. For a good introduction to projection, we refer the reader to [161] and the references therein.

4.4.7 Set-Difference

The set-difference of two polytopes $\mathcal{Y}$ and $\mathcal{R}_0$

$$\mathcal{R} = \mathcal{Y} \setminus \mathcal{R}_0 = \{x \in \mathbb{R}^n \ : \ x \in \mathcal{Y}, x \notin \mathcal{R}_0\}, \tag{4.22}$$

in general, can be a nonconvex and disconnected set and can be described as a P-collection $\mathcal{R} = \bigcup_{i=1}^m \mathcal{R}_i$, where $\mathcal{Y} = \bigcup_{i=1}^m \mathcal{R}_i \bigcup (\mathcal{R}_0 \bigcap \mathcal{Y})$. The P-collection $\mathcal{R} = \bigcup_{i=1}^m \mathcal{R}_i$ can be computed by consecutively inverting the half-spaces defining $\mathcal{R}_0$ as described in the following Theorem 4.2.

Note that here we use the term P-collection in the dual context of both P-collection and its underlying set (Definitions 4.3 and 4.4). The precise statement would say that $\underline{\mathcal{R}} = \mathcal{Y} \setminus \mathcal{R}_0$, where $\underline{\mathcal{R}}$ is the underlying set of the P-collection $\mathcal{R} = \{\mathcal{R}_i\}_{i=1}^m$. However, whenever it is clear from context, we will use the former, more compact form.

Theorem 4.2 *[44] Let $\mathcal{Y} \subseteq \mathbb{R}^n$ be a polyhedron, $\mathcal{R}_0 = \{x \in \mathbb{R}^n : Ax \leq b\}$, and $\bar{\mathcal{R}}_0 = \{x \in \mathcal{Y} : Ax \leq b\} = \mathcal{R}_0 \bigcap \mathcal{Y}$, where $b \in \mathbb{R}^m$, $\mathcal{R}_0 \neq \emptyset$ and $Ax \leq b$ is a minimal representation of $\mathcal{R}_0$. Also let*

$$\mathcal{R}_i = \left\{ x \in \mathcal{Y} : \begin{array}{c} A^i x > b^i \\ A^j x \leq b^j, \forall j < i \end{array} \right\}, \ i = 1, \ldots, m, \ j = 1, \ldots, m-1.$$

Let $\mathcal{R} = \bigcup_{i=1}^m \mathcal{R}_i$. Then, $\mathcal{R}$ is a P-collection and $\{\bar{\mathcal{R}}_0, \mathcal{R}_1, \ldots, \mathcal{R}_m\}$ is a strict polyhedral partition of $\mathcal{Y}$.

Proof: (i) We want to prove that given an $x \in \mathcal{Y}$, then either x belongs to $\bar{\mathcal{R}}_0$ or to $\mathcal{R}_i$ for some i but not both. If $x \in \bar{\mathcal{R}}_0$, we are done. Otherwise, there exists an index i such that $A^i x > b^i$. Let $i^* = \min_{i \leq m}\{i : A^i x > b^i\}$. Then $x \in \mathcal{R}_{i^*}$, as $A^{i^*} x > b^{i^*}$ and $A^j x \leq b^j$, $\forall j < i^*$, by definition of i^*.

(ii) Let $x \in \bar{\mathcal{R}}_0$. Then there does not exist any i such that $A^i x > b^i$, which implies that $x \notin \mathcal{R}_i$, $\forall i \leq m$. Let $x \in \mathcal{R}_i$ and take $i > j$. Because $x \in \mathcal{R}_i$, by definition of $\mathcal{R}_i$ $(i > j)$ $A^j x \leq b^j$, which implies that $x \notin \mathcal{R}_j$. ∎

As an illustration for the procedure proposed in Theorem 4.2 consider the two-dimensional case depicted in Figure 4.10(a). Here $\mathcal{Y}$ is defined by the inequalities $\{x_1^- < x_1 \leq x_1^+, x_2^- \leq x_2 \leq x_2^+\}$, and $\mathcal{R}_0$ by the inequalities $\{g_1 \leq 0, \ldots, g_5 \leq 0\}$ where $g_1, \ldots, g_5$ are linear in x. The procedure consists of considering one by one the inequalities which define $\mathcal{R}_0$. Considering, for example, the inequality $g_1 \leq 0$, the first set of the rest of the region $\mathcal{Y} \backslash \mathcal{R}_0$ is given by $\mathcal{R}_1 = \{g_1 \geq 0, x_1 \geq x_1^-, x_2^- \leq x_2 \leq x_2^+\}$, which is obtained by reversing the sign of the inequality $g_1 \leq 0$ and removing redundant constraints in $\mathcal{Y}$ (see Figure 4.10(b)). Thus, by considering the rest of the inequalities we get the partition of the rest of the parameter space $\mathcal{Y} \backslash \mathcal{R}_0 = \bigcup_{i=1}^5 \mathcal{R}_i$, as reported in Figure 4.10(d).

Remark 4.1 The set difference of two intersecting polytopes $\mathcal{P}$ and $\mathcal{Q}$ (or any closed sets) is not a closed set. This means that some borders of polytopes $\mathcal{R}_i$ from a P-collection $\mathcal{R} = \mathcal{P} \backslash \mathcal{Q}$ are open, while other borders are closed. Even though it is possible to keep track of the origin of particular borders of $\mathcal{R}_i$, thus specifying if they are open or closed, we are not doing so in the algorithms described in this book nor in MPT [149]. In computations, we will henceforth only consider the closure of the sets $\mathcal{R}_i$.

The set difference between two P-collections $\mathcal{P}$ and $\mathcal{Q}$ can be computed as described in [22, 134, 243].

4.4.8 Pontryagin Difference

The Pontryagin difference (also known as Minkowski difference) of two polytopes $\mathcal{P}$ and $\mathcal{Q}$ is a polytope

$$\mathcal{P} \ominus \mathcal{Q} = \{x \in \mathbb{R}^n : x + q \in \mathcal{P}, \forall q \in \mathcal{Q}\}. \tag{4.23}$$

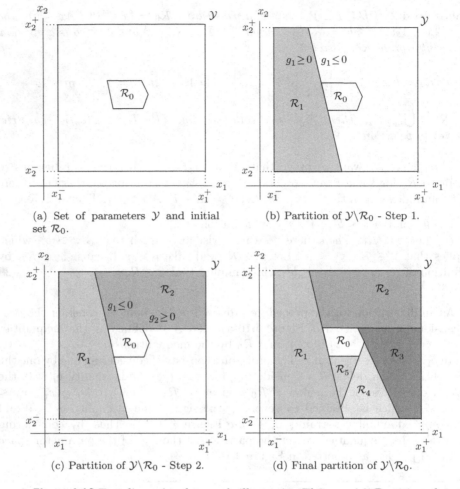

(a) Set of parameters $\mathcal{Y}$ and initial set $\mathcal{R}_0$.

(b) Partition of $\mathcal{Y}\backslash\mathcal{R}_0$ - Step 1.

(c) Partition of $\mathcal{Y}\backslash\mathcal{R}_0$ - Step 2.

(d) Final partition of $\mathcal{Y}\backslash\mathcal{R}_0$.

Figure 4.10 Two-dimensional example illustrating Theorem 4.2. Partition of the rest of the space $\mathcal{Y}\backslash\mathcal{R}_0$.

The Pontryagin difference can be efficiently computed for polytopes by solving a sequence of LPs as follows. Define the $\mathcal{P}$ and $\mathcal{Q}$ as

$$\mathcal{P} = \{y \in \mathbb{R}^n \; : \; P^y y \le P^b\}, \qquad \mathcal{Q} = \{z \in \mathbb{R}^n \; : \; Q^z z \le Q^b\}, \tag{4.24}$$

then

$$\mathcal{W} = \mathcal{P} \ominus \mathcal{Q} \tag{4.25a}$$

$$= \{x \in \mathbb{R}^n \; : \; P^y x \le P^b - H(P^y, \mathcal{Q})\}, \tag{4.25b}$$

where the i-th element of $H(P^y, \mathcal{Q})$ is

$$H_i(P^y, \mathcal{Q}) = \max_{x \in \mathcal{Q}} P_i^y x, \tag{4.26}$$

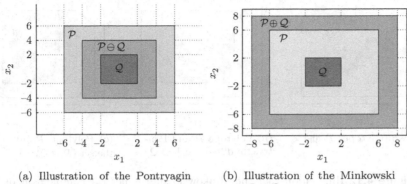

(a) Illustration of the Pontryagin difference $\mathcal{P} \ominus \mathcal{Q}$.

(b) Illustration of the Minkowski sum $\mathcal{P} \oplus \mathcal{Q}$.

Figure 4.11 Illustration of the Pontryagin difference and Minkowski sum operations.

and P_i^y is the i-th row of the matrix P^y. Note that for special cases (e.g., when $\mathcal{Q}$ is a hypercube), more efficient computational methods exist [175]. An illustration of the Pontryagin difference is given in Figure 4.11a.

4.4.9 Minkowski Sum

The Minkowski sum of two polytopes $\mathcal{P}$ and $\mathcal{Q}$ is a polytope

$$\mathcal{P} \oplus \mathcal{Q} = \{ y + z \in \mathbb{R}^n \; : \; y \in \mathcal{P}, \; z \in \mathcal{Q} \}. \tag{4.27}$$

The Minkowski sum is a computationally expensive operation which requires either vertex enumeration and convex hull computation in n-dimensions or a projection from $2n$ down to n dimensions. The implementation of the Minkowski sum via projection is described below. If the polytopes are defined by

$$P = \{ y \in \mathbb{R}^n \; : \; P^y y \le P^c \}, \quad \mathcal{Q} = \{ z \in \mathbb{R}^n \; : \; Q^z z \le Q^c \},$$

then it holds that

$$
\begin{aligned}
W &= P \oplus Q \\
&= \left\{ x \in \mathbb{R}^n \; : \; x = y + z, \; P^y y \le P^c, \; Q^z z \le Q^c, \; y, z \in \mathbb{R}^n \right\} \\
&= \left\{ x \in \mathbb{R}^n \; : \; \exists y \in \mathbb{R}^n, \text{ subj. to } P^y y \le P^c, \; Q^z(x - y) \le Q^c \right\} \\
&= \left\{ x \in \mathbb{R}^n \; : \; \exists y \in \mathbb{R}^n, \text{ subj. to } \begin{bmatrix} 0 & P^y \\ Q^z & -Q^z \end{bmatrix} \begin{bmatrix} x \\ y \end{bmatrix} \le \begin{bmatrix} P^c \\ Q^c \end{bmatrix} \right\} \\
&= \text{proj}_x \left(\left\{ \begin{bmatrix} x \\ y \end{bmatrix} \in \mathbb{R}^{n+n} \; : \; \begin{bmatrix} 0 & P^y \\ Q^z & -Q^z \end{bmatrix} \begin{bmatrix} x \\ y \end{bmatrix} \le \begin{bmatrix} P^c \\ Q^c \end{bmatrix} \right\} \right).
\end{aligned}
$$

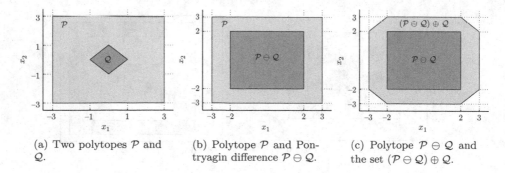

(a) Two polytopes $\mathcal{P}$ and $\mathcal{Q}$.

(b) Polytope $\mathcal{P}$ and Pontryagin difference $\mathcal{P} \ominus \mathcal{Q}$.

(c) Polytope $\mathcal{P} \ominus \mathcal{Q}$ and the set $(\mathcal{P} \ominus \mathcal{Q}) \oplus \mathcal{Q}$.

Figure 4.12 Illustration that the Minkowski sum is not the complement of the Pontryagin difference. $(\mathcal{P} \ominus \mathcal{Q}) \oplus \mathcal{Q} \subseteq \mathcal{P}$.

Both the projection and vertex enumeration-based methods are implemented in the MPT toolbox [149]. An illustration of the Minkowski sum is given in Figure 4.11b.

Remark 4.2   The Minkowski sum is **not** the complement of the Pontryagin difference. For two polytopes $\mathcal{P}$ and $\mathcal{Q}$, it holds that $(\mathcal{P} \ominus \mathcal{Q}) \oplus \mathcal{Q} \subseteq \mathcal{P}$. This is illustrated in Figure 4.12.

4.4.10 Polyhedra Union

Consider the following basic problem in polyhedral computation: given two polyhedra $\mathcal{P} \subset \mathbb{R}^n$ and $\mathcal{Q} \subset \mathbb{R}^n$, decide whether their union is convex, and, if so, compute it. There are three classes of algorithms for the given problem depending on their representation: (1) $\mathcal{P}$ and $\mathcal{Q}$ are given in $\mathcal{H}$-representation, (2) $\mathcal{P}$ and $\mathcal{Q}$ are given in $\mathcal{V}$-representation and (3) both $\mathcal{H}$- and $\mathcal{V}$-representations are available for $\mathcal{P}$ and $\mathcal{Q}$. Next we present an algorithm for case (1). Case (2), case (3) and the computational complexity of all three cases are discussed in [38].

Recall the definition of envelope in Section 4.4.3. By definition, it is easy to see that env$(\mathcal{P}, \mathcal{Q})$ is convex and that

$$\mathcal{P} \cup \mathcal{Q} \subseteq \text{env}(\mathcal{P}, \mathcal{Q}). \tag{4.28}$$

We have the following theorem.

Theorem 4.3 $\mathcal{P} \cup \mathcal{Q}$ *is convex* $\Leftrightarrow \mathcal{P} \cup \mathcal{Q} = \text{env}(\mathcal{P}, \mathcal{Q})$.

Algorithm 4.2 checks for points $x^* \in \text{env}(\mathcal{P}, \mathcal{Q})$ outside $\mathcal{P} \cup \mathcal{Q}$.

Algorithm 4.2 *Algorithm for recognizing the convexity of* $\mathcal{P} \cup \mathcal{Q}$

Input $\mathcal{P}$, $\mathcal{Q}$
Output Flag convex, if convex=true $\mathcal{H}$-representation of $\mathcal{P} \cup \mathcal{Q}$

    convex $\leftarrow$ true
    Construct env$(\mathcal{P}, \mathcal{Q})$ by removing nonvalid constraints (see Figure 4.7)
    Let $\tilde{A}x \leq \tilde{\alpha}$, $\tilde{B}x \leq \tilde{\beta}$ be the set of removed constraints from $\mathcal{P}$ and $\mathcal{Q}$
    Let env$(\mathcal{P}, \mathcal{Q}) = \{x : Cx \leq \gamma\}$ the resulting envelope
    Remove from env$(\mathcal{P}, \mathcal{Q})$ possible duplicates $(B_j, \beta_j) = (\sigma A_i, \sigma \alpha_i)$, $\sigma > 0$
    For each pair $\tilde{A}_i x \leq \tilde{\alpha}_i$, $\tilde{B}_j x \leq \tilde{\beta}_j$ **Do**
        Determine ϵ^* by solving the LP

$$\begin{aligned} \epsilon^* = \max_{(x,\epsilon)} \quad & \epsilon \\ \text{subj. to} \quad & \tilde{A}_i x \geq \tilde{\alpha}_i + \epsilon \\ & \tilde{B}_j x \geq \tilde{\beta}_j + \epsilon \\ & Cx \leq \gamma \end{aligned}$$

    If $\epsilon^* > 0$ **Stop**, **Return** convex=false
    Return env$(\mathcal{P}, \mathcal{Q})$

Note that if $\epsilon^* = 0$ for each i, j, then the union is convex and equals env$(\mathcal{P}, \mathcal{Q})$. On the other hand, $\epsilon^* > 0$ indicates the existence of a point $x \in$ env$(\mathcal{P}, \mathcal{Q})$ outside $\mathcal{P} \cup \mathcal{Q}$.

For recognizing convexity and computing the union of k polyhedra, the test can be modified by checking each k tuple of removed constraints. Let $\tilde{m}_1, \ldots, \tilde{m}_k$ be the number of removed constrains from the polyhedra $\mathcal{P}_1, \ldots, \mathcal{P}_k$, respectively. Then similarly for the loop in Algorithm 4.2, $\prod_{i=1}^{k} \tilde{m}_i$ linear programs need to be solved in the general case.

4.4.11 Affine Mappings and Polyhedra

This section deals with the composition of affine mappings and polyhedra. Consider a polyhedron $\mathcal{P} = \{x \in \mathbb{R}^n : P^x x \leq P^c\}$, with $P^x \in \mathbb{R}^{n_P \times n}$ and an affine mapping $f(z)$

$$f : z \in \mathbb{R}^m \mapsto Az + b, \quad A \in \mathbb{R}^{m_A \times m}, \quad b \in \mathbb{R}^{m_A}. \tag{4.29}$$

Let $m_A = n$. We define the composition of $\mathcal{P}$ and f as the following polyhedron

$$\mathcal{P} \circ f = \{z \in \mathbb{R}^m : P^x f(z) \leq P^c\} = \{z \in \mathbb{R}^m : P^x Az \leq P^c - P^x b\}. \tag{4.30}$$

Let $m = n$. We define the composition of f and $\mathcal{P}$ as the following polyhedron

$$f \circ \mathcal{P} = \{y \in \mathbb{R}^{m_A} : \exists x \in \mathcal{P} \text{ and } y = Ax + b \}. \tag{4.31}$$

The polyhedron $f \circ \mathcal{P}$ in (4.31) can be computed as follows. Let us write $\mathcal{P}$ in $\mathcal{V}$-representation

$$\mathcal{P} = \text{conv}(V), \tag{4.32}$$

and let us map the set of vertices $V - \{V_1, \ldots, V_k\}$ through the transformation f. Because the transformation is affine, the set $f \circ \mathcal{P}$ is simply the convex hull of the transformed vertices

$$f \circ \mathcal{P} = \operatorname{conv}(F), \quad F = \{AV_1 + b, \ldots, AV_k + b\}. \tag{4.33}$$

The polyhedron $f \circ \mathcal{P}$ in (4.31) can be computed immediately if $m_A = m = n$ and A is invertible. In this case, from the definition in (4.31), $x = A^{-1}y - A^{-1}b$ and therefore

$$f \circ \mathcal{P} = \{y \in \mathbb{R}^{m_A} \ : \ P^x A^{-1} y \leq P^c + P^x A^{-1} b\}. \tag{4.34}$$

Vertex enumeration can be avoided even if A is not invertible and $m_A \geq m = n$ by using a QR decomposition of the matrix A.

Remark 4.3   Often in the literature the symbol "$\circ$" is omitted for linear maps $f = Az$. Therefore, $A\mathcal{P}$ refers to the operation $A \circ \mathcal{P}$ and $\mathcal{P}A$ refers to the operation $\mathcal{P} \circ A$.

4.5 Operations on P-Collections

This section covers some results and algorithms which are specific to operations with P-collections. P-collections are unions of polytopes (see Definition 4.3) and therefore the set of points contained in a P-collection can be represented in an infinite number of ways, i.e., the P-collection representation is not unique. For example, one can subdivide any polytope $\mathcal{P}$ into a number of smaller polytopes whose union is a P-collection which covers $\mathcal{P}$. Note that the complexity of all subsequent computations depends strongly on the number of polytopes representing a P-collection. The smaller the cardinality of a P-collection, the more efficient the computations. The reader is referred to [243, 242] for proofs and comments on computational efficiency.

4.5.1 Set-Difference

The first two results given here show how the set difference of a P-collection and a P-collection (or polyhedron) may be computed.

Lemma 4.1   Let $\mathcal{C} = \bigcup_{j \in \{1,\ldots,J\}} \mathcal{C}_j$ be a P-collection, where all the $\mathcal{C}_j$, $j \in \{1,\ldots,J\}$, are nonempty polyhedra. If $\mathcal{D}$ is a nonempty polyhedron, then $\mathcal{C} \backslash \mathcal{D} = \bigcup_{j \in \{1,\ldots,J\}} (\mathcal{C}_j \backslash \mathcal{D})$ is a P-collection.

Lemma 4.2   Let the sets $\mathcal{C} = \bigcup_{j \in \{1,\ldots,J\}} \mathcal{C}_j$ and $\mathcal{D} = \bigcup_{y=1,\ldots,Y} \mathcal{D}_y$ be P-collections, where all the $\mathcal{C}_j$, $j \in \{1,\ldots,J\}$, and $\mathcal{D}_y$, $y \in \{1,\ldots,Y\}$, are nonempty polyhedra. If $\mathcal{E}_0 = \mathcal{C}$ and $\mathcal{E}_y = \mathcal{E}_{y-1} \backslash \mathcal{D}_y$, $y \in \{1,\ldots,Y\}$ then $\mathcal{C} \backslash \mathcal{D} = \mathcal{E}_Y$ is a P-collection.

The condition $\mathcal{C} \subseteq \mathcal{D}$ can be easily verified since $\mathcal{C} \subseteq \mathcal{D} \Leftrightarrow \mathcal{C} \backslash \mathcal{D} = \emptyset$. Similarly $\mathcal{C} = \mathcal{D}$ is also easily verified since

$$\mathcal{C} = \mathcal{D} \Leftrightarrow (\mathcal{C} \backslash \mathcal{D} = \emptyset \text{ and } \mathcal{D} \backslash \mathcal{C} = \emptyset).$$

Next, an algorithm for computing the Pontryagin difference of a P-collection and a polytope is presented. If $\mathcal{S}$ and $\mathcal{B}$ are two subsets of $\mathbb{R}^n$ then $\mathcal{S} \ominus \mathcal{B} = [\mathcal{S}^c \oplus (-\mathcal{B})]^c$ where $(\cdot)^c$ denotes the set complement. The following algorithm implements the computation of the Pontryagin difference of a P-collection $\mathcal{C} = \cup_{j \in \{1,\dots,J\}} \mathcal{C}_j$, where $\mathcal{C}_j, j \in \{1, \dots, J\}$ are polytopes in $\mathbb{R}^n$, and a polytope $\mathcal{B} \subset \mathbb{R}^n$.

Algorithm 4.3 *Pontryagin Difference for P-collections, $\mathcal{C} \ominus \mathcal{B}$*

Input: P-collection $\mathcal{C}$, polytope $\mathcal{B}$
Output: P-collection $\mathcal{G} \leftarrow \mathcal{C} \ominus \mathcal{B}$

$\quad \mathcal{H} \leftarrow \text{env}(\mathcal{C}) \ (\text{or } \mathcal{H} \leftarrow \text{conv}(\mathcal{C}))$
$\quad \mathcal{D} \leftarrow \mathcal{H} \ominus \mathcal{B}$
$\quad \mathcal{E} \leftarrow \mathcal{H} \backslash \mathcal{C}$
$\quad \mathcal{F} \leftarrow \mathcal{E} \oplus (-\mathcal{B})$
$\quad \mathcal{G} \leftarrow \mathcal{D} \backslash \mathcal{F}$

Remark 4.4 Note that $\mathcal{H}$ in Algorithm 4.3 can be any convex set containing the P-collection $\mathcal{C}$. The computation of $\mathcal{H}$ is generally more efficient if the envelope operation is used instead of convex hull.

Remark 4.5 It is important to note that $(\bigcup_{j \in \{1,\dots,J\}} \mathcal{C}_j) \ominus \mathcal{B} \neq \bigcup_{j \in \{1,\dots,J\}} (\mathcal{C}_j \ominus \mathcal{B})$, where $\mathcal{B}$ and $\mathcal{C}_j$ are polyhedra; hence, the relatively high computational effort of computing the Pontryagin difference of a P-collection and a polytope.

Theorem 4.4 (Computation of Pontryagin Difference [242]) *The output of Algorithm 4.3 is $\mathcal{G} = \mathcal{C} \ominus \mathcal{B}$.*

Proof: It holds by definition that

$$\mathcal{D} = \mathcal{H} \ominus \mathcal{B} = \{x \ : \ x + w \in \mathcal{H}, \ \forall w \in \mathcal{B}\},$$
$$\mathcal{E} = \mathcal{H} \backslash \mathcal{C} = \{x \ : \ x \in \mathcal{H} \text{ and } x \notin \mathcal{C}\}.$$

By the definition of the Minkowski sum:

$$\mathcal{F} = \mathcal{E} \oplus (-\mathcal{B}) = \{x \ : \ x = z + w, \ z \in \mathcal{E}, w \in (-\mathcal{B})\}$$
$$= \{x \ : \ \exists w \in (-\mathcal{B}), \text{ subj. to } x - w \in \mathcal{E}\}.$$

By definition of the set difference:

$$\mathcal{D} \backslash \mathcal{F} = \{x : \ x \in \mathcal{D} \text{ and } x \notin \mathcal{F}\}$$
$$= \{x \in \mathcal{D} \ : \ \nexists \, w \in \mathcal{B} \text{ s.t. } x + w \in \mathcal{E}\}$$
$$= \{x \in \mathcal{D} \ : \ x + w \notin \mathcal{E}, \ \forall w \in \mathcal{B}\}.$$

From the definition of the set $\mathcal{D}$:

$$\mathcal{D} \setminus \mathcal{F} = \{x \ : \ x + w \in \mathcal{H} \text{ and } x + w \notin \mathcal{E}, \ \forall w \in \mathcal{B}\}$$

and from the definition of the set $\mathcal{E}$ and because $\mathcal{C} \subseteq \mathcal{H}$:

$$\begin{aligned}
\mathcal{D} \setminus \mathcal{F} &= \{x \ : \ x + w \in \mathcal{H} \text{ and } (x + w \notin \mathcal{H} \text{ or } x + w \in \mathcal{C}) \ \forall w \in \mathcal{B}\} \\
&= \{x \ : \ x + w \in \mathcal{C}, \ \forall w \in \mathcal{B}\} \\
&= \mathcal{C} \ominus \mathcal{B}.
\end{aligned}$$

■

Algorithm 4.3 is illustrated on a sample P-collection in Figures 4.13(a) to 4.13(f).

Remark 4.6 It should be noted that Algorithm 4.3 for computation of the Pontryagin difference is conceptually similar to the algorithm proposed in [263, 172]. The envelope operation $\mathcal{H} = \text{env}(\mathcal{C})$ employed in Algorithm 4.3 might reduce the number of sets obtained when computing $\mathcal{H} \setminus \mathcal{C}$, which in turn results in fewer Minkowski set additions. Since the computation of a Minkowski set addition is expensive, a runtime improvement can be expected.

4.5.2 Polytope Covering

The problem of checking if some $\mathcal{P}$ polytope is covered with the union of other polytopes, i.e., a P-collection $\mathcal{Q} = \cup_i \mathcal{Q}_i$ is discussed in this section. We consider two related problems:

polycover: Check if $\mathcal{P} \subseteq \mathcal{Q}$, and

regiondiff: Compute P-collection $\mathcal{R} = \mathcal{P} \setminus \mathcal{Q}$.

Clearly, **polycover** is just a special case of **regiondiff**, where the resulting P-collection $\mathcal{R} = \emptyset$. Also, it is straightforward to extend the above problems to the case where both $\mathcal{P}$ and $\mathcal{Q}$ are both P-collections.

One idea of solving the polycover problem is inspired by the following observation

$$\mathcal{P} \subseteq \mathcal{Q} \quad \Leftrightarrow \quad \mathcal{P} = \cup_i (\mathcal{P} \cap \mathcal{Q}_i).$$

Therefore, we could create $\mathcal{R}_i = \mathcal{P} \cap \mathcal{Q}_i$, for $i = 1, \ldots, N_Q$ and then compute the union of the collection of polytopes $\{\mathcal{R}_i\}$ by using the polyunion algorithm for computing the convex union of $\mathcal{H}$-polyhedra reported discussed in Section 4.4.10. If polyunion succeeds (i.e., the union is a convex set) and the resulting polytope is equal to $\mathcal{P}$ then $\mathcal{P}$ is covered by $\mathcal{Q}$, otherwise it is not. However, this approach is computationally very expensive. More details can be found in [123].

4.5.3 Union of P-Collections

Consider a P-collection $\mathcal{P} = \{\mathcal{P}_i\}_{i=1}^{p}$. We study the problem of finding a minimal representation of $\mathcal{P}$ by merging one or more polyhedra belonging to the P-collection.

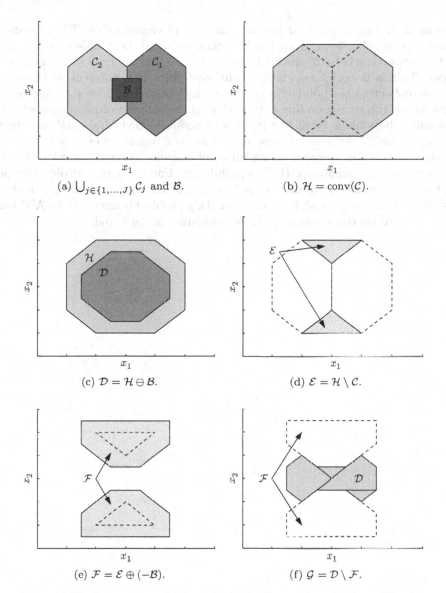

(a) $\bigcup_{j\in\{1,\dots,J\}} \mathcal{C}_j$ and $\mathcal{B}$.

(b) $\mathcal{H} = \mathrm{conv}(\mathcal{C})$.

(c) $\mathcal{D} = \mathcal{H} \ominus \mathcal{B}$.

(d) $\mathcal{E} = \mathcal{H} \setminus \mathcal{C}$.

(e) $\mathcal{F} = \mathcal{E} \oplus (-\mathcal{B})$.

(f) $\mathcal{G} = \mathcal{D} \setminus \mathcal{F}$.

Figure 4.13 Illustration of Algorithm 4.3. Computing the set difference $\mathcal{G} = \mathcal{C} \ominus \mathcal{B}$ between the P-collection $\mathcal{C} = \mathcal{C}_1 \cup \mathcal{C}_2$ and the polytope $\mathcal{B}$.

Clearly one could use the polyunion algorithm presented in Section 4.4.10 for all possible subsets of the P-collection and solve the problem by comparing all solutions. However this approach is not computationally efficient.

Our interest in this problem will be clear later in this book (Section 11.1) when computing the PWA state feedback control law to optimal control problems. Once the PWA state feedback control law has been derived, the memory requirement and the on-line computation time are linear in the number of polyhedra of the feedback law when using standard brute force search. Therefore, we will be interested in the problem of finding a *minimal representation* of piecewise affine (PWA) systems, or more specifically, for a given PWA system, we solve the problem of deriving a PWA system, that is both equivalent to the former and minimal in the number of regions. This is done by associating a different color to a different feedback law, and then collecting the polyhedra with the same color. Then, for a given color, we try to merge the corresponding P-collection. If the number of polyhedra with the same affine dynamic is large, the number of possible polyhedral combinations for merging explodes. As most of these unions are not convex or even not connected and thus cannot be merged, trying all combinations using standard techniques based on *linear programming* (LP) is prohibitive. Furthermore, our objective here is not only to reduce the number of polyhedra but rather to find the minimal and thus optimal number of disjoint polyhedra. This problem is known to be $\mathcal{NP}$-hard. In [123] details on the solutions of these problems can be found.

Part II

Multiparametric Programming

5

Multiparametric Nonlinear Programming

The operations research community has addressed parameter variations in mathematical programs at two levels: *sensitivity analysis*, which characterizes the change of the solution with respect to small perturbations of the parameters, and *parametric programming*, where the characterization of the solution for a full range of parameter values is studied. In this chapter we introduce the concept of multiparametric programming and recall the main results of nonlinear multiparametric programming. The main goal is to make the reader aware of the complexities of general multiparametric nonlinear programming. Later in this book we will use multiparametric programming to characterize and compute the state feedback solution of optimal control problems. There we will only make use of Corollary 5.1 for multiparametric linear programs and Corollary 5.2 for multiparametric quadratic programs, which show that these specific programs are "well behaved."

5.1 Introduction to Multiparametric Programs

Consider the mathematical program

$$J^*(x) = \inf_z \quad J(z, x)$$
$$\text{subj. to} \quad g(z, x) \leq 0$$

where z is the optimization vector and x is a vector of parameters. We are interested in studying the behavior of the value function $J^*(x)$ and the optimizer $z^*(x)$ as we vary the parameter x. Mathematical programs where x is a scalar are referred to as *parametric programs*, while programs where x is a vector are referred to as *multiparametric programs*.

There are several reasons to look for efficient solvers of multiparametric programs. Typically, mathematical programs are affected by uncertainties due to factors that are either unknown or that will be decided later. Parametric programming systematically subdivides the space of parameters into characteristic regions, which depict the feasibility and corresponding performance as a function of the uncertain parameters, and hence provide the decision maker with a complete map of various outcomes.

Our interest in multiparametric programming arises from the field of system theory and optimal control. For example, for discrete-time dynamical systems, finite time constrained optimal control problems can be formulated as mathematical programs where the cost function and the constraints are functions of the initial state of the dynamical system. In particular, Zadeh and Whalen [293] appear to have been the first ones to express the optimal control problem for constrained discrete-time linear systems as a linear program. We can interpret the initial state as a parameter. By using multiparametric programming we can characterize and compute the solution of the optimal control problem explicitly as a function of the initial state.

We are further motivated by the *model predictive control* (MPC) technique. MPC is very popular in the process industry for the automatic regulation of process units under operating constraints, and has attracted a considerable research effort in the last two decades. MPC requires an optimal control problem to be solved on-line in order to compute the next command action. This mathematical program depends on the current sensor measurements. The computation effort can be moved off-line by solving multiparametric programs, where the control inputs are the optimization variables and the measurements are the parameters. The solution of the parametric program problem is a *control law* describing the control inputs as function of the measurements. MPC and its multiparametric solution are discussed in Chapter 12.

In the following we will present several examples that illustrate the parametric programming problem and hint at some of the issues that need to be addressed by the solvers.

Example 5.1   Consider the parametric quadratic program

$$J^*(x) = \min_z \quad J(z,x) = \frac{1}{2}z^2 + 2xz + 2x^2$$
$$\text{subj. to} \quad z \le 1 + x,$$

where $x \in \mathbb{R}$. Our goals are:

1. to find $z^*(x) = \arg\min_z J(z,x)$,

2. to find all x for which the problem has a solution, and

3. to compute the value function $J^*(x)$.

The Lagrangian is

$$L(z,x,u) = \frac{1}{2}z^2 + 2xz + 2x^2 + u(z - x - 1)$$

and the KKT conditions are (see Section 2.3.3 for KKT conditions for quadratic programs)

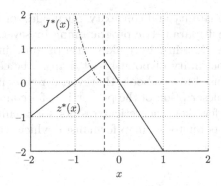

Figure 5.1 Example 5.1. Optimizer $z^*(x)$ and value function $J^*(x)$ as a function of the parameter x.

$$z + 2x + u = 0 \tag{5.1a}$$
$$u(z - x - 1) = 0 \tag{5.1b}$$
$$u \geq 0 \tag{5.1c}$$
$$z - x - 1 \leq 0. \tag{5.1d}$$

Consider (5.1) and the two strictly complementary cases:

$$
\text{A.} \quad
\begin{aligned}
z + 2x + u &= 0 \\
z - x - 1 &= 0 \\
u &\geq 0
\end{aligned}
\quad \Rightarrow \quad
\left\{
\begin{aligned}
z^* &= x + 1 \\
J^* &= \tfrac{9}{2}x^2 + 3x + \tfrac{1}{2} \\
x &\leq -\tfrac{1}{3}
\end{aligned}
\right.
$$

$$
\text{B.} \quad
\begin{aligned}
z + 2x + u &= 0 \\
z - x - 1 &< 0 \\
u &= 0
\end{aligned}
\quad \Rightarrow \quad
\left\{
\begin{aligned}
z^* &= -2x \\
J^* &= 0 \\
x &> -\tfrac{1}{3}
\end{aligned}
\right.
\tag{5.2}
$$

This solution is depicted in Figure 5.1.

The above simple procedure, which required nothing but the solution of the KKT conditions, yielded the optimizer $z^*(x)$ and the value function $J^*(x)$ for all values of the parameter x. The set of admissible parameter values was divided into two *critical regions*, defined by $x \leq -\frac{1}{3}$ and $x > -\frac{1}{3}$. In the region $x \leq -\frac{1}{3}$ the inequality constraint is active and the Lagrange multiplier is greater or equal than zero, in the other region $x > -\frac{1}{3}$ the inequality constraint is not active and the Lagrange multiplier is equal to zero.

In general, when there are more than one inequality constraints, a critical region is defined by the set of inequalities that are active in the region. Throughout a critical region the conditions for optimality derived from the KKT conditions do not change. For our example, in each critical region the optimizer $z^*(x)$ is affine and the value function $J^*(x)$ is quadratic. Thus, considering all x, $z^*(x)$ is piecewise affine and $J^*(x)$ is piecewise quadratic. Both $z^*(x)$ and $J^*(x)$ are continuous, but $z^*(x)$ is not continuously differentiable.

In much of this book we will be interested in two questions: how to find the value function $J^*(x)$ and the optimizer $z^*(x)$ and what are their structural properties, e.g., continuity, differentiability and convexity. Such questions have been addressed for general nonlinear multiparametric programming by several authors in the past (see [18] and references therein), by making use of quite involved mathematical theory based on the continuity of point-to-set maps. The concept of point-to-set maps will not be much used in this book. However, it represents a key element for a rigorous mathematical description of the properties of a nonlinear multiparametric program and hence a few key theoretical results for nonlinear multiparametric programs based on the point-to-set map formalism, which will be discussed in this chapter.

5.2   General Results for Multiparametric Nonlinear Programs

Consider the nonlinear mathematical program dependent on a parameter x appearing in the cost function and in the constraints

$$J^*(x) = \inf_z \quad J(z, x)$$
$$\text{subj. to} \quad g(z, x) \leq 0, \tag{5.3}$$

where $z \in \mathcal{Z} \subseteq \mathbb{R}^s$ is the optimization vector, $x \in \mathcal{X} \subseteq \mathbb{R}^n$ is the parameter vector, $J : \mathbb{R}^s \times \mathbb{R}^n \to \mathbb{R}$ is the cost function and $g : \mathbb{R}^s \times \mathbb{R}^n \to \mathbb{R}^{n_g}$ are the constraints. We denote by $g_i(z, x)$ the i-th component of the vector-valued function $g(z, x)$.

A small perturbation of the parameter x in the mathematical program (5.3) can cause a variety of results. Depending on the properties of the functions J and g the solution $z^*(x)$ may vary smoothly or change abruptly as a function of x. Denote by $2^{\mathcal{Z}}$ the set of subsets of $\mathcal{Z}$. We denote by R the point-to-set map which assigns to a parameter $x \in \mathcal{X}$ the (possibly empty) *set $R(x)$ of feasible variables* $z \in \mathcal{Z}$, $R : \mathcal{X} \mapsto 2^{\mathcal{Z}}$

$$R(x) = \{z \in \mathcal{Z} : \; g(z, x) \leq 0\}, \tag{5.4}$$

by $\mathcal{K}^*$ the set of feasible parameters

$$\mathcal{K}^* = \{x \in \mathcal{X} : \; R(x) \neq \emptyset\}, \tag{5.5}$$

by $J^*(x)$ the real-valued function that expresses the dependence of the minimum value of the objective function over $\mathcal{K}^*$ on x

$$J^*(x) = \inf_z \{J(z, x) : \; z \in R(x)\}, \tag{5.6}$$

and by $Z^*(x)$ the point-to-set map which assigns the (possibly empty) set of optimizers $z^* \in 2^{\mathcal{Z}}$ to a parameter $x \in \mathcal{X}$

$$Z^*(x) = \{z \in R(x) : \; J(z, x) = J^*(x)\}. \tag{5.7}$$

$J^*(x)$ will be referred to as optimal value function or simply *value function*, $Z^*(x)$ will be referred to as the *optimal set*. If $Z^*(x)$ is a singleton for all x, then $z^*(x) = Z^*(x)$ will be called *optimizer function*. We remark that R and Z^* are set-valued functions. As discussed in the notation section, with abuse of notation $J^*(x)$ and $Z^*(x)$ will denote both the functions and the value of the functions at the point x. The context will make clear which notation is being used.

The book by Bank and coauthors [18] and Chapter 2 of [107] describe conditions under which the solution of the nonlinear multiparametric program (5.3) is locally well behaved and establish properties of the optimal value function and of the optimal set. The description of such conditions requires the definition of continuity of point-to-set maps. Before introducing this concept we will show through two simple examples that continuity of the constraints $g_i(z, x)$ with respect to z and x is not enough to imply any "regularity" of the value function and the optimizer function.

Example 5.2 [18, p. 12]

Consider the following problem:

$$
\begin{aligned}
J^*(x) = \min_z \quad & x^2 z^2 - 2x(1-x)z \\
\text{subj. to} \quad & z \geq 0 \\
& 0 \leq x \leq 1.
\end{aligned}
\tag{5.8}
$$

Cost and constraints are continuous and continuously differentiable. For $0 < x \leq 1$ the optimizer function is $z^* = (1-x)/x$ and the value function is $J^*(x) = -(1-x)^2$. For $x = 0$, the value function is $J^*(x) = 0$ while the optimal set is $Z^* = \{z \in \mathbf{R} : z \geq 0\}$. Thus, the value function is *discontinuous* at 0 and the optimal set is single-valued for all $0 < x \leq 1$ and set-valued for $x = 0$.

Example 5.3 Consider the following problem:

$$
\begin{aligned}
J^*(x) = \inf_z \quad & z \\
\text{subj. to} \quad & zx \geq 0 \\
& -10 \leq z \leq 10 \\
& -10 \leq x \leq 10,
\end{aligned}
\tag{5.9}
$$

where $z \in \mathbb{R}$ and $x \in \mathbb{R}$. For each fixed x the set of feasible z is a segment. The point-to-set map $R(x)$ is plotted in Figure 5.2(a). The function $g_1 : (z, x) \mapsto zx$ is continuous. Nevertheless, the value function $J^*(x) = z^*(x)$ has a discontinuity at the origin as can be seen in Figure 5.2(b).

Example 5.4 Consider the following problem:

$$
\begin{aligned}
J^*(x) = \inf_{z_1, z_2} \quad & -z_1 \\
\text{subj. to} \quad & g_1(z_1, z_2) + x \leq 0 \\
& g_2(z_1, z_2) + x \leq 0,
\end{aligned}
\tag{5.10}
$$

where examples of the functions $g_1(z_1, z_2)$ and $g_2(z_1, z_2)$ are plotted in Figures 5.3(a)–5.3(c). Figures 5.3(a)–5.3(c) also depict the point-to-set map $R(x) = \{[z_1, z_2] \in \mathbb{R}^2 | g_1(z_1, z_2) + x \leq 0, \ g_2(z_1, z_2) + x \leq 0\}$ for three fixed x. Starting from $x = \bar{x}_1$, as x increases, the domain of feasibility in the space z_1, z_2 shrinks; at the beginning it is connected (Figure 5.3(a)), then it becomes disconnected (Figure 5.3(b)) and eventually connected again (Figure 5.3(c)). No matter how smooth one chooses the functions g_1 and g_2, the value function $J^*(x) = -z_1^*(x)$ will have a discontinuity at $x = \bar{x}_3$.

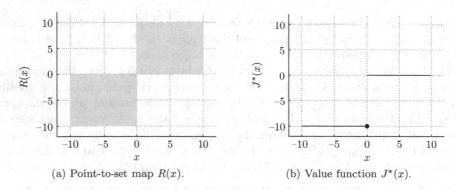

(a) Point-to-set map $R(x)$.   (b) Value function $J^*(x)$.

Figure 5.2 Example 5.3. Point-to-set map and value function.

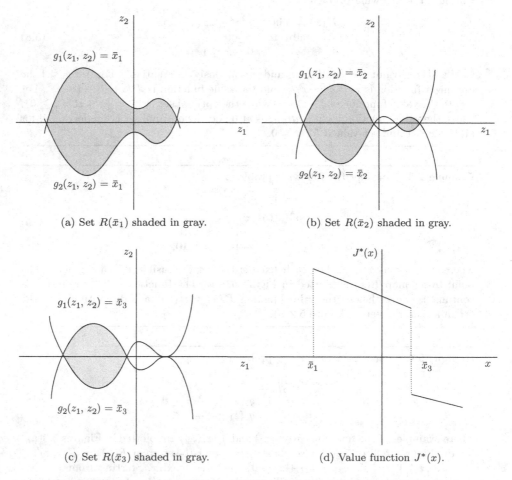

(a) Set $R(\bar{x}_1)$ shaded in gray.   (b) Set $R(\bar{x}_2)$ shaded in gray.

(c) Set $R(\bar{x}_3)$ shaded in gray.   (d) Value function $J^*(x)$.

Figure 5.3 Example 5.4. Problem (5.10). (a)–(c) Projections of the point-to-set map $R(x)$ for three values of the parameter x: $\bar{x}_1 < \bar{x}_2 < \bar{x}_3$; (d) Value function $J^*(x)$.

Examples 5.2, 5.3 and 5.4 show the case of simple and smooth constraints which lead to a discontinuous behavior of value function and in Examples 5.3 and 5.4 we also observe discontinuity of the optimizer function. The main causes are:

- in Example 5.2 the feasible vector space $\mathcal{Z}$ is unbounded ($z \geq 0$),

- in Examples 5.3 and 5.4 the feasible point-to-set map $R(x)$ (defined in (5.4)) is discontinuous, as defined precisely below.

In the next sections we discuss both cases in detail.

Continuity of Point-to-Set Maps

Consider a point-to-set map $R : x \in \mathcal{X} \mapsto R(x) \in 2^{\mathcal{Z}}$. We give the following definitions of open and closed maps according to Hogan [152]:

Definition 5.1   *The point-to-set map $R(x)$ is open at a point $\bar{x} \in \mathcal{K}^*$ if for all sequences $\{x^k\} \subset \mathcal{K}^*$ with $x^k \to \bar{x}$ and for all $\bar{z} \in R(\bar{x})$ there exists an integer m and a sequence $\{z^k\} \in \mathcal{Z}$ such that $z^k \in R(x^k)$ for $k \geq m$ and $z^k \to z$.*

Definition 5.2   *The point-to-set map $R(x)$ is closed at a point $\bar{x} \in \mathcal{K}^*$ if for each pair of sequences $\{x^k\} \subset \mathcal{K}^*$, and $z^k \in R(x^k)$ with the properties*

$$x^k \to \bar{x}, \ z^k \to \bar{z}.$$

it follows that $\bar{z} \in R(\bar{x})$.

We define the continuity of a point-to-set map according to Hogan [152] as follows:

Definition 5.3   *The point-to-set map $R(x)$ is continuous at a point $\bar{x}$ in $\mathcal{K}^*$ if it is both open and closed at $\bar{x}$. $R(x)$ is continuous in $\mathcal{K}^*$ if $R(x)$ is continuous at every point x in $\mathcal{K}^*$.*

The definitions above are illustrated through two examples.

Example 5.5   Consider

$$R(x) = \{z \in \mathbb{R} \mid z \in [0,1] \text{ if } x < 1, \ z \in [0, 0.5] \text{ if } x \geq 1.\}$$

The point-to-set map $R(x)$ is plotted in Figure 5.4. It is easy to see that $R(x)$ is not closed but open. In fact, if one considers a sequence $\{x^k\}$ that converges to $\bar{x} = 1$ from the left and extracts the sequence $\{z^k\}$ plotted in Figure 5.4 converging to $\bar{z} = 0.75$, then $\bar{z} \notin R(\bar{x})$ since $R(1) = [0, 0.5]$.

Example 5.6   Consider

$$R(x) = \{z \in \mathbb{R} \ : \ z \in [0,1] \ \text{ if } x \leq 1, \ z \in [0, 0.5] \text{ if } x > 1\}$$

The point-to-set map $R(x)$ is plotted in Figure 5.5. It is easy to verify that $R(x)$ is closed but not open. Choose $\bar{z} = 0.75 \in R(\bar{x})$. Then, for any sequence $\{x^k\}$ that converges to $\bar{x} = 1$ from the right, one is not able to construct a sequence $\{z^k\} \in \mathcal{Z}$ such that $z^k \in R(x^k)$ and $z^k \to \bar{z}$. In fact, such sequence z^k will always be bounded between 0 and 0.5.

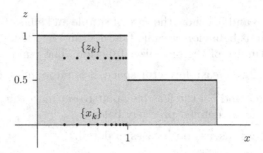

Figure 5.4 Example 5.5. Open and not closed point-to-set map $R(x)$.

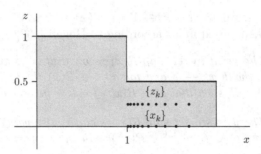

Figure 5.5 Example 5.6. Closed and not open point-to-set map $R(x)$.

Remark 5.1   We remark that "upper semicontinuous" and "lower semicontinuous" definitions of point-to-set map are sometimes preferred to open and closed definitions [47, p. 109]. In [18, p. 25], nine different definitions for the continuity of point-to-set maps are introduced and compared. We will not give any details on this subject and refer the interested reader to [18, p. 25].

The examples above are only illustrative. In general, it is difficult to test if a set is closed or open by applying the definitions. Several authors have proposed sufficient conditions on g_i which imply the continuity of $R(x)$. In the following we introduce a theorem which summarizes the main results of [254, 93, 152, 47, 18].

Theorem 5.1   *If $\mathcal{Z}$ is convex, if each component $g_i(z,x)$ of $g(z,x)$ is continuous on $\mathcal{Z} \times \bar{x}$ and convex in z for each $\bar{x} \in \mathcal{X}$ and if there exists a $\bar{z}$ such that $g(\bar{z},\bar{x}) < 0$, then $R(x)$ is continuous at $\bar{x}$.*

The proof is given in [152, Theorems 10 and 12]. An equivalent proof can be also derived from [18, Theorem 3.1.1 and Theorem 3.1.6]. ■

Remark 5.2   Note that convexity in z for each x is not enough to imply the continuity of $R(x)$ everywhere in $\mathcal{K}^*$. In [18, Example 3.3.1 on p. 53] an example illustrating this is presented. We remark that in Example 5.3 the origin does not satisfy the last hypothesis of Theorem 5.1.

Remark 5.3   If the assumptions of Theorem 5.1 hold at each $\bar{x} \in \mathcal{X}$ then one can extract a continuous single-valued function (often called a "continuous selection") $r : \mathcal{X} \mapsto \mathbb{R}$ such that $r(x) \in R(x)$, $\forall x \in \mathcal{X}$, provided that $\mathcal{Z}$ is finite-dimensional. Note that convexity of $R(x)$ is a critical assumption [18, Corollary 2.3.1]. The following example shows a point-to-set map $R(x)$ not convex for a fixed x which is continuous but has no continuous selection [18, p. 29]. Let Λ be the unit disk in $\mathbb{R}^2$, define $R(x)$ as

$$x \in \Lambda \mapsto R(x) = \left\{ z \in \Lambda : \; \|z - x\|_2 \geq \frac{1}{2} \right\}. \qquad (5.11)$$

It can be shown that the point-to-set map $R(x)$ in (5.11) is continuous according to Definition 5.3. In fact, for a fixed $\bar{x} \in \Lambda$ the set $R(\bar{x})$ is the set of points in the unit disk outside the disk centered in $\bar{x}$ and of radius 0.5 (next called the *half disk*); small perturbations of $\bar{x}$ yield small translations of the half disk inside the unit disk for all $\bar{x} \in \Lambda$. However $R(x)$ has no continuous selection. Assume that there exists a continuous selection $r : x \in \Lambda \mapsto r(x) \subset \Lambda$. Then, there exists a point x^* such that $x^* = r(x^*)$. Since $r(x) \in R(x)$, $\forall x \in \Lambda$, there exists a point x^* such that $x^* \in R(x^*)$. This is not possible since for all $x^* \in \Lambda$, $x^* \notin R(x^*)$ (recall that $R(x^*)$ is set of points in the unit disk outside the disk centered in x^* and of radius 0.5).

Remark 5.4   Let Λ be the unit disk in $\mathbb{R}$, define $R(x)$ as

$$x \in \Lambda \mapsto R(x) = \left\{ z \in \Lambda : \; |z - x| \geq \frac{1}{2} \right\}. \qquad (5.12)$$

$R(x)$ is closed and not open and it has no continuous selection.

Remark 5.5   Based on [18, Theorem 3.2.1-(I) and Theorem 3.3.3], the hypotheses of Theorem 5.1 can be relaxed for affine $g_i(z, x)$. In fact, affine functions are weakly analytic functions according to [18, p. 47]. Therefore, we can state that if $\mathcal{Z}$ is convex, if each component $g_i(z, x)$ of $g(z, x)$ is an affine function, then $R(x)$ is continuous at $\bar{x}$ for all $\bar{x} \in \mathcal{K}^*$.

Properties of the Value Function and Optimal Set

Consider the following definition

Definition 5.4   *A point-to-set map $R(x)$ is said to be* uniformly compact *near $\bar{x}$ if there exist a neighborhood N of $\bar{x}$ such that the closure of the set $\bigcup_{x \in N} R(x)$ is compact.*

Now we are ready to state the two main theorems on the continuity of the value function and of the optimizer function.

Theorem 5.2   *[152, Theorem 7] Consider problem (5.3)–(5.4). If $R(x)$ is a continuous point-to-set map at $\bar{x}$ and uniformly compact near $\bar{x}$ and if J is continuous on $\bar{x} \times R(\bar{x})$, then J^* is continuous at $\bar{x}$.* ∎

Theorem 5.3   *[152, Corollary 8.1] Consider problem (5.3)–(5.4). If $R(x)$ is a continuous point-to-set map at $\bar{x}$, J is continuous on $\bar{x} \times R(\bar{x})$, Z^* is nonempty and uniformly compact near $\bar{x}$, and $Z^*(\bar{x})$ is single valued, then Z^* is continuous at $\bar{x}$.*   ∎

Remark 5.6   Equivalent results of Theorems 5.2 and 5.3 can be found in [47, p. 116] and [18, Chapter 4.2].

Example 5.7   Example 5.2 revisited

Consider Example 5.2. The feasible map $R(x)$ is unbounded and therefore it does not satisfy the assumptions of Theorem 5.2 (since it is not uniformly compact). Modify Example 5.2 as follows:

$$
\begin{aligned}
J^*(x) = \min_z \quad & x^2 z^2 - 2x(1-x)z \\
\text{subj. to} \quad & 0 \le z \le M \\
& 0 \le x \le 1
\end{aligned}
\tag{5.13}
$$

with $M \ge 0$. The solution can be computed immediately. For $1/(1+M) < x \le 1$ the optimizer function is $z^* = (1-x)/x$ and the value function is $J^*(x) = -(1-x)^2$. For $0 < x \le 1/(1+M)$, the value function is $J^*(x) = x^2 M^2 - 2Mx(1-x)$ and the optimizer function is $z^* = M$. For $x = 0$, the value function is $J^*(x) = 0$ while the optimal set is $Z^* = \{z \in \mathbf{R} : 0 \le z \le M\}$.

No matter how large we choose M, the value function and the optimal set are continuous for all $x \in [0,1]$.

Example 5.8   Example 5.3 revisited

Consider Example 5.3. The feasible map $R(x)$ is not continuous at $x = 0$ and therefore it does not satisfy the assumptions of Theorem 5.2. Modify Example 5.3 as follows:

$$
\begin{aligned}
J^*(x) = \inf_z \quad & z \\
\text{subj. to} \quad & zx \ge -\varepsilon \\
& -10 \le z \le 10 \\
& -10 \le x \le 10,
\end{aligned}
\tag{5.14}
$$

where $\varepsilon > 0$. The value function and the optimal set are depicted in Figure 5.2 for $\varepsilon = 1$. No matter how small we choose ε, the value function and the optimal set are continuous for all $x \in [-10, 10]$

The following corollaries consider special classes of parametric problems.

Corollary 5.1 (mp-LP)   *Consider the special case of the multiparametric program (5.3). where the objective and the constraints are linear*

$$
\begin{aligned}
J^*(x) = \min_z \quad & c'z \\
\text{subj. to} \quad & Gz \le w + Sx,
\end{aligned}
\tag{5.15}
$$

and assume that there exists an $\bar{x}$ and $z^*(\bar{x})$ with a bounded cost $J^*(\bar{x})$. Then, $\mathcal{K}^*$ is a nonempty polyhedron, $J^*(x)$ is a continuous and convex function on $\mathcal{K}^*$ and the optimal set $Z^*(x)$ is a continuous point-to-set map on $\mathcal{K}^*$.

Proof: See Theorem 5.5.1 in [18] and the bottom of page 138 in [18].   ∎

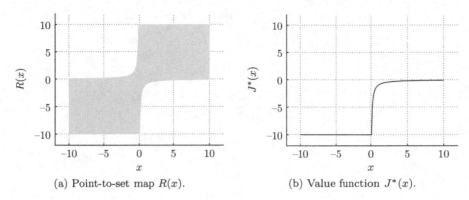

(a) Point-to-set map $R(x)$. (b) Value function $J^*(x)$.

Figure 5.6 Example 5.8. Point-to-set map and value function.

Corollary 5.2 (mp-QP) *Consider the special case of the multiparametric program (5.3). where the objective is quadratic and the constraints are linear*

$$J^*(x) = \min_{z} \quad \frac{1}{2} z'Hz + z'F$$
$$subj. \ to \quad Gz \le w + Sx,$$
(5.16)

and assume that $H \succ 0$ and that there exists $(\bar{z}, \bar{x})$ such that $G\bar{z} \le w + S\bar{x}$. Then, $\mathcal{K}^*$ is a nonempty polyhedron, $J^*(x)$ is a continuous and convex function on $\mathcal{K}^*$ and the optimizer function $z^*(x)$ is continuous in $\mathcal{K}^*$.

Proof: See Theorem 5.5.1 in [18] and the bottom of page 138 in [18]. ∎

Remark 5.7 We remark that Corollary 5.1 requires the existence of optimizer $z^*(x)$ with a bounded cost. This is implicitly guaranteed in the mp-QP case since in Corollary 5.2 the matrix H is assumed to be strictly positive definite. Moreover, the existence of an optimizer $z^*(\bar{x})$ with a bounded cost guarantees that $J^*(x)$ is bounded for all x in $\mathcal{K}^*$. This been proven in [115, p. 178, Theorem 1] for the mp-LP case and it is immediate to prove for the mp-QP case.

Remark 5.8 Both Corollary 5.1 (mp-LP) and Corollary 5.2 (mp-QP) could be formulated stronger: J^* and Z^* are even Lipschitz-continuous. J^* is also piecewise affine (mp-LP) or piecewise quadratic (mp-QP), and for the mp-QP $z^*(x)$ is piecewise affine. For the linear case, Lipschitz continuity is known from Walkup-Wets [283] as a consequence of Hoffman's theorem. For the quadratic case, Lipschitz continuity follows from Robinson [253], as e.g., shown by Klatte and Thiere [178]. The "piecewise" properties are consequences of local stability analysis of parametric optimization, e.g., [107, 18, 200] and are the main focus of the next chapter.

6

Multiparametric Programming: A Geometric Approach

In this chapter we will concentrate on multiparametric linear programs (mp-LP), multiparametric quadratic programs (mp-QP) and multiparametric mixed-integer linear programs (mp-MILP).

The main idea of the multiparametric algorithms presented in this chapter is to construct a critical region in a neighborhood of a given parameter by using necessary and sufficient conditions for optimality, and then to recursively explore the parameter space outside such a region. For this reason the methods are classified as "geometric." All the algorithms are easy to implement once standard solvers are available: linear programming, quadratic programming and mixed-integer linear programming for solving mp-LP, mp-QP and mp-MILP, respectively. A literature review is presented in Section 6.6.

6.1 Multiparametric Programs with Linear Constraints

6.1.1 Formulation

Consider the multiparametric program

$$
\begin{aligned}
J^*(x) = \min_z \quad & J(z, x) \\
\text{subj. to} \quad & Gz \leq w + Sx,
\end{aligned}
\tag{6.1}
$$

where $z \in \mathbb{R}^s$ are the optimization variables, $x \in \mathbb{R}^n$ is the vector of parameters, $J(z, x) : \mathbb{R}^{s+n} \to \mathbb{R}$ is the objective function and $G \in \mathbb{R}^{m \times s}$, $w \in \mathbb{R}^m$, and $S \in \mathbb{R}^{m \times n}$. Given a closed and bounded polyhedral set $\mathcal{K} \subset \mathbb{R}^n$ of parameters,

$$
\mathcal{K} = \{x \in \mathbb{R}^n : Tx \leq N\},
\tag{6.2}
$$

we denote by $\mathcal{K}^* \subseteq \mathcal{K}$ the region of parameters $x \in \mathcal{K}$ such that (6.1) is feasible:

$$
\mathcal{K}^* = \{x \in \mathcal{K} : \exists z \text{ satisfying } Gz \leq w + Sx\}.
\tag{6.3}
$$

In this book we assume that

1. the constraint $x \in \mathcal{K}$ is included in the constraints $Gz \leq w + Sx$.

2. the polytope $\mathcal{K}$ is full-dimensional. Otherwise we can reformulate the problem with a smaller set of parameters such that $\mathcal{K}$ becomes full-dimensional.

3. S has full column rank. Otherwise we can again reformulate the problem in a smaller set of parameters.

Theorem 6.1   *Consider the multiparametric problem (6.1). If the domain of $J(z, x)$ is $\mathbb{R}^{s+n}$ then $\mathcal{K}^*$ is a polytope.*

Proof: $\mathcal{K}^*$ is the projection of the set $Gz - Sx \leq w$ on the x space intersected with the polytope $\mathcal{K}$.   ∎

For any given $\bar{x} \in \mathcal{K}^*$, $J^*(\bar{x})$ denotes the minimum value of the objective function in problem (6.1) for $x = \bar{x}$. The function $J^* : \mathcal{K}^* \to \mathbb{R}$ expresses the dependence of the minimum value of the objective function on x, $J^*(x)$ is called the value function. The set-valued function $Z^* : \mathcal{K}^* \to 2^{\mathbb{R}^s}$, where $2^{\mathbb{R}^s}$ is the set of subsets of $\mathbb{R}^s$, describes for any fixed $x \in \mathcal{K}^*$ the set $Z^*(x)$ of optimizers $z^*(x)$ yielding $J^*(x)$.

We aim to determine the feasible set $\mathcal{K}^* \subseteq \mathcal{K}$ of parameters, the expression of the value function $J^*(x)$ and the expression of one of the optimizers $z^*(x) \in Z^*(x)$.

6.1.2   Definition of Critical Region

Consider the multiparametric program (6.1). Let $I = \{1, \ldots, m\}$ be the set of constraint indices. For any $A \subseteq I$, let G_A and S_A be the submatrices of G and S, respectively, comprising the rows indexed by A and denote with G_j, S_j and w_j the j-th row of G, S and w, respectively. We define CR_A as the set of parameters x for which the same set A of constraints is active at the optimum. More formally we have the following definitions.

Definition 6.1   *The optimal partition of I at x is the partition $(A(x), NA(x))$ where*

$$A(x) = \{j \in I : G_j z^*(x) - S_j x = w_j \text{ for all } z^*(x) \in Z^*(x)\}$$
$$NA(x) = \{j \in I : \text{exists } z^*(x) \in Z^*(x) \text{ satisfying } G_j z^*(x) - S_j x < w_j\}.$$

It is clear that $A(x)$ and $NA(x)$ are disjoint and their union is I.

Definition 6.2   *Consider a set $A \subseteq I$. The critical region associated with the set of active constraints A is defined as*

$$CR_A = \{x \in \mathcal{K}^* : A(x) = A\}. \tag{6.4}$$

The set CR_A is the set of all parameters x such that the constraints indexed by A are active at the optimum of problem (6.1). Our first objective is to work with full-dimensional critical regions. For this reason, we discuss next how the dimension of the parameter space can be reduced in case it is not full-dimensional.

6.1.3 Reducing the Dimension of the Parameter Space

It may happen that the set of inequality constraints in (6.1) contains some "hidden" or "implicit" equality constraints as the following example shows.

Example 6.1

$$\min_z \quad J(z, x)$$

$$\text{subj. to} \quad \begin{cases} z_1 + z_2 \leq 9 - x_1 - x_2 \\ z_1 - z_2 \leq 1 - x_1 - x_2 \\ z_1 + z_2 \leq 7 + x_1 + x_2 \\ z_1 - z_2 \leq -1 + x_1 + x_2 \\ -z_1 \leq -4 \\ -z_2 \leq -4 \\ z_1 \leq 20 - x_2, \end{cases} \tag{6.5}$$

where $\mathcal{K} = \{[x_1, x_2]' \subset \mathbb{R}^2 : -100 \leq x_1 \leq 100, -100 \leq x_2 \leq 100\}$. The reader can check that *all* feasible values of x_1, x_2, z_1, z_2 satisfy

$$\begin{aligned} z_1 + z_2 &= 9 - x_1 - x_2 \\ z_1 - z_2 &= 1 - x_1 - x_2 \\ z_1 + z_2 &= 7 + x_1 + x_2 \\ z_1 - z_2 &= -1 + x_1 + x_2 \\ z_1 &= 4 \\ z_2 &= 4 \\ z_1 &\leq 20 - x_2, \end{aligned} \tag{6.6}$$

where we have identified many of the inequalities to be hidden equalities. This can be simplified to

$$\begin{aligned} x_1 + x_2 &= 1 \\ x_2 &\leq 16. \end{aligned} \tag{6.7}$$

Thus

$$\mathcal{K}^* = \left\{ [x_1, x_2]' \in \mathbb{R}^2 : \begin{array}{c} x_1 + x_2 = 1 \\ -100 \leq x_1 \leq 100 \\ -100 \leq x_2 \leq 16 \end{array} \right\}. \tag{6.8}$$

The example shows that the polytope $\mathcal{K}^*$ is contained in a lower dimensional subspace of $\mathcal{K}$, namely a line segment in $\mathbb{R}^2$.

Our goal is to identify the hidden equality constraints (as in (6.6)) and use them to reduce the dimension of the parameter space (as in (6.7)) for which the multiparametric program needs to be solved.

Definition 6.3 *A hidden equality of the polyhedron $\mathcal{C} = \{\xi \in \mathbb{R}^s : B\xi \leq v\}$ is an inequality $B_i \xi \leq v_i$ such that $B_i \bar{\xi} = v_i \quad \forall \bar{\xi} \in \mathcal{C}$.*

To find hidden equalities we need to solve

$$\begin{aligned} v_i^* = \min \quad & B_i \xi \\ \text{subj. to} \quad & B\xi \leq v, \end{aligned}$$

for all constraints $i = 1, \ldots, m$. If $v_i^* = v_i$, then $B_i \xi = v_i$ is a hidden equality.

We can apply this procedure with $\xi = \begin{bmatrix} x \\ z \end{bmatrix}$ to identify the hidden equalities in the set of inequalities in (6.1)

$$Gz \leq w + Sx,$$

to obtain

$$G_{nh}z \leq w_{nh} + S_{nh}x \qquad (6.9a)$$

$$G_h z = w_h + S_h x, \qquad (6.9b)$$

where we have partitioned G, w and S to reflect the hidden equalities. In order to find the equality constraints involving only the parameter x that allow us to reduce the dimension of x, we need to project the equalities (6.9b) onto the parameter space. Let the singular value decomposition of G_h be

$$G_h = [U_1 \ U_2] \, \Sigma \begin{bmatrix} V_1' \\ V_2' \end{bmatrix},$$

where the columns of U_2 are the singular vectors associated with the zero singular values. Then the matrix U_2' defines the projection of (6.9b) onto the parameter space, i.e.,

$$U_2'G_h z = 0 = U_2'w_h + U_2'S_h x. \qquad (6.10)$$

We can use this set of equalities to replace the parameter $x \in \mathbb{R}^n$ with a set of $n' = n - \mathrm{rank}(U_2'S_h)$ new parameters in (6.9a) which simplifies the parametric program (6.1).

In the rest of this book we always assume that the multiparametric program has been preprocessed using the ideas of this section so that the number of independent parameters is reduced as much as possible and $\mathcal{K}^*$ is full-dimensional.

6.2   Multiparametric Linear Programming

6.2.1   Formulation

Consider the special case of the multiparametric program (6.1) where the objective is linear

$$\begin{aligned} J^*(x) = \min_{z} \quad & c'z \\ \text{subj. to} \quad & Gz \leq w + Sx. \end{aligned} \qquad (6.11)$$

All the variables were defined in Section 6.1.1. Our goal is to find the value function $J^*(x)$ and an optimizer function $z^*(x)$ for $x \in \mathcal{K}^*$. Note that $\mathcal{K}^*$ can be determined as discussed in Theorem 6.1. As suggested through Example 5.1 our search for these functions proceeds by partitioning the set of feasible parameters into critical regions. This is shown through a simple example next.

Example 6.2   Consider the parametric linear program

$$\begin{aligned} J^*(x) = \min_{z} \quad & z + 1 \\ \text{subj. to} \quad & z \geq 1 + x \\ & z \geq 0, \end{aligned}$$

where $z \in \mathbb{R}$ and $x \in \mathbb{R}$. Our goals are:

1. to find $z^*(x) = \arg\min_{z,\ z \geq 0,\ z \geq 1+x} z + 1$,
2. to find all x for which the problem has a solution, and
3. to compute the value function $J^*(x)$.

The Lagrangian is

$$L(z, x, u_1, u_2) = z + u_1(-z + x + 1) + u_2(-z)$$

and the KKT conditions are (see Section 2.2.3 for KKT conditions for linear programs)

$$- u_1 - u_2 = -1 \tag{6.12a}$$

$$u_1(-z + x + 1) = 0 \tag{6.12b}$$

$$u_2(-z) = 0 \tag{6.12c}$$

$$u_1 \geq 0 \tag{6.12d}$$

$$u_2 \geq 0 \tag{6.12e}$$

$$-z + x + 1 \leq 0 \tag{6.12f}$$

$$-z \leq 0. \tag{6.12g}$$

Consider (6.12) and the three complementary cases:

$$
\text{A.} \quad
\begin{aligned}
u_1 + u_2 &= 1 \\
-z + x + 1 &= 0 \\
-z &< 0 \\
u_1 &> 0 \\
u_2 &= 0
\end{aligned}
\quad \Rightarrow \quad
\left\{
\begin{aligned}
z^* &= 1 + x \\
u_1^* &= 1,\ u_2^* = 0 \\
J^* &= 2 + x \\
x &> -1
\end{aligned}
\right.
$$

$$
\text{B.} \quad
\begin{aligned}
u_1 + u_2 &= 1 \\
-z + x + 1 &< 0 \\
-z &= 0 \\
u_1 &= 0 \\
u_2 &> 0
\end{aligned}
\quad \Rightarrow \quad
\left\{
\begin{aligned}
z^* &= 0 \\
u_1^* &= 0,\ u_2^* = 1 \\
J^* &= 1 \\
x &< -1
\end{aligned}
\right.
\tag{6.13}
$$

$$
\text{C.} \quad
\begin{aligned}
u_1 + u_2 &= 1 \\
-z + x + 1 &= 0 \\
-z &= 0 \\
u_1 &\geq 0 \\
u_2 &\geq 0
\end{aligned}
\quad \Rightarrow \quad
\left\{
\begin{aligned}
z^* &= 0 \\
u_1^* &\geq 0,\ u_2 \geq 0,\ u_1^* + u_2^* = 1 \\
J^* &= 1 \\
x &= -1
\end{aligned}
\right.
$$

This solution is depicted in Figure 6.1.

The above simple procedure, which required nothing but the solution of the KKT conditions, yielded the optimizer $z^*(x)$ and the value function $J^*(x)$ for all values of the parameter x. The set of admissible parameters values was divided into three *critical regions*, defined by $x < -1$, $x > -1$ and $x = -1$. In the region $x > -1$ the first inequality constraint is active ($z = 1 + x$) and the Lagrange multiplier u_1 is greater than zero, in the second region $x < -1$ the second inequality constraint is active ($z = 0$) and the Lagrange multiplier u_2 is greater than zero. In the third region $x = -1$ both constraints are active and the Lagrange multipliers belong to the set $u_1^* \geq 0$, $u_2 \geq 0$, $u_1^* + u_2^* = 1$. Throughout a critical region the conditions for optimality derived from the KKT conditions do not change. For our example, in each critical region the optimizer $z^*(x)$ is affine and the value function $J^*(x)$ is also affine.

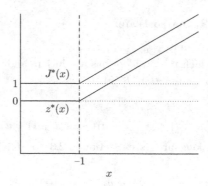

Figure 6.1 Example 6.2. Optimizer $z^*(x)$ and value function $J^*(x)$ as a function of the parameter x.

6.2.2   Critical Regions, Value Function and Optimizer: Local Properties

Consider Definition 6.2 of a critical region. In this section we show that critical regions of mp-LP are polyhedra. We use primal feasibility to derive the $\mathcal{H}$-polyhedral representation of the critical regions, the complementary slackness conditions to compute an optimizer $z^*(x)$, and the dual problem of (6.11) to derive the optimal value function $J^*(x)$ inside each critical region.

Consider the dual problem of (6.11):

$$\begin{aligned}
\min_{u} \quad & (w + Sx)'u \\
\text{subj. to} \quad & G'u = -c \\
& u \geq 0.
\end{aligned} \tag{6.14}$$

The dual feasibility, complementary slackness and primal feasibility conditions for problems (6.11), (6.14) are

$$G'u + c = 0, \ u \geq 0 \tag{6.15a}$$

$$u_i(G_i z - w_i - S_i x) = 0, \ i = 1, \dots, m \tag{6.15b}$$

$$Gz - w - Sx \leq 0. \tag{6.15c}$$

Let us assume that we have determined the optimal partition

$$(A, NA) = (A(x^*), NA(x^*))$$

for some $x^* \in \mathcal{K}^*$. The primal feasibility condition (6.15c) can be rewritten as

$$G_A z^* - S_A x = w_A \tag{6.16a}$$

$$G_{NA} z^* - S_{NA} x < w_{NA}. \tag{6.16b}$$

Here (6.16a) assigns to a parameter x an optimizer $z^*(x)$ (in general, a set of optimizers $z^*(x)$) and (6.16b) defines the *critical region* CR_A in the parameter space $\mathcal{K}^*$ for which this assignment is valid. In the simplest case assume that G_A is square and of full rank. From (6.16a) we find

$$z^* = G_A^{-1}(S_A x + w_A), \tag{6.17}$$

and substituting in (6.16b) we obtain the set of inequalities that defines the critical region CR_A of parameters x for which the optimizer is (6.17):

$$\mathrm{CR}_A = \left\{ x : \; G_{NA}G_A^{-1}(S_A x + w_A) - S_{NA}x < w_{NA} \right\}. \tag{6.18}$$

The following derivation deals with the general case when the optimizer $z^*(x)$ is not necessarily unique and G_A is not invertible. We know from (6.16a) that for $x \in CR_A$

$$G_A z^* - S_A x = w_A.$$

For another $\bar{x} \in CR_A$ we require

$$G_A \bar{z} - S_A \bar{x} = w_A,$$

or for the perturbation $\tilde{x} = x - \bar{x}$

$$G_A \tilde{z} - S_A \tilde{x} = 0.$$

This equation will have a solution $\tilde{z}$ for arbitrary values of $\tilde{x}$, i.e., a full dimensional CR_A if

$$\mathrm{rank}[G_A] = \mathrm{rank}[G_A \; S_A].$$

Otherwise the dimension over which $\tilde{x}$ can vary will be restricted to

$$n - (\mathrm{rank}[G_A \; S_A] - \mathrm{rank}[G_A]).$$

Next we will show how to construct these restrictions on the variation of x.

Let $l = \mathrm{rank}\, G_A$ and consider the QR decomposition of G_A

$$G_A = Q \begin{bmatrix} U_1 & U_2 \\ \mathbf{0}_{|A|-l \times l} & \mathbf{0}_{|A|-l \times |A|-l} \end{bmatrix},$$

where $Q \in \mathbb{R}^{|A| \times |A|}$ is a unitary matrix, $U_1 \in \mathbb{R}^{l \times l}$ is a full-rank square matrix and $U_2 \in \mathbb{R}^{l \times |A|-l}$. Let $\begin{bmatrix} P \\ D \end{bmatrix} = -Q^{-1}S_A$ and $\begin{bmatrix} q \\ r \end{bmatrix} = Q^{-1}w_A$. From (6.16a) we obtain

$$\begin{bmatrix} U_1 & U_2 & P \\ \mathbf{0}_{|A|-l \times l} & \mathbf{0}_{|A|-l \times |A|-l} & D \end{bmatrix} \begin{bmatrix} z_1^* \\ z_2^* \\ x \end{bmatrix} = \begin{bmatrix} q \\ r \end{bmatrix}. \tag{6.19}$$

We partition (6.16b) accordingly

$$\begin{bmatrix} E & F \end{bmatrix} \begin{bmatrix} z_1^* \\ z_2^* \end{bmatrix} - S_{NA}x < w_{NA}. \tag{6.20}$$

Calculating z_1^* from (6.19) we obtain

$$z_1^* = U_1^{-1}(-U_2 z_2^* - Px + q), \tag{6.21}$$

which substituted in (6.20) gives:

$$(F - EU_1^{-1}U_2)z_2^* + (S_{NA} - EU_1^{-1}P)x < w_{NA} - EU_1^{-1}q. \tag{6.22}$$

The critical region CR_A can be equivalently written as:

$$CR_A = \left\{ x \in \mathcal{P}_x \ : \ \exists z_2 \text{ such that } \begin{bmatrix} z_2 \\ x \end{bmatrix} \in \mathcal{P}_{z_2 x} \right\}, \qquad (6.23)$$

where:

$$\mathcal{P}_x = \{x \ : \ Dx = r\}, \qquad (6.24)$$

$$\mathcal{P}_{z_2 x} = \left\{ \begin{bmatrix} z_2 \\ x \end{bmatrix} \ : \ (F - EU_1^{-1}U_2)z_2 + (S_{NA} - EU_1^{-1}P)x < w_{NA} - EU_1^{-1}q \right\}. \qquad (6.25)$$

In other words:

$$CR_A = \mathcal{P}_x \cap \operatorname{proj}_x(\mathcal{P}_{z_2 x}). \qquad (6.26)$$

In summary, if the optimizer $z^*(x)$ is not unique then it can be expressed as $z_1^*(x)$ (6.21), where the set of optimizers is characterized by $z_2^*(x)$, which is allowed to vary within the set $\mathcal{P}_{z_2 x}$ (6.25). We have also constructed the critical region CR_A (6.26), where $z_1^*(x)$ remains optimal. If G_A is not invertible and the matrix D is not empty then the critical region is not full dimensional.

We can now state some fundamental properties of the critical regions, value function and optimizer inside a critical region.

Theorem 6.2   *Let* $(A, NA) = (A(x^*), NA(x^*))$ *for some* $x^* \in \mathcal{K}^*$

i) *CR_A is an open polyhedron of dimension d where $d = n - \operatorname{rank} \begin{bmatrix} G_A & S_A \end{bmatrix} + \operatorname{rank} G_A$. If $d = 0$ then $CR_A = \{x^*\}$.*

ii) *If $\operatorname{rank} G_A = s$ (recall that $z \in \mathbb{R}^s$) then the optimizer $z^*(x)$ is unique and given by an affine function of the state inside CR_A, i.e., $z^*(x) = F_i x + g_i$ for all $x \in CR_A$.*

iii) *If the optimizer is not unique in CR_A then $Z^*(x)$ is an open polyhedron for all $x \in CR_A$.*

iv) *$J^*(x)$ is an affine function of the state, i.e., $J^*(x) = c_i' x + d_i$ for all $x \in CR_A$.*

Proof:

i) Polytope $\mathcal{P}_{z_2 x}$ (6.25) is open and nonempty, therefore it is full-dimensional in the (z_2, x) space and $\dim \operatorname{proj}_x(\mathcal{P}_{z_2 x}) = n$. Also,

$$\dim \mathcal{P}_x = n - \operatorname{rank} D = n - (\operatorname{rank} \begin{bmatrix} G_A & S_A \end{bmatrix} - \operatorname{rank} G_A).$$

Since the intersection of $\mathcal{P}_x$ and $\operatorname{proj}_x(\mathcal{P}_{z_2 x})$ is nonempty (it contains at least the point x^*) we can conclude that

$$\dim CR_A = n - \operatorname{rank} \begin{bmatrix} G_A & S_A \end{bmatrix} + \operatorname{rank} G_A.$$

Since we assumed that the set $\mathcal{K}$ in (6.2) is bounded, CR_A is bounded. This implies that CR_A is an open polytope since it is the intersection of an open polytope and the subspace $Dx = r$. In general, if we allow $\mathcal{K}$ in (6.2) to be unbounded, then the critical region CR_A can be unbounded.

ii) Consider (6.21) and recall that $l = \operatorname{rank} G_A$. If $l = s$, then the primal optimizer is unique, U_2 is an empty matrix and

$$z^*(x) = z_1^*(x) = U_1^{-1}(-Px + q). \tag{6.27}$$

iii) If the primal optimizer is not unique in CR_A then $Z^*(x)$ in CR_A is the following point-to-set map: $Z^*(x) = \{[z_1, \ z_2] : z_2 \in \mathcal{P}_{z_2 x}, \ U_1 z_1 + U_2 z_2 + Px = q)\}$. $Z^*(x)$ is an open polyhedron since $\mathcal{P}_{z_2 x}$ is open.

iv) Consider the dual problem (6.14) and one of its optimizer u_0^* for $x = x^*$. By definition of a critical region u_0^* remains optimal for all $x \in CR_A$. Therefore the value function in CR_A is

$$J^*(x) = (w + Sx)' u_0^*, \tag{6.28}$$

which is an affine function of x on CR_A.

∎

Remark 6.1 If the optimizer is unique then the computation of CR_A in (6.26) does not require the projection of the set $\mathcal{P}_{z_2 x}$ in (6.25). In fact, U_2 and F are empty matrices and

$$CR_A = \mathcal{P}_{z_2 x} = \left\{ x : Dx = r, \ (S_{NA} - EU^{-1}P)x < w_{NA} - EU^{-1}q \right\}. \tag{6.29}$$

If D is also empty, then the critical region is full dimensional.

6.2.3 Propagation of the Set of Active Constraints

The objective of this section is to briefly describe the propagation of the set of active constraints when moving from one full-dimensional critical region to a neighboring full-dimensional critical region. We will use a simple example in order to illustrate the main points.

Example 6.3 Consider the mp-LP problem

$$\min_{z_1, z_2, z_3, z_4} \quad z_1 + z_2$$

$$\text{subj. to} \quad \begin{cases} -z_1 \leq x_1 + x_2 \\ -z_1 - z_3 \leq x_2 \\ -z_1 \leq -x_1 - x_2 \\ -z_1 + z_3 \leq -x_2 \\ -z_2 - z_3 \leq x_1 + 2x_2 \\ -z_2 - z_3 - z_4 \leq x_2 \\ -z_2 + z_3 \leq -x_1 - 2x_2 \\ -z_2 + z_3 + z_4 \leq -x_2 \\ z_3 \leq 1 \\ -z_3 \leq 1 \\ z_4 \leq 1 \\ -z_4 \leq 1, \end{cases} \tag{6.30}$$

Table 6.1 Example 6.3. Critical regions and corresponding value function.

Critical Region	Value function
$CR1 = CR_{\{1,4,5,6,7,8\}}$	$-x_1 - x_2$
$CR2 = CR_{\{2,3,5,6,7,8\}}$	$x_1 + x_2$
$CR3 = CR_{\{1,5,6,9\}}$	$-2x_1 - 3x_2 - 1$
$CR4 = CR_{\{4,6,7,10\}}$	$x_1 + 3x_2 - 2$
$CR5 = CR_{\{1,5,8,12\}}$	$-1.5x_1 - 1.5x_2 + 0.5$
$CR6 = CR_{\{2,5,6,9\}}$	$-x_1 - 3x_2 - 2$
$CR7 = CR_{\{3,6,7,10\}}$	$2x_1 + 3x_2 - 1$
$CR8 = CR_{\{3,6,7,11\}}$	$1.5x_1 + 1.5x_2 - 0.5$
$CR9 = CR_{\{1,5,9,12\}}$	$-2x_1 - 3x_2 - 1$
$CR10 = CR_{\{4,7,10,12\}}$	$x_1 + 3x_2 - 2$
$CR11 = CR_{\{2,5,9,12\}}$	$-x_1 - 3x_2 - 2$
$CR12 = CR_{\{3,7,10,12\}}$	$2x_1 + 3x_2 - 1$

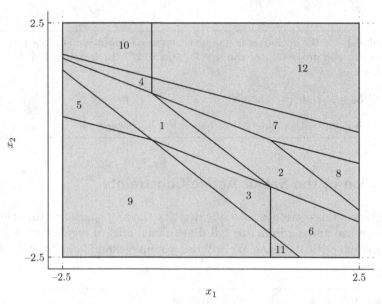

Figure 6.2 Example 6.3. Polyhedral partition of the parameter space.

where $\mathcal{K}$ is given by

$$-2.5 \leq x_1 \leq 2.5$$
$$-2.5 \leq x_2 \leq 2.5. \tag{6.31}$$

A solution to the mp-LP problem is shown in Figure 6.2 and the constraints, which are active in each associated critical region are reported in Table 6.1. Clearly, as $z \in \mathbb{R}^4$, $CR1 = CR_{\{1,4,5,6,7,8\}}$ and $CR2 = CR_{\{2,3,5,6,7,8\}}$ are primal degenerate full-dimensional critical regions.

By observing Figure 6.2 and Table 6.1 we notice the following. Under no primal and dual degeneracy, *(i)* full critical regions are described by a set of active constraints of dimension n; *(ii)* two neighboring full-dimensional critical regions CR_{A_i} and CR_{A_j} have A_i and A_j differing only in one constraint; *(iii)* CR_{A_i} and CR_{A_j} will share a facet which is a primal degenerate critical region CR_{A_p} of dimension $n-1$ with $A_p = A_i \cup A_j$. In Example 6.3, CR7 and CR12 are two full-dimensional and neighboring critical regions and the corresponding active set differs only in one constraint (constraint 6 in CR7 and constraint 12 in CR12). They share a facet which is a one-dimensional critical region (an open line), CR13 $= CR_{\{3,6,7,10,12\}}$. (The solution depicted in Figure 6.2 and detailed in Table 6.1 contains only full-dimensional critical regions.)

If primal and/or dual degeneracy occur, then the situation becomes more complex. In particular, in the case of primal degeneracy, it might happen that full-dimensional critical regions are described by more than s active constraints (CR1 or CR2 in Example 6.3). In case of dual degeneracy, it might happen that full-dimensional critical regions are described by fewer than s active constraints. Details on this case are provided in the next section.

6.2.4 Nonunique Optimizer

If $\text{rank}[G_A] < s$ then $Z^*(x)$ is not a singleton and the projection of the set $\mathcal{P}_{z_2 x}$ in (6.25) is required in order to compute the critical region $CR_{A(x^*)}$, which is "expensive" (see Section 6.1.1 for a discussion of polyhedra projection). It is preferable to move on the optimal facet to a vertex and to construct the critical region starting with this optimizer. This is explained next.

If one needs to determine one possible optimizer $z^*(\cdot)$ in the dual degenerate region $CR_{A(x^*)}$ the following simple method can be used. Choose a particular optimizer which lies on a vertex of the feasible set, i.e., determine set $\widehat{A}(x^*) \supset A(x^*)$ of active constraints for which $\text{rank}(G_{\widehat{A}(x^*)}) = s$, and compute a subset $\widehat{CR}_{\widehat{A}(x^*)}$ of the dual degenerate critical region (namely, the subset of parameters x such that the constraints $\widehat{A}(x^*)$ are active at the optimizer, which is not a critical region in the sense of Definition 6.2). Within $\widehat{CR}_{\widehat{A}(x^*)}$, the piecewise linear expression of an optimizers $z^*(x)$ is available from (6.27) and $\widehat{CR}_{\widehat{A}(x^*)}$ is defined by (6.29).

The algorithm proceeds by exploring the space surrounding $\widehat{CR}_{\widehat{A}(x^*)}$ until $CR_{A(x^*)}$ is covered. The arbitrariness in choosing an optimizer leads to different ways of partitioning $CR_{A(x^*)}$, where the partitions, in general, may overlap. Nevertheless, in each region a unique optimizer is defined. The storing of overlapping regions can be avoided by intersecting each new region (inside the dual degenerate region) with the current partition computed so far. This procedure is illustrated in the following example.

Example 6.4 Consider the following mp-LP reported in [115, p. 152]

$$
\begin{aligned}
\min \quad & -2z_1 - z_2 \\
\text{subj. to} \quad & \left\{
\begin{array}{l}
z_1 + 3z_2 \leq 9 - 2x_1 + x_2 \\
2z_1 + z_2 \leq 8 + x_1 - 2x_2 \\
z_1 \leq 4 + x_1 + x_2 \\
-z_1 \leq 0 \\
-z_2 \leq 0,
\end{array}
\right.
\end{aligned}
\tag{6.32}
$$

Table 6.2 Example 6.4. Critical regions and corresponding optimal value.

Region	Optimizer	Value Function
$CR_{\{2\}}$	not single valued	$-x_1 + 2x_2 - 8$
$CR_{\{1,5\}}$	$z_1^* = -2x_1 + x_2 + 9,\ z_2^* = 0$	$4x_1 - 2x_2 - 18$
$CR_{\{1,3\}}$	$z_1^* = x_1 + x_2 + 4,\ z_2^* = -x_1 + 5/3$	$-x_1 - 2x_2 - 29/3$

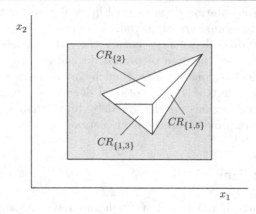

Figure 6.3 Example 6.4. Polyhedral partition of the parameter space corresponding to the solution.

where $\mathcal{K}$ is given by:
$$-10 \leq x_1 \leq 10$$
$$-10 \leq x_2 \leq 10. \tag{6.33}$$

The solution is represented in Figure 6.3 and the critical regions are listed in Table 6.2.

The critical region $CR_{\{2\}}$ is related to a dual degenerate solution with non-unique optimizers. The analytical expression of $CR_{\{2\}}$ is obtained by projecting the $\mathcal{H}$-polyhedron

$$
\begin{aligned}
z_1 + 3z_2 + 2x_1 - x_2 &< 9 \\
2z_1 + z_2 - x_1 + 2x_2 &= 8 \\
z_1 - x_1 - x_2\quad\ &< 4 \\
-z_1\qquad\qquad\ &< 0 \\
-z_2\qquad\qquad\ &< 0
\end{aligned}
\tag{6.34}
$$

on the parameter space to obtain:

$$
\overline{CR}_{\{2\}} = \left\{ [x_1, x_2] : \begin{array}{r} 2.5x_1 - 2x_2 \leq 5 \\ -0.5x_1 + x_2 \leq 4 \\ -12x_2 \leq 5 \\ -x_1 - x_2 \leq 4 \end{array} \right\}, \tag{6.35}
$$

which is effectively the result of (6.26). For all $x \in CR_{\{2\}}$, only one constraint is active at the optimum, which makes the optimizer not unique.

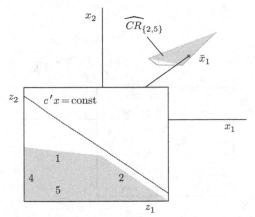

(a) First region $\widehat{CR}_{\{2,5\}} \subset CR_{\{2\}}$, and below the feasible set in the z-space corresponding to $\bar{x}_1 \in \widehat{CR}_{\{2,5\}}$.

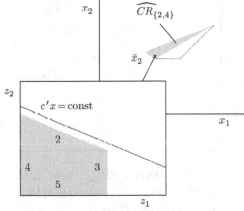

(b) Second region $\widehat{CR}_{\{2,4\}} \subset CR_{\{2\}}$, and below the feasible set in the z-space corresponding to $\bar{x}_2 \in \widehat{CR}_{\{2,4\}}$.

Figure 6.4 Example 6.4. A possible sub-partitioning of the degenerate region CR_2 where the regions $\widehat{CR}_{\{2,5\}}$ (Figure 6.4(a)), $\widehat{CR}_{\{2,4\}}$ (Figure 6.4(b)) and $\widehat{CR}_{\{2,1\}}$ (Figure 6.5(a)) are overlapping. Note that below each picture the feasible set and the level set of the value function in the z-space are depicted for a particular choice of the parameter x indicated by a point marked with ×.

Figures 6.4, 6.5 and 6.6 show two possible ways of covering $CR_{\{2\}}$ without using projection. The generation of overlapping regions is avoided by intersecting each new region with the current partition computed so far, as shown in Figure 6.7 where $\widetilde{CR}_{\{2,4\}}$ and $\widetilde{CR}_{\{2,1\}}$ represent the intersected critical regions. In Figures 6.4 and 6.5 the regions are overlapping, and in Figure 6.7 artificial cuts are introduced at the boundaries inside the degenerate critical region $CR_{\{2\}}$. No artificial cuts are introduced in Figure 6.6 because the $\widetilde{CR}_{\{2,3\}}$ and $\widetilde{CR}_{\{2,5\}}$ happen to be nonoverlapping.

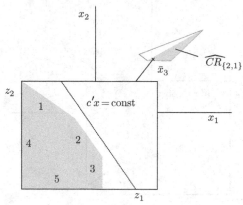

(a) Third region $\widehat{CR}_{\{2,1\}} \subset CR_{\{2\}}$, and below the feasible set in the z-space corresponding to $\bar{x}_3 \in \widehat{CR}_{\{2,1\}}$.

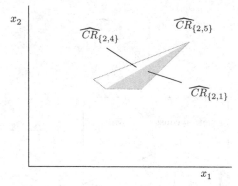

(b) Final partition of $CR_{\{2\}}$. Note that the region $\widehat{CR}_{\{2,5\}}$ is hidden by region $\widehat{CR}_{\{2,4\}}$ and region $\widehat{CR}_{\{2,1\}}$.

Figure 6.5 Example 6.4. A possible sub-partitioning of the degenerate region CR_2 where the regions $\widehat{CR}_{\{2,5\}}$ (Figure 6.4(a)), $\widehat{CR}_{\{2,4\}}$ (Figure 6.4(b)) and $\widehat{CR}_{\{2,1\}}$ (Figure 6.5(a)) are overlapping. Note that below each picture the feasible set and the level set of the value function in the z-space are depicted for a particular choice of the parameter x indicated by a point marked with $\times$.

6.2.5 Value Function and Optimizer: Global Properties

In this section we discuss global properties of the value function $J^*(x)$, optimizer $z^*(x)$, and of the set $\mathcal{K}^*$.

Theorem 6.3 *Assume that for a fixed $x^0 \in \mathcal{K}$ there exists a finite optimal solution $z^*(x_0)$ of (6.11). Then, for all $x \in \mathcal{K}$, (6.11) has either a finite optimum or no feasible solution.*

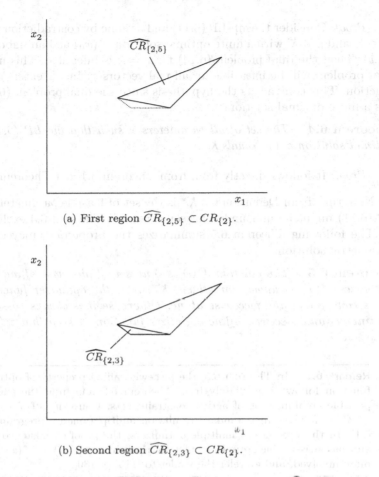

(a) First region $\widehat{CR}_{\{2,5\}} \subset CR_{\{2\}}$.

(b) Second region $\widehat{CR}_{\{2,3\}} \subset CR_{\{2\}}$.

Figure 6.6 Example 6.4. A possible solution where the regions $\widehat{CR}_{\{2,5\}}$ and $\widehat{CR}_{\{2,3\}}$ are nonoverlapping.

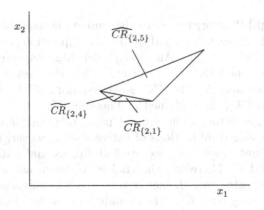

Figure 6.7 Example 6.4. A possible solution where $\widetilde{CR}_{\{2,4\}}$ is obtained by intersecting $\widehat{CR}_{\{2,4\}}$ with the complement of $\widehat{CR}_{\{2,5\}}$, and $\widehat{CR}_{\{2,1\}}$ by intersecting $\widehat{CR}_{\{2,1\}}$ with the complement of $\widehat{CR}_{\{2,5\}}$ and $\widehat{CR}_{\{2,4\}}$.

Proof: Consider the mp-LP (6.11) and assume by contradiction that there exist $x_0 \in \mathcal{K}$ and $\bar{x} \in \mathcal{K}$ with a finite optimal solution $z^*(x_0)$ and an unbounded solution $z^*(\bar{x})$. Then the dual problem (6.14) for $x = \bar{x}$ is infeasible. This implies that the dual problem will be infeasible for all real vectors x since x enters only in the cost function. This contradicts the hypothesis since the dual problem (6.14) for $x = x_0$ has a finite optimal solution. ∎

Theorem 6.4   *The set of all parameters x such that the LP (6.11) has a finite optimal solution $z^*(x)$ equals $\mathcal{K}^*$.*

Proof: It follows directly from from Theorem 6.1 and Theorem 6.3. ∎

Note that from Definition 6.3 $\mathcal{K}^*$ is the set of feasible parameters. However the LP (6.11) might be unbounded for some $x \in \mathcal{K}^*$. Theorem 6.4 excludes this case.

The following Theorem 6.5 summarizes the properties enjoyed by the multi-parametric solution.

Theorem 6.5   *The function $J^*(\cdot)$ is convex and piecewise affine over $\mathcal{K}^*$. If the optimizer $z^*(x)$ is unique for all $x \in \mathcal{K}^*$, then the optimizer function $z^* : \mathcal{K}^* \to \mathbb{R}^s$ is continuous and piecewise affine. Otherwise it is always possible to define a continuous and piecewise affine optimizer function z^* such that $z^*(x) \in Z^*(x)$ for all $x \in \mathcal{K}^*$.*

Remark 6.2   In Theorem 6.5, the piecewise affine property of optimizer and value function follows immediately from Theorem 6.2 and from the enumeration of all possible combinations of active constraint sets. Convexity of $J^*(\cdot)$ and continuity of $Z^*(x)$ follows from standard results on multiparametric programs (see Corollary 5.1). In the presence of multiple optimizers, the proof of existence of a continuous and piecewise affine optimizer function z^* such that $z^*(x) \in Z^*(x)$ for all $z \in \mathcal{K}^*$ is more involved, and we refer the reader to [115, p. 180].

6.2.6   mp-LP Algorithm

The goal of an mp-LP algorithm is to determine the partition of $\mathcal{K}^*$ into full dimensional critical regions CR_{A_i}, and to find the expression of the functions $J^*(\cdot)$ and $z^*(\cdot)$ for each critical region. An mp-LP algorithm has two components: the "active set generator" and the "KKT solver." The active set generator computes the set of active constraints A_i. The KKT solver computes CR_{A_i} and the expression of $J^*(\cdot)$ and $z^*(\cdot)$ in CR_{A_i} as explained in Theorem 6.2.

The active set generator is the critical part. In principle, one could simply generate all the possible combinations of active sets. However, in many problems only a few active constraints sets generate full-dimensional critical regions inside the region of interest $\mathcal{K}$. Therefore, the goal is to design an active set generator algorithm which computes only the active sets A_i with associated full-dimensional critical regions covering only $\mathcal{K}^*$. This avoids the combinatorial explosion of a complete enumeration. Also, we will use the technique described in Section 6.2.4 in order to avoid expressions for nonunique optimizers (6.21) with U_2 nonempty. Next we will describe one possible implementation of an mp-LP algorithm.

In order to start solving the mp-LP problem, we need an initial vector x_0 inside the polyhedral set $\mathcal{K}^*$ of feasible parameters. A possible choice for x_0 is the Chebyshev center (see Section 4.4.5) of $\mathcal{K}^*$, i.e., x_0 solving the following LP:

$$\begin{aligned}
\max_{x,z,\epsilon} \quad & \epsilon \\
\text{subj. to} \quad & T_i x + \epsilon \|T_i\|_2 \le N_i, \quad i = 1, \ldots, n_T \\
& Gz - Sx \le w,
\end{aligned} \qquad (6.36)$$

where n_T is the number of rows T_i of the matrix T defining the set $\mathcal{K}$ in (6.2). If $\epsilon \le 0$, then the LP problem (6.11) is infeasible for all x in the interior of $\mathcal{K}$. Otherwise, we solve the primal and dual problems (6.11), (6.14) for $x = x_0$. Let z_0^* and u_0^* be the optimizers of the primal and the dual problem, respectively. The value z_0^* defines the following optimal partition

$$\begin{aligned}
\mathrm{A}(x_0) &= \{j \in I : G_j z_0^* - S_j x_0 - w_j = 0\} \\
\mathrm{NA}(x_0) &= \{j \in I : G_j z_0^* - S_j x_0 - w_j < 0\}
\end{aligned} \qquad (6.37)$$

and consequently the critical region $CR_{A(x_0)}$. Once the critical region $CR_{A(x_0)}$ has been defined, the rest of the space $R^{\text{rest}} = \mathcal{K} \backslash CR_{A(x_0)}$ has to be explored and new critical regions generated. An approach for generating a polyhedral partition $\{R_1, \ldots, R_{n_{rest}}\}$ of the rest of the space R^{rest} is described in Theorem 4.2 in Section 4.4.7. The procedure proposed in Theorem 4.2 for partitioning the set of parameters allows one to recursively explore the parameter space. Such an iterative procedure terminates after a finite time, as the number of possible combinations of active constraints decreases with each iteration. The following two issues need to be considered:

1. The partitioning in Theorem 4.2 defines new polyhedral regions R_k to be explored that are not related to the critical regions which still need to be determined. This may split some of the critical regions, due to the artificial cuts induced by Theorem 4.2. Postprocessing can be used to join cut critical regions [44]. As an example, in Figure 6.8 the critical region $CR_{\{3,7\}}$ is discovered twice, one part during the exploration of R_1 and the second part during the exploration of R_2.

   Although algorithms exist for convexity recognition and computation of the union of polyhedra, the postprocessing operation is computationally expensive. Therefore, it is more efficient not to intersect the critical region obtained by (6.29) with halfspaces generated by Theorem 4.2, which is only used to drive the exploration of the parameter space. Then, no postprocessing is needed to join subpartitioned critical regions.

   On the other hand, some critical regions may appear more than once. Duplicates can be uniquely identified by the set of active constraints $A(x)$ and can be easily eliminated. To this aim, in the implementation of the algorithm we keep a list of all the critical regions which have already been generated in order to avoid duplicates. In Figure 6.8 the critical region $CR_{\{3,7\}}$ is discovered twice but stored only once.

2. If a region is generated which is not full-dimensional we want to avoid further recursion of the algorithm not producing any full-dimensional critical region,

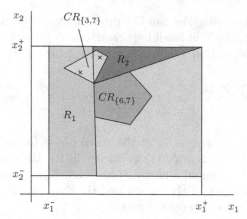

Figure 6.8 Example of a critical region explored twice.

and therefore lengthening the number of steps required to determine the solution to the mp-LP. We perturb the parameter x_0 by a random vector with length smaller than the Chebyshev radius of the polyhedral region R_k where we are looking for a new critical region. This will ensure that the perturbed vector is still contained in R_k.

Based on the above discussion, the mp-LP solver can be summarized in the following recursive Algorithm 6.1. Note that the algorithm generates a partition of the state space which is not strict. The algorithm could be modified to store the critical regions as defined in (6.4) which are open sets, instead of storing their closure. In this case the algorithm would have to explore and store all the critical regions that are not full-dimensional in order to generate a strict polyhedral partition of the set of feasible parameters. From a practical point of view such a procedure is not necessary since the value function and the optimizer are continuous functions of x.

Algorithm 6.1   *mp-LP Algorithm (nondegenerate case)*

Input: Matrices c, G, w, S of the mp-LP (6.11) and set $\mathcal{K}$ in (6.2)
Output: Multiparametric solution to the mp-LP (6.11)

    Execute partition($\mathcal{K}$)
    End

    Function partition(Y)
        Let $x_0 \in Y$ and ϵ be the solution to the LP (6.36)
        If $\epsilon \leq 0$ **Then** exit (no full-dimensional CR is in Y)
        Solve the LP (6.11), (6.14) for $x = x_0$
        Let $A(x_0)$ be the set of active constraints as in (6.37)
        Determine $z^*(x)$ from (6.17) and $CR_{A(x_0)}$ from (6.18)
        Partition the rest of the region as in Theorem 4.2
        For each new sub-region R_i, **Do** partition(R_i)
    End function

Remark 6.3 In the degenerate case $z^*(x)$ and $CR_{A(x_0)}$ are given by (6.21) and (6.26), respectively. As remarked in Section 6.2.2, if rank(D) > 0 the region $CR_{A(x_0)}$ is not full-dimensional and therefore not of interest. To use Algorithm 6.1 after computing U, P, D if $D \neq 0$ one should compute a random vector $\epsilon \in \mathbb{R}^n$ smaller than the Chebyshev radius of Y and such that the LP (6.11) is feasible for $x_0 + \epsilon$ and then repeat step where $A(x_0)$ is computed with $x_0 \leftarrow x_0 + \epsilon$.

Remark 6.4 The algorithm determines the partition of $\mathcal{K}$ recursively. After the first critical region is found, the rest of the region in $\mathcal{K}$ is partitioned into polyhedral sets $\{R_i\}$ as in Theorem 4.2. By using the same method, each set R_i is further partitioned, and so on.

6.3 Multiparametric Quadratic Programming

6.3.1 Formulation

In this section we investigate multiparametric quadratic programs (mp-QP), a special case of the multiparametric program (6.1) where the objective is a quadratic function

$$J^*(x) = \min_z \quad J(z, x) = \tfrac{1}{2} z' H z$$
$$\text{subj. to} \quad Gz \leq w + Sx. \tag{6.38}$$

All the variables were defined in Section 6.1.1. We assume $H \succ 0$. Our goal is to find the value function $J^*(x)$ and the optimizer function $z^*(x)$ in $\mathcal{K}^*$. Note that $\mathcal{K}^*$ can be determined by projection as discussed in Theorem 6.1. As suggested through Example 5.1 our search for these functions proceeds by partitioning the set of feasible parameters into critical regions. Note that the more general problem with $J(z, x) = \tfrac{1}{2} z' H z + x' F z$ can always be transformed into an mp-QP of form (6.38) by using the variable substitution $\tilde{z} = z + H^{-1} F' x$.

As in the previous sections, we denote with the subscript j the j-th row of a matrix or j-th element of a vector. Also, $J = \{1, \ldots, m\}$ is the set of constraint indices and for any $A \subseteq J$, G_A, w_A and S_A are the submatrices of G, w and S, respectively, consisting of the rows indexed by A. Without loss of generality we will assume that $\mathcal{K}^*$ is full-dimensional (if it is not, then the procedure described in Section 6.1.3 can be used to obtain a full-dimensional $\mathcal{K}^*$ in a reduced parameter space).

Example 6.5 Consider the parametric quadratic program

$$J^*(x) = \min_z \quad J(z, x) = \tfrac{1}{2} z^2$$
$$\text{subj. to} \quad z \leq 1 + 3x,$$

where $z \in \mathbb{R}$ and $x \in \mathbb{R}$. Our goals are:

1. to find $z^*(x) = \arg\min_{z,\, z \leq 1+3x} J(z, x)$,

2. to find all x for which the problem has a solution, and

3. to compute the value function $J^*(x)$.

The Lagrangian function is

$$L(z, x, u) = \frac{1}{2}z^2 + u(z - 3x - 1)$$

and the KKT conditions are (see Section 2.3.3 for KKT conditions for quadratic programs)

$$z + u = 0 \tag{6.39a}$$
$$u(z - 3x - 1) = 0 \tag{6.39b}$$
$$u \geq 0 \tag{6.39c}$$
$$z - 3x - 1 \leq 0. \tag{6.39d}$$

Consider (6.39) and the two strictly complementary cases:

$$
\text{A.} \quad
\begin{aligned}
z + u &= 0 \\
z - 3x - 1 &= 0 \\
u &\geq 0
\end{aligned}
\Rightarrow
\left\{
\begin{aligned}
z^* &= 3x + 1 \\
u^* &= -3x - 1 \\
J^* &= \frac{9}{2}x^2 + 3x + \frac{1}{2} \\
x &\leq -\frac{1}{3}
\end{aligned}
\right.
$$

$$
\text{B.} \quad
\begin{aligned}
z + u &= 0 \\
z - 3x - 1 &< 0 \\
u &= 0
\end{aligned}
\Rightarrow
\left\{
\begin{aligned}
z^* &= 0 \\
u^* &= 0 \\
J^* &= 0 \\
x &> -\frac{1}{3}.
\end{aligned}
\right.
\tag{6.40}
$$

This solution is depicted in Figure 6.9. The above simple procedure, which required nothing but the solution of the KKT conditions, yielded the optimizer $z^*(x)$ and the value function $J^*(x)$ for all values of the parameter x. The set of admissible parameters values was divided into two *critical regions*, defined by $x \leq -\frac{1}{3}$ and $x > -\frac{1}{3}$. In the region $x \leq -\frac{1}{3}$ the inequality constraint is active and the Lagrange multiplier is greater or equal than zero, in the other region $x > -\frac{1}{3}$ the inequality constraint is not active and the Lagrange multiplier is equal to zero. Note that for

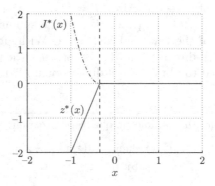

Figure 6.9 Example 6.5. Optimizer $z^*(x)$ and value function $J^*(x)$ as a function of the parameter x.

$x = -\frac{1}{3}$ the inequality constraint $z - 3x - 1 \le 0$ is active at z^* and the Lagrange multiplier is equal to zero.

Throughout a critical region the conditions for optimality derived from the KKT conditions do not change. In each critical region the optimizer $z^*(x)$ is affine and the value function $J^*(x)$ is quadratic. Both $z^*(x)$ and $J^*(x)$ are continuous.

6.3.2 Critical Regions, Value Function and Optimizer: Local Properties

Consider the definition of a critical region given in Section 6.1.2. We show next that critical regions of mp-QP are polyhedra. We use the KKT conditions (Section 1.6) to derive the $\mathcal{H}$-polyhedral representation of the critical regions and to compute the optimizer function $z^*(x)$ and the value function $J^*(x)$ inside each critical region.

The following theorem introduces fundamental properties of critical regions, the value function and the optimizer inside a critical region.

Theorem 6.6   *Let* $(A, NA) = (A(\bar{x}), NA(\bar{x}))$ *for some* $\bar{x} \in \mathcal{K}^*$, *Then*

  i)   the closure of CR_A *is a polyhedron.*

  ii) $z^*(x)$ is an affine function of the state inside CR_A, i.e., $z^*(x) = F_i x + g_i$ for all $x \in CR_A$.

  iii) $J^*(x)$ is a quadratic function of the state inside CR_A, i.e., $J^*(x) = x' M_i x + c_i' x + d_i$ for all $x \in CR_A$.

Proof: The first-order Karush Kuhn-Tucker (KKT) optimality conditions (see Section 2.3.3) for the mp-QP are given by

$$Hz^* + G'u^* = 0, \ u \in \mathbb{R}^m \tag{6.41a}$$

$$u_i^*(G_i z^* - w_i - S_i x) = 0, \ i = 1, \dots, m \tag{6.41b}$$

$$u^* \ge 0 \tag{6.41c}$$

$$Gz^* - w - Sx \le 0. \tag{6.41d}$$

Let us assume that we have determined the optimal partition

$$(A, NA) = (A(\bar{x}), NA(\bar{x}))$$

for some $\bar{x} \in \mathcal{K}^*$. The primal feasibility condition (6.41d) can be rewritten as

$$G_A z^* - S_A x = w_A \tag{6.42a}$$

$$G_{NA} z^* - S_{NA} x < w_{NA}. \tag{6.42b}$$

We solve (6.41a) for z^*

$$z^* = -H^{-1}G'u^* \tag{6.43}$$

and substitute the result into (6.41b) to obtain the complementary slackness conditions

$$u_i^*(-G_i H^{-1}G'u^* - w_i - S_i x) = 0, \ i = 1, \dots, m. \tag{6.44}$$

Let u_{NA}^* and u_A^* denote the Lagrange multipliers corresponding to inactive and active constraints, respectively. For inactive constraints $u_{NA}^* = 0$. For active constraints:

$$(-G_A H^{-1} G_A')u_A^* - w_A - S_A x = 0. \tag{6.45}$$

If the set of active constraint A is empty, then $u^* = u_{NA}^* = 0$ and therefore $z^* = 0$ which implies that the critical region CR_A is

$$CR_A = \{x : \ Sx + w > 0\}. \tag{6.46}$$

Otherwise we distinguish two cases.

Case 1: LICQ holds, i.e., the rows of G_A are linearly independent. This implies that $(G_A H^{-1} G_A')$ is a square full rank matrix and therefore

$$u_A^* = -(G_A H^{-1} G_A')^{-1}(w_A + S_A x), \tag{6.47}$$

where G_A, w_A, S_A correspond to the set of active constraints A. Thus u^* is an affine function of x. We can substitute u_A^* from (6.47) into (6.43) to obtain

$$z^* = H^{-1} G_A'(G_A H^{-1} G_A')^{-1}(w_A + S_A x) \tag{6.48}$$

and note that z^* is also an affine function of x. $J^*(x) = \frac{1}{2}z^*(x)'Hz^*(x)$ and therefore it is a quadratic function of x. The critical region CR_A is computed by substituting z^* from (6.48) in the primal feasibility conditions (6.42b)

$$\mathcal{P}_p = \{x : G_{NA} H^{-1} G_A'(G_A H^{-1} G_A')^{-1}(w_A + S_A x) < w_{NA} + S_{NA}x\}, \tag{6.49}$$

and the Lagrange multipliers from (6.47) in the dual feasibility conditions (6.41c)

$$\mathcal{P}_d = \{x : -(G_A H^{-1} G_A')^{-1}(w_A + S_A x) \geq 0\}. \tag{6.50}$$

In conclusion, the critical region CR_A is the intersection of $\mathcal{P}_p$ and $\mathcal{P}_d$:

$$CR_A = \{x : \ x \in \mathcal{P}_p, \ x \in \mathcal{P}_d\}. \tag{6.51}$$

Obviously, the closure of CR_A is a polyhedron in the x-space.

The polyhedron $\mathcal{P}_p$ is open and non empty (it contains at least the point $\bar{x}$). Therefore it is full-dimensional in the x space. This implies that $\dim CR_A = \dim \mathcal{P}_d$.

Case 2: LICQ does not hold, the rows of G_A are not linearly independent. For instance, this happens when more than s constraints are active at the optimizer $z^*(\bar{x}) \in \mathbb{R}^s$, i.e., in a case of *primal degeneracy*. In this case the vector of Lagrange multipliers u^* might not be uniquely defined, as the dual problem of (6.38) is not strictly convex. Note that *dual degeneracy* and nonuniqueness of $z^*(\bar{x})$ cannot occur, as $H \succ 0$.

Using the same arguments as in Section 6.2.2, equation (6.45) will allow a full-dimensional critical region only if $\text{rank}[(G_A H^{-1} G_A') \ S_A] = \text{rank}[G_A H^{-1} G_A']$. Indeed the dimension of the critical region will be

$$n - (\text{rank}[(G_A H^{-1} G_A') \ S_A] - \text{rank}[G_A H^{-1} G_A']).$$

We will now show how to compute the critical region.

Let $l = \operatorname{rank} G_A$ and consider the QR decomposition of $-G_A H^{-1} G_A'$

$$-G_A H^{-1} G_A' = Q \begin{bmatrix} U_1 & U_2 \\ \mathbf{0}_{|A|-l\times l} & \mathbf{0}_{|A|-l\times|A|-l} \end{bmatrix},$$

where $Q \in \mathbb{R}^{|A|\times|A|}$ is a unitary matrix, $U_1 \in \mathbb{R}^{l\times l}$ is a full-rank square matrix and $U_2 \in \mathbb{R}^{l\times|A|-l}$. Let $\begin{bmatrix} P \\ D \end{bmatrix} = -Q^{-1} S_A$ and $\begin{bmatrix} q \\ r \end{bmatrix} = Q^{-1} w_A$. From (6.45) we obtain

$$\begin{bmatrix} U_1 & U_2 & P \\ \mathbf{0}_{|A|-l\times l} & \mathbf{0}_{|A|-l\times|A|-l} & D \end{bmatrix} \begin{bmatrix} u^*_{A,1} \\ u^*_{A,2} \\ x \end{bmatrix} = \begin{bmatrix} q \\ r \end{bmatrix}. \tag{6.52}$$

We compute $u^*_{A,1}$ from (6.52)

$$u^*_{A,1} = U_1^{-1}(-U_2 u^*_{A,2} - Px + q). \tag{6.53}$$

Finally, we can substitute $u^*_A = \begin{bmatrix} u^*_{A,1}{}' & u^*_{A,2}{}' \end{bmatrix}'$ from (6.53) into (6.43) to obtain

$$z^* = H^{-1} G_A{}' \begin{bmatrix} U_1^{-1}(U_2 u^*_{A,2} + Px - q) \\ u^*_{A,2} \end{bmatrix}. \tag{6.54}$$

The optimizer z^* is unique and therefore independent of the choice of $u^*_{A,2} \geq 0$. Setting $u^*_{A,2} = 0$ we obtain

$$z^* = H^{-1} G_{A,1}{}' U_1^{-1}(Px - q), \tag{6.55}$$

where we have partitioned G_A as $G_A = \begin{bmatrix} G_{A,1}' & G_{A,2}' \end{bmatrix}'$ and $G_{A,1} \in \mathbb{R}^{l\times s}$.

Note that z^* is an affine function of x. $J^*(x) = \frac{1}{2} z^*(x)' H z^*(x)$ and therefore is a quadratic function of x.

The critical region CR_A is computed by substituting z^* from (6.55) in the primal feasibility conditions (6.42b)

$$\mathcal{P}_p = \{x : G_{NA}(H^{-1} G_{A,1}{}' U_1^{-1}(Px - q)) < w_{NA} + S_{NA}x\}, \tag{6.56}$$

and the Lagrange multipliers from (6.53) in the dual feasibility conditions (6.41c)

$$\mathcal{P}_{u^*_{A,2},x} = \{[\, u^*_{A,2}, x] : U_1^{-1}(-U_2 u^*_{A,2} - Px + q) \geq 0, \ u^*_{A,2} \geq 0\}. \tag{6.57}$$

The critical region CR_A is the intersection of the sets $Dx = r$, $\mathcal{P}_p$ and $\operatorname{proj}_x(\mathcal{P}_{u^*_{A,2},x})$:

$$CR_A = \{x : \ Dx = r, \ x \in \mathcal{P}_p, \ x \in \operatorname{proj}_x(\mathcal{P}_{u^*_{A,2},x})\}. \tag{6.58}$$

The closure of CR_A is a polyhedron in the x-space. ∎

Remark 6.5 In general, the critical region polyhedron (6.49)–(6.51) is open on facets arising from primal feasibility and closed on facets arising from dual feasibility.

Remark 6.6 If D in (6.52) is nonzero, then from (6.58), CR_A is a lower dimensional region, which, in general, corresponds to a common boundary between two or more full-dimensional regions.

6.3.3   Propagation of the Set of Active Constraints

The objective of this section is to briefly describe the propagation of the set of active constraints when moving from one full-dimensional critical region to a neighboring full-dimensional critical region. We will use a simple example in order to illustrate the main points.

Example 6.6   Consider the mp-QP problem

$$J^*(x) = \min_{z} \quad \tfrac{1}{2} z' H z + x' F z$$
$$\text{subj. to} \quad Gz \le w + Sx, \tag{6.59}$$

with

$$H = \begin{bmatrix} 1 & 0 \\ 0 & 1 \end{bmatrix}, \quad F = \begin{bmatrix} 1 & 1 \\ 1 & -1 \end{bmatrix} \tag{6.60}$$

and

$$G = \begin{bmatrix} 1 & 0 \\ -1 & 0 \\ 0 & 1 \\ 0 & -1 \\ 1 & -1 \\ -1 & 1 \end{bmatrix}, \quad S = \begin{bmatrix} 0 & 0 \\ 0 & 0 \\ 0 & 0 \\ 0 & 0 \\ 1 & 1 \\ -2 & 1 \end{bmatrix}, \quad w = \begin{bmatrix} 1 \\ 1 \\ 1 \\ 1 \\ 0 \\ 0 \end{bmatrix}, \tag{6.61}$$

where $\mathcal{K}$ is given by

$$-1 \le x_1 \le 1$$
$$-1 \le x_2 \le 1. \tag{6.62}$$

A solution to the mp-QP problem is shown in Figure 6.10 and the constraints which are active in each associated full-dimensional critical region are reported in Table 6.3.

Since A_3 is empty in CR3 we can conclude from (6.46) that the facets of CR3 are facets of primal feasibility and therefore do not belong to CR3. In general, as discussed in Remark 6.5 critical regions are open on facets arising from primal

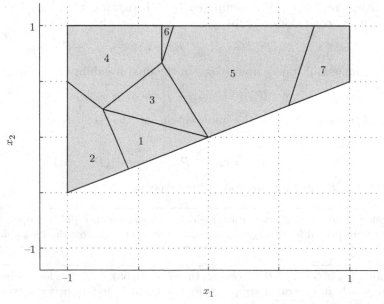

Figure 6.10 Example 6.6. Polyhedral partition of the parameter space.

Table 6.3 Example 6.6. Critical regions and corresponding set of active constraints.

Critical Region	Active Constraints
CR1	{5}
CR2	{3,5}
CR3	{}
CR4	{3}
CR5	{6}
CR6	{3,6}
CR7	{4,6}

feasibility and closed on facets arising from dual feasibility. Next we focus on the closure of the critical regions.

By observing Figure 6.10 and Table 6.3 we notice that as we move away from region CR3 (corresponding to no active constraints), the number of active constraints increases. In particular, for any two neighboring full-dimensional critical regions CR_{A_i} and CR_{A_j} we have $A_i \subset A_j$ and $|A_i| = |A_j| - 1$ or $A_j \subset A_i$ and $|A_i| = |A_j| + 1$. This means that as one moves from one full-dimensional region to a neighboring full-dimensional region, one constraint is either added to the list of active constraints or removed from it. This happens for instance when moving from CR3 to CR4, CR5, CR1 or from CR1 to CR2, or from CR5 to CR7.

The situation is more complex if LICQ does not hold everywhere in $\mathcal{K}^*$. In particular, there might exist two neighboring full-dimensional critical regions CR_{A_i} and CR_{A_j} where A_i and A_j do not share any constraint and with $|A_i| = |A_j|$. Also, CR_{A_i} might have multiple neighboring region on the same facet. In other words, it can happen that the intersection of the closures of two adjacent full-dimensional critical regions is a not a facet of both regions but only a subset of it. We refer the reader to [268] for more details on such a degenerate condition.

6.3.4 Value Function and Optimizer: Global Properties

The convexity of the value function $J^*(x)$ and the continuity of the solution $z^*(x)$ follow from the general results on multiparametric programming (Corollary 5.2). In the following we present an alternate simple proof.

Theorem 6.7 *Consider the multiparametric quadratic program (6.38) and let* $H \succ 0$. *Then the optimizer* $z^*(x) : \mathcal{K}^* \to \mathbb{R}^s$ *is continuous and piecewise affine on polyhedra, in particular it is affine in each critical region, and the optimal solution* $J^*(x) : \mathcal{K}^* \to \mathbb{R}$ *is continuous, convex and piecewise quadratic on polyhedra.*

Proof: We first prove convexity of $J^*(x)$. Take generic x_1, $x_2 \in \mathcal{K}^*$, and let $J^*(x_1)$, $J^*(x_2)$ and z_1, z_2 the corresponding optimal values and minimizers. Let $x_\alpha = \alpha x_1 + (1 - \alpha)x_2$ and $z_\alpha = \alpha z_1 + (1 - \alpha)z_2$. By optimality of $J^*(x_\alpha)$,

$$J^*(x_\alpha) \leq \frac{1}{2} z'_\alpha H z_\alpha.$$

Hence

$$J^*(x_\alpha) - \frac{1}{2}[\alpha z_1' H z_1 + (1-\alpha) z_2' H z_2] \leq \frac{1}{2} z_\alpha' H z_\alpha - \frac{1}{2}[\alpha z_1' H z_1 + (1-\alpha) z_2' H z_2].$$

We can upperbound the latter term as follows:

$$\frac{1}{2} z_\alpha' H z_\alpha - \frac{1}{2}[\alpha z_1' H z_1 + (1-\alpha) z_2' H z_2] = \frac{1}{2}[\alpha^2 z_1' H z_1 + (1-\alpha)^2 z_2' H z_2 +$$

$$+ 2\alpha(1-\alpha) z_2' H z_1 - \alpha z_1' H z_1 - (1-\alpha) z_2' H z_2] =$$

$$= -\frac{1}{2}\alpha(1-\alpha)(z_1 - z_2)' H(z_1 - z_2) \leq 0.$$

In conclusion

$$J^*(\alpha x_1 + (1-\alpha)x_2) \leq \alpha J^*(x_1) + (1-\alpha) J^*(x_2), \ \forall x_1, x_2 \in \mathcal{K}, \ \forall \alpha \in [0,1],$$

which proves the convexity of $J^*(x)$ on $\mathcal{K}^*$.

Within the closed polyhedral regions CR_i in $\mathcal{K}^*$ the solution $z^*(x)$ is affine (6.48) by Theorem 6.6. The boundary between two regions belongs to both closed regions. Since $H \succ 0$, the optimum is unique and hence the solution must be continuous across the boundary. Therefore $z^*(x) : \mathcal{K}^* \to \mathbb{R}^s$ is continuous and piecewise affine on polyhedra. The fact that $J^*(x)$ is continuous and piecewise quadratic follows trivially. ∎

We can easily derive a result similar to Theorem 6.7 for the dual function $d^*(x)$ (see (6.38) and (2.28))

$$d^*(x) = \max_{u \geq 0} \left[d(u) = -\frac{1}{2} u'(GH^{-1}G')u - u'(w + Sx) \right]. \tag{6.63}$$

Theorem 6.8   *Consider the multiparametric quadratic program (6.38) with $H \succ 0$ and assume that LICQ holds $\forall \ x \in \mathcal{K}^*$. Then the optimizer $u^*(x) : \mathcal{K}^* \to \mathbb{R}^m$ of the dual (6.63) is continuous and piecewise affine on polyhedra, in particular it is affine in each critical region, and the optimal solution $d^*(x) : \mathcal{K}^* \to \mathbb{R}$ is continuous, convex and piecewise quadratic on polyhedra.*

Proof: The proof follows along the lines of the proof of Theorem 6.7 once we know that the optimizer $u^*(x)$ is unique because of the LICQ assumptions. ∎

Theorem 6.9   *Assume that the mp-QP problem (6.38) is not degenerate, then the value function $J^*(x)$ in (6.38) is continuously differentiable ($C^{(1)}$).*

Proof: The dual of (6.38) is

$$d^*(x) = \max_{u \geq 0} -\frac{1}{2} u'(GH^{-1}G')u - u'(w + Sx). \tag{6.64}$$

By strong duality we have

$$J^*(x) = d^*(x) = -\frac{1}{2} u^*(x)'(GH^{-1}G')u^*(x) - u^*(x)'(w + Sx). \tag{6.65}$$

For the moment let us restrict our attention to the interior of a critical region. From (6.47) we know that $u^*(x)$ is continuously differentiable. Therefore

$$\nabla_x J(x) = -\left[\frac{\partial u^*}{\partial x}(x)\right]' GH^{-1}G'u^*(x) - \left[\frac{\partial u^*}{\partial x}(x)\right]'(w+Sx) - S'u^*(x) =$$

$$= \left[\frac{\partial u^*}{\partial x}(x)\right]'(-GH^{-1}G'u^*(x) - w - Sx) - S'u^*(x). \tag{6.66}$$

Using (6.43)

$$z^* = -H^{-1}G'u^*, \tag{6.67}$$

we get

$$\nabla_x J(x) = -S'u^*(x) + \left[\frac{\partial u^*}{\partial x}(x)\right]'(Gz^* - w - Sx) =$$

$$= -S'u^*(x) + \sum_{i \in A}\left[\frac{\partial u_i^*}{\partial x}(x)\right]'\underbrace{(G_i z^* - w_i - S_i x)}_{0} +$$

$$+ \sum_{i \in NA}\left[\frac{\partial u_i^*}{\partial x}(x)\right]'\underbrace{(G_i z^* - w_i - S_i x)}_{0} = \tag{6.68}$$

$$= -S'u^*(x).$$

Continuous differentiability of J^* in the interior of a critical region follows from (6.47). Continuous differentiability of J^* at the boundary follows from the fact that $u^*(x)$ is continuous (Theorem 6.8) for all x in the interior of $\mathcal{K}^*$[20]. ∎

Note that in case of degeneracy the value function $J^*(x)$ in (6.38) may not be $C^{(1)}$. The following counterexample was given in [48].

Example 6.7

Consider the mp-QP (6.38) with

$$H = \begin{bmatrix} 3 & 3 & -1 \\ 3 & 11 & 23 \\ -1 & 23 & 75 \end{bmatrix}$$

$$G = \begin{bmatrix} 1 & 1 & 1 \\ 1 & 3 & 5 \\ -1 & -1 & -1 \\ -1 & -3 & -5 \\ -1 & 0 & 0 \\ 0 & -1 & 0 \\ 0 & 0 & -1 \end{bmatrix} \quad w = \begin{bmatrix} 1 \\ 0 \\ -1 \\ 0 \\ 0 \\ 0 \\ 0 \end{bmatrix} \quad S = \begin{bmatrix} 0 \\ 1 \\ 0 \\ -1 \\ 0 \\ 0 \\ 0 \end{bmatrix} \tag{6.69}$$

and $\mathcal{K} = \{x \in \mathbb{R} | 1 \le x \le 5\}$.

The problem was solved by using Algorithm 6.2 described next. The solution comprises five critical regions. The critical regions and the expression of the value function are reported in Table 6.4. The reader can verify that the value function is not continuously differentiable at $x = 3$. Indeed, at $x = 3$ the LICQ condition does not hold and therefore, the hypothesis of Theorem 6.9 is not fulfilled.

Table 6.4 Critical regions and value function corresponding to the solution of Example 6.7.

Region	Optimal value
$CR_{\{1,2,3,4,6\}} = \{x \ : \ 1 \le x < 1.5\}$	$2.5x^2 - 6x + 5$
$CR_{\{1,2,3,4\}} = \{x \ : \ 1.5 \le x \le 2\}$	$0.5x^2 + 0.5$
$CR_{\{1,2,3,4,7\}} = \{x \ : \ 2 < x < 3\}$	$x^2 - 2x + 2.5$
$CR_{\{1,2,3,4,5,7\}} = \{x \ : \ x = 3\}$	$x^2 - 2x + 2.5$
$CR_{\{1,2,3,4,5\}} = \{x \ : \ 3 < x \le 5\}$	$5x^2 - 24x + 32.5$

6.3.5 mp-QP Algorithm

The goal of an mp-QP algorithm is to determine the partition of $\mathcal{K}^*$ into critical regions CR_i, and find the expression of the functions $J^*(\cdot)$ and $z^*(\cdot)$ for each critical region. An mp-QP algorithm has two components: the "active set generator" and the "KKT solver." The active set generator computes the set of active constraints A_i. The KKT solver computes CR_{A_i} and the expression of $J^*(\cdot)$ and $z^*(\cdot)$ in CR_{A_i} as explained in Theorem 6.6. The active set generator is the critical part. In principle, one could simply generate all the possible combinations of active sets. However, in many problems only a few active constraints sets generate full-dimensional critical regions inside the region of interest $\mathcal{K}$. Therefore, the goal is to design an active set generator algorithm which computes only the active sets A_i with the associated full-dimensional critical regions covering only $\mathcal{K}^*$.

Next an implementation of an mp-QP algorithm is described. See Section 6.6 for a literature review on alternative approaches to the solution of an mp-QP.

In order to start solving the mp-QP problem, we need an initial vector x_0 inside the polyhedral set $\mathcal{K}^*$ of feasible parameters. A possible choice for x_0 is the Chebyshev center (see Section 4.4.5) of $\mathcal{K}^*$ obtained as a solution of

$$\begin{aligned} \max_{x,\bar{z},\epsilon} \quad & \epsilon \\ \text{subj. to} \quad & T_i x + \epsilon \|T_i\|_2 \le N_i, \quad i = 1, \ldots, n_T \\ & G\bar{z} - Sx \le w \end{aligned} \qquad (6.70)$$

where n_T is the number of rows T_i of the matrix T defining the set $\mathcal{K}$ in (6.2). If $\epsilon \le 0$, then the QP problem (6.38) is infeasible for all x in the interior of $\mathcal{K}$. Otherwise, we set $x = x_0$ and solve the QP problem (6.38), in order to obtain the corresponding optimal solution z_0^*. Such a solution is unique, because $H \succ 0$. The value of z_0^* defines the following optimal partition

$$\begin{aligned} \mathrm{A}(x_0) &= \{j \in J : G_j z_0^* - S_j x_0 - w_j = 0\} \\ \mathrm{NA}(x_0) &= \{j \in J : G_j z_0^* - S_j x_0 - w_j < 0\} \end{aligned} \qquad (6.71)$$

and consequently the critical region $CR_{A(x_0)}$. Once the critical region $CR_{A(x_0)}$ has been defined, the rest of the space $R^{\mathrm{rest}} = \mathcal{K} \backslash CR_{A(x_0)}$ has to be explored and new critical regions generated. An approach for generating a polyhedral partition $\{R_1, \ldots, R_{n_{rest}}\}$ of the rest of the space R^{rest} is described in Theorem 4.2.

Theorem 4.2 provides a way of partitioning the nonconvex set $\mathcal{K} \setminus CR_0$ into polyhedral subsets R_i. For each R_i, a new vector x_i is determined by solving the LP (6.70), and, correspondingly, an optimum z_i^*, a set of active constraints A_i, and a critical region CR_i. The procedure proposed in Theorem 4.2 for partitioning the set of parameters allows one to recursively explore the parameter space. Such an iterative procedure terminates after a finite time, as the number of possible combinations of active constraints decreases with each iteration. Two main elements need to be considered:

1. As for the mp-LP algorithm, the partitioning in Theorem 4.2 defines new polyhedral regions R_k to be explored that are not related to the critical regions which still need to be determined. This may split some of the critical regions, due to the artificial cuts induced by Theorem 4.2. Postprocessing can be used to join cut critical regions [44]. As an example, in Figure 6.8 the critical region $CR_{\{3,7\}}$ is discovered twice, one part during the exploration of R_1 and the second part during the exploration of R_2.

   Although algorithms exist for convexity recognition and computation of the union of polyhedra, the postprocessing operation is computationally expensive. Therefore, it is more efficient not to intersect the critical region obtained by (6.29) with halfspaces generated by Theorem 4.2, which is only used to drive the exploration of the parameter space. Then, no postprocessing is needed to join subpartitioned critical regions. On the other hand, some critical regions may appear more than once. Duplicates can be uniquely identified by the set of active constraints $A(x)$ and can be easily eliminated. To this aim, in the implementation of the algorithm we keep a list of all the critical regions which have already been generated in order to avoid duplicates. In Figure 6.8 the critical region $CR_{\{3,7\}}$ is discovered twice but stored only once.

2. If Case 2 occurs in Section 6.3.2 and D is nonzero, $CR_{\mathcal{A}}$ is a lower dimensional critical region (see Remark 6.6). Therefore we do not need to explore the actual combination G_A, S_A, w_A. On the other hand, if $D = 0$ the KKT conditions do not lead directly to (6.49)–(6.50). In this case, a full-dimensional critical region can be obtained from (6.58) by projecting the set $\mathcal{P}_{x, u_{A,2}^*}$ in (6.57).

Based on the discussion and results above, the main steps of the mp-QP solver are outlined in the following algorithm.

Algorithm 6.2 *mp-QP Algorithm*

Input: Matrices H, G, w, S of problem (6.38) and set $\mathcal{K}$ in (6.2)
Output: Multiparametric solution to problem (6.38)

    Execute partition($\mathcal{K}$)
    end

    Function partition(Y)
        Let $x_0 \in Y$ and ϵ be the solution to the LP (6.70);
        If $\epsilon \leq 0$ **Then** exit (no full-dimensional CR is in Y)

Solve the QP (6.38) for $x = x_0$ to obtain (z_0^*, u_0^*)

Determine the set of active constraints A when $z = z_0^*$, $x = x_0$, and build G_A, w_A, S_A

If G_A has full row rank **Then**

Determine $u_A^*(x)$, $z^*(x)$ from (6.47) and (6.48)

Characterize the CR from (6.49) and (6.50)

Else

Determine $z^*(x)$ from (6.55)

Characterize the CR from (6.58)

End

Partition the rest of the region as in Theorem 4.2

For each new sub-region R_i, **Do** partition(R_i)

End function

Remark 6.7 If rank(D) > 0 in Algorithm 6.2, the region $CR_{A(x_0)}$ is not full-dimensional. To avoid further recursion in the algorithm which does not produce any full-dimensional critical region, one should compute a random vector $\epsilon \in \mathbb{R}^n$ smaller than the Chebyshev radius of Y and such that the QP (6.38) is feasible for $x_0 + \epsilon$ and then repeat step where $A(x_0)$ is computed with $x_0 \leftarrow x_0 + \epsilon$.

Remark 6.8 The algorithm solves the mp-QP problem by partitioning the given parameter set $\mathcal{K}$ into N_r closed polyhedral regions. Note that the algorithm generates a partition of the state space which is not strict. The algorithm could be modified to store the critical regions as defined in Section 6.1.2 (which are neither closed nor open as proven in Theorem 6.6) instead of storing their closure. This can be done by keeping track of which facet belongs to a certain critical region and which not. From a practical point of view, such a procedure is not necessary since the value function and the optimizer are continuous functions of x.

Remark 6.9 The proposed algorithm does not apply to the case when $H \succeq 0$ and when the optimizer may not be unique. In this case one may resort to "regularization," i.e., adding an appropriate "small" quadratic term, or follow the ideas of [231].

6.4 Multiparametric Mixed-Integer Linear Programming

6.4.1 Formulation and Properties

Consider the mp-LP

$$J^*(x) = \min_z \quad \{J(z,x) = c'z\}$$
$$\text{subj. to} \quad Gz \leq w + Sx, \tag{6.72}$$

where $z \in \mathbb{R}^s$ are the optimization variables, $x \in \mathbb{R}^n$ is the vector of parameters, $G \in \mathbb{R}^{m \times s}$, $w \in \mathbb{R}^m$, and $S \in \mathbb{R}^{m \times n}$. When we restrict some of the optimization variables to be 0 or 1, $z = \{z_c, z_d\}$, $z_c \in \mathbb{R}^{s_c}$, $z_d \in \{0,1\}^{s_d}$ and $s = s_c + s_d$, we refer to (6.72) as a (right-hand side) *multiparametric mixed-integer linear program* (mp-MILP).

6.4.2 Geometric Algorithm for mp-MILP

Consider the mp-MILP (6.72). Given a closed and bounded polyhedral set $\mathcal{K} \subset \mathbb{R}^n$ of parameters,

$$\mathcal{K} = \{x \in \mathbb{R}^n : Tx \leq N\}, \tag{6.73}$$

we denote by $\mathcal{K}^* \subseteq \mathcal{K}$ the region of parameters $x \in \mathcal{K}$ such that the MILP (6.72) is feasible and the optimum $J^*(x)$ is finite. For any given $\bar{x} \in \mathcal{K}^*$, $J^*(\bar{x})$ denotes the minimum value of the objective function in problem (6.72) for $x = \bar{x}$. The function $J^* : \mathcal{K}^* \to \mathbb{R}$ will denote the function which expresses the dependence on x of the minimum value of the objective function over $\mathcal{K}^*$, J^* will be called the value function. The set-valued function $Z^* : \mathcal{K}^* \to 2^{\mathbb{R}^{s_c}} \times 2^{\{0,1\}^{s_d}}$ will describe for any fixed $x \in \mathcal{K}^*$ the set of optimizers $z^*(x)$ related to $J^*(x)$.

We aim to determine the region $\mathcal{K}^* \subseteq \mathcal{K}$ of feasible parameters x and to find the expression of the value function $J^*(x)$ and the expression of an optimizer function $z^*(x) \in Z^*(x)$.

Two main approaches have been proposed for solving mp-MILP problems. In [1], the authors develop an algorithm based on branch and bound (B&B) methods. At each node of the B&B tree an mp-LP is solved. The solution at the root node where all the binary variables are relaxed to the interval [0,1] represents a valid lower bound, while the solution at a node where all the integer variables have been fixed represents a valid upper bound. As in standard B&B methods, the complete enumeration of combinations of 0–1 integer variables is avoided by comparing the multiparametric solutions, and by fathoming the nodes where there is no improvement of the value function.

In [101] an alternative algorithm was proposed, which will be detailed in this section. Problem (6.72) is alternatively decomposed into an mp-LP and an MILP subproblem. In one step the values of the binary variable are fixed for a region and an mp-LP is solved. Its solution provides a parametric upper bound to the value function $J^*(x)$ in the region. In the other step, the parameters x are treated as additional free variables and an MILP is solved. In this way a parameter x is found and an associated new integer vector which improves the value function at this point.

The algorithm is composed of an *initialization step*, and a recursion between the solution of an *mp-LP subproblem* and an *MILP subproblem*.

Initialization

Solve the following MILP problem

$$\begin{aligned}
\min_{\{z,x\}} \quad & c'z \\
\text{subj. to} \quad & Gz - Sx \leq w \\
& x \in \mathcal{K},
\end{aligned} \tag{6.74}$$

where x is treated as an independent variable. If the MILP (6.74) is infeasible then the mp-MILP (6.72) admits no solution, i.e., $\mathcal{K}^* = \emptyset$; otherwise its solution z^*, x^* provides an integer variable z^* that is feasible at point x^*.

At step $j = 0$, set: $N_0 = 1$, $CR_1 = \mathcal{K}$, $Z_1 = \emptyset$, $\bar{J}_1 = +\infty$, $N_{b_i} = 0$, $\bar{z}_{d_1}^1 = \bar{z}_d$.

mp-LP Subproblem

For each CR_i we solve the following mp-LP problem

$$
\begin{aligned}
\tilde{J}_i(x) = \quad &\min_z \quad c'z \\
&\text{subj. to} \quad Gz \leq w + Sx \\
&\qquad\qquad z_d = \bar{z}_{d_i}^{Nb_i+1} \\
&\qquad\qquad x \in CR_i.
\end{aligned}
\tag{6.75}
$$

By Theorem 6.5, the solution of mp-LP (6.75) provides a partition of CR_i into polyhedral regions R_i^k, $k = 1, \ldots, N_{R_i}$ and a PWA value function

$$
\tilde{J}_i(x) = (\tilde{J}R_i^k(x) = c_i^{k'}x + p_i^k) \text{ if } x \in R_i^k, \ k = 1, \ldots, N_{R_i}
\tag{6.76}
$$

where $\tilde{J}R_i^j(x) = +\infty$ in R_i^j if the integer variable $\bar{z}_{d_i}$ is not feasible in R_i^j and a PWA continuous optimizer $z^*(x)$ ($z^*(x)$ is not defined in R_i^j if $\tilde{J}R_i^j(x) = +\infty$).

The function $\tilde{J}_i(x)$ will be an upper bound of $J^*(x)$ for all $x \in CR_i$. Such a bound $\tilde{J}_i(x)$ on the value function has to be compared with the current bound $\bar{J}_i(x)$ in CR_i in order to obtain the lowest of the two parametric value functions and to update the bound.

While updating $\bar{J}_i(x)$ three cases are possible:

1. $\bar{J}_i(x) = \tilde{J}R_i^k(x) \ \forall x \in R_i^k$ if $(\tilde{J}R_i^k(x) \leq \bar{J}_i(x) \ \forall \ x \in R_i^k)$.

2. $\bar{J}_i(x) = \bar{J}_i(x) \ \forall x \in R_i^k$ (if $\tilde{J}R_i^k(x) \geq \bar{J}_i(x) \ \forall \ x \in R_i^k$).

3. $\bar{J}_i(x) = \begin{cases} \bar{J}_i(x) & \forall x \in (R_i^k)_1 = \{x \in R_i^k : \ \tilde{J}R_i^k(x) \geq \bar{J}_i(x)\} \\ \tilde{J}R_i^k(x) & \forall x \in (R_i^k)_2 = \{x \in R_i^k : \ \tilde{J}R_i^k(x) \leq \bar{J}_i(x)\}. \end{cases}$

The three cases above can be distinguished by using a simple linear program. We add the constraint $\tilde{J}R_i^k(x) \leq \bar{J}_i(x)$ to the constraints defining R_i^k and tests its redundancy (Section 4.4.1). If it is redundant we have Case 1, if it is infeasible we have Case 2. Otherwise $\tilde{J}R_i^k(x) = \bar{J}_i(x)$ defines the facet separating $(R_i^k)_1$ and $(R_i^k)_2$. In the third case, the region R_i^k is partitioned into two regions $(R_i^k)_1$ and $(R_i^k)_2$ which are convex polyhedra since $\tilde{J}R_i^k(x)$ and $\bar{J}_i(x)$ are affine functions of x.

After the mp-LP (6.75) has been solved for all $i = 1, \ldots, N_j$ (the subindex j denotes that we are at step j of the recursion) and the value function has been updated, each initial region CR_i has been subdivided into at most $2N_{R_i}$ polyhedral regions R_i^k and possibly $(R_i^k)_1$ and $(R_i^k)_2$ with a corresponding updated parametric bound on the value function $\bar{J}_i(x)$. For each R_i^k, $(R_i^k)_1$ and $(R_i^k)_2$ we define the set of integer variables already explored as $Z_i = Z_i \bigcup \bar{z}_{d_i}^{Nb_i+1}$, $Nb_i = Nb_i + 1$. In the sequel the polyhedra of the new partition will be referred to as CR_i.

MILP Subproblem

At step j for each critical region CR_i (note that these CR_i are the output of the previous phase) we solve the following MILP problem

$$\min_{\{z,x\}} \quad c'z \qquad\qquad\qquad (6.77)$$

$$\text{subj. to} \quad Gz - Sx \le w \qquad\qquad\qquad (6.78)$$

$$c'z \le \bar{J}_i(x) \qquad\qquad\qquad (6.79)$$

$$z_d \ne \bar{z}_{d_i}^k, \ \ k = 1,\dots,Nb_i \qquad\qquad (6.80)$$

$$x \in CR_i, \qquad\qquad\qquad (6.81)$$

where constraints (6.80) prohibit integer solutions that have been already analyzed in CR_i from appearing again and constraint (6.79) excludes integer solutions with higher values than the current upper bound. If problem (6.81) is infeasible then the region CR_i is excluded from further recursion and the current upper bound represents the final solution. If problem (6.81) is feasible, then the discrete optimal component $z_{d_i}^*$ is stored and represents a feasible integer variable that is optimal at least in one point of CR_i.

Recursion

At step j we have stored

1. A list of N_j polyhedral regions CR_i and for each of them an associated parametric affine upper bound $\bar{J}_i(x)$ ($\bar{J}_i(x) = +\infty$ if no integer solution has been found yet in CR_i).

2. For each CR_i a set of integer variables $Z_i = \bar{z}_{d_i}^0,\dots,\bar{z}_{d_i}^{Nb_i}$, that have already been explored in the region CR_i.

3. For each CR_i an integer feasible variable $\bar{z}_{d_i}^{Nb_i+1} \notin Z_i$ such that there exists z_c and $\hat{x} \in CR_i$ for which $Gz \le w + S\hat{x}$ and $c'z < \bar{J}_i(\hat{x})$ where $z = \{z_c, \bar{z}_{d_i}^{Nb_i+1}\}$. That is, $\bar{z}_{d_i}^{Nb_i+1}$ is an integer variable that improves the current bound for at least one point of the current polyhedron.

For all the regions CR_i not excluded from the MILP's subproblem (6.77)–(6.81) the algorithm continues to iterate between the mp-LP (6.75) with $\bar{z}_{d_i}^{Nb_i+1} = z_{d_i}^*$ and the MILP (6.77)–(6.81). The algorithm terminates when all the MILPs (6.77)–(6.81) are infeasible.

Note that the algorithm generates a partition of the state space. Some parameter x could belong to the boundary of several regions. Differently from the LP and QP case, the value function may be discontinuous and therefore such a case has to be treated carefully. If a point x belongs to different critical regions, the expressions of the value function associated with such regions have to be compared in order to assign to x the right optimizer. Such a procedure can be avoided by keeping track of which facet belongs to a certain critical region and which not. Moreover, if the value functions associated with the regions containing the same parameter x coincide this may imply the presence of multiple optimizers.

6.4.3 Solution Properties

The following properties of $J^*(x)$ and $Z^*(x)$ follow easily from the algorithm described above.

Theorem 6.10   *Consider the mp-MILP (6.72). The set $\mathcal{K}^*$ is the union of a finite number of (possibly open) polyhedra and the value function J^* is piecewise affine on polyhedra. If the optimizer $z^*(x)$ is unique for all $x \in \mathcal{K}^*$, then the optimizer functions $z_c^* : \mathcal{K}^* \to \mathbb{R}^{s_c}$ and $z_d^* : \mathcal{K}^* \to \{0,1\}^{s_d}$ are piecewise affine and piecewise constant, respectively, on polyhedra. Otherwise, it is always possible to define a piecewise affine optimizer function $z^*(x) \in Z^*(x)$ for all $x \in \mathcal{K}^*$.*

Note that, differently from the mp-LP case, the set $\mathcal{K}^*$ can be nonconvex and even disconnected.

6.5 Multiparametric Mixed-Integer Quadratic Programming

6.5.1 Formulation and Properties

Consider the mp-QP

$$
\begin{aligned}
J^*(x) = \min_z \quad & J(z,x) = z'H_1 z + c_1' z \\
\text{subj. to} \quad & Gz \leq w + Sx,
\end{aligned}
\tag{6.82}
$$

When we restrict some of the optimization variables to be 0 or 1, $z = [z_c', z_d']'$, where $z_c \in \mathbb{R}^{s_c}$, $z_d \in \{0,1\}^{s_d}$, we refer to (6.82) as a *multiparametric mixed-integer quadratic program* (mp-MIQP). Given a closed and bounded polyhedral set $\mathcal{K} \subset \mathbb{R}^n$ of parameters,

$$
\mathcal{K} = \{x \in \mathbb{R}^n : Tx \leq N\},
\tag{6.83}
$$

we denote by $\mathcal{K}^* \subseteq \mathcal{K}$ the region of parameters $x \in \mathcal{K}$ such that the MIQP (6.82) is feasible and the optimum $J^*(x)$ is finite. For any given $\bar{x} \in \mathcal{K}^*$, $J^*(\bar{x})$ denotes the minimum value of the objective function in problem (6.82) for $x = \bar{x}$. The value function $J^* : \mathcal{K}^* \to \mathbb{R}$ denotes the function which expresses the dependence on x of the minimum value of the objective function over $\mathcal{K}^*$. The set-valued function $Z^* : \mathcal{K}^* \to 2^{\mathbb{R}^{s_c}} \times 2^{\{0,1\}^{s_d}}$ describes for any fixed $x \in \mathcal{K}^*$ the set of optimizers $z^*(x)$ related to $J^*(x)$.

We aim at determining the region $\mathcal{K}^* \subseteq \mathcal{K}$ of feasible parameters x and at finding the expression of the value function $J^*(x)$ and the expression of an optimizer function $z^*(x) \in Z^*(x)$.

We show with a simple example that the geometric approach discussed in this chapter cannot be used for solving mp-MIQPs.

Example 6.8   Suppose $z_1, z_2, x_1, x_2 \in \mathbb{R}$ and $\delta \in \{0, 1\}$, then the following mp-MIQP

$$J^*(x_1, x_2) = \min_{z_1, z_2, \delta} \quad z_1^2 + z_2^2 - 25\delta + 100$$

$$\text{subj. to} \quad
\begin{bmatrix}
1 & 0 & 10 \\
-1 & 0 & 10 \\
0 & 1 & 10 \\
0 & -1 & 10 \\
1 & 0 & -10 \\
-1 & 0 & -10 \\
0 & 1 & -10 \\
0 & -1 & -10 \\
0 & 0 & 0 \\
0 & 0 & 0 \\
0 & 0 & 0 \\
0 & 0 & 0
\end{bmatrix}
\begin{bmatrix}
z_1 \\ z_2 \\ \delta
\end{bmatrix}
\leq
\begin{bmatrix}
1 & 0 \\
-1 & 0 \\
0 & 1 \\
0 & -1 \\
0 & 0 \\
0 & 0 \\
0 & 0 \\
0 & 0 \\
1 & 0 \\
-1 & 0 \\
0 & 1 \\
0 & -1
\end{bmatrix}
\begin{bmatrix}
x_1 \\ x_2
\end{bmatrix}
+
\begin{bmatrix}
10 \\ 10 \\ 10 \\ 10 \\ 0 \\ 0 \\ 0 \\ 0 \\ 10 \\ 10 \\ 10 \\ 10
\end{bmatrix}$$

$$(6.84)$$

can be simply solved by noting that for $\delta = 1$ $z_1 = x_1$ and $z_2 = x_2$ while for $\delta = 0$ $z_1 = z_2 = 0$. By comparing the value functions associated with $\delta = 0$ and $\delta = 1$ we obtain two critical regions

$$
\begin{aligned}
CR_1 &= \{x_1, x_2 \in \mathbb{R} : \ x_1^2 + x_2^2 \leq 25\} \\
CR_2 &= \{x_1, x_2 \in \mathbb{R} : \ -10 \leq x_1 \leq 10, \ -10 \leq x_2 \leq 10, \ x_1^2 + x_2^2 > 25\},
\end{aligned}
\qquad (6.85)
$$

$$
z_1^*(x_1, x_2) = \begin{cases} x_1 & \text{if } [x_1, x_2] \in CR_1 \\ 0 & \text{if } [x_1, x_2] \in CR_2 \end{cases}
$$
$$
z_2^*(x_1, x_2) = \begin{cases} x_2 & \text{if } [x_1, x_2] \in CR_1 \\ 0 & \text{if } [x_1, x_2] \in CR_2, \end{cases}
\qquad (6.86)
$$

and the parametric value function

$$
J^*(x_1, x_2) = \begin{cases} x_1^2 + x_2^2 + 75 & \text{if } [x_1, x_2] \in CR_1 \\ 100 & \text{if } [x_1, x_2] \in CR_2. \end{cases}
\qquad (6.87)
$$

The two critical regions and the value function are depicted in Figure 6.11.

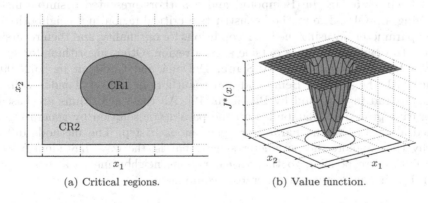

(a) Critical regions.   (b) Value function.

Figure 6.11 Example 6.8. Solution to the mp-MIQP (6.84).

Example 6.8 demonstrates that, in general, the critical regions of an mp-MIQP cannot be decomposed into convex polyhedra. Therefore the method of partitioning the rest of the space presented in Theorem 4.2 cannot be applied here. In Chapter 17 we will present an algorithm that efficiently solves specific mp-MIQPs that stem from the optimal control of discrete-time hybrid systems.

6.6 Literature Review

Many of the theoretical results on parametric programming can be found in [18, 107, 47, 116].

The first method for solving *parametric linear programs* was proposed by Gass and Saaty [120], and since then extensive research has been devoted to sensitivity and multiparametric linear analysis, as attested by the hundreds of references in [115] (see also [116] for recent advances in the field). One of the first methods for solving multiparametric linear programs (mp-LPs) was formulated by Gal and Nedoma [117]. The method constructs the critical regions iteratively, by visiting the graph of bases associated with the LP tableau of the original problem. Many of the results on mp-LP presented in this book can be found in [115, p. 178–180].

Note that in [117, 115] a critical region is defined as a subset of the parameter space on which a certain basis of the linear program is optimal. The algorithm proposed in [117] for solving multiparametric linear programs generates nonoverlapping critical regions by generating and exploring the graph of bases. In the graph of bases the nodes represent optimal bases of the given multiparametric problem and two nodes are connected by an edge if it is possible to pass from one basis to another by one pivot step (in this case the bases are called neighbors). In this book we use the definition (6.4) of critical regions which is not associated with the bases but with the set of active constraints and it is directly related to the definition given in [2, 207, 116].

The solution to *multiparametric quadratic programs* has been studied in detail in [18, Chapter 5]. In [44] Bemporad and coauthors presented a simple method for solving mp-QPs. The method constructs a critical region in a neighborhood of a given parameter, by using the KKT conditions for optimality, and then recursively explores the parameter space outside such a region. Other algorithms for solving mp-QPs have been proposed by Seron, DeDoná and Goodwin in [262, 99] in parallel with the study of Bemporad and coauthors in [44], by Tøndel, Johansen and Bemporad in [274] and by Baotic in [19]. All these algorithms are based on an iterative procedure that builds up the parametric solution by generating new polyhedral regions of the parameter space at each step. The methods differ in the way they explore the parameter space, that is, the way they identify active constraints corresponding to the critical regions neighboring to a given critical region, i.e., in the "active set generator" component.

In [262, 99] the authors construct the unconstrained critical region and then generate neighboring critical regions by enumerating all possible combinations of active constraints.

In [274] the authors explore the parameter space outside a given region CR_i by examining its set of active constraints A_i. The critical regions neighboring to CR_i are constructed by elementary operations on the active constraints set A_i that can be seen as an equivalent "pivot" for the quadratic program. For this reason the method can be considered as an extension of the method of Gal [115] to multiparametric quadratic programming.

In [19] the author uses a direct exploration of the parameter space as in [44] but he avoids the partition of the state space described in Theorem 4.2. Given a polyhedral critical region CR_i, the procedure goes through all its facets and generates the Chebyshev center of each facet. For each facet $\mathcal{F}_i$ a new parameter x_ε^i is generated, by moving from the center of the facet in the direction of the normal to the facet by a small step. If such parameter x_ε^i is infeasible or is contained in a critical region already stored, then the exploration in the direction of $\mathcal{F}_i$ stops. Otherwise, the set of active constraints corresponding to the critical region sharing the facet $\mathcal{F}_i$ with the region CR_i is found by solving a QP for the new parameter x_ε^i.

In [1, 101] two approaches were proposed for solving mp-MILP problems. In both methods the authors use an mp-LP algorithm and a branch and bound strategy that avoids the complete enumeration of combinations of 0–1 integer variables by comparing the available bounds on the multiparametric solutions.

Part III

Optimal Control

7

General Formulation and Discussion

In this chapter we introduce the optimal control problem we will be studying in a very general form. We want to communicate the basic definitions and essential concepts. We will sacrifice mathematical precision for the sake of simplicity. In later chapters we will study specific versions of this problem for specific cost functions and system classes in greater detail.

7.1 Problem Formulation

We consider the nonlinear time-invariant system

$$x(t+1) = g(x(t), u(t)), \tag{7.1}$$

subject to the constraints

$$h(x(t), u(t)) \leq 0 \tag{7.2}$$

at all time instants $t \geq 0$. In (7.1)–(7.2), $x(t) \subset \mathbb{R}^n$ and $u(t) \in \mathbb{R}^m$ are the state and input vector, respectively. Inequality (7.2) with $h : \mathbb{R}^n \times \mathbb{R}^m \to \mathbb{R}^{n_c}$ expresses the n_c constraints imposed on the input and the states. These may be simple upper and lower bounds or more complicated expressions. We assume that the origin is an equilibrium point $(g(0,0) = 0)$ in the interior of the feasible set, i.e., $h(0,0) < 0$.

We assumed the system to be specified in discrete time. One reason is that we are looking for solutions to engineering problems. In practice, the controller will almost always be implemented through a digital computer by sampling the variables of the system and transmitting the control action to the system at discrete time points. Another reason is that for the solution of the optimal control problems for discrete-time systems we will be able to make ready use of powerful mathematical programming software.

We want to caution the reader, however, that in many instances the discrete time model is an approximation of the continuous time model. It is generally difficult to derive "good" discrete time models from nonlinear continuous time models, and especially so when the nonlinear system has discontinuities as would be the case for switched systems. We also note that continuous time switched systems can exhibit behavioral characteristics not found in discrete-time systems, for example, an ever increasing number of switches in an ever decreasing time interval (*Zeno behavior* [127]).

We define the following *performance objective* or *cost function* from time instant 0 to time instant N

$$J_{0 \to N}(x_0, U_{0 \to N}) = p(x_N) + \sum_{k=0}^{N-1} q(x_k, u_k), \qquad (7.3)$$

where N is the time *horizon* and x_k denotes the state vector at time k obtained by starting from the measured state $x_0 = x(0)$ and applying to the system model

$$x_{k+1} = g(x_k, u_k), \qquad (7.4)$$

the input sequence $u_0, \ldots, u_{k-1}$. From this sequence we define the vector of future inputs $U_{0 \to N} = [u_0', \ldots, u_{N-1}']' \in \mathbb{R}^s$, $s = mN$. The terms $q(x_k, u_k)$ and $p(x_N)$ are referred to as *stage cost* and *terminal cost*, respectively, and are assumed to be positive definite ($q \succ 0$, $p \succ 0$):

$$p(x, u) > 0 \ \forall x \neq 0, \ u \neq 0, \ \ p(0, 0) = 0$$
$$q(x, u) > 0 \ \forall x \neq 0, \ u \neq 0, \ \ q(0, 0) = 0.$$

The form of the cost function (7.3) is very general. If a practical control objective can be expressed as a scalar function then this function usually takes the indicated form. Specifically, we consider the following constrained finite time optimal control (CFTOC) problem.

$$
\begin{aligned}
J_{0 \to N}^*(x_0) = \min_{U_{0 \to N}} \quad & J_{0 \to N}(x_0, U_{0 \to N}) \\
\text{subj. to} \quad & x_{k+1} = g(x_k, u_k), \ k = 0, \ldots, N-1 \\
& h(x_k, u_k) \leq 0, \ k = 0, \ldots, N-1 \\
& x_N \in \mathcal{X}_f \\
& x_0 = x(0).
\end{aligned}
\qquad (7.5)
$$

Here $\mathcal{X}_f \subseteq \mathbb{R}^n$ is a *terminal region* that we want the system states to reach at the end of the horizon. The terminal region could be the origin, for example. We define $\mathcal{X}_{0 \to N} \subseteq \mathbb{R}^n$ to be the set of initial conditions $x(0)$ for which there exists an input vector $U_{0 \to N}$ so that the inputs $u_0, \ldots, u_{N-1}$ and the states $x_0, \ldots, x_N$ satisfy the model $x_{k+1} = g(x_k, u_k)$ and the constraints $h(x_k, u_k) \leq 0$ and that the state x_N lies in the terminal set $\mathcal{X}_f$.

We can determine this set of feasible initial conditions in a recursive manner. Let us denote with $\mathcal{X}_{j \to N}$ the set of states x_j at time j which can be steered into $\mathcal{X}_f$ at time N, i.e., for which the model $x_{k+1} = g(x_k, u_k)$ and the constraints $h(x_k, u_k) \leq 0$ are feasible for $k = j, \ldots, N-1$ and $x_N \in \mathcal{X}_f$. This set can be defined recursively by

$$\mathcal{X}_{j \to N} = \{x \in \mathbb{R}^n : \ \exists u \text{ such that } (h(x, u) \leq 0, \text{ and } g(x, u) \in \mathcal{X}_{j+1 \to N})\},$$
$$j = 0, \ldots, N-1 \qquad (7.6)$$
$$\mathcal{X}_{N \to N} = \mathcal{X}_f. \qquad (7.7)$$

The set $\mathcal{X}_{0 \to N}$ is the final result of these iterations starting with $\mathcal{X}_f$.

The optimal cost $J_{0 \to N}^*(x_0)$ is also called *value function*. In general, the problem (7.3)–(7.5) may not have a minimum. We will assume that there exists a minimum. This is the case, for example, when the set of feasible input vectors $U_{0 \to N}$ (defined by h and $\mathcal{X}_f$) is compact and when the functions g, p and q are continuous. Also, there might be several input vectors $U_{0 \to N}^*$ which yield the minimum ($J_{0 \to N}^*(x_0) = J_{0 \to N}(x_0, U_{0 \to N}^*)$). In this case we will define one of them as the minimizer $U_{0 \to N}^*$.

Note that throughout the book we will distinguish between the *current* state $x(k)$ of system (7.1) at time k and the variable x_k in the optimization problem (7.5), that is the *predicted* state of system (7.1) at time k obtained by starting from the state x_0 and applying to system (7.4) the input sequence $u_0, \ldots, u_{k-1}$. Analogously, $u(k)$ is the input applied to system (7.1) at time k while u_k is the k-th optimization variable of the optimization problem (7.5). Clearly, $x(k) = x_k$ for any k if $u(k) = u_k$ for all k (under the assumption that our model is perfect).

In the rest of this chapter we will be interested in the following questions related to the general optimal control problem (7.3)–(7.5).

- *Solution.* We will show that the problem can be expressed and solved either as one general nonlinear programming problem, or in a recursive manner by invoking Bellman's Principle of Optimality.

- *Infinite horizon.* We will investigate if a solution exists as $N \to \infty$, the properties of this solution and how it can be obtained or at least approximated by using a *receding horizon* technique.

7.2 Solution via Batch Approach

If we write the equality constraints appearing in (7.5) explicitly

$$
\begin{aligned}
x_1 &= g(x(0), u_0) \\
x_2 &= g(x_1, u_1) \\
&\vdots \\
x_N &= g(x_{N-1}, u_{N-1}),
\end{aligned}
\tag{7.8}
$$

then the optimal control problem (7.3)–(7.5), rewritten below

$$
J_{0 \to N}^*(x_0) = \min_{U_{0 \to N}} \quad p(x_N) + \sum_{k=0}^{N-1} q(x_k, u_k)
$$

$$
\begin{aligned}
\text{subj. to} \quad &x_1 = g(x_0, u_0) \\
&x_2 = g(x_1, u_1) \\
&\vdots \\
&x_N = g(x_{N-1}, u_{N-1}) \\
&h(x_k, u_k) \le 0, \ k = 0, \ldots, N-1 \\
&x_N \in \mathcal{X}_f \\
&x_0 = x(0)
\end{aligned}
\tag{7.9}
$$

is recognized more easily as a general nonlinear programming problem with variables $u_0, \ldots, u_{N-1}$ and $x_1, \ldots, x_N$.

As an alternative we may try to eliminate the state variables and equality constraints (7.8) by successive substitution so that we are left with $u_0, \ldots, u_{N-1}$ as the only decision variables. For example, we can express x_2 as a function of $x(0)$, u_0 and u_1 only, by eliminating the intermediate state x_1

$$
\begin{aligned}
x_2 &= g(x_1, u_1) \\
x_2 &= g(g(x(0), u_0), u_1).
\end{aligned}
\tag{7.10}
$$

Except when the state equations are linear this successive substitution may become complex. Even when they are linear it may be bad from a numerical point of view.

Either with or without successive substitution the solution of the nonlinear programming problem is a sequence of present and future inputs $U_{0 \to N}^* = [u_0^{*'}, \ldots, u_{N-1}^{*'}]'$ determined for the particular initial state $x(0)$.

7.3 Solution via Recursive Approach

The recursive approach, Bellman's dynamic programming technique, rests on a simple idea, the *principle of optimality*. It states that for a trajectory $x_0, x_1^*, \ldots, x_N^*$ to be optimal, the trajectory starting from any intermediate point x_j^*, i.e., $x_j^*, x_{j+1}^*, \ldots, x_N^*$, $0 \le j \le N-1$, must be optimal.

Consider the following example to provide an intuitive justification [53]. Suppose that the fastest route from Los Angeles to Boston passes through Chicago. Then the principle of optimality formalizes the obvious fact that the Chicago to Boston portion of the route is also the fastest route for a trip that starts from Chicago and ends in Boston.

We can utilize the principle of optimality for the optimal control problem we are investigating. We define the cost over the reduced horizon from j to N

$$
J_{j \to N}(x_j, u_j, u_{j+1}, \ldots, u_{N-1}) = p(x_N) + \sum_{k=j}^{N-1} q(x_k, u_k),
\tag{7.11}
$$

also called the *cost-to-go*. Then the *optimal cost-to-go* $J_{j \to N}^*$ is

$$
\begin{aligned}
J_{j \to N}^*(x_j) = \min_{u_j, u_{j+1}, \ldots, u_{N-1}} \quad & J_{j \to N}(x_j, u_j, u_{j+1}, \ldots, u_{N-1}) \\
\text{subj. to} \quad & x_{k+1} = g(x_k, u_k), \ k = j, \ldots, N-1 \\
& h(x_k, u_k) \le 0, \ k = j, \ldots, N-1 \\
& x_N \in \mathcal{X}_f.
\end{aligned}
\tag{7.12}
$$

Note that the optimal cost-to-go $J_{j \to N}^*(x_j)$ depends only on the initial state x_j.

The principle of optimality implies that the optimal cost-to-go $J_{j-1 \to N}^*$ from time $j-1$ to the final time N can be found by minimizing the sum of the stage cost $q(x_{j-1}, u_{j-1})$ and the optimal cost-to-go $J_{j \to N}^*(x_j)$ from time j onwards:

$$
\begin{aligned}
J_{j-1 \to N}^*(x_{j-1}) = \min_{u_{j-1}} \quad & q(x_{j-1}, u_{j-1}) + J_{j \to N}^*(x_j) \\
\text{subj. to} \quad & x_j = g(x_{j-1}, u_{j-1}) \\
& h(x_{j-1}, u_{j-1}) \le 0 \\
& x_j \in \mathcal{X}_{j \to N}.
\end{aligned}
\tag{7.13}
$$

Here the only decision variable left for the optimization is u_{j-1}, the input at time $j-1$. All the other inputs $u_j^*, \ldots, u_{N-1}^*$ have already been selected optimally to yield the optimal cost-to-go $J_{j \to N}^*(x_j)$. We can rewrite (7.13) as

$$
\begin{aligned}
J_{j-1 \to N}^*(x_{j-1}) = \min_{u_{j-1}} \quad & q(x_{j-1}, u_{j-1}) + J_{j \to N}^*(g(x_{j-1}, u_{j-1})) \\
\text{subj. to} \quad & h(x_{j-1}, u_{j-1}) \le 0 \\
& g(x_{j-1}, u_{j-1}) \in \mathcal{X}_{j \to N},
\end{aligned}
\tag{7.14}
$$

making the dependence of x_j on the initial state x_{j-1} explicit.

The optimization problem (7.14) suggests the following recursive algorithm backwards in time to determine the optimal control law. We start with the terminal cost and constraint

$$
J_{N \to N}^*(x_N) = p(x_N) \tag{7.15}
$$
$$
\mathcal{X}_{N \to N} = \mathcal{X}_f, \tag{7.16}
$$

and then proceed backwards

$$
\begin{aligned}
J_{N-1 \to N}^*(x_{N-1}) = \min_{u_{N-1}} \quad & q(x_{N-1}, u_{N-1}) + J_{N \to N}^*(g(x_{N-1}, u_{N-1})) \\
\text{subj. to} \quad & h(x_{N-1}, u_{N-1}) \le 0, \\
& g(x_{N-1}, u_{N-1}) \in \mathcal{X}_{N \to N}
\end{aligned}
$$

$$
\vdots
$$
$$
\begin{aligned}
J_{0 \to N}^*(x_0) = \min_{u_0} \quad & q(x_0, u_0) + J_{1 \to N}^*(g(x_0, u_0)) \\
\text{subj. to} \quad & h(x_0, u_0) \le 0, \\
& g(x_0, u_0) \in \mathcal{X}_{1 \to N} \\
& x_0 = x(0).
\end{aligned}
\tag{7.17}
$$

This algorithm, popularized by Bellman, is referred to as *dynamic programming*. The dynamic programming problem is appealing because it can be stated compactly and because at each step the optimization takes place over one element u_j of the optimization vector only. This optimization is rather complex, however. It is not a standard nonlinear programming problem, since we have to construct the optimal cost-to-go $J_{j \to N}^*(x_j)$, a *function* defined over the subset $\mathcal{X}_{j \to N}$ of the state space.

In a few special cases we know the type of function and we can find it efficiently. For example, in the next chapter we will cover the case when the system is linear and the cost is quadratic. Then the optimal cost-to-go is also quadratic and can be constructed rather easily. Later in the book we will show that, when constraints are added to this problem, the optimal cost-to-go becomes piecewise quadratic and efficient algorithms for its construction are also available.

In general, however, we may have to resort to a "brute force" approach to construct the cost-to-go function $J_{j-1 \to N}^*$ and to solve the dynamic program. Let us assume that at time $j-1$ the cost-to-go $J_{j \to N}^*$ is known and discuss how to construct an approximation of $J_{j-1 \to N}^*$. With $J_{j \to N}^*$ known, for a fixed x_{j-1} the optimization problem (7.14) becomes a standard nonlinear programming problem. Thus, we can define a grid in the set $\mathcal{X}_{j-1 \to N}$ of the state space and compute the optimal cost-to-go function on each grid point. We can then define an approximate value function $\tilde{J}^*_{j-1 \to N}(x_{j-1})$ at intermediate points via interpolation. The complexity of constructing the cost-to-go function in this manner increases rapidly with the dimension of the state space ("curse of dimensionality").

The extra benefit of solving the optimal control problem via dynamic programming is that we do not only obtain the vector of optimal inputs $U^*_{0 \to N}$ for a particular initial state $x(0)$ as with the batch approach. At each time j the optimal cost-to-go function defines implicitly a nonlinear feedback control law.

$$u^*_j(x_j) = \arg \min_{u_j} \quad q(x_j, u_j) + J^*_{j+1 \to N}(g(x_j, u_j))$$
$$\text{subj. to} \quad h(x_j, u_j) \leq 0, \qquad\qquad (7.18)$$
$$g(x_j, u_j) \in \mathcal{X}_{j+1 \to N}.$$

For a fixed x_j this nonlinear programming problem can be solved quite easily in order to find $u^*_j(x_j)$. Because the optimal cost-to-go function $J^*_{j \to N}(x_j)$ changes with time j, the nonlinear feedback control law is time-varying.

7.4 Optimal Control Problem with Infinite Horizon

We are interested in the optimal control problem (7.3)–(7.5) as the horizon N approaches infinity.

$$J^*_{0 \to \infty}(x_0) = \min_{u_0, u_1, \dots} \quad \sum_{k=0}^{\infty} q(x_k, u_k)$$
$$\text{subj. to} \quad x_{k+1} = g(x_k, u_k), \ k = 0, \dots, \infty \qquad (7.19)$$
$$h(x_k, u_k) \leq 0, \ k = 0, \dots, \infty$$
$$x_0 = x(0).$$

We define the set of initial conditions for which this problem has a solution.

$$\mathcal{X}_{0 \to \infty} = \{x(0) \in \mathbb{R}^n : \text{ Problem (7.19) is feasible and } J^*_{0 \to \infty}(x(0)) < +\infty\}.$$
$$(7.20)$$

For the value function $J^*_{0 \to \infty}(x_0)$ to be finite it must hold that

$$\lim_{k \to \infty} q(x_k, u_k) = 0,$$

and because $q(x_k, u_k) > 0$ for all $(x_k, u_k) \neq 0$

$$\lim_{k \to \infty} x_k = 0$$

and

$$\lim_{k \to \infty} u_k = 0.$$

Thus the sequence of control actions generated by the solution of the infinite horizon problem drives the system to the origin. For this solution to exists the system must be, loosely speaking, stabilizable.

Using the recursive dynamic programming approach we can seek the solution of the infinite horizon optimal control problem by increasing N until we observe convergence. If the dynamic programming algorithm converges as $N \to \infty$ then (7.14) becomes the Bellman equation

$$
\begin{aligned}
J^*(x) = \min_u \quad & q(x,u) + J^*(g(x,u)) \\
\text{subj. to} \quad & h(x,u) \leq 0 \\
& g(x,u) \in \mathcal{X}_{0 \to \infty}
\end{aligned}
\tag{7.21}
$$

This procedure of simply increasing N may not be well behaved numerically and it may also be difficult to define a convergence criterion that is meaningful for the control problem. We will describe a method, called *Value Function Iteration*, in the next section.

An alternative is *receding horizon control* which can yield a time invariant controller guaranteeing convergence to the origin without requiring $N \to \infty$. We will describe this important idea later in this chapter.

7.4.1   Value Function Iteration

Once the value function $J^*(x)$ is known, the nonlinear feedback control law $u^*(x)$ is defined implicitly by (7.21)

$$
\begin{aligned}
u^*(x) = \arg\min_u \quad & q(x,u) + J^*(g(x,u)) \\
\text{subj. to} \quad & h(x,u) \leq 0 \\
& g(x,u) \subset \mathcal{X}_{0 \to \infty}.
\end{aligned}
\tag{7.22}
$$

It is *time invariant* and guarantees convergence to the origin for all states in $\mathcal{X}_{0 \to \infty}$. For a given $x \in \mathcal{X}_{0 \to \infty}$, $u^*(x)$ can be found from (7.21) by solving a standard nonlinear programming problem.

In order to find the value function $J^*(x)$ we need to solve (7.21). We can start with some initial guess $\tilde{J}_0^*(x)$ for the value function and an initial guess $\tilde{\mathcal{X}}_0$ for the region in the state space where we expect the infinite horizon problem to converge and iterate. Then at iteration $i+1$ solve

$$
\begin{aligned}
\tilde{J}_{i+1}^*(x) = \min_u \quad & q(x,u) + \tilde{J}_i^*(g(x,u)) \\
\text{subj. to} \quad & h(x,u) \leq 0 \\
& g(x,u) \in \tilde{\mathcal{X}}_i
\end{aligned}
\tag{7.23}
$$

$$
\tilde{\mathcal{X}}_{i+1} = \{x \in \mathbb{R}^n : \ \exists u \ (h(x,u) \leq 0, \text{ and } g(x,u) \in \tilde{\mathcal{X}}_i)\}.
\tag{7.24}
$$

Again, here i is the iteration index and does not denote time. This iterative procedure is called *value function iteration*. It can be executed as follows. Let us assume that at iteration step i we gridded the set $\tilde{\mathcal{X}}_i$ and that $\tilde{J}_i^*(x)$ is known at each grind point from the previous iteration. We can approximate $\tilde{J}_i^*(x)$ at intermediate points via interpolation. For a fixed point $\bar{x}$ the optimization problem (7.23) is a nonlinear programming problem yielding $\tilde{J}_i^*(\bar{x})$. In this manner the approximate value function $\tilde{J}_i^*(x)$ can be constructed at all grid points and we can proceed to the next iteration step $i+1$.

7.4.2   Receding Horizon Control

Receding Horizon Control will be covered in detail in Chapter 12. Here we illustrate the main idea and discuss the fundamental properties.

Assume that at time $t = 0$ we determine the control action u_0 by solving the finite horizon optimal control problem (7.3)–(7.5). If $J^*_{0 \to N}(x_0)$ converges to $J^*_{0 \to \infty}(x_0)$ as $N \to \infty$ then the effect of increasing N on the value of u_0 should diminish as $N \to \infty$. Thus, intuitively, instead of making the horizon infinite we can get a similar behavior when we use a long, but finite horizon N, and repeat this optimization at each time step, in effect moving the horizon forward (*moving horizon* or *receding horizon* control). We can use the batch or the dynamic programming approach.

Batch approach. We solve an optimal control problem with horizon N yielding a sequence of optimal inputs $u^*_0, \ldots, u^*_{N-1}$, but we would implement only the first one of these inputs u^*_0. At the next time step we would measure the current state and then again solve the N-step problem with the current state as new initial condition x_0. If the horizon N is long enough then we expect that this approximation of the infinite horizon problem should not matter and the implemented sequence should drive the states to the origin.

Dynamic programming approach. We always implement the control u_0 obtained from the optimization problem

$$
\begin{aligned}
J^*_{0 \to N}(x_0) = \min_{u_0} \quad & q(x_0, u_0) + J^*_{1 \to N}(g(x_0, u_0)) \\
\text{subj. to} \quad & h(x_0, u_0) \leq 0, \\
& g(x_0, u_0) \in \mathcal{X}_{1 \to N}, \\
& x_0 = x(0)
\end{aligned}
\tag{7.25}
$$

where $J^*_{1 \to N}(g(x_0, u_0))$ is the optimal cost-to-go from the state $x_1 = g(x_0, u_0)$ at time 1 to the end of the horizon N.

If the dynamic programming iterations converge as $N \to \infty$, then for a long, but finite horizon N we expect that this receding horizon approximation of the infinite horizon problem should not matter and the resulting controller will drive the system asymptotically to the origin.

In both the batch and the recursive approach, however, it is not obvious how long N must be for the receding horizon controller to inherit these desirable convergence characteristics. Indeed, for computational simplicity we would like to keep N small. We will argue next that the proposed control scheme guarantees convergence just like the infinite horizon variety if we impose a specific terminal constraint, for example, if we require the terminal region to be the origin $\mathcal{X}_f = 0$.

From the principle of optimality we know that

$$
J^*_{0 \to N}(x_0) = \min_{u_0} \quad q(x_0, u_0) + J^*_{1 \to N}(x_1).
\tag{7.26}
$$

Assume that we are at $x(0)$ at time 0 and implement the optimal u^*_0 that takes us to the next state $x_1 = g(x(0), u^*_0)$. At this state at time 1 we postulate to use over the next N steps the sequence of optimal moves determined at the previous step followed by zero: $u^*_1, \ldots, u^*_{N-1}, 0$. This sequence is not optimal but the associated cost over the shifted horizon from 1 to $N + 1$ can be easily determined. It consists of three parts: (1) the optimal cost $J_{0 \to N}^*(x_0)$ from time 0 to N computed at time 0, minus (2) the stage cost $q(x_0, u_0)$ at time 0 plus (3) the cost at time $N + 1$. But this last cost is zero because we imposed the terminal constraint $x_N = 0$ and assumed $u_N = 0$. Thus the cost over the shifted horizon for the assumed sequence of control moves is

$$J_{0 \to N}^*(x_0) - q(x_0, u_0).$$

Because this postulated sequence of inputs is not optimal at time 1

$$J_{1 \to N+1}^*(x_1) \leq J_{0 \to N}^*(x_0) - q(x_0, u_0).$$

Because the system and the objective are time invariant $J_{1 \to N+1}^*(x_1) = J_{0 \to N}^*(x_1)$ so that

$$J_{0 \to N}^*(x_1) \leq J_{0 \to N}^*(x_0) - q(x_0, u_0).$$

As $q \succ 0$ for all $(x, u) \neq (0, 0)$, the sequence of optimal costs $J_{0 \to N}^*(x_0)$, $J_{0 \to N}^*(x_1), \ldots$ is strictly decreasing for all $(x, u) \neq (0, 0)$. Because the cost $J_{0 \to N}^* \geq 0$ the sequence $J_{0 \to N}^*(x_0), J_{0 \to N}^*(x_1), \ldots$ (and thus the sequence $x_0, x_1, \ldots$) is converging. Thus we have established the following important theorem.

Theorem 7.1 (Convergence of Receding Horizon Control) *At time step j consider the cost function*

$$J_{j \to j+N}(x_j, u_j, u_{j+1}, \ldots, u_{j+N-1}) = \sum_{k=j}^{j+N} q(x_k, u_k), \quad q \succ 0 \qquad (7.27)$$

and the CFTOC problem

$$\begin{aligned}
J_{j \to j+N}^*(x_j) = \min_{u_j, u_{j+1}, \ldots, u_{j+N-1}} \quad & J_{j \to j+N}(x_j, u_j, u_{j+1}, \ldots, u_{j+N-1}) \\
\text{subj. to} \quad & x_{k+1} = g(x_k, u_k) \\
& h(x_k, u_k) \leq 0, \quad k = j, \ldots, j + N - 1 \\
& x_N = 0
\end{aligned}$$

$$(7.28)$$

Assume that only the optimal u_j^* is implemented. At the next time step $j + 1$ the CFTOC problem is solved again starting from the resulting state $x_{j+1} = g(x_j, u_j^*)$. Assume that the CFTOC problem (7.28) has a solution for every state $x_j, x_{j+1}, \ldots$ resulting from the control policy. Then the system will converge to the origin as $j \to \infty$. ∎

Thus we have established that a receding horizon controller with terminal constraint $x_N = 0$ has the same desirable convergence characteristics as the infinite horizon controller. At first sight the theorem appears very general and powerful. It is based on the implicit assumption, however, that at every time step the CFTOC problem has a solution. Infeasibility would occur, for example, if the underlying system is not stabilizable. It could also happen that the constraints on the inputs which restrict the control action prevent the system from reaching the terminal state in N steps. In Chapter 12 we will present special formulations of problem (7.28) such that feasibility at the initial time guarantees feasibility for all future times. Furthermore, in addition to asymptotic convergence to the origin we will establish stability for the closed-loop system with the receding horizon controller.

Remark 7.1   For the sake of simplicity in the rest of the book we will use the following shorter notation

$$J_j^*(x_j) = J_{j \to N}^*(x_j), \ j = 0, \ldots, N$$
$$J_\infty^*(x_0) = J_{0 \to \infty}^*(x_0)$$
$$\mathcal{X}_j = \mathcal{X}_{j \to N}, \ j = 0, \ldots, N \tag{7.29}$$
$$\mathcal{X}_\infty = \mathcal{X}_{0 \to \infty}$$
$$U_0 = U_{0 \to N}$$

and use the original notation only if needed.

7.5 Lyapunov Stability

While asymptotic convergence $\lim_{k \to \infty} x_k = 0$ is a desirable property, it is generally not sufficient in practice. We would also like a system to stay in a small neighborhood of the origin when it is disturbed slightly. Formally, this is expressed as Lyapunov stability.

7.5.1 General Stability Conditions

Consider the autonomous system

$$x_{k+1} = g(x_k) \tag{7.30}$$

with $g(0) = 0$.

Definition 7.1 (Lyapunov Stability)   *The equilibrium point $x = 0$ of system (7.30) is*

- stable *(in the sense of Lyapunov) if, for each $\varepsilon > 0$, there is $\delta > 0$ such that*

$$\|x_0\| < \delta \Rightarrow \|x_k\| < \varepsilon, \ \forall k \geq 0 \tag{7.31}$$

- unstable *if not stable*

- asymptotically stable *in $\Omega \subseteq \mathbb{R}^n$ if it is stable and*

$$\lim_{k \to \infty} x_k = 0, \ \forall x_0 \in \Omega \tag{7.32}$$

- globally asymptotically stable *if it is asymptotically stable and $\Omega = \mathbb{R}^n$*

- exponentially stable *if it is stable and there exist constants $\alpha > 0$ and $\gamma \in (0, 1)$ such that*

$$\|x_0\| < \delta \Rightarrow \|x_k\| \leq \alpha \|x_0\| \gamma^k, \ \forall k \geq 0. \tag{7.33}$$

The ε-δ requirement for stability (7.31) takes a challenge–answer form. To demonstrate that the origin is stable, for any value of ε that a challenger may chose (however small), we must produce a value of δ such that a trajectory starting in a δ neighborhood of the origin will never leave the ε neighborhood of the origin.

Remark 7.2   If in place of system (7.30), we consider the time-varying system $x_{k+1} = g(x_k, k)$, then δ in Definition 7.1 is a function of ε and k, i.e., $\delta = \delta(\varepsilon, k) > 0$. In this case, we introduce the concept of "uniform stability." The equilibrium point $x = 0$ is *uniformly stable* if, for each $\varepsilon > 0$, there is $\delta = \delta(\varepsilon) > 0$ (independent from k) such that

$$\|x_0\| < \delta \Rightarrow \|x_k\| < \varepsilon, \ \forall k \geq 0. \tag{7.34}$$

The following example shows that Lyapunov stability and convergence are, in general, different properties.

Example 7.1   Consider the following system with one state $x \in \mathbb{R}$:

$$x_{k+1} = x_k(x_k - 1 - |x_k - 1|). \tag{7.35}$$

The state $x = 0$ is an equilibrium for the system. For any state $x \in [-1, 1]$ we have $(x - 1) \leq 0$ and the system dynamics (7.35) become

$$x_{k+1} = 2x_k^2 - 2x_k. \tag{7.36}$$

System (7.36) generates oscillating and diverging trajectories for any $x_0 \in (-1, 1) \setminus \{0\}$. Any such trajectory will enter in finite time T the region with $x \geq 1$. In this region the system dynamics (7.35) become

$$x_{k+1} = 0, \ \forall \, k \geq T. \tag{7.37}$$

Therefore the origin is not Lyapunov stable, however the system converges to the origin for all $x_0 \in (-\infty, +\infty)$.

Usually, to show Lyapunov stability of the origin for a particular system one constructs a so called *Lyapunov function*, i.e., a function satisfying the conditions of the following theorem.

Theorem 7.2   *Consider the equilibrium point $x = 0$ of system (7.30). Let $\Omega \subset \mathbb{R}^n$ be a closed and bounded set containing the origin. Assume there exists a function $V : \mathbb{R}^n \to \mathbb{R}$ continuous at the origin, finite for every $x \in \Omega$, and such that*

$$V(0) = 0 \ and \ V(x) > 0, \ \forall x \in \Omega \setminus \{0\} \tag{7.38a}$$

$$V(x_{k+1}) - V(x_k) \leq -\alpha(x_k) \ \forall x_k \in \Omega \setminus \{0\} \tag{7.38b}$$

where $\alpha : \mathbb{R}^n \to \mathbb{R}$ is a continuous positive definite function. Then $x = 0$ is asymptotically stable in Ω.

Definition 7.2   *A function $V(x)$ satisfying conditions (7.38a)–(7.38b) is called a* Lyapunov Function.

The main idea of Theorem 7.2 can be explained as follows. We aim to find a scalar function $V(x)$ that captures qualitative characteristics of the system response, and, in particular, its stability. We can think of V as an energy function that is zero at the origin and positive elsewhere (condition (7.38a)). Condition (7.38b) of Theorem 7.2 requires that for any state $x_k \in \Omega$, $x_k \neq 0$ the energy decreases as the system evolves to x_{k+1}.

A proof of Theorem 7.2 for continuous dynamical systems can be found in [184]. The continuity assumptions on the dynamical system g is not used in the proof in [247, p. 609]. Assumption in equation (7.38b), however, together with the continuity of $V(\cdot)$ at the origin, implies that $g(\cdot)$ must be continuous at the origin.

Theorem 7.2 states that if we find an energy function which satisfies the two conditions (7.38a)–(7.38b), then the system states starting from any initial state $x_0 \in \Omega$ will eventually settle to the origin.

Note that Theorem 7.2 is only sufficient. If condition (7.38b) is not satisfied for a particular choice of V nothing can be said about stability of the origin. Condition (7.38b) of Theorem 7.2 can be relaxed to allow α to be a continuous positive semi-definite (psd) function:

$$V(x_{k+1}) - V(x_k) \leq -\alpha(x_k), \ \forall x_k \neq 0, \ \alpha \text{ continuous and psd.} \tag{7.39}$$

Condition (7.39) along with condition (7.38a) are sufficient to guarantee stability of the origin as long as the set $\{x_k : V(g(x_k)) - V(x_k) = 0\}$ contains no trajectory of the system $x_{k+1} = g(x_k)$ except for $x_k = 0$ for all $k \geq 0$. This relaxation of Theorem 7.2 is the so called Barbashin-Krasovski-LaSalle principle [183]. It basically means that $V(x_k)$ may stay constant and non zero at one or more time instants as long as it does not do so at an equilibrium point or periodic orbit of the system.

A similar result as Theorem 7.2 can be derived for *global* asymptotic stability, i.e., $\Omega = \mathbb{R}^n$.

Theorem 7.3   *Consider the equilibrium point $x = 0$ of system (7.30). Assume there exists a function $V : \mathbb{R}^n \to \mathbb{R}$ continuous at the origin, finite for every $x \in \mathbb{R}^n$, and such that*

$$\|x\| \to \infty \Rightarrow V(x) \to \infty \tag{7.40a}$$

$$V(0) = 0 \text{ and } V(x) > 0, \ \forall x \neq 0 \tag{7.40b}$$

$$V(x_{k+1}) - V(x_k) \leq -\alpha(x_k) \ \forall x_k \neq 0 \tag{7.40c}$$

where $\alpha : \mathbb{R}^n \to \mathbb{R}$ is a continuous positive definite function. Then $x = 0$ is globally asymptotically stable. ∎

Definition 7.3   *A function $V(x)$ satisfying condition (7.40a) is said to be* radially unbounded.

Definition 7.4   *A radially unbounded Lyapunov function is called a* Global Lyapunov function.

Note that it was not enough just to restate Theorem 7.2 with $\Omega = \mathbb{R}^n$ but we also have to require $V(x)$ to be radially unbounded to guarantee global asymptotic stability. To motivate this condition consider the candidate Lyapunov function for a system in $\mathbb{R}^2$[176]

$$V(x) = \frac{x_1^2}{1 + x_1^2} + x_2^2, \tag{7.41}$$

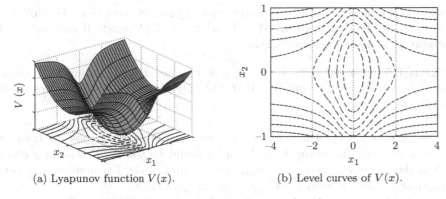

(a) Lyapunov function $V(x)$.   (b) Level curves of $V(x)$.

Figure 7.1 Lyapunov function (7.41).

which is depicted in Figure 7.1, where x_1 and x_2 denote the first and second components of the state vector x, respectively. $V(x)$ in (7.41) is not radially unbounded as for $x_2 = 0$

$$\lim_{x_1 \to \infty} V(x) = 1.$$

For this Lyapunov function even if condition (7.40c) is satisfied, the state x may escape to infinity. Condition (7.40c) of Theorem 7.3 guarantees that the level sets Ω_c of $V(x)$ ($\Omega_c = \{x \in \mathbb{R}^n : V(x) \le c\}$) are closed.

The construction of suitable Lyapunov functions is a challenge except for linear systems. First of all one can quite easily show that for linear systems Lyapunov stability agrees with the notion of stability based on eigenvalue location.

Theorem 7.4 *A linear system $x_{k+1} = Ax_k$ is globally asymptotically stable in the sense of Lyapunov if and only if all its eigenvalues are strictly inside the unit circle.*

We also note that stability is always "global" for linear systems.

7.5.2 Quadratic Lyapunov Functions for Linear Systems

A simple effective Lyapunov function for linear systems is

$$V(x) = x'Px, \; P \succ 0 \tag{7.42}$$

which satisfies conditions (7.40a)–(7.40b) of Theorem 7.3. In order to test condition (7.40c) we compute

$$V(x_{k+1}) - V(x_k) = x'_{k+1}Px_{k+1} - x'_kPx_k = x'_kA'PAx_k - x'_kPx_k = x'_k(A'PA - P)x_k. \tag{7.43}$$

Therefore condition (7.40c) is satisfied if $P \succ 0$ can be found such that

$$A'PA - P = -Q, \; Q \succ 0. \tag{7.44}$$

Equation (7.44) is referred to as discrete-time Lyapunov equation. The following Theorem [75, p. 211] shows that $P \succ 0$ satisfying (7.44) exists if and only if the linear system is asymptotically stable.

Theorem 7.5 *Consider the linear system $x_{k+1} = Ax_k$. Equation (7.44) has a unique solution $P \succ 0$ for any $Q \succ 0$ if and only if A has all eigenvalues strictly inside the unit circle.*

Thus, a quadratic form $x'Px$ is always a suitable Lyapunov function for linear systems and an appropriate P can be found by solving (7.44) for a chosen $Q \succ 0$ iff the system's eigenvalues lie inside the unit circle. For nonlinear systems, determining a suitable form for $V(x)$ is generally difficult.

For a stable linear system $x_{k+1} = Ax_k$, P turns out to be the infinite time cost matrix

$$J_\infty(x_0) = \sum_{k=0}^{\infty} x'_k Q x_k = x'_0 P x_0 \tag{7.45}$$

as we can easily show. From

$$J_\infty(x_1) - J_\infty(x_0) = x'_1 P x_1 - x'_0 P x_0 = x'_0 A' P A x_0 - x'_0 P x_0 = -x'_0 Q x_0 \tag{7.46}$$

we recognize that P is the solution of the Lyapunov equation (7.44). In other words the infinite time cost (7.45) is a Lyapunov function for the linear system $x_{k+1} = Ax_k$.

The conditions of Theorem 7.5 can be relaxed as follows.

Theorem 7.6 *Consider the linear system $x_{k+1} = Ax_k$. Equation (7.44) has a unique solution $P \succ 0$ for any $Q = C'C \succeq 0$ if and only if A has all eigenvalues inside the unit circle and (C, A) is observable.*

We can prove Theorem 7.6 in the same way as Callier and Desoer [75, p. 211] proved Theorem 7.5. In Theorem 7.6 we do not require that the Lyapunov function decreases at every time step, i.e., we allow Q to be positive *semi*definite. To understand this, let us assume that for a particular system state $\bar{x}$, V does not decrease, i.e., $\bar{x}'Q\bar{x} = (C\bar{x})'(C\bar{x}) = 0$. Then at the next time steps we have the rate of decrease $(CA\bar{x})'(CA\bar{x}), (CA^2\bar{x})'(CA^2\bar{x}), \ldots$. If the system (C, A) is observable, then for all $\bar{x} \neq 0$

$$\bar{x}' \left[C \; (CA)' \; (CA^2)' \cdots (CA^{n-1})' \right] \neq 0, \tag{7.47}$$

which implies that after at most $(n-1)$ steps the rate of decrease will become nonzero. This is a special case of the Barbashin-Krasovski-LaSalle principle. Note that for C square and nonsingular Theorem 7.6 reduces to Theorem 7.5.

Similarly, we can analyze the controlled system $x_{k+1} = Ax_k + Bu_k$ with $u_k = Fx_k$ and the infinite time cost

$$J_\infty(x_0) = \sum_{k=0}^{\infty} x'_k Q x_k + u'_k R u_k \tag{7.48}$$

with $Q = C'C$ and $R = D'D$ with $\det(D) \neq 0$. We can rewrite the cost as

$$J_\infty(x_0) = \sum_{k=0}^{\infty} x_k'(Q + F'RF)x_k = \sum_{k=0}^{\infty} x_k' \begin{bmatrix} C \\ DF \end{bmatrix}' \begin{bmatrix} C & DF \end{bmatrix} x_k \qquad (7.49)$$

for the controlled system $x_{k+1} = (A + BF)x_k$. The infinite time cost matrix P can now be found from the Lyapunov equation

$$(A + BF)'P(A + BF) - P = \begin{bmatrix} C \\ DF \end{bmatrix}' \begin{bmatrix} C & DF \end{bmatrix}. \qquad (7.50)$$

According to Theorem 7.6 the solution P is unique and positive definite iff $(A+BF)$ is stable and $\left[\begin{pmatrix} C \\ DF \end{pmatrix}, (A + BF) \right]$ is observable. This follows directly from the observability of (C, A). If (C, A) is observable, then so is $(C, A + BF)$ because feedback does not affect observability. Observability is also not affected by adding the observed outputs DFx.

From (7.44) it follows that for stable systems and for a chosen $Q \succ 0$ one can always find $P \succ 0$ solving

$$A'PA - P + Q \preceq 0. \qquad (7.51)$$

This *Lyapunov inequality* shows that for a stable system we can always find a P such that $V(x) = x'Px$ decreases at a desired "rate" indicated by Q. We will need this result later to prove stability of receding horizon control schemes.

7.5.3 $1/\infty$ Norm Lyapunov Functions for Linear Systems

For $p = \{1, \infty\}$ the function

$$V(x) = \|Px\|_p$$

with $P \in \mathbb{R}^{l \times n}$ of full column rank satisfies the requirements (7.40a), (7.40b) of a Lyapunov function. It can be shown that a matrix P can be found such that condition (7.40c) is satisfied for the system $x_{k+1} = Ax_k$ if and only if the eigenvalues of A are inside the unit circle. The number of rows l necessary in P depends on the system. The techniques to construct P are based on the following theorem [177, 237].

Theorem 7.7   *Let $P \in \mathbb{R}^{l \times n}$ with $\text{rank}(P) = n$ and $p \in \{1, \infty\}$. The function*

$$V(x) = \|Px\|_p \qquad (7.52)$$

is a Lyapunov function for the discrete-time system

$$x_{k+1} = Ax_k, \qquad k \geq 0, \qquad (7.53)$$

if and only if there exists a matrix $H \in \mathbb{R}^{l \times l}$, such that

$$PA = HP, \qquad (7.54a)$$

$$\|H\|_p < 1. \qquad (7.54b)$$

An effective method to find both H and P was proposed by Christophersen and Morari in [88].

To prove the stability of receding horizon control, later in this book, we will need to find a $\tilde{P}$ such that

$$\|\tilde{P}Ax\|_\infty - \|\tilde{P}x\|_\infty + \|Qx\|_\infty \le 0, \ \forall x \in \mathbb{R}^n. \tag{7.55}$$

Note that the inequality (7.55) is equivalent to the Lyapunov inequality (7.51) when the squared two-norm is replaced by the $1-$ or $\infty-$norm. Once we have constructed a P and H to fulfill the conditions of Theorem 7.7 we can easily find $\tilde{P}$ to satisfy (7.55) according to the following lemma.

Lemma 7.1   *Let P and H be matrices satisfying conditions (7.54), with P full column rank. Let $\sigma = 1 - \|H\|_\infty$, $\rho = \|QP^{\#}\|_\infty$, where $P^{\#} = (P'P)^{-1}P'$ is the left pseudoinverse of P. Then, the square matrix*

$$\tilde{P} = \frac{\rho}{\sigma}P \tag{7.56}$$

satisfies condition (7.55).

Proof: Since $\tilde{P}$ satisfies $\tilde{P}A = H\tilde{P}$, we obtain $-\|\tilde{P}x\|_\infty + \|\tilde{P}Ax\|_\infty + \|Qx\|_\infty = -\|\tilde{P}x\|_\infty + \|H\tilde{P}x\|_\infty + \|Qx\|_\infty \le (\|H\|_\infty - 1)\|\tilde{P}x\|_\infty + \|Qx\|_\infty \le (\|H\|_\infty - 1)\|\tilde{P}x\|_\infty + \|QP^{\#}\|_\infty\|Px\|_\infty = 0$. Therefore, (7.55) is satisfied.   ∎

8

Linear Quadratic Optimal Control

In this chapter we study the finite time and infinite time optimal control problem for unconstrained linear systems with quadratic objective functions. We derive the structure of the optimal control law by using two approaches: the batch approach and the dynamic programming approach. For problems with quadratic objective functions we obtain the well-known Algebraic Riccati Equations.

8.1 Problem Formulation

We consider a special case of the problem stated in the last chapter, where the system is linear and time-invariant

$$x(t+1) = Ax(t) + Bu(t). \tag{8.1}$$

Again, $x(t) \in \mathbb{R}^n$ and $u(t) \in \mathbb{R}^m$ are the state and input vectors respectively.

We define the following quadratic cost function over a finite horizon of N steps

$$J_0(x_0, U_0) = x_N' P x_N + \sum_{k=0}^{N-1} x_k' Q x_k + u_k' R u_k, \tag{8.2}$$

where x_k denotes the state vector at time k obtained by starting from the state $x_0 = x(0)$ and applying to the system model

$$x_{k+1} = Ax_k + Bu_k \tag{8.3}$$

the input sequence $u_0, \ldots, u_{k-1}$. Consider the finite time optimal control problem

$$
\begin{aligned}
J_0^*(x(0)) = \min_{U_0} \quad & J_0(x(0), U_0) \\
\text{subj. to} \quad & x_{k+1} = Ax_k + Bu_k, \ k = 0, 1, \ldots, N-1 \\
& x_0 = x(0).
\end{aligned} \tag{8.4}
$$

In (8.4) $U_0 = [u_0', \ldots, u_{N-1}']' \in \mathbb{R}^s$, $s = mN$ is the decision vector containing all future inputs. We will assume that the state penalty is positive semi-definite $Q = Q' \succeq 0$, $P = P' \succeq 0$ and the input penalty is positive definite $R = R' \succ 0$.

As introduced in the previous chapter we will present two alternate approaches to solve problem (8.4), the batch approach and the recursive approach using dynamic programming.

8.2 Solution via Batch Approach

First we write the equality constraints (8.4) explicitly to express all future states $x_1, x_2, \ldots$ as a function of the future inputs $u_0, u_1, \ldots$ and then we eliminate all intermediate states by successive substitution to obtain

$$
\underbrace{\begin{bmatrix} x(0) \\ x_1 \\ \vdots \\ \vdots \\ x_N \end{bmatrix}}_{\mathcal{X}} = \underbrace{\begin{bmatrix} I \\ A \\ \vdots \\ \vdots \\ A^N \end{bmatrix}}_{\mathcal{S}^x} x(0) + \underbrace{\begin{bmatrix} 0 & \cdots & \cdots & 0 \\ B & 0 & \cdots & 0 \\ AB & \ddots & \ddots & \vdots \\ \vdots & \ddots & \ddots & \vdots \\ A^{N-1}B & \cdots & \cdots & B \end{bmatrix}}_{\mathcal{S}^u} \begin{bmatrix} u_0 \\ \vdots \\ \vdots \\ u_{N-1} \end{bmatrix}. \tag{8.5}
$$

Here all future states are explicit functions of the present state $x(0)$ and the future inputs $u_0, u_1, u_2, \ldots$ only. By defining the appropriate quantities we can rewrite this expression compactly as

$$
\mathcal{X} = \mathcal{S}^x x(0) + \mathcal{S}^u U_0. \tag{8.6}
$$

Using the same notation the objective function can be rewritten as

$$
J(x(0), U_0) = \mathcal{X}' \bar{Q} \mathcal{X} + U_0' \bar{R} U_0, \tag{8.7}
$$

where $\bar{Q} = \text{blockdiag}\{Q, \ldots, Q, P\}, \bar{Q} \succeq 0$, and $\bar{R} = \text{blockdiag}\{R, \ldots, R\}, \bar{R} \succ 0$. Substituting (8.6) into the objective function (8.7) yields

$$
\begin{aligned}
J_0(x(0), U_0) &= (\mathcal{S}^x x(0) + \mathcal{S}^u U_0)' \bar{Q} (\mathcal{S}^x x(0) + \mathcal{S}^u U_0) + U_0' \bar{R} U_0 \\
&= U_0' \underbrace{(\mathcal{S}^{u'} \bar{Q} \mathcal{S}^u + \bar{R})}_{H} U_0 + 2 x'(0) \underbrace{(\mathcal{S}^{x'} \bar{Q} \mathcal{S}^u)}_{F} U_0 + x'(0) \underbrace{(\mathcal{S}^{x'} \bar{Q} \mathcal{S}^x)}_{Y} x(0) \\
&= U_0' H U_0 + 2 x'(0) F U_0 + x'(0) Y x(0).
\end{aligned} \tag{8.8}
$$

Because $\bar{R} \succ 0$, also $H \succ 0$. Thus $J_0(x(0), U_0)$ is a positive definite quadratic function of U_0. Therefore, its minimum can be found by computing its gradient and setting it to zero. This yields the optimal vector of future inputs

$$
\begin{aligned}
U_0^*(x(0)) &= -H^{-1} F' x(0) \\
&= -\left(\mathcal{S}^{u'} \bar{Q} \mathcal{S}^u + \bar{R}\right)^{-1} \mathcal{S}^{u'} \bar{Q} \mathcal{S}^x x(0).
\end{aligned} \tag{8.9}
$$

With this choice of U_0 the optimal cost is

$$
\begin{aligned}
J_0^*(x(0)) &= -x(0)'FH^{-1}F'x(0) + x(0)'Yx(0) \\
&= x(0)'\left[\mathcal{S}^{x'}\bar{Q}\mathcal{S}^x - \mathcal{S}^{x'}\bar{Q}\mathcal{S}^u\left(\mathcal{S}^{u'}\bar{Q}\mathcal{S}^u + \bar{R}\right)^{-1}\mathcal{S}^{u'}\bar{Q}\mathcal{S}^x\right]x(0).
\end{aligned}
\tag{8.10}
$$

Note that the optimal vector of future inputs $U_0^*(x(0))$ is a linear function (8.9) of the initial state $x(0)$ and the optimal cost $J_0^*(x(0))$ is a quadratic function (8.10) of the initial state $x(0)$.

8.3 Solution via Recursive Approach

Alternatively, we can use dynamic programming to solve the same problem in a recursive manner. We define the optimal cost $J_j^*(x_j)$ for the $N-j$ step problem starting from state x_j by

$$
J_j^*(x_j) \triangleq \min_{u_j,\dots,u_{N-1}} x_N'Px_N + \sum_{k=j}^{N-1} x_k'Qx_k + u_k'Ru_k.
\tag{8.11}
$$

According to the principle of optimality, the optimal one-step cost-to-go can be obtained from

$$
J_{N-1}^*(x_{N-1}) = \min_{u_{N-1}} x_N'P_Nx_N + x_{N-1}'Qx_{N-1} + u_{N-1}'Ru_{N-1}
\tag{8.12}
$$

$$
\text{subj. to} \quad
\begin{aligned}
x_N &= Ax_{N-1} + Bu_{N-1} \\
P_N &= P.
\end{aligned}
\tag{8.13}
$$

Substituting (8.13) into the objective function (8.12),

$$
\begin{aligned}
J_{N-1}^*(x_{N-1}) = \min_{u_{N-1}} \{ & x_{N-1}'(A'P_NA + Q)x_{N-1} \\
&+ 2x_{N-1}'A'P_NBu_{N-1} \\
&+ u_{N-1}'(B'P_NB + R)u_{N-1} \}.
\end{aligned}
\tag{8.14}
$$

We note that the cost-to-go $J_{N-1}(x_{N-1})$ is a positive definite quadratic function of the decision variable u_{N-1}. We find the optimum by setting the gradient to zero and obtain the optimal input

$$
u_{N-1}^* = \underbrace{-(B'P_NB + R)^{-1}B'P_NA}_{F_{N-1}}x_{N-1}
\tag{8.15}
$$

and the one-step optimal cost-to-go

$$
J_{N-1}^*(x_{N-1}) = x_{N-1}'P_{N-1}x_{N-1},
\tag{8.16}
$$

where we have defined

$$
P_{N-1} = A'P_NA + Q - A'P_NB(B'P_NB + R)^{-1}B'P_NA.
\tag{8.17}
$$

At the next stage, consider the two-step problem from time $N-2$ forward:

$$J_{N-2}^*(x_{N-2}) = \min_{u_{N-2}} x_{N-1}' P_{N-1} x_{N-1} + x_{N-2}' Q x_{N-2} + u_{N-2}' R u_{N-2} \qquad (8.18)$$

$$x_{N-1} = Ax_{N-2} + Bu_{N-2}. \qquad (8.19)$$

We recognize that (8.18), (8.19) has the same form as (8.12), (8.13). Therefore, we can state the optimal solution directly.

$$u_{N-2}^* = \underbrace{-(B'P_{N-1}B + R)^{-1}B'P_{N-1}A}_{F_{N-2}} x_{N-2}. \qquad (8.20)$$

The optimal two-step cost-to-go is

$$J_{N-2}^*(x_{N-2}) = x_{N-2}' P_{N-2} x_{N-2}, \qquad (8.21)$$

where we defined

$$P_{N-2} = A'P_{N-1}A + Q - A'P_{N-1}B(B'P_{N-1}B + R)^{-1}B'P_{N-1}A. \qquad (8.22)$$

Continuing in this manner, at some arbitrary time k the optimal control action is

$$\begin{aligned} u^*(k) &= -(B'P_{k+1}B + R)^{-1}B'P_{k+1}Ax(k), \\ &= F_k x(k), \qquad \text{for } k = 0, \dots, N-1, \end{aligned} \qquad (8.23)$$

where

$$P_k = A'P_{k+1}A + Q - A'P_{k+1}B(B'P_{k+1}B + R)^{-1}B'P_{k+1}A \qquad (8.24)$$

and the optimal cost-to-go starting from the measured state $x(k)$ is

$$J_k^*(x(k)) = x'(k)P_k x(k). \qquad (8.25)$$

Equation (8.24) (called *Discrete Time Riccati Equation* or *Riccati Difference Equation* – RDE) is initialized with $P_N = P$ and is solved backwards, i.e., starting with P_N and solving for P_{N-1}, etc. Note from (8.23) that the optimal control action $u^*(k)$ is obtained in the form of a feedback law as a linear function of the measured state $x(k)$ at time k. The optimal cost-to-go (8.25) is found to be a quadratic function of the state at time k.

Remark 8.1   According to Section 7.4.2, the receding horizon control policy is obtained by solving problem (8.4) at each time step t with $x_0 = x(t)$. Consider the state feedback solution $u^*(k)$ in (8.23) to problem (8.4). Then, the receding horizon control policy is:

$$u^*(t) = F_0 x(t), \quad t \geq 0 \qquad (8.26)$$

8.4   Comparison of the Two Approaches

We will compare the batch and the recursive dynamic programming approach in terms of the results and the methods used to obtain the results.

Most importantly we observe that the results obtained by the two methods are fundamentally different. The batch approach yields a formula for the *sequence of inputs* as a function of the initial state.

$$U_0^* = -\left(\mathcal{S}^{u\prime}\bar{Q}\mathcal{S}^u + \bar{R}\right)^{-1}\mathcal{S}^{u\prime}\bar{Q}\mathcal{S}^x x(0). \tag{8.27}$$

The recursive dynamic programming approach yields a feedback policy, i.e., a *sequence of feedback laws* expressing at each time step the control action as a function of the state at that time.

$$u^*(k) = F_k x(k), \text{ for } k = 0, \dots, N-1. \tag{8.28}$$

As this expression implies, we determine $u(k)$ at each time k as a function of the current state $x(k)$ rather than use a $u(k)$ precomputed at $k = 0$ as in the batch method. If the state evolves exactly according to the linear model (8.3) then the sequence of control actions $u(k)$ obtained from the two approaches is identical. In practice, the result of applying the sequence (8.27) in an open-loop fashion may be rather different from applying the time-varying feedback law (8.28) because the model (8.1) for predicting the system states may be inaccurate and the system may be subject to disturbances not included in the model. We expect the application of the feedback law to be more robust because at each time step the *observed* state $x(k)$ is used to determine the control action rather than the state x_k *predicted* at time $t = 0$.

We note that we can get the same feedback effect with the batch approach if we recalculate the optimal open-loop sequence at each time step j with the current measurement as initial condition. In this case we need to solve the following optimization problem

$$J_j^*(x(j)) = \min_{u_j,\dots,u_{N-1}} \quad x_N' P x_N + \sum_{k=j}^{N-1} x_k' Q x_k + u_k' R u_k \tag{8.29}$$
$$\text{subj. to} \qquad x_j = x(j),$$

where we note that the horizon length is shrinking at each time step.

As seen from (8.27) the solution to (8.29) relates the sequence of inputs $u_j^*, u_{j+1}^*, \dots$ to the state $x(j)$ through a linear expression. The first part of this expression yields again the optimal feedback law (8.28) at time j, $u^*(j) = F_j x(j)$.

Here the dynamic programming approach is clearly a more efficient way to generate the feedback policy because it only uses a simple matrix recursion (8.24). Repeated application of the batch approach, on the other hand, requires the repeated inversion of a potentially large matrix in (8.27). For such inversion, however, one can take advantage of the fact that only a small part of the matrix H changes at every time step.

What makes dynamic programming so effective here is that in this special case, where the system is linear and the objective is quadratic, the optimal cost-to-go, the value function $J_j^*(x(j))$ has a very simple form: it is quadratic. If we make the problem only slightly more complicated, e.g., if we add constraints on the inputs or states, the value function can still be constructed, but it is much more complex. In general, the value function can only be approximated as discussed in the previous chapter. Then a repeated application of the batch policy, where we resolve the optimization problem at each time step is an attractive alternative.

8.5 Infinite Horizon Problem

For continuous processes operating over a long time period it would be interesting to solve the following infinite horizon problem.

$$J_\infty^*(x(0)) = \min_{u_0,u_1,\dots} \sum_{k=0}^\infty x_k'Qx_k + u_k'Ru_k. \qquad (8.30)$$

Since the prediction must be carried out to infinity, application of the batch method becomes impossible. On the other hand, derivation of the optimal feedback law via dynamic programming remains viable. We can initialize the RDE (8.24)

$$P_k = A'P_{k+1}A + Q - A'P_{k+1}B(B'P_{k+1}B + R)^{-1}B'P_{k+1}A \qquad (8.31)$$

with the terminal cost matrix $P_0 = Q$ and solve it backwards for $k \to -\infty$. Let us assume for the moment that the iterations converge to a solution P_∞. Such P_∞ would then satisfy the *Algebraic Riccati Equation* (ARE)

$$P_\infty = A'P_\infty A + Q - A'P_\infty B(B'P_\infty B + R)^{-1}B'P_\infty A. \qquad (8.32)$$

Then the optimal feedback control law is

$$u^*(k) = \underbrace{-(B'P_\infty B + R)^{-1}B'P_\infty A}_{F_\infty}\, x(k), \quad k = 0, \cdots, \infty \qquad (8.33)$$

and the optimal infinite horizon cost is

$$J_\infty^*(x(0)) = x(0)'P_\infty x(0). \qquad (8.34)$$

Controller (8.33) is referred to as the asymptotic form of the *Linear Quadratic Regulator* (LQR) or the ∞-horizon LQR.

Convergence of the RDE has been studied extensively. A nice summary of the various results can be found in Appendix E of the book by Goodwin and Sin [129]. Intuitively we expect that the system (8.3) must be controllable so that all states can be affected by the control and that the cost function should capture the behavior of all the states, e.g., that $Q \succ 0$. These conditions are indeed sufficient for the RDE to converge and to yield a stabilizing feedback control law. Less restrictive conditions are possible as stated in the following theorem.

Theorem 8.1   *[190, Theorem 2.4-2] If (A, B) is a stabilizable pair and $(Q^{1/2}, A)$ is an observable pair, then the Riccati difference equation (8.31) with $P_0 \succeq 0$ converges to the unique positive definite solution P_∞ of the ARE (8.32) and all the eigenvalues of $(A + BF_\infty)$ lie strictly inside the unit circle.*

The first condition is clearly necessary for J_∞^* (and P_∞) to be finite. To understand the second condition, we write the state dependent term in the objective function as $x'Qx = (x'Q^{1/2})(Q^{1/2}x)$. Thus not the state but the "output" $(Q^{1/2}x)$ is penalized in the objective. Therefore the second condition $((Q^{1/2}, A)$ observable) requires that this output captures all system modes. In this manner convergence of the output $(Q^{1/2}x)$ implies convergence of the state to zero.

From Section 7.5.2 we know that the optimal infinite horizon cost (8.34) is a Lyapunov function for the system $x_{k+1} = Ax_k + Bu_k$ with $u_k = F_\infty x_k$ and satisfies

$$(A + BF_\infty)'P_\infty(A + BF_\infty) - P_\infty = Q + F'_\infty RF_\infty.$$

The reader can verify that substituting F_∞ from (8.33) we recover the Riccati equation (8.32).

9

Linear 1/∞ Norm Optimal Control

In this chapter, we study the finite time and infinite time optimal control problem for unconstrained linear systems with convex piecewise linear objective functions. We derive the structure of the optimal control law by using two approaches: the Batch approach and the Dynamic Programming approach.

9.1 Problem Formulation

We consider a special case of the problem stated in Chapter 7, where the system is linear and time-invariant

$$x(t + 1) = Ax(t) + Bu(t). \tag{9.1}$$

Again, $x(t) \in \mathbb{R}^n$ and $u(t) \in \mathbb{R}^m$ are the state and input vector respectively.

We define the following piecewise linear cost function over a finite horizon of N steps

$$J_0(x_0, U_0) = \|Px_N\|_p + \sum_{k=0}^{N-1} \|Qx_k\|_p + \|Ru_k\|_p \tag{9.2}$$

with $p = 1$ or $p = \infty$ and where x_k denotes the state vector at time k obtained by starting from the state $x_0 = x(0)$ and applying to the system model

$$x_{k+1} = Ax_k + Bu_k \tag{9.3}$$

the input sequence $u_0, \ldots, u_{k-1}$. The weighting matrices in (9.2) could have an arbitrary number of rows. For simplicity of notation we will assume $Q \in \mathbb{R}^{n \times n}$, $R \in \mathbb{R}^{m \times m}$ and $P \in \mathbb{R}^{r \times n}$. Consider the finite time optimal control problem

$$
\begin{aligned}
J_0^*(x(0)) = \min_{U_0} \quad & J_0(x(0), U_0) \\
\text{subj. to} \quad & x_{k+1} = Ax_k + Bu_k, \ k = 0, 1, \ldots, N-1 \\
& x_0 = x(0).
\end{aligned}
\tag{9.4}
$$

In (9.4) $U_0 = [u_0', \ldots, u_{N-1}']' \in \mathbb{R}^s$, $s = mN$ is the decision vector containing all future inputs.

We will present two different approaches to solve problem (9.2)–(9.4), the batch approach and the recursive approach using dynamic programming. Unlike in the 2-norm case presented in the previous chapter, there does not exist a simple closed-form solution of problem (9.2)–(9.4). In this chapter we will show how to use multiparametric linear programming to compute the solution to problem (9.2)–(9.4). We will concentrate on the use of the ∞-norm, the results can be extended easily to cost functions based on the 1-norm or mixed 1/∞ norms.

9.2 Solution via Batch Approach

First we write the equality constraints (9.4) explicitly to express all future states $x_1, x_2, \ldots$ as a function of the future inputs $u_1, u_2, \ldots$ and then we eliminate all intermediate states by using

$$
x_k = A^k x_0 + \sum_{j=0}^{k-1} A^j B u_{k-1-j}
\tag{9.5}
$$

so that all future states are explicit functions of the present state $x(0)$ and the future inputs $u_0, u_1, u_2, \ldots$ only.

The optimal control problem (9.4) with $p = \infty$ can be rewritten as a linear program by using the following standard approach (see e.g., [79]). The sum of components of any vector $\{\varepsilon_0^x, \ldots, \varepsilon_N^x, \varepsilon_0^u, \ldots, \varepsilon_{N-1}^u\}$ that satisfies

$$
\begin{aligned}
-\mathbf{1}_n \varepsilon_k^x &\leq Q x_k, \ k = 0, 1, \ldots, N-1 \\
-\mathbf{1}_n \varepsilon_k^x &\leq -Q x_k, \ k = 0, 1, \ldots, N-1 \\
-\mathbf{1}_r \varepsilon_N^x &\leq P x_N, \\
-\mathbf{1}_r \varepsilon_N^x &\leq -P x_N, \\
-\mathbf{1}_m \varepsilon_k^u &\leq R u_k, \ k = 0, 1, \ldots, N-1 \\
-\mathbf{1}_m \varepsilon_k^u &\leq -R u_k, \ k = 0, 1, \ldots, N-1
\end{aligned}
\tag{9.6}
$$

forms an upper bound on $J_0(x(0), U_0)$, where $\mathbf{1}_k = [\underbrace{1 \ \ldots \ 1}_{k}]'$, and the inequalities (9.6) hold componentwise. It is easy to prove that the vector $z_0 = \{\varepsilon_0^x, \ldots, \varepsilon_N^x, \varepsilon_0^u, \ldots, \varepsilon_{N-1}^u, u_0', \ldots, u_{N-1}'\} \in \mathbb{R}^s$, $s = (m+2)N+1$, that satisfies equations (9.6) and simultaneously minimizes $J(z_0) = \varepsilon_0^x + \cdots + \varepsilon_N^x + \varepsilon_0^u + \cdots + \varepsilon_{N-1}^u$ also solves the original problem (9.4), i.e., the same optimum $J_0^*(x(0))$ is achieved [293, 79]. Therefore, problem (9.4) can be reformulated as the following LP problem

$$\min_{z_0} \quad \varepsilon_0^x + \cdots + \varepsilon_N^x + \varepsilon_0^u + \cdots + \varepsilon_{N-1}^u \tag{9.7a}$$

$$\text{subj. to} \quad -\mathbf{1}_n \varepsilon_k^x \leq \pm Q \left[A^k x_0 + \sum_{j=0}^{k-1} A^j B u_{k-1-j} \right] \tag{9.7b}$$

$$-\mathbf{1}_r \varepsilon_N^x \leq \pm P \left[A^N x_0 + \sum_{j=0}^{N-1} A^j B u_{N-1-j} \right] \tag{9.7c}$$

$$-\mathbf{1}_m \varepsilon_k^u \leq \pm R u_k \tag{9.7d}$$

$$k = 0, \ldots, N-1$$

$$x_0 = x(0), \tag{9.7e}$$

where constraints (9.7c)–(9.7d) are componentwise, and $\pm$ means that the constraint appears once with each sign, as in (9.6).

Remark 9.1 The cost function (9.2) with $p = \infty$ can be interpreted as a special case of a cost function with 1-norm over time and ∞-norm over space. For instance, the dual choice (∞-norm over time and 1-norm over space) leads to the following cost function

$$J_0(x(0), U_0) = \max_{k=0,\ldots,N} \{ \|Qx_k\|_1 + \|Ru_k\|_1 \}. \tag{9.8}$$

We remark that any combination of 1- and ∞-norms leads to a linear program. In general, ∞-norm over time could result in a poor closed-loop performance (only the largest state deviation and the largest input would be penalized over the prediction horizon), while 1-norm over space leads to an LP with a larger number of variables.

The results of this chapter hold for any combination of 1- and ∞-norms over time and space. Clearly the LP formulation will differ from the one in (9.7). For instance, the 1$-$norm in space requires the introduction of nN slack variables for the terms $\|Qx_k\|_1$, $\varepsilon_{k,i} \geq \pm Q^i x_k$ $k = 0, 2, \ldots, N-1$, $i = 1, 2, \ldots, n$, plus r slack variables for the terminal penalty $\|Px_N\|_1$, $\varepsilon_{N,i} \geq \pm P_i x_N$ $i = 1, 2, \ldots, r$, plus mN slack variables for the input terms $\|Ru_k\|_1$, $\varepsilon_{k,i}^u \geq \pm R_i u_k$ $k = 0, 1, \ldots, N-1$, $i = 1, 2, \ldots, m$. Here we have used the notation M_i to denote the i-th row of matrix M.

Problem (9.7) can be rewritten in the more compact form

$$\begin{aligned} \min_{z_0} \quad & c_0' z_0 \\ \text{subj. to} \quad & G_\varepsilon z_0 \leq w_\varepsilon + S_\varepsilon x(0), \end{aligned} \tag{9.9}$$

where $c_0 \in \mathbb{R}^s$ and $G_\varepsilon \in \mathbb{R}^{q \times s}$, $S_\varepsilon \in \mathbb{R}^{q \times n}$ and $w_\varepsilon \in \mathbb{R}^q$ are

$$
c_0 = [\overbrace{1 \ldots 1}^{N+1} \; \overbrace{1 \ldots 1}^{N} \; \overbrace{0 \ldots 0}^{mN}]'
$$

$$
G_\epsilon =
\left[
\begin{array}{ccccccc}
\multicolumn{4}{c}{\overbrace{\hspace{4cm}}^{N+1}} & \multicolumn{3}{c}{\overbrace{\hspace{3cm}}^{N}} \\
-\mathbf{1}_n & 0 & \ldots & 0 & 0 & \ldots & 0 \\
-\mathbf{1}_n & 0 & \ldots & 0 & 0 & \ldots & 0 \\
0 & -\mathbf{1}_n & \ldots & 0 & 0 & \ldots & 0 \\
0 & -\mathbf{1}_n & \ldots & 0 & 0 & \ldots & 0 \\
\ldots & \ldots & & \ldots & & \ldots & \ldots \\
0 & \ldots & -\mathbf{1}_n & 0 & 0 & \ldots & 0 \\
0 & \ldots & -\mathbf{1}_n & 0 & 0 & \ldots & 0 \\
0 & \ldots & 0 & -\mathbf{1}_r & 0 & \ldots & 0 \\
0 & \ldots & 0 & -\mathbf{1}_r & 0 & \ldots & 0 \\
0 & 0 & \ldots & 0 & -\mathbf{1}_m & \ldots & 0 \\
0 & 0 & \ldots & 0 & -\mathbf{1}_m & \ldots & 0 \\
\ldots & \ldots & \ldots & \ldots & & \ldots & \ldots \\
0 & 0 & \ldots & 0 & 0 & \ldots & -\mathbf{1}_m \\
0 & 0 & \ldots & 0 & 0 & \ldots & -\mathbf{1}_m
\end{array}
\right.
$$

$$
\left.
\begin{array}{cccc}
\multicolumn{4}{c}{\overbrace{\hspace{5cm}}^{mN}} \\
0 & 0 & \ldots & 0 \\
0 & 0 & \ldots & 0 \\
QB & 0 & \ldots & 0 \\
-QB & 0 & \ldots & 0 \\
\ldots & & \ldots & \ldots \\
QA^{N-2}B & QA^{N-3}B & \ldots & 0 \\
-QA^{N-2}B & -QA^{N-3}B & \ldots & 0 \\
PA^{N-1}B & PA^{N-2}B & \ldots & PB \\
-PA^{N-1}B & -PA^{N-2}B & \ldots & -PB \\
R & 0 & \ldots & 0 \\
-R & 0 & \ldots & 0 \\
\ldots & & \ldots & \ldots \\
0 & 0 & \ldots & R \\
0 & 0 & \ldots & -R
\end{array}
\right]
$$

$$
w_\epsilon = [\overbrace{0 \ldots 0}^{2nN+2r} \; \overbrace{0 \ldots 0}^{2mN}]'
$$

$$
S_\epsilon = [\overbrace{-Q' \quad Q' \quad (-QA)' \quad (QA)' \quad (-QA^2)' \quad \ldots \quad (-QA^{N-1})' \quad (QA^{N-1})'}^{2nN}{}'
$$

$$
\overbrace{(-PA^N)' \; (PA^N)'}^{2r} \; \overbrace{0'_m \; \ldots \; 0'_m}^{2mN}]'.
$$

$$(9.10)$$

Note that in (9.10) we include the zero vector w_ϵ to make the notation consistent with the one used in Section 6.2.

By treating $x(0)$ as a vector of parameters, the problem (9.9) becomes a *multiparametric linear program* (mp-LP) that can be solved as described in Section 6.2. Once the multiparametric problem (9.7) has been solved, the explicit solution $z_0^*(x(0))$ of (9.9) is available as a piecewise affine function of $x(0)$, and the optimal control law U_0^* is also available explicitly, as the optimal input U_0^* consists simply of the last part of $z_0^*(x(0))$

$$U_0^*(x(0)) = [0 \ \ldots \ 0 \ I_m \ I_m \ \ldots \ I_m]z_0^*(x(0)). \tag{9.11}$$

Theorem 6.5 states that there exists a continuous and PPWA solution $z_0^*(x)$ of the mp-LP problem (9.9). Clearly the same properties are inherited by the controller. The following Corollaries of Theorem 6.5 summarize the analytical properties of the optimal control law and the value function.

Corollary 9.1 *There exists a control law $U_0^* = \bar{f}_0(x(0))$, $\bar{f}_0 : \mathbb{R}^n \to \mathbb{R}^m$, obtained as a solution of the optimal control problem (9.2)–(9.4) with $p = 1$ or $p = \infty$, which is continuous and PPWA*

$$\bar{f}_0(x) = \bar{F}_0^i x \quad if \quad x \in CR_0^i, \ i = 1, \ldots, N_0^r, \tag{9.12}$$

where the polyhedral sets $CR_0^i = \{H_0^i x \leq 0\}$, $i = 1, \ldots, N_0^r$, are a partition of $\mathbb{R}^n$.

Note that in Corollary 9.1 the control law is linear (not affine) and the critical regions have a conic shape ($CR_0^i = \{H_0^i x \leq 0\}$). This can be proven immediately from the results in Section 6.2 by observing that the constant term w_ϵ at the right-hand side on the mp-LP problem (9.9) is zero.

Corollary 9.2 *The value function $J^*(x)$ obtained as a solution of the optimal control problem (9.2)–(9.4) is convex and PPWA.*

Remark 9.2 Note that if the optimizer of problem (9.4) is unique for all $x(0) \in \mathbb{R}^n$, then Corollary 9.1 reads: "**The** control law $U^*(0) = \bar{f}_0(x(0))$, $\bar{f}_0 : \mathbb{R}^n \to \mathbb{R}^m$, obtained as a solution of the optimal control problem (9.2)–(9.4) with $p = 1$ or $p = \infty$, is continuous and PPWA,…". From the results of Section 6.2 we know that in case of multiple optimizers for some $x(0) \in \mathbb{R}^n$, a control law of the form (9.12) can always be computed.

9.3 Solution via Recursive Approach

Alternatively we can use dynamic programming to solve the same problem in a recursive manner. We define the optimal cost $J_j^*(x_j)$ for the $N - j$ step problem starting from state x_j by

$$J_j^*(x_j) = \min_{u_j, \cdots, u_{N-1}} \|Px_N\|_\infty + \sum_{k=j}^{N-1} \|Qx_k\|_\infty + \|Ru_k\|_\infty.$$

According to the principle of optimality the optimal one step cost-to-go can be obtained from

$$J^*_{N-1}(x_{N-1}) = \min_{u_{N-1}} \|P_N x_N\|_\infty + \|Q x_{N-1}\|_\infty + \|R u_{N-1}\|_\infty \qquad (9.13)$$

$$x_N = A x_{N-1} + B u_{N-1}$$

$$P_N = P. \qquad (9.14)$$

Substituting (9.14) into the objective function (9.13), we have

$$J^*_{N-1}(x_{N-1}) = \min_{u_{N-1}} \|P_N(A x_{N-1} + B u_{N-1})\|_\infty + \|Q x_{N-1}\|_\infty + \|R u_{N-1}\|_\infty. \quad (9.15)$$

We find the optimum by solving the mp-LP

$$\min_{\varepsilon^x_{N-1}, \varepsilon^x_N, \varepsilon^u_{N-1}, u_{N-1}} \quad \varepsilon^x_{N-1} + \varepsilon^x_N + \varepsilon^u_{N-1} \qquad (9.16a)$$

$$\text{subj. to} \quad -\mathbf{1}_n \varepsilon^x_{N-1} \le \pm Q x_{N-1} \qquad (9.16b)$$

$$-\mathbf{1}_{r_N} \varepsilon^x_N \le \pm P_N [A x_{N-1} + B u_{N-1}] \qquad (9.16c)$$

$$-\mathbf{1}_m \varepsilon^u_{N-1} \le \pm R u_{N-1}, \qquad (9.16d)$$

where r_N is the number of rows of the matrix P_N. By Theorem 6.5, J^*_{N-1} is a convex and piecewise affine function of x_{N-1}, the corresponding optimizer u^*_{N-1} is piecewise affine and continuous, and the feasible set $\mathcal{X}_{N-1}$ is $\mathbb{R}^n$. We use the equivalence of representation between convex and PPWA functions and infinity norm (see Section 2.2.5) to write the one-step optimal cost-to-go as

$$J^*_{N-1}(x_{N-1}) = \|P_{N-1} x_{N-1}\|_\infty \qquad (9.17)$$

with P_{N-1} defined appropriately. At the next stage, consider the two-step problem from time $N-2$ forward:

$$J^*_{N-2}(x_{N-2}) = \min_{u_{N-2}} \|P_{N-1} x_{N-1}\|_\infty + \|Q x_{N-2}\|_\infty + \|R u_{N-2}\|_\infty \quad (9.18)$$

$$x_{N-1} = A x_{N-2} + B u_{N-2}. \qquad (9.19)$$

We recognize that (9.18), (9.19) has the same form as (9.13), (9.14). Therefore we can compute the optimal solution again by solving the mp-LP

$$\min_{\varepsilon^x_{N-2}, \varepsilon^x_{N-1}, \varepsilon^u_{N-2}, u_{N-2}} \quad \varepsilon^x_{N-2} + \varepsilon^x_{N-1} + \varepsilon^u_{N-2} \qquad (9.20a)$$

$$\text{subj. to} \quad -\mathbf{1}_n \varepsilon^x_{N-2} \le \pm Q x_{N-2} \qquad (9.20b)$$

$$-\mathbf{1}_{r_{N-1}} \varepsilon^x_{N-1} \le \pm P_{N-1} [A x_{N-2} + B u_{N-2}] \qquad (9.20c)$$

$$-\mathbf{1}_m \varepsilon^u_{N-2} \le \pm R u_{N-2}, \qquad (9.20d)$$

where r_{N-1} is the number of rows of the matrix P_{N-1}. The optimal two-step cost-to-go is

$$J^*_{N-2}(x_{N-2}) = \|P_{N-2} x_{N-2}\|_\infty, \qquad (9.21)$$

Continuing in this manner, at some arbitrary time k the optimal control action is

$$u^*(k) = f_k(x(k)), \qquad (9.22)$$

where $f_k(x)$ is continuous and PPWA

$$f_k(x) = F_k^i x \quad \text{if} \quad H_k^i x \leq 0, \ i = 1, \dots, N_k^r, \tag{9.23}$$

where the polyhedral sets $\{H_k^i x \leq 0\}$, $i = 1, \dots, N_k^r$, are a partition of $\mathbb{R}^n$. The optimal cost-to-go starting from the measured state $x(k)$ is

$$J_k^*(x(k)) = \|P_k x(k)\|_\infty. \tag{9.24}$$

Here we have introduced the notation P_k to express the optimal cost-to-go $J_k^*(x(k)) = \|P_k x(k)\|_\infty$ from time k to the end of the horizon N. We also remark that the rows of P_k correspond to the different affine functions constituting J_k^* and thus their number varies with the time index k. Clearly, we do not have a closed form as for the 2-norm with the Riccati Difference Equation (8.24) linking cost and control law at time k given their value at time $k - 1$.

9.4 Comparison of the two Approaches

We will compare the batch and the recursive dynamic programming approach in terms of the results and the methods used to obtain the results. Most importantly we observe that the results obtained by the two methods are fundamentally different. The batch approach yields a formula for the *sequence of inputs* as a function of the initial state.

$$U_0^* = \bar{F}_0^i x(0) \quad \text{if} \quad \bar{H}_0^i x(0) \leq 0, \ i = 1, \dots, \bar{N}_0^r. \tag{9.25}$$

The recursive dynamic programming approach yields a feedback policy, i.e., a *sequence of feedback laws* expressing at each time step the control action as a function of the state at that time.

$$u^*(k) = F_k^i x(k) \quad \text{if} \quad H_k^i x(k) \leq 0, \ i = 1, \dots, N_k^r \text{ for } k = 0, \dots, N - 1. \tag{9.26}$$

As this expression implies, we determine $u(k)$ at each time k as a function of the current state $x(k)$ rather than use a $u(k)$ precomputed at $k = 0$ as in the batch method. If the state evolves exactly according to the linear model (9.3) then the sequence of control actions $u(k)$ obtained from the two approaches is identical. In practice, the result of applying the sequence (9.25) in an open-loop fashion may be rather different from applying the time-varying feedback law (9.26) because the model (9.3) for predicting the system states may be inaccurate and the system may be subject to disturbances not included in the model. We expect the application of the feedback law to be more robust because at each time step the *observed* state $x(k)$ is used to determine the control action rather than the state x_k *predicted* at time $t = 0$.

We note that we can get the same feedback effect with the batch approach if we recalculate the optimal open-loop sequence at each time step j with the current measurement as initial condition. In this case we need to solve the following optimization problem

$$J_j^*(x(j)) = \min_{u_j,\ldots,u_{N-1}} \quad \|Px_N\|_\infty + \sum_{k=j}^{N-1} \|Qx_k\|_\infty + \|Ru_k\|_\infty \tag{9.27}$$
$$\text{subj. to} \qquad x_j = x(j),$$

where we note that the horizon length is shrinking at each time step.

As seen from (9.25) the solution to (9.27) relates the sequence of inputs $u_j^*, u_{j+1}^*, \ldots$ to the state $x(j)$ through a linear expression. The first part of this expression yields again the optimal feedback law (9.26) at time j, $u^*(j) = f_j(x(j))$.

Here the dynamic programming approach is clearly a more efficient way to generate the feedback policy because it requires the solution of a small mp-LP problem (9.7) for each time step. Repeated application of the batch approach, on the other hand, requires repeatedly the solution of a larger mp-LP for each time step.

9.5 Infinite Horizon Problem

For continuous processes operating over a long time period it would be interesting to solve the following infinite horizon problem.

$$J_\infty^*(x(0)) = \min_{u(0),u(1),\ldots} \sum_{k=0}^{\infty} \|Qx_k\|_\infty + \|Ru_k\|_\infty. \tag{9.28}$$

Since the prediction must be carried out to infinity, application of the batch method becomes impossible. On the other hand, derivation of the optimal feedback law via dynamic programming remains viable. We can use the dynamic programming formulation

$$\|P_j x_j\|_\infty = \min_{u_j} \|P_{j+1} x_{j+1}\|_\infty + \|Qx_j\|_\infty + \|Ru_j\|_\infty \tag{9.29}$$

$$x_{j+1} = Ax_j + Bu_j \tag{9.30}$$

with the terminal cost matrix $P_0 = Q$ and solve it backwards for $k \to -\infty$. Let us assume for the moment that the iterations converge to a solution P_∞ in a finite number of iterations. Then the optimal feedback control law is time-invariant and piecewise linear

$$u^*(k) = F^i x(k) \quad \text{if} \quad H^i x \leq 0, \ i = 1,\ldots,N^r \tag{9.31}$$

and the optimal infinite horizon cost is

$$J_\infty^*(x(0)) = \|P_\infty x(0)\|_\infty. \tag{9.32}$$

In general, the infinite time optimal cost $J_\infty^*(x(0))$ and the optimal feedback control law are not necessarily piecewise linear (with a finite number of regions). Convergence of the recursive scheme (9.29) has been studied in detail in [87]. If this recursive scheme converges and Q and R are of full column rank, then the resulting control law (9.31) stabilizes the system (see Section 7.4).

Example 9.1 Consider the double integrator system

$$\left\{ \; x(t+1) = \begin{bmatrix} 1 & 1 \\ 0 & 1 \end{bmatrix} x(t) + \begin{bmatrix} 0 \\ 1 \end{bmatrix} u(t) \right. \tag{9.33}$$

The aim is to compute the infinite horizon optimal controller that solves the optimization problem (9.28) with $Q = \begin{bmatrix} 1 & 0 \\ 0 & 1 \end{bmatrix}$ and $R = 20$.

The dynamic programming iteration (9.29) converges after 18 iterations to the following optimal solution:

$$
u = \begin{cases}
\begin{bmatrix} 9.44 & 29.44 \end{bmatrix} x & \text{if} & \begin{bmatrix} -0.10 & -1.00 \\ -0.71 & -0.71 \end{bmatrix} x \leq \begin{bmatrix} 0 \\ 0 \end{bmatrix} & \text{(Region \#1)} \\[2mm]
\begin{bmatrix} 9.00 & 25.00 \end{bmatrix} x & \text{if} & \begin{bmatrix} 0.10 & 1.00 \\ -0.11 & -0.99 \end{bmatrix} x \leq \begin{bmatrix} 0 \\ 0 \end{bmatrix} & \text{(Region \#2)} \\[2mm]
\begin{bmatrix} -1.00 & 19.00 \end{bmatrix} x & \text{if} & \begin{bmatrix} -0.45 & -0.89 \\ 0.71 & 0.71 \end{bmatrix} x \leq \begin{bmatrix} 0 \\ 0 \end{bmatrix} & \text{(Region \#3)} \\[2mm]
\begin{bmatrix} 8.00 & 16.00 \end{bmatrix} x & \text{if} & \begin{bmatrix} 0.11 & 0.99 \\ -0.12 & -0.99 \end{bmatrix} x \leq \begin{bmatrix} 0 \\ 0 \end{bmatrix} & \text{(Region \#4)} \\[2mm]
\begin{bmatrix} -2.00 & 17.00 \end{bmatrix} x & \text{if} & \begin{bmatrix} -0.32 & -0.95 \\ 0.45 & .89 \end{bmatrix} x \leq \begin{bmatrix} 0 \\ 0 \end{bmatrix} & \text{(Region \#5)} \\[2mm]
\begin{bmatrix} 7.00 & 8.00 \end{bmatrix} x & \text{if} & \begin{bmatrix} 0.12 & 0.99 \\ -0.14 & -0.99 \end{bmatrix} x \leq \begin{bmatrix} 0 \\ 0 \end{bmatrix} & \text{(Region \#6)} \\[2mm]
\begin{bmatrix} -3.00 & 14.00 \end{bmatrix} x & \text{if} & \begin{bmatrix} 0.32 & 0.95 \\ -0.24 & -0.97 \end{bmatrix} x \leq \begin{bmatrix} 0 \\ 0 \end{bmatrix} & \text{(Region \#7)} \\[2mm]
\begin{bmatrix} 6.00 & 1.00 \end{bmatrix} x & \text{if} & \begin{bmatrix} 0.14 & 0.99 \\ -0.16 & -0.99 \end{bmatrix} x \leq \begin{bmatrix} 0 \\ 0 \end{bmatrix} & \text{(Region \#8)} \\[2mm]
\begin{bmatrix} -4.00 & 10.00 \end{bmatrix} x & \text{if} & \begin{bmatrix} 0.24 & 0.97 \\ -0.20 & -0.98 \end{bmatrix} x \leq \begin{bmatrix} 0 \\ 0 \end{bmatrix} & \text{(Region \#9)} \\[2mm]
\begin{bmatrix} 5.00 & -5.00 \end{bmatrix} x & \text{if} & \begin{bmatrix} 0.16 & 0.99 \\ -0.20 & -0.98 \end{bmatrix} x < \begin{bmatrix} 0 \\ 0 \end{bmatrix} & \text{(Region \#10)} \\[2mm]
\begin{bmatrix} -5.00 & 5.00 \end{bmatrix} x & \text{if} & \begin{bmatrix} 0.20 & 0.98 \\ -0.16 & -0.99 \end{bmatrix} x \leq \begin{bmatrix} 0 \\ 0 \end{bmatrix} & \text{(Region \#11)} \\[2mm]
\begin{bmatrix} 4.00 & -10.00 \end{bmatrix} x & \text{if} & \begin{bmatrix} 0.20 & 0.98 \\ -0.24 & -0.97 \end{bmatrix} x \leq \begin{bmatrix} 0 \\ 0 \end{bmatrix} & \text{(Region \#12)} \\[2mm]
\begin{bmatrix} -6.00 & -1.00 \end{bmatrix} x & \text{if} & \begin{bmatrix} 0.16 & 0.99 \\ -0.14 & -0.99 \end{bmatrix} x \leq \begin{bmatrix} 0 \\ 0 \end{bmatrix} & \text{(Region \#13)} \\[2mm]
\begin{bmatrix} 3.00 & -14.00 \end{bmatrix} x & \text{if} & \begin{bmatrix} 0.24 & 0.97 \\ -0.32 & -0.95 \end{bmatrix} x \leq \begin{bmatrix} 0 \\ 0 \end{bmatrix} & \text{(Region \#14)} \\[2mm]
\begin{bmatrix} -7.00 & -8.00 \end{bmatrix} x & \text{if} & \begin{bmatrix} 0.14 & 0.99 \\ -0.12 & -0.99 \end{bmatrix} x \leq \begin{bmatrix} 0 \\ 0 \end{bmatrix} & \text{(Region \#15)} \\[2mm]
\begin{bmatrix} 2.00 & -17.00 \end{bmatrix} x & \text{if} & \begin{bmatrix} 0.32 & 0.95 \\ -0.45 & -0.89 \end{bmatrix} x \leq \begin{bmatrix} 0 \\ 0 \end{bmatrix} & \text{(Region \#16)} \\[2mm]
\begin{bmatrix} -8.00 & -16.00 \end{bmatrix} x & \text{if} & \begin{bmatrix} 0.12 & 0.99 \\ -0.11 & -0.99 \end{bmatrix} x \leq \begin{bmatrix} 0 \\ 0 \end{bmatrix} & \text{(Region \#17)} \\[2mm]
\begin{bmatrix} 1.00 & -19.00 \end{bmatrix} x & \text{if} & \begin{bmatrix} 0.45 & 0.89 \\ -0.71 & -0.71 \end{bmatrix} x \leq \begin{bmatrix} 0 \\ 0 \end{bmatrix} & \text{(Region \#18)} \\[2mm]
\begin{bmatrix} -9.00 & -25.00 \end{bmatrix} x & \text{if} & \begin{bmatrix} 0.11 & 0.99 \\ -0.10 & -1.00 \end{bmatrix} x \leq \begin{bmatrix} 0 \\ 0 \end{bmatrix} & \text{(Region \#19)} \\[2mm]
\begin{bmatrix} -9.44 & -29.44 \end{bmatrix} x & \text{if} & \begin{bmatrix} 0.10 & 1.00 \\ 0.71 & 0.71 \end{bmatrix} x \leq \begin{bmatrix} 0 \\ 0 \end{bmatrix} & \text{(Region \#20)}
\end{cases}
$$

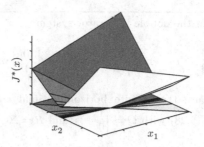

Figure 9.1 Example 9.1. ∞-norm objective function, ∞-horizon controller. Piecewise linear optimal cost (value function) and corresponding polyhedral partition.

with P_∞ equal to

$$
P_\infty = \begin{bmatrix}
9.44 & 29.44 \\
9.00 & 25.00 \\
-1.00 & 19.00 \\
8.00 & 16.00 \\
-2.00 & 17.00 \\
7.00 & 8.00 \\
-3.00 & 14.00 \\
6.00 & 1.00 \\
-4.00 & 10.00 \\
5.00 & -5.00 \\
-5.00 & 5.00 \\
4.00 & -10.00 \\
-6.00 & -1.00 \\
3.00 & -14.00 \\
-7.00 & -8.00 \\
2.00 & -17.00 \\
-8.00 & -16.00 \\
1.00 & -19.00 \\
-9.00 & -25.00 \\
-9.44 & -29.44
\end{bmatrix}
\tag{9.34}
$$

Note that P_∞ in (9.34) has 20 rows corresponding to the 20 linear terms (or pieces) of the piecewise linear value function $J_\infty^*(x) = \|P_\infty x\|_\infty$ for $x \in \mathbb{R}^2$. For instance, $J_\infty^*(x)$ in region 1 is $J_\infty^*(x) = 9.44x_1 + 29.44x_2$, where x_1 and x_2 denote the first and second component of the state vector x, respectively. Note that each linear term appears twice, with positive and negative sign. Therefore $J_\infty^*(x)$ can be written in minimal form as the infinity norm of a matrix $\|\tilde{P}_\infty x\|_\infty$ with $\tilde{P}_\infty$ being a matrix with ten rows. The value function $J_\infty^*(x) = \|P_\infty x\|_\infty$ is plotted in Figure 9.1.

Part IV

Constrained Optimal Control of Linear Systems

10

Controllability, Reachability and Invariance

This chapter is a self-contained introduction to controllability, reachability and invariant set theory. N-steps reachable sets are defined for autonomous systems. They represent the set of states which a system can evolve to after N steps. N-steps controllable sets are defined for systems with inputs. They represent the set of states which a system can be steered to after N steps.

Invariant sets are the infinite time versions of N-steps controllable and reachable sets. *Invariant sets* are computed for autonomous systems. These types of sets are useful to answer questions such as: "For a given feedback controller $u = k(x)$, find the set of states whose trajectory will never violate the system constraints." *Control invariant sets* are defined for systems subject to external inputs. These types of sets are useful to answer questions such as: "Find the set of states for which there exists a controller such that the system constraints are never violated."

This chapter focuses on computational tools for constrained linear systems and constrained linear systems subject to additive and parametric uncertainty. A thorough presentation of the basic notions and algorithms presented in this chapter can be found in the book by Blanchini and Miani [59].

10.1 Controllable and Reachable Sets

In this section we deal with two types of systems, namely, autonomous systems:

$$x(t + 1) = g_a(x(t)), \qquad (10.1)$$

and systems subject to external inputs:

$$x(t + 1) = g(x(t), u(t)). \qquad (10.2)$$

Both systems are subject to state and input constraints

$$x(t) \in \mathcal{X}, \ u(t) \in \mathcal{U}, \ \forall \, t \geq 0. \qquad (10.3)$$

The sets $\mathcal{X}$ and $\mathcal{U}$ are polyhedra.

Definition 10.1 *For the autonomous system (10.1) we denote the* precursor set *to the set $\mathcal{S}$ as*

$$\text{Pre}(\mathcal{S}) = \{x \in \mathbb{R}^n \;:\; g_a(x) \in \mathcal{S}\}. \tag{10.4}$$

$\text{Pre}(\mathcal{S})$ is the set of states which evolve into the target set $\mathcal{S}$ in one time step.

Definition 10.2 *For the system (10.2) we denote the* precursor set *to the set $\mathcal{S}$ as*

$$\text{Pre}(\mathcal{S}) = \{x \in \mathbb{R}^n \;:\; \exists u \in \mathcal{U} \ s.t. \ g(x,u) \in \mathcal{S}\}. \tag{10.5}$$

For a system with inputs, $\text{Pre}(\mathcal{S})$ is the set of states which can be driven into the target set $\mathcal{S}$ in one time step while satisfying input and state constraints.

Definition 10.3 *For the autonomous system (10.1) we denote the* successor set *from the set $\mathcal{S}$ as*

$$\text{Suc}(\mathcal{S}) = \{x \in \mathbb{R}^n \;:\; \exists \ x(0) \in \mathcal{S} \ s.t. \ x = g_a(x(0))\}.$$

Definition 10.4 *For the system (10.2) with inputs we will denote the* successor set *from the set $\mathcal{S}$ as*

$$\text{Suc}(\mathcal{S}) = \{x \in \mathbb{R}^n \;:\; \exists \ x(0) \in \mathcal{S}, \ \exists \ u(0) \in \mathcal{U} \ s.t. \ x = g(x(0), u(0))\}.$$

Therefore, all the states contained in $\mathcal{S}$ are mapped into the set $\text{Suc}(\mathcal{S})$ under the map g_a or under the map g for some input $u \in \mathcal{U}$.

Remark 10.1   The sets $\text{Pre}(\mathcal{S})$ and $\text{Suc}(\mathcal{S})$ are also denoted as 'one-step backward-reachable set' and 'one-step forward-reachable set', respectively, in the literature.

N-step controllable and reachable sets are defined by iterating $\text{Pre}(\cdot)$ and $\text{Suc}(\cdot)$ computations, respectively.

Definition 10.5 (N-Step Controllable Set $\mathcal{K}_N(\mathcal{S})$) *For a given target set $\mathcal{S} \subseteq \mathcal{X}$, the N-step controllable set $\mathcal{K}_N(\mathcal{S})$ of the system (10.1) or (10.2) subject to the constraints (10.3) is defined recursively as:*

$$\mathcal{K}_j(\mathcal{S}) = \text{Pre}(\mathcal{K}_{j-1}(\mathcal{S})) \cap \mathcal{X}, \quad \mathcal{K}_0(\mathcal{S}) = \mathcal{S}, \quad j \in \{1, \dots, N\} \tag{10.6}$$

From Definition 10.5, all states x_0 of the system (10.1) belonging to the N-Step Controllable Set $\mathcal{K}_N(\mathcal{S})$ will evolve to the target set $\mathcal{S}$ in N steps, while satisfying state constraints.

Also, all states x_0 of the system (10.2) belonging to the N-Step Controllable Set $\mathcal{K}_N(\mathcal{S})$ can be driven, by a suitable control sequence, to the target set $\mathcal{S}$ in N steps, while satisfying input and state constraints.

Definition 10.6 (N-Step Reachable Set $\mathcal{R}_N(\mathcal{X}_0)$) *For a given initial set $\mathcal{X}_0 \subseteq \mathcal{X}$, the N-step reachable set $\mathcal{R}_N(\mathcal{X}_0)$ of the system (10.1) or (10.2) subject to the constraints (10.3) is defined as:*

$$\mathcal{R}_{i+1}(\mathcal{X}_0) = \text{Suc}(\mathcal{R}_i(\mathcal{X}_0)) \cap \mathcal{X}, \quad \mathcal{R}_0(\mathcal{X}_0) = \mathcal{X}_0, \quad i = 0, \dots, N-1 \tag{10.7}$$

From Definition 10.6, all states x_0 belonging to $\mathcal{X}_0$ will evolve to the N-step reachable set $\mathcal{R}_N(\mathcal{X}_0)$ in N steps.

10.1.1 Computation of Controllable and Reachable Sets

Next, we will show through simple examples the main steps involved in the computation of controllable and reachable sets for constrained linear systems. Later in this section we will provide compact formulas based on polyhedral operations.

Example 10.1   Consider the second order autonomous stable system

$$x(t+1) = Ax(t) = \begin{bmatrix} 0.5 & 0 \\ 1 & -0.5 \end{bmatrix} x(t) \tag{10.8}$$

subject to the state constraints

$$x(t) \in \mathcal{X} = \left\{ x \ : \ \begin{bmatrix} -10 \\ -10 \end{bmatrix} \leq x \leq \begin{bmatrix} 10 \\ 10 \end{bmatrix} \right\}, \ \forall t \geq 0. \tag{10.9}$$

The set $\mathrm{Pre}(\mathcal{X})$ can be obtained as follows: Since the set $\mathcal{X}$ is a polytope, it can be represented as a $\mathcal{H}$-polytope (Section 4.2)

$$\mathcal{X} = \{x \ : \ Hx \leq h\}, \tag{10.10}$$

where

$$H = \begin{bmatrix} 1 & 0 \\ 0 & 1 \\ -1 & 0 \\ 0 & -1 \end{bmatrix} \text{ and } h = \begin{bmatrix} 10 \\ 10 \\ 10 \\ 10 \end{bmatrix}.$$

By using this $\mathcal{H}$-presentation and the system equation (10.8), the set $\mathrm{Pre}(\mathcal{X})$ can be derived:

$$\mathrm{Pre}(\mathcal{X}) = \{x \ : \ Hg_a(x) \leq h\} \tag{10.11}$$

$$= \{x \ : \ HAx \leq h\}. \tag{10.12}$$

The set (10.12) may contain redundant inequalities which can be removed by using Algorithm 4.1 in Section 4.4.1 to obtain its minimal representation. Note that by using the notation in Section 4.4.11, the set $\mathrm{Pre}(\mathcal{X})$ in (10.12) is simply $\mathcal{X} \circ A$.

The set $\mathrm{Pre}(\mathcal{X})$ is

$$\mathrm{Pre}(\mathcal{X}) = \left\{ x \ : \ \begin{bmatrix} 1 & 0 \\ 1 & -0.5 \\ -1 & 0 \\ -1 & -0.5 \end{bmatrix} x < \begin{bmatrix} 20 \\ 10 \\ 20 \\ 10 \end{bmatrix} \right\}.$$

The one-step controllable set to $\mathcal{X}$, $\mathcal{K}_1(\mathcal{X}) = \mathrm{Pre}(\mathcal{X}) \cap \mathcal{X}$ is

$$\mathrm{Pre}(\mathcal{X}) \cap \mathcal{X} = \left\{ x \ : \ \begin{bmatrix} 1 & 0 \\ 0 & 1 \\ -1 & 0 \\ 0 & -1 \\ 1 & -0.5 \\ -1 & 0.5 \end{bmatrix} x \leq \begin{bmatrix} 10 \\ 10 \\ 10 \\ 10 \\ 10 \\ 10 \end{bmatrix} \right\}$$

and it is depicted in Figure 10.1.

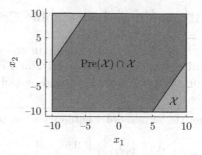

Figure 10.1 Example 10.1. One-step controllable set $\mathrm{Pre}(\mathcal{X}) \cap \mathcal{X}$ for system (10.8) under constraints (10.9).

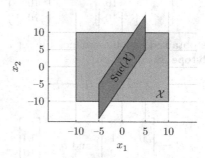

Figure 10.2 Example 10.1. Successor set for system (10.8).

The set $\mathrm{Suc}(\mathcal{X})$ is obtained by applying the map A to the set $\mathcal{X}$. Let us write $\mathcal{X}$ in $\mathcal{V}$-representation (see Section 4.1)

$$\mathcal{X} = \mathrm{conv}(V), \tag{10.13}$$

and let us map the set of vertices V through the transformation A. Because the transformation is linear, the successor set is simply the convex hull of the transformed vertices

$$\mathrm{Suc}(\mathcal{X}) = A \circ \mathcal{X} = \mathrm{conv}(AV). \tag{10.14}$$

We refer the reader to Section 4.4.11 for a detailed discussion on linear transformations of polyhedra.

The set $\mathrm{Suc}(\mathcal{X})$ in $\mathcal{H}$-representation is

$$\mathrm{Suc}(\mathcal{X}) = \left\{ x \ : \ \begin{bmatrix} 1 & 0 \\ -1 & 0 \\ 1 & -0.5 \\ -1 & 0.5 \end{bmatrix} x \leq \begin{bmatrix} 5 \\ 5 \\ 2.5 \\ 2.5 \end{bmatrix} \right\}$$

and is depicted in Figure 10.2.

Example 10.2 Consider the second order unstable system

$$x(t+1) = Ax + Bu = \begin{bmatrix} 1.5 & 0 \\ 1 & -1.5 \end{bmatrix} x(t) + \begin{bmatrix} 1 \\ 0 \end{bmatrix} u(t) \qquad (10.15)$$

subject to the input and state constraints

$$u(t) \in \mathcal{U} = \{u \ : \ -5 \le u \le 5\}, \ \forall t \ge 0 \qquad (10.16a)$$

$$x(t) \in \mathcal{X} = \left\{ x \ : \ \begin{bmatrix} -10 \\ -10 \end{bmatrix} \le x \le \begin{bmatrix} 10 \\ 10 \end{bmatrix} \right\}, \ \forall t \ge 0. \qquad (10.16b)$$

For the nonautonomous system (10.15), the set $\text{Pre}(\mathcal{X})$ can be computed using the $\mathcal{H}$-representation of $\mathcal{X}$ and $\mathcal{U}$,

$$\mathcal{X} = \{x \ : \ Hx \le h\}, \quad \mathcal{U} = \{u \ : \ H_u u \le h_u\}, \qquad (10.17)$$

to obtain

$$\text{Pre}(\mathcal{X}) = \{x \in \mathbb{R}^2 \ : \ \exists u \in \mathcal{U} \text{ s.t. } g(x,u) \in \mathcal{X}, \} \qquad (10.18)$$

$$= \left\{ x \in \mathbb{R}^2 \ : \ \exists u \in \mathbb{R} \text{ s.t. } \begin{bmatrix} HA & HB \\ 0 & H_u \end{bmatrix} \begin{pmatrix} x \\ u \end{pmatrix} \le \begin{bmatrix} h \\ h_u \end{bmatrix} \right\}. \qquad (10.19)$$

The half-spaces in (10.19) define a polytope in the state-input space, and a projection operation (see Section 4.4.6) is used to derive the half-spaces which define $\text{Pre}(\mathcal{X})$ in the state space. The one-step controllable set $\text{Pre}(\mathcal{X}) \cap \mathcal{X}$

$$\begin{bmatrix} 1 & 0 \\ 0 & 1 \\ -1 & 0 \\ 0 & -1 \\ 1 & -1.5 \\ -1 & 1.5 \end{bmatrix} x \le \begin{bmatrix} 10 \\ 10 \\ 10 \\ 10 \\ 10 \\ 10 \end{bmatrix}$$

is depicted in Figure 10.3.

Note that by using the definition of the Minkowski sum given in Section 4.4.9 and the affine operation on polyhedra in Section 4.4.11 we can write the operations in (10.19) compactly as follows:

$$\begin{aligned} \text{Pre}(\mathcal{X}) &= \{x \ : \ \exists u \in \mathcal{U} \text{ s.t. } Ax + Bu \in \mathcal{X}\} \\ &\{x \ : \ y = Ax + Bu, \ y \in \mathcal{X}, \ u \in \mathcal{U}\} \\ &\{x \ : \ Ax = y + (-Bu), \ y \in \mathcal{X}, \ u \in \mathcal{U}\} \\ &\{x \ : \ Ax \in \mathcal{C}, \ \mathcal{C} = \mathcal{X} \oplus (-B) \circ \mathcal{U}\} \\ &\{x \ : \ x \in \mathcal{C} \circ A, \ \mathcal{C} = \mathcal{X} \oplus (-B) \circ \mathcal{U}\} \\ &\{x \ : \ x \in (\mathcal{X} \oplus (-B) \circ \mathcal{U}) \circ A\}. \end{aligned} \qquad (10.20)$$

The set $\text{Suc}(\mathcal{X}) = \{Ax + Bu \in \mathbb{R}^2 \ : \ x \in \mathcal{X}, \ u \in \mathcal{U}\}$ is obtained by applying the map A to the set $\mathcal{X}$ and then considering the effect of the input $u \in \mathcal{U}$. As shown before,

$$A \circ \mathcal{X} = \text{conv}(AV) \qquad (10.21)$$

and therefore

$$\text{Suc}(\mathcal{X}) = \{y + Bu \ : \ y \in A \circ \mathcal{X}, \ u \in \mathcal{U}\}.$$

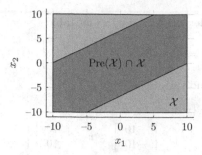

Figure 10.3 Example 10.2. One-step controllable set $\mathrm{Pre}(\mathcal{X}) \cap \mathcal{X}$ for system (10.15) under constraints (10.16).

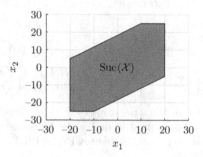

Figure 10.4 Example 10.2. Successor set for system (10.15) under constraints (10.16).

We can use the definition of the Minkowski sum given in Section 4.4.9 and rewrite the set $\mathrm{Suc}(\mathcal{X})$ as

$$\mathrm{Suc}(\mathcal{X}) = (A \circ \mathcal{X}) \oplus (B \circ \mathcal{U}).$$

We can compute the Minkowski sum via projection or vertex enumeration as explained in Section 4.4.9 and obtain the set $\mathrm{Suc}(\mathcal{X})$ in $\mathcal{H}$-representation

$$\mathrm{Suc}(\mathcal{X}) = \left\{ x \; : \; \begin{bmatrix} 1 & 0 \\ -1 & 0 \\ 0 & 1 \\ 0 & -1 \\ 1 & -1.5 \\ -1 & 1.5 \end{bmatrix} x \leq \begin{bmatrix} 20 \\ 20 \\ 25 \\ 25 \\ 27.5 \\ 27.5 \end{bmatrix} \right\},$$

which is depicted in Figure 10.4.

In summary, the sets $\mathrm{Pre}(\mathcal{X})$ and $\mathrm{Suc}(\mathcal{X})$ are the results of linear operations on the polyhedra $\mathcal{X}$ and $\mathcal{U}$ and therefore are polyhedra. By using the definition of the Minkowski sum given in Section 4.4.9 and of affine operation on polyhedra in Section 4.4.11 we can compactly summarize the Pre and Suc operations on linear systems in Table 10.1.

Table 10.1 Pre and Suc operations for linear systems subject to polyhedral state and input constraints $x(t) \in \mathcal{X}$, $u(t) \in \mathcal{U}$

	$x(t+1) = Ax(t)$	$x(t+1) = Ax(t) + Bu(t)$
$\text{Pre}(\mathcal{X})$	$\mathcal{X} \circ A$	$(\mathcal{X} \oplus (-B \circ \mathcal{U})) \circ A$
$\text{Suc}(\mathcal{X})$	$A \circ \mathcal{X}$	$(A \circ \mathcal{X}) \oplus (B \circ \mathcal{U})$

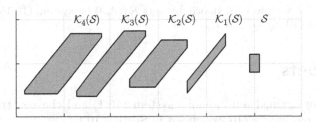

Figure 10.5 Example 10.3. Controllable sets $\mathcal{K}_j(\mathcal{S})$ for system (10.15) under constraints (10.16) for $j = 1, 2, 3, 4$. Note that the sets are shifted along the x-axis for a clearer visualization.

The N-step controllable set $\mathcal{K}_N(\mathcal{S})$ and the N-step reachable set $\mathcal{R}_N(\mathcal{X}_0)$ can be computed by using their recursive formulas (10.6), (10.7) and computing the Pre and Suc operations as in Table 10.1.

Example 10.3 Consider the second order unstable system (10.15) subject to the input and state constraints (10.16). Consider the target set

$$S = \left\{ x : \begin{bmatrix} -1 \\ -1 \end{bmatrix} \leq x \leq \begin{bmatrix} 1 \\ 1 \end{bmatrix} \right\}.$$

The N-step controllable set $\mathcal{K}_N(\mathcal{S})$ of the system (10.15) subject to the constraints (10.16) can be computed by using the recursive formula (10.6)

$$\mathcal{K}_j(\mathcal{S}) = \text{Pre}(\mathcal{K}_{j-1}(\mathcal{S})) \cap \mathcal{X}, \quad \mathcal{K}_0(\mathcal{S}) = \mathcal{S}, \quad j = 1, \dots, N$$

and the steps described in Example 10.2 to compute the Pre($\cdot$) set.

The sets $\mathcal{K}_j(\mathcal{S})$ for $j = 1, 2, 3, 4$ are depicted in Figure 10.5.

Example 10.4 Consider the second order unstable system (10.15) subject to the input and state constraints (10.16). Consider the initial set

$$\mathcal{X}_0 = \left\{ x : \begin{bmatrix} -1 \\ -1 \end{bmatrix} \leq x \leq \begin{bmatrix} 1 \\ 1 \end{bmatrix} \right\}.$$

The N-step reachable set $\mathcal{R}_N(\mathcal{X}_0)$ of the system (10.15) subject to the constraints (10.16) can be computed by using the recursively formula (10.7)

$$\mathcal{R}_{j+1}(\mathcal{X}_0) = \text{Suc}(\mathcal{R}_j(\mathcal{X}_0)) \cap \mathcal{X}, \quad \mathcal{R}_0(\mathcal{X}_0) = \mathcal{X}_0, \quad j = 0, \dots, N-1$$

and the steps described in Example 10.2 to compute the Suc($\cdot$) set.

The sets $\mathcal{R}_j(\mathcal{X}_0))$ for $j = 1, 2, 3, 4$ are depicted in Figure 10.6. The sets are shifted along the x-axis for a clearer visualization.

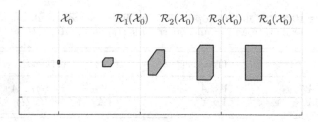

Figure 10.6 Example 10.3. Reachable sets $\mathcal{R}_j(\mathcal{X}_0)$) for system (10.15) under constraints (10.16) for $j = 1, 2, 3, 4$.

10.2 Invariant Sets

Consider the constrained autonomous system (10.1) and the constrained system subject to external inputs (10.2) defined in Section 10.1.

Two different types of sets are considered in this section: *invariant sets* and *control invariant sets*. We will first discuss invariant sets.

Positive Invariant Sets

Invariant sets are used for characterizing the behavior of autonomous systems. These types of sets are useful to answer questions such as: "For a *given* feedback controller $u = f(x)$, find the set of initial states whose trajectory will never violate the system constraints." The following definitions, derived from [172, 58, 54, 49, 179, 137, 139, 140], introduce the different types of invariant sets.

Definition 10.7 (Positive Invariant Set) *A set $\mathcal{O} \subseteq \mathcal{X}$ is said to be a positive invariant set for the autonomous system (10.1) subject to the constraints in (10.3), if*

$$x(0) \in \mathcal{O} \quad \Rightarrow \quad x(t) \in \mathcal{O}, \quad \forall t \in \mathbb{N}_+.$$

Definition 10.8 (Maximal Positive Invariant Set $\mathcal{O}_\infty$) *The set $\mathcal{O}_\infty \subseteq \mathcal{X}$ is the maximal invariant set of the autonomous system (10.1) subject to the constraints in (10.3) if $\mathcal{O}_\infty$ is invariant and $\mathcal{O}_\infty$ contains all the invariant sets contained in $\mathcal{X}$.*

Remark 10.2   The maximal invariant sets defined here are often referred to as "maximal admissible sets" or "maximal output admissible sets" in the literature (e.g., [124]), depending on whether the system state or output is constrained.

Remark 10.3   Note that, in general, the nonlinear system (10.1) may have multiple equilibrium points, and thus $\mathcal{O}_\infty$ might be the union of disconnected sets each containing an equilibrium point.

Theorem 10.1 (Geometric condition for invariance [100])   *A set $\mathcal{O} \subseteq \mathcal{X}$ is a positive invariant set for the autonomous system (10.1) subject to the constraints in (10.3), if and only if*

$$\mathcal{O} \subseteq \text{Pre}(\mathcal{O}). \tag{10.22}$$

Proof: We prove both the necessary and sufficient parts by contradiction. ($\Leftarrow$:) If $\mathcal{O} \nsubseteq \text{Pre}(\mathcal{O})$ then $\exists \bar{x} \in \mathcal{O}$ such that $\bar{x} \notin \text{Pre}(\mathcal{O})$. From the definition of $\text{Pre}(\mathcal{O})$, $g_a(\bar{x}) \notin \mathcal{O}$ and thus $\mathcal{O}$ is not positive invariant. ($\Rightarrow$:) If $\mathcal{O}$ is not a positive invariant set then $\exists \bar{x} \in \mathcal{O}$ such that $g_a(\bar{x}) \notin \mathcal{O}$. This implies that $\bar{x} \in \mathcal{O}$ and $\bar{x} \notin \text{Pre}(\mathcal{O})$ and thus $\mathcal{O} \nsubseteq \text{Pre}(\mathcal{O})$ ∎

It is immediate to prove that condition (10.22) of Theorem 10.1 is equivalent to the following condition

$$\text{Pre}(\mathcal{O}) \cap \mathcal{O} = \mathcal{O}. \tag{10.23}$$

Based on condition (10.23), the following algorithm provides a procedure for computing the maximal positive invariant subset $\mathcal{O}_\infty$ for system (10.1),(10.3) [10, 49, 172, 124].

Algorithm 10.1   *Computation of $\mathcal{O}_\infty$*

Input: g_a, $\mathcal{X}$
Output: $\mathcal{O}_\infty$

$\quad \Omega_0 \leftarrow \mathcal{X}$, $k \leftarrow -1$
$\quad$ **Repeat**
$\quad\quad k \leftarrow k + 1$
$\quad\quad \Omega_{k+1} \leftarrow \text{Pre}(\Omega_k) \cap \Omega_k$
$\quad$ **Until** $\Omega_{k+1} = \Omega_k$
$\quad \mathcal{O}_\infty \leftarrow \Omega_k$

Algorithm 10.1 generates the set sequence $\{\Omega_k\}$ satisfying $\Omega_{k+1} \subseteq \Omega_k, \forall k \in \mathbb{N}$ and it terminates when $\Omega_{k+1} = \Omega_k$. If it terminates, then Ω_k is the maximal positive invariant set $\mathcal{O}_\infty$ for the system (10.1)–(10.3). If $\Omega_k = \emptyset$ for some integer k then the simple conclusion is that $\mathcal{O}_\infty = \emptyset$.

In general, Algorithm 10.1 may never terminate. If the algorithm does not terminate in a finite number of iterations, it can be proven that [179]

$$\mathcal{O}_\infty = \lim_{k \to +\infty} \Omega_k.$$

Conditions for finite time termination of Algorithm 10.1 can be found in [124]. A simple sufficient condition for finite time termination of Algorithm 10.1 requires the system $g_a(x)$ to be linear and stable, and the constraint set $\mathcal{X}$ to be bounded and to contain the origin.

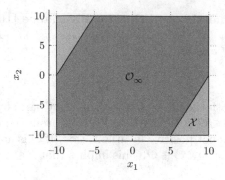

Figure 10.7 Example 10.5. Maximal Positive Invariant Set of system (10.8) under constraints (10.9).

Example 10.5 Consider the second order stable system in Example 10.1. The maximal positive invariant set of system (10.8) subject to constraints (10.9)

$$
\begin{bmatrix}
1 & 0 \\
0 & 1 \\
-1 & 0 \\
0 & -1 \\
1 & -0.5 \\
-1 & 0.5
\end{bmatrix}
x \leq
\begin{bmatrix}
10 \\
10 \\
10 \\
10 \\
10 \\
10
\end{bmatrix}
$$

is depicted in Figure 10.7.

Note from the previous discussion of the example and from Figure 10.1 that here the maximal positive invariant set $\mathcal{O}_\infty$ is obtained after a single step of Algorithm 10.1, i.e.,

$$\mathcal{O}_\infty = \Omega_1 = \mathrm{Pre}(\mathcal{X}) \cap \mathcal{X}.$$

Control Invariant Sets

Control invariant sets are defined for systems subject to external inputs. These types of sets are useful to answer questions such as: "Find the set of initial states for which *there exists* a controller such that the system constraints are never violated." The following definitions, adopted from [172, 58, 54, 49, 179], introduce the different types of control invariant sets.

Definition 10.9 (Control Invariant Set)   *A set $\mathcal{C} \subseteq \mathcal{X}$ is said to be a control invariant set for the system (10.2) subject to the constraints in (10.3), if*

$$x(t) \in \mathcal{C} \quad \Rightarrow \quad \exists u(t) \in \mathcal{U} \text{ such that } g(x(t), u(t)) \in \mathcal{C}, \quad \forall t \in \mathbb{N}_+.$$

Definition 10.10 (Maximal Control Invariant Set $\mathcal{C}_\infty$)   *The set $\mathcal{C}_\infty \subseteq \mathcal{X}$ is said to be the maximal control invariant set for the system (10.2) subject to the constraints in (10.3), if it is control invariant and contains all control invariant sets contained in $\mathcal{X}$.*

Remark 10.4 The geometric conditions for invariance (10.22), (10.23) hold for control invariant sets.

The following algorithm provides a procedure for computing the maximal control invariant set $\mathcal{C}_\infty$ for system (10.2),(10.3) [10, 49, 172, 124].

Algorithm 10.2 *Computation of $\mathcal{C}_\infty$*

Input: $g, \mathcal{X}, \mathcal{U}$
Output: $\mathcal{C}_\infty$

$\quad \Omega_0 \leftarrow \mathcal{X}, k \leftarrow -1$
$\quad$ **Repeat**
$\quad\quad k \leftarrow k + 1$
$\quad\quad \Omega_{k+1} \leftarrow \mathrm{Pre}(\Omega_k) \cap \Omega_k$
$\quad$ **Until** $\Omega_{k+1} = \Omega_k$
$\quad \mathcal{C}_\infty \leftarrow \Omega_{k+1}$

Algorithm 10.2 generates the set sequence $\{\Omega_k\}$ satisfying $\Omega_{k+1} \subseteq \Omega_k, \forall k \in \mathbb{N}$. Algorithm 10.2 terminates when $\Omega_{k+1} = \Omega_k$. If it terminates, then Ω_k is the maximal control invariant set $\mathcal{C}_\infty$ for the system (10.2)–(10.3). In general, Algorithm 10.2 may never terminate [10, 49, 172, 164]. If the algorithm does not terminate in a finite number of iterations, in general, convergence to the maximal control invariant set is not guaranteed

$$\mathcal{C}_\infty \neq \lim_{k \to +\infty} \Omega_k. \tag{10.24}$$

The work in [50] reports examples of nonlinear systems where (10.24) can be observed. A sufficient condition for the convergence of Ω_k to $\mathcal{C}_\infty$ as $k \to +\infty$ requires the polyhedral sets $\mathcal{X}$ and $\mathcal{U}$ to be bounded and the system $g(x, u)$ to be continuous [50].

Example 10.6 Consider the second order unstable system in Example 10.2. Algorithm 10.2 is used to compute the maximal control invariant set of system (10.15) subject to constraints (10.16). Algorithm 10.2 terminates after 45 iterations and the maximal control invariant set $\mathcal{C}_\infty$ is:

$$\begin{bmatrix} 0 & 1 \\ 0 & -1 \\ 0.55 & -0.83 \\ -0.55 & 0.83 \\ 1 & 0 \\ -1 & 0 \end{bmatrix} x \leq \begin{bmatrix} 4 \\ 4 \\ 2.22 \\ 2.22 \\ 10 \\ 10 \end{bmatrix}.$$

The results of the iterations and $\mathcal{C}_\infty$ are depicted in Figure 10.8.

Definition 10.11 (Finitely determined set) *Consider Algorithm 10.1 (Algorithm 10.2). The set $\mathcal{O}_\infty$ ($\mathcal{C}_\infty$) is finitely determined if and only if $\exists\ i \in \mathbb{N}$ such that $\Omega_{i+1} = \Omega_i$. The smallest element $i \in \mathbb{N}$ such that $\Omega_{i+1} = \Omega_i$ is called the determinedness index.*

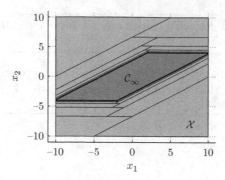

Figure 10.8 Example 10.6. Maximal Control Invariant Set of system (10.15) subject to constraints (10.16).

Remark 10.5   From the results in Section 10.1.1, for linear system with linear constraints the sets $\mathcal{O}_\infty$ and $\mathcal{C}_\infty$ are polyhedra if they are finitely determined.

For all states contained in the maximal control invariant set $\mathcal{C}_\infty$ there exists a control law such that the system constraints are never violated. This does not imply that there exists a control law which can drive the state into a user-specified target set. This issue is addressed in the following by introducing the concepts of maximal controllable sets and stabilizable sets.

Definition 10.12 (Maximal Controllable Set $\mathcal{K}_\infty(\mathcal{O})$)   *For a given target set $\mathcal{O} \subseteq \mathcal{X}$, the maximal controllable set $\mathcal{K}_\infty(\mathcal{O})$ for system (10.2) subject to the constraints in (10.3) is the union of all N-step controllable sets $\mathcal{K}_N(\mathcal{O})$ contained in $\mathcal{X}$ ($N \in \mathbb{N}$).*

We will often deal with controllable sets $\mathcal{K}_N(\mathcal{O})$ where the target $\mathcal{O}$ is a control invariant set. They are special sets, since in addition to guaranteeing that from $\mathcal{K}_N(\mathcal{O})$ we reach $\mathcal{O}$ in N steps, one can ensure that once it has reached $\mathcal{O}$, the system can stay there at all future time instants.

Definition 10.13 (N-step (Maximal) Stabilizable Set)   *For a given control invariant set $\mathcal{O} \subseteq \mathcal{X}$, the N-step (maximal) stabilizable set of the system (10.2) subject to the constraints (10.3) is the N-step (maximal) controllable set $\mathcal{K}_N(\mathcal{O})$ ($\mathcal{K}_\infty(\mathcal{O})$).*

The set $\mathcal{K}_\infty(\mathcal{O})$ contains all states which can be steered into the control invariant set $\mathcal{O}$ and hence $\mathcal{K}_\infty(\mathcal{O}) \subseteq \mathcal{C}_\infty$. The set $\mathcal{K}_\infty(\mathcal{O}) \subseteq \mathcal{C}_\infty$ can be computed as follows [58, 50]:

Algorithm 10.3   *Computation of $\mathcal{K}_\infty(\mathcal{O})$*

Input: $g, \mathcal{X}, \mathcal{U}$
Output: $\mathcal{K}_\infty(\mathcal{O})$

    $\mathcal{K}_0 \leftarrow \mathcal{O}$, where $\mathcal{O}$ is a control invariant set

$c \leftarrow -1$
Repeat
$\quad c \leftarrow c + 1$
$\quad \mathcal{K}_{c+1} \leftarrow \text{Pre}(\mathcal{K}_c) \bigcap \mathcal{X}$
Until $\mathcal{K}_{c+1} = \mathcal{K}_c$
$\mathcal{K}_{\infty}(\mathcal{O}) \leftarrow \mathcal{K}_c$

Since $\mathcal{O}$ is control invariant, it holds $\forall c \in \mathbb{N}$ that $\mathcal{K}_c(\mathcal{O})$ is control invariant and $\mathcal{K}_c \subseteq \mathcal{K}_{c+1}$. Note that Algorithm 10.3 is not guaranteed to terminate in finite time.

Remark 10.6 In general, the maximal stabilizable set $\mathcal{K}_{\infty}(\mathcal{O})$ is not equal to the maximal control invariant set $\mathcal{C}_{\infty}$, even for linear systems. $\mathcal{K}_{\infty}(\mathcal{O}) \subseteq \mathcal{C}_{\infty}$ for all control invariant sets $\mathcal{O}$. The set $\mathcal{C}_{\infty} \backslash \mathcal{K}_{\infty}(\mathcal{O})$ includes all initial states from which it is not possible to steer the system to the stabilizable region $\mathcal{K}_{\infty}(\mathcal{O})$ and hence $\mathcal{O}$.

Example 10.7 Consider the simple constrained one-dimensional system

$$x(t+1) = 2x(t) + u(t) \tag{10.25a}$$

$$|x(t)| \leq 1, \text{ and } |u(t)| \leq 1 \tag{10.25b}$$

and the state feedback control law

$$u(t) = \begin{cases} 1 & \text{if } x(t) \in \left[-1, -\frac{1}{2}\right] \\ -2x(t) & \text{if } x(t) \in \left[-\frac{1}{2}, \frac{1}{2}\right] \\ -1 & \text{if } x(t) \in \left[\frac{1}{2}, 1\right]. \end{cases} \tag{10.26}$$

The closed-loop system has three equilibria at -1, 0, and 1 and system (10.25) is always feasible for all initial states in $[-1, 1]$ and therefore $\mathcal{C}_{\infty} = [-1, 1]$. The equilibrium point 0 is, however, asymptotically stable only for the open set $(-1, 1)$. In fact, $u(t)$ and any other feasible control law cannot stabilize the system from $x = 1$ and from $x = -1$ and therefore when $\mathcal{O} = \emptyset$ then $\mathcal{K}_{\infty}(\mathcal{O}) = (-1, 1) \subset \mathcal{C}_{\infty}$. We note in this example that the maximal stabilizable set is open. One can easily argue that, in general, if the maximal stabilizable set is closed then it is equal to the maximal control invariant set.

10.3 Robust Controllable and Reachable Sets

In this section we deal with two types of systems, namely, autonomous systems

$$x(k+1) = g_a(x(k), w(k)), \tag{10.27}$$

and systems subject to external controllable inputs

$$x(k+1) = g(x(k), u(k), w(k)). \tag{10.28}$$

Both systems are subject to the disturbance $w(k)$ and to the constraints

$$x(k) \in \mathcal{X}, \ u(k) \in \mathcal{U}, \ w(k) \in \mathcal{W} \ \forall \ k \geq 0. \tag{10.29}$$

The sets $\mathcal{X}, \mathcal{U}$ and $\mathcal{W}$ are polyhedra.

Definition 10.14   *For the autonomous system (10.27) we will denote the* robust precursor set to the set $\mathcal{S}$ *as*

$$\text{Pre}(\mathcal{S}, \mathcal{W}) = \{x \in \mathbb{R}^n \ : \ g_a(x, w) \in \mathcal{S}, \ \forall w \in \mathcal{W}\}. \tag{10.30}$$

$\text{Pre}(\mathcal{S}, \mathcal{W})$ defines the set of states of system (10.27) which evolve into the target set $\mathcal{S}$ in one time step for all possible disturbances $w \in \mathcal{W}$.

Definition 10.15   *For the system (10.28) we will denote the* robust precursor set to the set $\mathcal{S}$ *as*

$$\text{Pre}(\mathcal{S}, \mathcal{W}) = \{x \in \mathbb{R}^n \ : \ \exists u \in \mathcal{U} \ s.t. \ g(x, u, w) \subseteq \mathcal{S}, \ \forall w \in \mathcal{W}\}. \tag{10.31}$$

For a system with inputs, $\text{Pre}(\mathcal{S}, \mathcal{W})$ is the set of states which can be robustly driven into the target set $\mathcal{S}$ in one time step for all admissible disturbances.

Definition 10.16   *For the autonomous system (10.27) we will denote the* robust successor set from the set $\mathcal{S}$ *as*

$$\text{Suc}(\mathcal{S}, \mathcal{W}) = \{x \in \mathbb{R}^n \ : \ \exists \ x(0) \in \mathcal{S}, \ \exists \ w \in \mathcal{W} \ such \ that \ x = g_a(x(0), w)\}.$$

Definition 10.17   *For the system (10.28) with inputs we will denote the* robust successor set from the set $\mathcal{S}$ *as*

$$\text{Suc}(\mathcal{S}, \mathcal{W}) = \{x \in \mathbb{R}^n : \exists \, x(0) \in \mathcal{S}, \ \exists \, u \in \mathcal{U}, \ \exists \, w \in \mathcal{W}, \ such \ that \ x = g(x(0), u, w)\}.$$

Thus, all the states contained in $\mathcal{S}$ are mapped into the set $\text{Suc}(\mathcal{S}, \mathcal{W})$ under the map g_a for all disturbances $w \in \mathcal{W}$, and under the map g for all inputs $u \in \mathcal{U}$ and for all disturbances $w \in \mathcal{W}$.

Remark 10.7   The sets $\text{Pre}(\mathcal{S}, \mathcal{W})$ and $\text{Suc}(\mathcal{S}, \mathcal{W})$ are also denoted as "one-step robust backward-reachable set" and "one-step robust forward-reachable set," respectively, in the literature.

N-step robust controllable and robust reachable sets are defined by iterating $\text{Pre}(\cdot, \cdot)$ and $\text{Suc}(\cdot, \cdot)$ computations, respectively.

Definition 10.18 (N-Step Robust Controllable Set $\mathcal{K}_N(\mathcal{S}, \mathcal{W})$)   *For a given target set $\mathcal{S} \subseteq \mathcal{X}$, the N-step robust controllable set $\mathcal{K}_N(\mathcal{S}, \mathcal{W})$ of the system (10.27) or (10.28) subject to the constraints (10.29) is defined recursively as:*

$$\mathcal{K}_j(\mathcal{S}, \mathcal{W}) = \text{Pre}(\mathcal{K}_{j-1}(\mathcal{S}, \mathcal{W}), \mathcal{W}) \cap \mathcal{X}, \ \ \mathcal{K}_0(\mathcal{S}, \mathcal{W}) = \mathcal{S}, \ \ \ j \in \{1, \dots, N\}. \tag{10.32}$$

From Definition 10.18, all states x_0 belonging to the N-Step Robust Controllable Set $\mathcal{K}_N(\mathcal{S}, \mathcal{W})$ can be robustly driven, through a time-varying control law, to the target set $\mathcal{S}$ in N steps, while satisfying input and state constraints for all possible disturbances.

N-step robust reachable sets are defined analogously to N-step robust controllable set.

Definition 10.19 (N-Step Robust Reachable Set $\mathcal{R}_N(\mathcal{X}_0, \mathcal{W})$) *For a given initial set $\mathcal{X}_0 \subseteq \mathcal{X}$, the N-step robust reachable set $\mathcal{R}_N(\mathcal{X}_0, \mathcal{W})$ of the system (10.27) or (10.28) subject to the constraints (10.29) is defined recursively as:*

$$\mathcal{R}_{i+1}(\mathcal{X}_0, \mathcal{W}) = \text{Suc}(\mathcal{R}_i(\mathcal{X}_0, \mathcal{W}), \mathcal{W}) \cap \mathcal{X}, \quad \mathcal{R}_0(\mathcal{X}_0, \mathcal{W}) = \mathcal{X}_0, \quad i = 0, \ldots, N-1. \tag{10.33}$$

From Definition 10.19, all states x_0 belonging to $\mathcal{X}_0$ will evolve to the N-step robust reachable set $\mathcal{R}_N(\mathcal{X}_0, \mathcal{W})$ in N steps.

Next, we will show through simple examples the main steps involved in the computation of robust controllable and robust reachable sets for certain classes of uncertain constrained linear systems.

10.3.1 Linear Systems with Additive Uncertainty and without Inputs

Example 10.8 Consider the second order autonomous system

$$x(t+1) = Ax(t) + w(t) = \begin{bmatrix} 0.5 & 0 \\ 1 & -0.5 \end{bmatrix} x(t) + w(t) \tag{10.34}$$

subject to the state constraints

$$x(t) \in \mathcal{X} = \left\{ x : \begin{bmatrix} -10 \\ -10 \end{bmatrix} \leq x \leq \begin{bmatrix} 10 \\ 10 \end{bmatrix} \right\}, \ \forall t \geq 0, \tag{10.35}$$

and where the additive disturbance belongs to the set

$$w(t) \in \mathcal{W} = \left\{ w : \begin{bmatrix} -1 \\ -1 \end{bmatrix} \leq w \leq \begin{bmatrix} 1 \\ 1 \end{bmatrix} \right\}, \ \forall t \geq 0. \tag{10.36}$$

The set $\text{Pre}(\mathcal{X}, \mathcal{W})$ can be obtained as described next. Since the set $\mathcal{X}$ is a polytope, it can be represented as an $\mathcal{H}$-polytope (Section 4.2)

$$\mathcal{X} = \{ x : Hx \leq h \}, \tag{10.37}$$

where

$$H = \begin{bmatrix} 1 & 0 \\ 0 & 1 \\ -1 & 0 \\ 0 & -1 \end{bmatrix} \text{ and } h = \begin{bmatrix} 10 \\ 10 \\ 10 \\ 10 \end{bmatrix}.$$

By using this $\mathcal{H}$-presentation and the system equation (10.34), the set $\text{Pre}(\mathcal{X}, \mathcal{W})$ can be rewritten as

$$\text{Pre}(\mathcal{X}, \mathcal{W}) = \{ x : Hg_a(x, w) \leq h, \ \forall w \in \mathcal{W} \} \tag{10.38a}$$

$$= \{ x : HAx \leq h - Hw, \ \forall w \in \mathcal{W} \}. \tag{10.38b}$$

The set (10.38) can be represented as a the following polyhedron

$$\text{Pre}(\mathcal{X}, \mathcal{W}) = \{ x \in \mathbb{R}^n : HAx \leq \tilde{h} \} \tag{10.39}$$

with

$$\tilde{h}_i = \min_{w \in \mathcal{W}} (h_i - H_i w). \tag{10.40}$$

In general, a linear program is required to solve problems (10.40). In this example H_i and $\mathcal{W}$ have simple expressions and we get $\tilde{h} = \begin{bmatrix} 9 \\ 9 \\ 9 \\ 9 \end{bmatrix}$. The set (10.39) might contain redundant inequalities which can be removed by using Algorithm 4.1 in Section 4.4.1 to obtain its minimal representation.

The set $\text{Pre}(\mathcal{X}, \mathcal{W})$ is

$$\text{Pre}(\mathcal{X}, \mathcal{W}) = \left\{ x \ : \ \begin{bmatrix} 1 & 0 \\ 1 & -0.5 \\ -1 & 0 \\ -1 & 0.5 \end{bmatrix} x \leq \begin{bmatrix} 18 \\ 9 \\ 18 \\ 9 \end{bmatrix} \right\}.$$

The one-step robust controllable set $\mathcal{K}_1(\mathcal{X}, \mathcal{W}) = \text{Pre}(\mathcal{X}, \mathcal{W}) \cap \mathcal{X}$ is

$$\text{Pre}(\mathcal{X}, \mathcal{W}) \cap \mathcal{X} = \left\{ x \ : \ \begin{bmatrix} 1 & 0 \\ 0 & 1 \\ -1 & 0 \\ 0 & -1 \\ 1 & -0.5 \\ -1 & 0.5 \end{bmatrix} x \leq \begin{bmatrix} 10 \\ 10 \\ 10 \\ 10 \\ 9 \\ 9 \end{bmatrix} \right\}$$

and is depicted in Figure 10.9.

Note that by using the definition of the Pontryagin difference given in Section 4.4.8 and affine operations on polyhedra in Section 4.4.11 we can compactly summarize the operations in (10.38) and write the set Pre in (10.30) as

$$\text{Pre}(\mathcal{X}, \mathcal{W}) = \{x \in \mathbb{R}^n \ : \ Ax + w \in \mathcal{X}, \ \forall w \in \mathcal{W}\} = \{x \in \mathbb{R}^n \ : \ Ax \in \mathcal{X} \ominus \mathcal{W}\} =$$
$$= (\mathcal{X} \ominus \mathcal{W}) \circ A.$$

The set

$$\text{Suc}(\mathcal{X}, \mathcal{W}) = \{y \ : \ \exists x \in \mathcal{X}, \ \exists w \in \mathcal{W} \text{ such that } y = Ax + w\} \tag{10.41}$$

is obtained by applying the map A to the set $\mathcal{X}$ and then considering the effect of the disturbance $w \in \mathcal{W}$. Let us write $\mathcal{X}$ in $\mathcal{V}$-representation (see Section 4.1)

$$\mathcal{X} = \text{conv}(V), \tag{10.42}$$

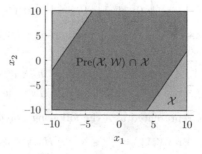

Figure 10.9 Example 10.8. One-step robust controllable set $\text{Pre}(\mathcal{X}, \mathcal{W}) \cap \mathcal{X}$ for system (10.34) under constraints (10.35)–(10.36).

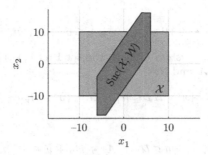

Figure 10.10 Example 10.8. Robust successor set for system (10.34) under constraints (10.36).

and let us map the set of vertices V through the transformation A. Because the transformation is linear, the composition of the map A with the set $\mathcal{X}$, denoted as $A \circ \mathcal{X}$, is simply the convex hull of the transformed vertices

$$A \circ \mathcal{X} = \mathrm{conv}(AV). \qquad (10.43)$$

We refer the reader to Section 4.4.11 for a detailed discussion on linear transformations of polyhedra. Rewrite (10.41) as

$$\mathrm{Suc}(\mathcal{X}, \mathcal{W}) = \{y \in \mathbb{R}^n \ : \ \exists \, z \in A \circ \mathcal{X}, \ \exists \, w \in \mathcal{W} \text{ such that } y = z + w\}.$$

We can use the definition of the Minkowski sum given in Section 4.4.9 and rewrite the Suc set as

$$\mathrm{Suc}(\mathcal{X}, \mathcal{W}) = (A \circ \mathcal{X}) \oplus \mathcal{W}.$$

We can compute the Minkowski sum via projection or vertex enumeration as explained in Section 4.4.9. The set $\mathrm{Suc}(\mathcal{X}, \mathcal{W})$ in $\mathcal{H}$-representation is

$$\mathrm{Suc}(\mathcal{X}, \mathcal{W}) = \left\{ x \ : \ \begin{bmatrix} 1 & -0.5 \\ 0 & -1 \\ -1 & 0 \\ -1 & 0.5 \\ 0 & 1 \\ 1 & 0 \end{bmatrix} x \leq \begin{bmatrix} 4 \\ 16 \\ 6 \\ 4 \\ 16 \\ 6 \end{bmatrix} \right\},$$

and is depicted in Figure 10.10.

10.3.2 Linear Systems with Additive Uncertainty and Inputs

Example 10.9 Consider the second order unstable system

$$\left\{ \ x(t+1) = Ax + Bu = \begin{bmatrix} 1.5 & 0 \\ 1 & -1.5 \end{bmatrix} x(t) + \begin{bmatrix} 1 \\ 0 \end{bmatrix} u(t) + w(t) \qquad (10.44)$$

subject to the input and state constraints

$$u(t) \in \mathcal{U} = \{u \ : \ -5 \leq u \leq 5\}, \ \forall t \geq 0 \qquad (10.45a)$$

$$x(t) \in \mathcal{X} = \left\{ x \ : \ \begin{bmatrix} -10 \\ -10 \end{bmatrix} \leq x \leq \begin{bmatrix} 10 \\ 10 \end{bmatrix} \right\}, \ \forall t \geq 0, \qquad (10.45b)$$

where

$$w(t) \in \mathcal{W} = \{w \; : \; -1 \leq w \leq 1\}, \; \forall t \geq 0. \tag{10.46}$$

For the nonautonomous system (10.44), the set $\text{Pre}(\mathcal{X}, \mathcal{W})$ can be computed using the $\mathcal{H}$-presentation of $\mathcal{X}$ and $\mathcal{U}$,

$$\mathcal{X} = \{x \; : \; Hx \leq h\}, \quad \mathcal{U} = \{u \; : \; H_u u \leq h_u\}, \tag{10.47}$$

to obtain

$$\text{Pre}(\mathcal{X}, \mathcal{W}) = \left\{ x \in \mathbb{R}^2 \; : \; \exists u \in \mathcal{U} \text{ s.t. } Ax + Bu + w \in \mathcal{X}, \; \forall \, w \in \mathcal{W} \right\} \tag{10.48a}$$

$$= \left\{ x \in \mathbb{R}^2 \; : \; \exists u \in \mathbb{R} \text{ s.t. } \begin{bmatrix} HA & HB \\ 0 & H_u \end{bmatrix} \begin{pmatrix} x \\ u \end{pmatrix} \leq \begin{bmatrix} h - Hw \\ h_u \end{bmatrix}, \; \forall \, w \in \mathcal{W} \right\}. \tag{10.48b}$$

As in Example 10.8, the set $\text{Pre}(\mathcal{X}, \mathcal{W})$ can be compactly written as

$$\text{Pre}(\mathcal{X}, \mathcal{W}) = \left\{ x \in \mathbb{R}^2 \; : \; \exists u \in \mathbb{R} \text{ s.t. } \begin{bmatrix} HA & HB \\ 0 & H_u \end{bmatrix} \begin{pmatrix} x \\ u \end{pmatrix} \leq \begin{bmatrix} \tilde{h} \\ h_u \end{bmatrix} \right\}, \tag{10.49}$$

where

$$\tilde{h}_i = \min_{w \in \mathcal{W}} (h_i - H_i w). \tag{10.50}$$

In general, a linear program is required to solve problems (10.50). In this example H_i and $\mathcal{W}$ have simple expressions and we get $\tilde{h} = \begin{bmatrix} 9 \\ 9 \\ 9 \\ 9 \end{bmatrix}$.

The halfspaces in (10.49) define a polytope in the state-input space, and a projection operation (see Section 4.4.6) is used to derive the halfspaces which define $\text{Pre}(\mathcal{X}, \mathcal{W})$ in the state space. The set $\text{Pre}(\mathcal{X}, \mathcal{W}) \cap \mathcal{X}$ is depicted in Figure 10.11 and reported below:

$$\begin{bmatrix} 1 & 0 \\ -1 & 0 \\ 1 & -1.5 \\ -1 & 1.5 \\ 0 & 1 \\ 0 & -1 \end{bmatrix} x \leq \begin{bmatrix} 9.3 \\ 9.3 \\ 9 \\ 9 \\ 10 \\ 10 \end{bmatrix}.$$

Note that by using the definition of a Minkowski sum given in Section 4.4.9 and the affine operation on polyhedra in Section 4.4.11 we can compactly write the operations in (10.48) as follows:

$$\begin{aligned} \text{Pre}(\mathcal{X}, \mathcal{W}) &= \{x \; : \; \exists u \in \mathcal{U} \text{ s.t. } Ax + Bu + w \in \mathcal{X}, \; \forall \, w \in \mathcal{W}\} \\ &= \{x \; : \; \exists y \in \mathcal{X}, \; \exists u \in \mathcal{U} \text{ s.t. } y = Ax + Bu + w, \; \forall \, w \in \mathcal{W}\} \\ &= \{x \; : \; \exists y \in \mathcal{X}, \; \exists u \in \mathcal{U} \text{ s.t. } Ax = y + (-Bu) - w, \; \forall \, w \in \mathcal{W}\} \\ &= \{x \; : \; Ax \in \mathcal{C} \text{ and } \mathcal{C} = \mathcal{X} \oplus (-B) \circ \mathcal{U} \ominus \mathcal{W}\} \\ &= \{x \; : \; x \in \mathcal{C} \circ A, \; \mathcal{C} = \mathcal{X} \oplus (-B) \circ \mathcal{U} \ominus \mathcal{W}\} \\ &= \{x \; : \; x \in ((\mathcal{X} \ominus \mathcal{W}) \oplus (-B \circ \mathcal{U})) \circ A\}. \end{aligned} \tag{10.51}$$

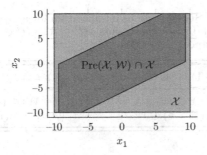

Figure 10.11 Example 10.9. One-step robust controllable set $\mathrm{Pre}(\mathcal{X}, \mathcal{W}) \cap \mathcal{X}$ for system (10.44) under constraints (10.45)–(10.46).

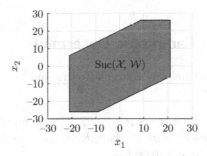

Figure 10.12 Example 10.9. Robust successor set for system (10.44) under constraints (10.45)–(10.46).

Remark 10.8   Note that in (10.51) we have used the fact that if a set $\mathcal{S}$ is described as $\mathcal{S} = \{v \ : \ \exists z \in \mathcal{Z}, \text{ s.t. } v = z - w, \ \forall \, w \in \mathcal{W}\}$, then $\mathcal{S} = \{v \ : \ \exists z \in \mathcal{Z}, \text{ s.t. } z = v + w, \ \forall \, w \in \mathcal{W}\}$ or $\mathcal{S} = \{v \ : \ v + w \in \mathcal{Z}, \ \forall \, w \in \mathcal{W}\} = \mathcal{Z} \ominus \mathcal{W}$. Also, to derive the last equation of (10.51) we have used the associative property of the Pontryagin difference.

The set $\mathrm{Suc}(\mathcal{X}, \mathcal{W}) = \{y \ : \ \exists x \in \mathcal{X}, \ \exists u \in \mathcal{U}, \ \exists w \in \mathcal{W} \text{ s.t. } y = Ax + Bu + w\}$ is obtained by applying the map A to the set $\mathcal{X}$ and then considering the effect of the input $u \in \mathcal{U}$ and of the disturbance $w \in \mathcal{W}$. We can use the definition of Minkowski sum given in Section 4.4.9 and rewrite $\mathrm{Suc}(\mathcal{X}, \mathcal{W})$ as

$$\mathrm{Suc}(\mathcal{X}, \mathcal{W}) = (A \circ \mathcal{X}) \oplus (B \circ \mathcal{U}) \oplus \mathcal{W}.$$

The set $\mathrm{Suc}(\mathcal{X}, \mathcal{W})$ is depicted in Figure 10.12.

In summary, for linear systems with additive disturbances the sets $\mathrm{Pre}(\mathcal{X}, \mathcal{W})$ and $\mathrm{Suc}(\mathcal{X}, \mathcal{W})$ are the results of linear operations on the polytopes $\mathcal{X}$, $\mathcal{U}$ and $\mathcal{W}$ and therefore are polytopes. By using the definition of Minkowski sum given in Section 4.4.9, Pontryagin difference given in Section 4.4.8 and affine operation on polyhedra in Section 4.4.11 we can compactly summarize the operations in Table 10.2. Note that the summary in Table 10.2 applies also to the class of systems $x(k+1) = Ax(t) + Bu(t) + E\tilde{d}(t)$ where $\tilde{d} \in \tilde{\mathcal{W}}$. This can be transformed into $x(k+1) = Ax(t) + Bu(t) + w(t)$ where $w \in \mathcal{W} = E \circ \tilde{\mathcal{W}}$.

Table 10.2 Pre and Suc operations for uncertain linear systems subject to polyhedral input and state constraints $x(t) \in \mathcal{X}$, $u(t) \in \mathcal{U}$ with additive polyhedral disturbances $w(t) \in \mathcal{W}$.

	$x(t+1) = Ax(t) + w(t)$	$x(t+1) = Ax(t) + Bu(t) + w(t)$
$\mathrm{Pre}(\mathcal{X}, \mathcal{W})$	$(\mathcal{X} \ominus \mathcal{W}) \circ A$	$((\mathcal{X} \ominus \mathcal{W}) \oplus (-B \circ \mathcal{U})) \circ A$
$\mathrm{Suc}(\mathcal{X}, \mathcal{W})$	$(A \circ \mathcal{X}) \oplus \mathcal{W}$	$(A \circ \mathcal{X}) \oplus (B \circ \mathcal{U}) \oplus \mathcal{W}$

The N-step robust controllable set $\mathcal{K}_N(\mathcal{S}, \mathcal{W})$ and the N-step robust reachable set $\mathcal{R}_N(\mathcal{X}_0, \mathcal{W})$ can be computed by using their recursive formulas (10.32), (10.33) and computing the Pre and Suc operations as described in Table 10.2.

10.3.3 Linear Systems with Parametric Uncertainty

The next Lemma 10.1 will help us computing Pre and Suc sets for linear systems with parametric uncertainty.

Lemma 10.1   *Let* $g : \mathbb{R}^{n_z} \times \mathbb{R}^n \times \mathbb{R}^{n_w} \to \mathbb{R}^{n_g}$ *be a function of* (z, x, w) *convex in* w *for each* (z, x). *Assume that the variable* w *belongs to the polytope* $\mathcal{W}$ *with vertices* $\{\bar{w}_i\}_{i=1}^{n_{\mathcal{W}}}$. *Then, the constraint*

$$g(z, x, w) \leq 0 \;\; \forall w \in \mathcal{W} \tag{10.52}$$

is satisfied if and only if

$$g(z, x, \bar{w}_i) \leq 0, \;\; i = 1, \dots, n_{\mathcal{W}}. \tag{10.53}$$

Proof: Easily follows from the fact that the maximum of a convex function over a compact convex set is attained at an extreme point of the set. ∎

Lemma 10.2 shows how to reduce the number of constraints in (10.53) for a specific class of constraint functions.

Lemma 10.2   *Assume* $g(z, x, w) = g^1(z, x) + g^2(w)$. *Then the constraint (10.52) can be replaced by* $g^1(z, x) \leq -\bar{g}$, *where* $\bar{g} = \left[\bar{g}_1, \dots, \bar{g}_{n_g}\right]'$ *is a vector whose i-th component is*

$$\bar{g}_i = \max_{w \in \mathcal{W}} \; g_i^2(w), \tag{10.54}$$

and $g_i^2(w)$ *denotes the i-th component of* $g^2(w)$.

Example 10.10   Consider the second order autonomous system

$$x(t+1) = A(w^p(t))x(t) + w^a(t) = \begin{bmatrix} 0.5 + w^p(t) & 0 \\ 1 & -0.5 \end{bmatrix} x(t) + w^a(t) \tag{10.55}$$

subject to the constraints

$$x(t) \in \mathcal{X} = \left\{ x \ : \ \begin{bmatrix} -10 \\ -10 \end{bmatrix} \le x \le \begin{bmatrix} 10 \\ 10 \end{bmatrix} \right\}, \ \forall t \ge 0$$
$$w^a(t) \in \mathcal{W}^a = \left\{ w^a \ : \ \begin{bmatrix} -1 \\ -1 \end{bmatrix} \le w^a \le \begin{bmatrix} 1 \\ 1 \end{bmatrix} \right\}, \ \forall t \ge 0 \qquad (10.56)$$
$$w^p(t) \in \mathcal{W}^p = \{w^p \ : \ 0 \le w^p \le 0.5\}, \ \forall t \ge 0.$$

Let $w = [w^a; \ w^p]$ and $\mathcal{W} = \mathcal{W}^a \times \mathcal{W}^p$. The set $\mathrm{Pre}(\mathcal{X}, \mathcal{W})$ can be obtained as follows. The set $\mathcal{X}$ is a polytope and it can be represented as an $\mathcal{H}$-polytope (Section 4.2)

$$\mathcal{X} = \{x : \ Hx \le h\}, \qquad (10.57)$$

where

$$H = \begin{bmatrix} 1 & 0 \\ 0 & 1 \\ -1 & 0 \\ 0 & -1 \end{bmatrix} \text{ and } h = \begin{bmatrix} 10 \\ 10 \\ 10 \\ 10 \end{bmatrix}.$$

By using this $\mathcal{H}$-presentation and the system equation (10.55), the set $\mathrm{Pre}(\mathcal{X}, \mathcal{W})$ can be rewritten as

$$\mathrm{Pre}(\mathcal{X}, \mathcal{W}) = \{x \ : \ Hg_a(x, w) \le h, \ \forall w \in \mathcal{W}\} \qquad (10.58)$$
$$= \{x \ : \ HA(w^p)x \le h - Hw^a, \ \forall w^a \in \mathcal{W}^a, \ w^p \in \mathcal{W}^p\}. \qquad (10.59)$$

By using Lemmas 10.1 and 10.2, the set (10.59) can be rewritten as a the polytope

$$x \in \mathrm{Pre}(\mathcal{X}, \mathcal{W}) = \left\{ x \in \mathbb{R}^n \ : \ \begin{bmatrix} HA(0) \\ HA(0.5) \end{bmatrix} x < \begin{bmatrix} \tilde{h} \\ \tilde{h} \end{bmatrix} \right\} \qquad (10.60)$$

with

$$\tilde{h}_i = \min_{w^a \in \mathcal{W}^a} (h_i - H_i w^a), \ i = 1, \dots, 4. \qquad (10.61)$$

The set $\mathrm{Pre}(\mathcal{X}, \mathcal{W}) \cap \mathcal{X}$ is depicted in Figure 10.13 and reported below:

$$\begin{bmatrix} 1 & 0 \\ 0.89 & -0.44 \\ 1 & 0 \\ -0.89 & 0.44 \\ 0 & 1 \\ 0 & 1 \end{bmatrix} x \le \begin{bmatrix} 9 \\ 8.049 \\ 9 \\ 8.049 \\ 10 \\ 10 \end{bmatrix}.$$

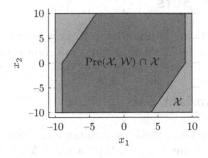

Figure 10.13 Example 10.10. One-step robust controllable set $\mathrm{Pre}(\mathcal{X}, \mathcal{W}) \cap \mathcal{X}$ for system (10.55) under constraints (10.56).

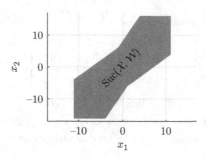

Figure 10.14 Example 10.11. Robust successor set $\mathrm{Suc}(\mathcal{X}, \mathcal{W})$ for system (10.62) under constraints (10.63).

Example 10.11 Consider the second order autonomous system

$$x(t+1) = A(w^p(t))x(t) = \begin{bmatrix} 0.5 + w^p(t) & 0 \\ 1 & -0.5 \end{bmatrix} x(t) \tag{10.62}$$

subject to the constraints

$$\begin{aligned} x(t) \in \mathcal{X} &= \left\{ x \; : \; \begin{bmatrix} -10 \\ -10 \end{bmatrix} \le x \le \begin{bmatrix} 10 \\ 10 \end{bmatrix} \right\}, \; \forall t \ge 0 \\ w^p(t) \in \mathcal{W}^p &= \{ w^p \; : \; 0 \le w^p \le 0.5 \}, \; \forall t \ge 0. \end{aligned} \tag{10.63}$$

Let $w = [w^p]$ and $\mathcal{W} = \mathcal{W}^p$. The set $\mathrm{Suc}(\mathcal{X}, \mathcal{W})$ can be written as infinite union of reachable sets

$$\mathrm{Suc}(\mathcal{X}, \mathcal{W}) = \bigcup_{\bar{w} \in \mathcal{W}} \mathrm{Suc}(\mathcal{X}, \bar{w}) \tag{10.64}$$

where $\mathrm{Suc}(\mathcal{X}, \bar{w})$ is computed as described in Example 10.1 for the system $A(\bar{w})$. In general, the union in equation (10.64) generates a nonconvex set, as can be seen in Figure 10.14. Nonconvexity of successor sets for parametric linear systems is also discussed in [59][Section 6.1.2].

10.4   Robust Invariant Sets

Two different types of sets are considered in this chapter: *robust invariant sets* and *robust control invariant sets*. We will first discuss robust invariant sets.

Robust Positive Invariant Sets

Robust invariant sets are computed for autonomous systems. These types of sets are useful to answer questions such as: "For a *given* feedback controller $u = f(x)$, find the set of states whose trajectory will never violate the system constraints for all possible disturbances." The following definitions introduce the different types of robust invariant sets.

Definition 10.20 (Robust Positive Invariant Set) *A set $\mathcal{O} \subseteq \mathcal{X}$ is said to be a robust positive invariant set for the autonomous system (10.27) subject to the constraints (10.29), if*

$$x(0) \in \mathcal{O} \quad \Rightarrow \quad x(t) \in \mathcal{O}, \quad \forall w(t) \in \mathcal{W}, \ t \in \mathbb{N}_+.$$

Definition 10.21 (Maximal Robust Positive Invariant Set $\mathcal{O}_\infty$) *The set $\mathcal{O}_\infty \subseteq \mathcal{X}$ is the maximal robust invariant set of the autonomous system (10.27) subject to the constraints (10.29) if $\mathcal{O}_\infty$ is a robust invariant set and $\mathcal{O}_\infty$ contains all the robust positive invariant sets contained in $\mathcal{X}$.*

Theorem 10.2 (Geometric condition for invariance) *A set $\mathcal{O} \subseteq \mathcal{X}$ is a robust positive invariant set for the autonomous system (10.27) subject to the constraints (10.29), if and only if*

$$\mathcal{O} \subseteq \mathrm{Pre}(\mathcal{O}, \mathcal{W}). \tag{10.65}$$

The proof of Theorem 10.2 follows the same lines of the proof of Theorem 10.1. ∎

It is immediate to prove that condition (10.65) of Theorem 10.2 is equivalent to the following condition

$$\mathrm{Pre}(\mathcal{O}, \mathcal{W}) \cap \mathcal{O} = \mathcal{O}. \tag{10.66}$$

Based on condition (10.66), the following algorithm provides a procedure for computing the maximal robust positive invariant subset $\mathcal{O}_\infty$ for system (10.27)–(10.29) (for reference to proofs and literature see Section 10.1).

Algorithm 10.4 *Computation of $\mathcal{O}_\infty$*

Input: g_a, $\mathcal{X}$, $\mathcal{W}$
Output: $\mathcal{O}_\infty$

$\quad \Omega_0 \leftarrow \mathcal{X}, \ k \leftarrow -1$
$\quad$ **Repeat**
$\quad\quad k \leftarrow k+1$
$\quad\quad \Omega_{k+1} \leftarrow \mathrm{Pre}(\Omega_k, \mathcal{W}) \cap \Omega_k$
$\quad$ **Until** $\Omega_{k+1} = \Omega_k$
$\quad \mathcal{O}_\infty \leftarrow \Omega_k$

Algorithm 10.4 generates the set sequence $\{\Omega_k\}$ satisfying $\Omega_{k+1} \subseteq \Omega_k, \forall k \in \mathbb{N}$ and it terminates when $\Omega_{k+1} = \Omega_k$. If it terminates, then Ω_k is the maximal robust positive invariant set $\mathcal{O}_\infty$ for system (10.27)–(10.29). If $\Omega_k = \emptyset$ for some integer k then the simple conclusion is that $\mathcal{O}_\infty = \emptyset$.

In general, Algorithm 10.4 may never terminate. If the algorithm does not terminate in a finite number of iterations, it can be proven that [179]

$$\mathcal{O}_\infty = \lim_{k \to +\infty} \Omega_k.$$

Conditions for finite time termination of Algorithm 10.4 can be found in [124]. A simple sufficient condition for finite time termination of Algorithm 10.1 requires the system $g_a(x, w)$ to be linear and stable, and the constraint set $\mathcal{X}$ and disturbance set $\mathcal{W}$ to be bounded and to contain the origin.

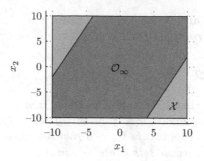

Figure 10.15 Example 10.12. Maximal Robust Positive Invariant Set of system (10.67) subject to constraints (10.68).

Example 10.12 Consider the second order stable system in Example 10.8

$$x(t+1) = Ax(t) + w(t) = \begin{bmatrix} 0.5 & 0 \\ 1 & -0.5 \end{bmatrix} x(t) + w(t) \qquad (10.67)$$

subject to the constraints

$$x(t) \in \mathcal{X} = \left\{ x \ : \ \begin{bmatrix} -10 \\ -10 \end{bmatrix} \le x \le \begin{bmatrix} 10 \\ 10 \end{bmatrix} \right\}, \ \forall t \ge 0$$
$$w(t) \in \mathcal{W} = \left\{ w \ : \ \begin{bmatrix} -1 \\ -1 \end{bmatrix} \le w \le \begin{bmatrix} 1 \\ 1 \end{bmatrix} \right\}, \ \forall t \ge 0. \qquad (10.68)$$

The maximal robust positive invariant set of system (10.67) subject to constraints (10.68) is depicted in Figure 10.15 and reported below:

$$\begin{bmatrix} 0.89 & -0.44 \\ -0.89 & 0.44 \\ -1 & 0 \\ 0 & -1 \\ 1 & 0 \\ 0 & 1 \end{bmatrix} x \le \begin{bmatrix} 8.04 \\ 8.04 \\ 10 \\ 10 \\ 10 \\ 10 \end{bmatrix}.$$

Robust Control Invariant Sets

Robust control invariant sets are defined for systems subject to controllable inputs. These types of sets are useful to answer questions such as: "Find the set of states for which *there exists* a controller such that the system constraints are never violated for all possible disturbances." The following definitions introduce the different types of robust control invariant sets.

Definition 10.22 (Robust Control Invariant Set) *A set $\mathcal{C} \subseteq \mathcal{X}$ is said to be a robust control invariant set for the system (10.28) subject to the constraints (10.29), if*

$$x(t) \in \mathcal{C} \quad \Rightarrow \quad \exists u(t) \in \mathcal{U} \text{ such that } g(x(t), u(t), w(t)) \in \mathcal{C}, \ \forall \, w(t) \in \mathcal{W}, \ \forall \, t \in \mathbb{N}_+.$$

Definition 10.23 (Maximal Robust Control Invariant Set $\mathcal{C}_\infty$)   *The set $\mathcal{C}_\infty \subseteq \mathcal{X}$ is said to be the maximal robust control invariant set for the system (10.28) subject to the constraints (10.29), if it is robust control invariant and contains all robust control invariant sets contained in $\mathcal{X}$.*

Remark 10.9   The geometric conditions for invariance (10.65)–(10.66) hold for control invariant sets.

The following algorithm provides a procedure for computing the maximal robust control invariant set $\mathcal{C}_\infty$ for system (10.28)–(10.29).

Algorithm 10.5   *Computation of $\mathcal{C}_\infty$*

Input: $g, \mathcal{X}, \mathcal{U}, \mathcal{W}$
Output: $\mathcal{C}_\infty$
$\quad$ $\Omega_0 \leftarrow \mathcal{X}, k \leftarrow -1$
$\quad$ **Repeat**
$\qquad$ $k \leftarrow k + 1$
$\qquad$ $\Omega_{k+1} \leftarrow \text{Pre}(\Omega_k, \mathcal{W}) \cap \Omega_k$
$\quad$ **Until** $\Omega_{k+1} = \Omega_k$
$\quad$ $\mathcal{C}_\infty \leftarrow \Omega_{k+1}$

Algorithm 10.5 generates the set sequence $\{\Omega_k\}$ satisfying $\Omega_{k+1} \subseteq \Omega_k, \forall k \in \mathbb{N}$. Algorithm 10.5 terminates when $\Omega_{k+1} = \Omega_k$. If it terminates, then Ω_k is the maximal robust control invariant set $\mathcal{C}_\infty$ for the system (10.28)–(10.29).

In general, Algorithm 10.5 may never terminate [10, 49, 172, 164]. If the algorithm does not terminate in a finite number of iterations, in general, convergence to the maximal robust control invariant set is not guaranteed

$$\mathcal{C}_\infty \neq \lim_{k \to +\infty} \Omega_k. \tag{10.69}$$

The work in [50] reports examples of nonlinear systems where (10.69) can be observed. A sufficient condition for the convergence of Ω_k to $\mathcal{C}_\infty$ as $k \to +\infty$ requires the polyhedral sets $\mathcal{X}, \mathcal{U}$ and $\mathcal{W}$ to be bounded and the system $g(x, u, w)$ to be continuous [50].

Example 10.13   Consider the second order unstable system in Example 10.9. The maximal robust control invariant set of system (10.44) subject to constraints (10.45)–(10.46) is an empty set. If the uncertain set (10.46) is replaced with

$$w(t) \in \mathcal{W} = \{w : -0.1 \leq w \leq 0.1\}, \ \forall t \geq 0$$

the maximal robust control invariant set is

$$\begin{bmatrix} 0 & 1 \\ 0 & -1 \\ 0.55 & -0.83 \\ -0.55 & 0.83 \end{bmatrix} x \leq \begin{bmatrix} 3.72 \\ 3.72 \\ 2.0 \\ 2.0 \end{bmatrix}$$

which is depicted in Figure 10.16.

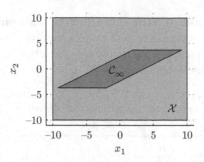

Figure 10.16 Example 10.13. Maximal Robust Control Invariant Set of system (10.44) subject to constraints (10.45).

Definition 10.24 (Finitely determined set)   *Consider Algorithm 10.4 (10.5). The set $\mathcal{O}_\infty$ $(\mathcal{C}_\infty)$ is finitely determined if and only if $\exists\ i \in \mathbb{N}$ such that $\Omega_{i+1} = \Omega_i$. The smallest element $i \in \mathbb{N}$ such that $\Omega_{i+1} = \Omega_i$ is called the determinedness index.*

For all states contained in the maximal robust control invariant set $\mathcal{C}_\infty$ there exists a control law, such that the system constraints are never violated for all feasible disturbances. This does not imply that there exists a control law which can drive the state into a user-specified target set. This issue is addressed in the following by introducing the concept of robust controllable and stabilizable sets.

Definition 10.25 (Maximal Robust Controllable Set $\mathcal{K}_\infty(\mathcal{O}, \mathcal{W})$)   *For a given target set $\mathcal{O} \subseteq \mathcal{X}$, the maximal robust controllable set $\mathcal{K}_\infty(\mathcal{O}, \mathcal{W})$ for the system (10.28) subject to the constraints (10.29) is the union of all N-step robust controllable sets contained in $\mathcal{X}$ for $N \in \mathbb{N}$.*

Robust controllable sets $\mathcal{K}_N(\mathcal{O}, \mathcal{W})$ where the target $\mathcal{O}$ is a robust control invariant set are special sets, since in addition to guaranteeing that from $\mathcal{K}_N(\mathcal{O}, \mathcal{W})$ we robustly reach $\mathcal{O}$ in N steps, one can ensure that once reached $\mathcal{O}$, the system can stay there at all future time instants and for all possible disturbance realizations.

Definition 10.26 (N-step (Maximal) Robust Stabilizable Set)   *For a given robust control invariant set $\mathcal{O} \subseteq \mathcal{X}$, the N-step (maximal) robust stabilizable set of the system (10.28) subject to the constraints (10.29) is the N-step (maximal) robust controllable set $\mathcal{K}_N(\mathcal{O}, \mathcal{W})$ $(\mathcal{K}_\infty(\mathcal{O}, \mathcal{W}))$.*

The set $\mathcal{K}_\infty(\mathcal{O}, \mathcal{W})$ contains all states which can be robustly steered into the robust control invariant set $\mathcal{O}$ and hence $\mathcal{K}_\infty(\mathcal{O}, \mathcal{W}) \subseteq \mathcal{C}_\infty$. The set $\mathcal{K}_\infty(\mathcal{O}, \mathcal{W}) \subseteq \mathcal{C}_\infty$ can be computed as follows:

Algorithm 10.6   *Computation of $\mathcal{K}_\infty(\mathcal{O}, \mathcal{W})$*

Input: g_a, $\mathcal{X}$, $\mathcal{W}$
Output: $\mathcal{K}_\infty(\mathcal{O}, \mathcal{W})$

    $\mathcal{K}_0 \leftarrow \mathcal{O}$, where $\mathcal{O}$ is a robust control invariant set
    $c \leftarrow -1$

Repeat
$\quad c \leftarrow c + 1$
$\quad \mathcal{K}_{c+1} \leftarrow \text{Pre}(\mathcal{K}_c, \mathcal{W}) \bigcap \mathcal{X}$
Until $\mathcal{K}_{c+1} = \mathcal{K}_c$
$\mathcal{K}_\infty(\mathcal{O}, \mathcal{W}) = \mathcal{K}_c$

Since $\mathcal{O}$ is robust control invariant, it holds $\forall c \in \mathbb{N}$ that $\mathcal{K}_c$ is robust control invariant and $\mathcal{K}_c \subseteq \mathcal{K}_{c+1}$. Note that Algorithm 10.6 is not guaranteed to terminate in finite time.

11

Constrained Optimal Control

In this chapter we study the finite time and infinite time optimal control problem for linear systems with linear constraints on inputs and state variables. We establish the structure of the optimal control law and derive algorithms for its computation. For finite time problems with linear and quadratic objective functions we show that the time varying feedback law is piecewise affine and continuous. The value function is a convex piecewise linear for linear objective functions and convex piecewise quadratic for quadratic objective functions.

We describe how the optimal control action for a given initial state can be computed by means of linear or quadratic programming. We also describe how the optimal control law can be computed by means of multiparametric linear or quadratic programming. Finally, we show how to compute the infinite time optimal controller for linear and quadratic objective functions and prove that, when it exists, the infinite time controller inherits all the structural properties of the finite time optimal controller.

11.1 Problem Formulation

Consider the linear time-invariant system

$$x(t+1) = Ax(t) + Bu(t), \tag{11.1}$$

where $x(t) \in \mathbb{R}^n$, $u(t) \in \mathbb{R}^m$ are the state and input vectors, respectively, subject to the constraints

$$x(t) \in \mathcal{X}, \ u(t) \in \mathcal{U}, \ \forall t \geq 0. \tag{11.2}$$

The sets $\mathcal{X} \subseteq \mathbb{R}^n$ and $\mathcal{U} \subseteq \mathbb{R}^m$ are polyhedra.

Remark 11.1 The results of this chapter also hold for more general forms of linear constraints such as mixed input and state constraints

$$[x(t)', u(t)'] \in \mathcal{P}_{x,u}, \tag{11.3}$$

where $\mathcal{P}_{x,u}$ is a polyhedron in $\mathbb{R}^{n+m}$ of mixed input and state constraints over a finite time, or, even more general, constraints of the type:

$$[x(0)', \ldots, x(N-1)', u(0)', \ldots, u(N-1)'] \in \mathcal{P}_{x,u,N}, \qquad (11.4)$$

where $\mathcal{P}_{x,u,N}$ is a polyhedron in $\mathbb{R}^{N(n+m)}$. Note that constraints of the type (11.4) can arise, for example, from constraints on the input rate $\Delta u(t) = u(t) - u(t-1)$. In this chapter, for the sake of simplicity, we will use the less general form (11.2).

Define the cost function

$$J_0(x(0), U_0) = p(x_N) + \sum_{k=0}^{N-1} q(x_k, u_k), \qquad (11.5)$$

where x_k denotes the state vector at time k obtained by starting from the state $x_0 = x(0)$ and applying to the system model

$$x_{k+1} = Ax_k + Bu_k \qquad (11.6)$$

the input sequence $u_0, \ldots, u_{k-1}$.

If the 1-norm or ∞-norm is used in the cost function (11.5), then we set $p(x_N) = \|Px_N\|_p$ and $q(x_k, u_k) = \|Qx_k\|_p + \|Ru_k\|_p$ with $p = 1$ or $p = \infty$ and P, Q, R full column rank matrices. Cost (11.5) is rewritten as

$$J_0(x(0), U_0) = \|Px_N\|_p + \sum_{k=0}^{N-1} \|Qx_k\|_p + \|Ru_k\|_p. \qquad (11.7)$$

If the squared Euclidian norm is used in the cost function (11.5), then we set $p(x_N) = x_N' P x_N$ and $q(x_k, u_k) = x_k' Q x_k + u_k' R u_k$ with $P \succeq 0$, $Q \succeq 0$ and $R \succ 0$. Cost (11.5) is rewritten as

$$J_0(x(0), U_0) = x_N' P x_N + \sum_{k=0}^{N-1} x_k' Q x_k + u_k' R u_k. \qquad (11.8)$$

Consider the constrained finite time optimal control problem (CFTOC)

$$\begin{aligned}
J_0^*(x(0)) = \min_{U_0} \quad & J_0(x(0), U_0) \\
\text{subj. to} \quad & x_{k+1} = Ax_k + Bu_k, \ k = 0, \ldots, N-1 \\
& x_k \in \mathcal{X}, \ u_k \in \mathcal{U}, \ k = 0, \ldots, N-1 \\
& x_N \in \mathcal{X}_f \\
& x_0 = x(0),
\end{aligned} \qquad (11.9)$$

where N is the time horizon and $\mathcal{X}_f \subseteq \mathbb{R}^n$ is a terminal polyhedral region. In (11.5)–(11.9) $U_0 = [u_0', \ldots, u_{N-1}']' \in \mathbb{R}^s$, $s = mN$ is the optimization vector. We denote with $\mathcal{X}_0 \subseteq \mathcal{X}$ the set of initial states $x(0)$ for which the optimal control problem (11.5)–(11.9) is feasible, i.e.,

$$\begin{aligned}
\mathcal{X}_0 = \{x_0 \in \mathbb{R}^n : \ & \exists(u_0, \ldots, u_{N-1}) \text{ such that } x_k \in \mathcal{X}, \ u_k \in \mathcal{U}, \ k = 0, \ldots, N-1, \\
& x_N \in \mathcal{X}_f \text{ where } x_{k+1} = Ax_k + Bu_k, \ k = 0, \ldots, N-1\}.
\end{aligned}$$

$$(11.10)$$

Remark 11.2   Note that we distinguish between the *current* state $x(k)$ of system (11.1) at time k and the variable x_k in the optimization problem (11.9), that is the *predicted* state of system (11.1) at time k obtained by starting from the state $x_0 = x(0)$ and applying to system (11.6) the input sequence $u_0, \ldots, u_{k-1}$. Analogously, $u(k)$ is the input applied to system (11.1) at time k while u_k is the k-th optimization variable of the optimization problem (11.9).

If we use cost (11.8) with the squared Euclidian norm and set

$$\{(x,u) \in \mathbb{R}^{n+m} \ : \ x \in \mathcal{X}, \ u \in \mathcal{U}\} = \mathbb{R}^{n+m}, \ \mathcal{X}_f = \mathbb{R}^n, \tag{11.11}$$

problem (11.9) becomes the standard unconstrained finite time optimal control problem (Chapter 8) whose solution (under standard assumptions on A, B, P, Q and R) can be expressed through the time varying state feedback control law (8.28)

$$u^*(k) = F_k x(k) \quad k = 0, \ldots, N-1. \tag{11.12}$$

From (8.25) the optimal cost is given by

$$J_0^*(x(0)) = x(0)' P_0 x(0). \tag{11.13}$$

If we let $N \to \infty$ as discussed in Section 8.5, then problem (11.8), (11.9), (11.11) becomes the standard infinite horizon linear quadratic regulator (LQR) problem whose solution (under standard assumptions on A, B, P, Q and R) can be expressed as the state feedback control law (see (8.33))

$$u^*(k) = F_\infty x(k), \quad k = 0, 1, \ldots \tag{11.14}$$

In the following chapters we will show that the solution to problem (11.9) can again be expressed in feedback form where now $u^*(k)$ is a continuous piecewise affine function on polyhedra of the state $x(k)$, i.e., $u^*(k) = f_k(x(k))$ where

$$f_k(x) = F_k^j x + g_k^j \quad \text{if} \quad H_k^j x \le K_k^j, \ j = 1, \ldots, N_k^r. \tag{11.15}$$

Matrices H_k^j and K_k^j in equation (11.15) describe the j-th polyhedron $CR_k^j = \{x \in \mathbb{R}^n : H_k^j x \le K_k^j\}$ inside which the feedback optimal control law $u^*(k)$ at time k has the affine form $F_k^j x + g_k^j$. The set of polyhedra CR_k^j, $j = 1, \ldots, N_k^r$ is a *polyhedral partition* of the set of feasible states $\mathcal{X}_k$ of problem (11.9) at time k. The sets $\mathcal{X}_k$ are discussed in detail in the next section. Since the functions $f_k(x(k))$ are continuous, the use of polyhedral partitions rather than strict polyhedral partitions (Definition 4.5) will not cause any problem, indeed it will simplify the exposition.

In the rest of this chapter we will characterize the structure of the value function and describe how the optimal control law can be efficiently computed by means of multiparametric linear and quadratic programming. We will distinguish the cases 1- or ∞-norm and squared 2-norm.

11.2 Feasible Solutions

We denote with $\mathcal{X}_i$ the set of states x_i at time i for which (11.9) is feasible, for $i = 0, \ldots, N$. The sets $\mathcal{X}_i$ for $i = 0, \ldots, N$ play an important role in the solution of (11.9). They are independent of the cost function (as long as it guarantees the existence of a minimum) and of the algorithm used to compute the solution to problem (11.9). There are two ways to rigorously define and compute the sets $\mathcal{X}_i$: the *batch approach* and the *recursive approach*. In the batch approach

$$\mathcal{X}_i = \{x_i \in \mathcal{X} : \exists (u_i, \ldots, u_{N-1}) \text{ such that } x_k \in \mathcal{X}, \ u_k \in \mathcal{U}, \ k = i, \ldots, N-1,$$
$$x_N \in \mathcal{X}_f \text{ where } x_{k+1} = Ax_k + Bu_k, \ k = i, \ldots, N-1\}.$$
$$(11.16)$$

The definition of $\mathcal{X}_i$ in (11.16) requires that for any initial state $x_i \in \mathcal{X}_i$ there exists a feasible sequence of inputs $U_i = [u'_i, \ldots, u'_{N-1}]$ which keeps the state evolution in the feasible set $\mathcal{X}$ at future time instants $k = i+1, \ldots, N-1$ and forces x_N into $\mathcal{X}_f$ at time N. Clearly $\mathcal{X}_N = \mathcal{X}_f$. Next we show how to compute $\mathcal{X}_i$ for $i = 0, \ldots, N-1$. Let the state and input constraint sets $\mathcal{X}$, $\mathcal{X}_f$ and $\mathcal{U}$ be the $\mathcal{H}$-polyhedra $A_x x \leq b_x$, $A_f x_N \leq b_f$, $A_u u \leq b_u$, respectively. Define the polyhedron $\mathcal{P}_i$ for $i = 0, \ldots, N-1$ as follows

$$\mathcal{P}_i = \{(U_i, x_i) \in \mathbb{R}^{m(N-i)+n} : \ G_i U_i - E_i x_i \leq w_i\}, \qquad (11.17)$$

where G_i, E_i and w_i are defined as follows

$$G_i = \begin{bmatrix} A_u & 0 & \ldots & 0 \\ 0 & A_u & \ldots & 0 \\ \vdots & \vdots & \vdots & \vdots \\ 0 & 0 & \ldots & A_u \\ 0 & 0 & \ldots & 0 \\ A_x B & 0 & \ldots & 0 \\ A_x AB & A_x B & \ldots & 0 \\ \vdots & \vdots & \vdots & \vdots \\ A_f A^{N-i-1}B & A_f A^{N-i-2}B & \ldots & A_f B \end{bmatrix}, \ E_i = \begin{bmatrix} 0 \\ 0 \\ \vdots \\ 0 \\ -A_x \\ -A_x A \\ -A_x A^2 \\ \vdots \\ -A_f A^{N-i} \end{bmatrix}, \ w_i = \begin{bmatrix} b_u \\ b_u \\ \vdots \\ b_u \\ b_x \\ b_x \\ b_x \\ \vdots \\ b_f \end{bmatrix}.$$
$$(11.18)$$

The set $\mathcal{X}_i$ is a polyhedron as it is the projection of the polyhedron $\mathcal{P}_i$ in (11.17)–(11.18) on the x_i space.

In the *recursive approach*,

$$\mathcal{X}_i = \{x \in \mathcal{X} : \exists u \in \mathcal{U} \text{ such that } Ax + Bu \in \mathcal{X}_{i+1}\},$$
$$i = 0, \ldots, N-1$$
$$\mathcal{X}_N = \mathcal{X}_f. \qquad (11.19)$$

The definition of $\mathcal{X}_i$ in (11.19) is recursive and requires that for any feasible initial state $x_i \in \mathcal{X}_i$ there exists a feasible input u_i which keeps the next state $Ax_i + Bu_i$ in the feasible set $\mathcal{X}_{i+1}$. It can be compactly written as

$$\mathcal{X}_i = \text{Pre}(\mathcal{X}_{i+1}) \cap \mathcal{X}. \qquad (11.20)$$

Initializing $\mathcal{X}_N$ to $\mathcal{X}_f$ and solving (11.19) backward in time yields *the same* sets $\mathcal{X}_i$ as the batch approach. This recursive formulation, however, leads to an alternative approach for computing the sets $\mathcal{X}_i$. Let $\mathcal{X}_i$ be the $\mathcal{H}$-polyhedra $A_{\mathcal{X}_i}x \leq b_{\mathcal{X}_i}$. Then the set $\mathcal{X}_{i-1}$ is the projection of the following polyhedron

$$\begin{bmatrix} A_u \\ 0 \\ A_{\mathcal{X}_i}B \end{bmatrix} u_i + \begin{bmatrix} 0 \\ A_x \\ A_{\mathcal{X}_i}A \end{bmatrix} x_i \leq \begin{bmatrix} b_u \\ b_x \\ b_{\mathcal{X}_i} \end{bmatrix} \qquad (11.21)$$

on the x_i space.

Consider problem (11.9). The set $\mathcal{X}_0$ is the set of all initial states x_0 for which (11.9) is feasible. The sets $\mathcal{X}_i$ with $i = 1, \ldots, N-1$ are hidden. A given $\bar{U}_0 = [\bar{u}_0, \ldots, \bar{u}_{N-1}]$ is feasible for problem (11.9) if and only if at all time instants i, the state x_i obtained by applying $\bar{u}_0, \ldots, \bar{u}_{i-1}$ to the system model $x_{k+1} = Ax_k + Bu_k$ with initial state $x_0 \in \mathcal{X}_0$ belongs to $\mathcal{X}_i$. Also, $\mathcal{X}_i$ is the set of feasible initial states for problem

$$\begin{aligned} J_i^*(x(0)) = \min_{U_i} \quad & p(x_N) + \sum_{k=i}^{N-1} q(x_k, u_k) \\ \text{subj. to} \quad & x_{k+1} = Ax_k + Bu_k, \ k = i, \ldots, N-1 \\ & x_k \in \mathcal{X}, \ u_k \in \mathcal{U}, \ k = i, \ldots, N-1 \\ & x_N \in \mathcal{X}_f. \end{aligned} \qquad (11.22)$$

Next, we provide more insights into the set $\mathcal{X}_i$ by using the invariant set theory of Section 10.2. We consider two cases: (1) $\mathcal{X}_f = \mathcal{X}$ which corresponds to effectively "removing" the terminal constraint set and (2) $\mathcal{X}_f$ chosen to be a control invariant set.

Theorem 11.1   *[172, Theorem 5.3]. Let the terminal constraint set $\mathcal{X}_f$ be equal to $\mathcal{X}$. Then,*

1. *The feasible set $\mathcal{X}_i$, $i = 0, \ldots, N-1$ is equal to the $(N-i)$-step controllable set:*

$$\mathcal{X}_i = \mathcal{K}_{N-i}(\mathcal{X}).$$

2. *The feasible set $\mathcal{X}_i$, $i = 0, \ldots, N-1$ contains the maximal control invariant set:*

$$\mathcal{C}_\infty \subseteq \mathcal{X}_i.$$

3. *The feasible set $\mathcal{X}_i$ is control invariant if and only if the maximal control invariant set is finitely determined and $N-i$ is equal to or greater than its determinedness index $\bar{N}$, i.e.,*

$$\mathcal{X}_i \subseteq Pre(\mathcal{X}_i) \Leftrightarrow \mathcal{C}_\infty = \mathcal{K}_{N-i}(\mathcal{X}) \quad \text{for all } i \leq N - \bar{N}.$$

4. *$\mathcal{X}_i \subseteq \mathcal{X}_j$ if $i < j$ for $i = 0, \ldots, N-1$. The size of the feasible set $\mathcal{X}_i$ stops decreasing (with decreasing i) if and only if the maximal control invariant set is finitely determined and $N-i$ is larger than its determinedness index, i.e.,*

$$\mathcal{X}_i \subset \mathcal{X}_j \text{ if } N - \bar{N} < i < j < N.$$

Furthermore,

$$\mathcal{X}_i = \mathcal{C}_\infty \ \text{if } i \le N - \bar{N}.$$

Theorem 11.2 *[172, Theorem 5.4]. Let the terminal constraint set $\mathcal{X}_f$ be a control invariant subset of $\mathcal{X}$. Then,*

1. *The feasible set $\mathcal{X}_i$, $i = 0, \ldots, N - 1$ is equal to the $(N - i)$-step stabilizable set:*

$$\mathcal{X}_i = \mathcal{K}_{N-i}(\mathcal{X}_f).$$

2. *The feasible set $\mathcal{X}_i$, $i = 0, \ldots, N - 1$ is control invariant and contained within the maximal control invariant set:*

$$\mathcal{X}_i \subseteq \mathcal{C}_\infty.$$

3. *$\mathcal{X}_i \supseteq \mathcal{X}_j$ if $i < j$, $i = 0, \ldots, N - 1$. The size of the feasible $\mathcal{X}_i$ set stops increasing (with decreasing i) if and only if the maximal stabilizable set is finitely determined and $N - i$ is larger than its determinedness index, i.e.,*

$$\mathcal{X}_i \supset \mathcal{X}_j \ \text{if } N - \bar{N} < i < j < N.$$

Furthermore,

$$\mathcal{X}_i = \mathcal{K}_\infty(\mathcal{X}_f) \ \text{if } i \le N - \bar{N}.$$

Remark 11.3   Theorems 11.1 and 11.2 help us understand how the feasible sets $\mathcal{X}_i$ propagate backward in time as a function of the terminal set $\mathcal{X}_f$. In particular, when $\mathcal{X}_f = \mathcal{X}$ the set $\mathcal{X}_i$ shrinks as i becomes smaller and stops shrinking when it becomes the maximal control invariant set. Also, depending on i, either it is not a control invariant set or it is the maximal control invariant set. We have the opposite if a control invariant set is chosen as terminal constraint $\mathcal{X}_f$. The set $\mathcal{X}_i$ grows as i becomes smaller and stops growing when it becomes the maximal stabilizable set. Both cases are shown in the Example 11.1 below.

Remark 11.4   In this section we investigated the behavior of $\mathcal{X}_i$ as i varies for a fixed horizon N. Equivalently, we could study the behavior of $\mathcal{X}_0$ as the horizon N varies. Specifically, the sets $\mathcal{X}_{0 \to N_1}$ and $\mathcal{X}_{0 \to N_2}$ with $N_2 > N_1$ are equal to the sets $\mathcal{X}_{N_2 - N_1 \to N}$ and $\mathcal{X}_{0 \to N}$, respectively, with $N = N_2$.

Example 11.1   Consider the double integrator

$$\begin{cases} x(t+1) = \begin{bmatrix} 1 & 1 \\ 0 & 1 \end{bmatrix} x(t) + \begin{bmatrix} 0 \\ 1 \end{bmatrix} u(t) \\ \quad y(t) = \begin{bmatrix} 1 & 0 \end{bmatrix} x(t) \end{cases} \tag{11.23}$$

subject to the input constraints

$$-1 \le u(k) \le 1 \ \text{for all } k \ge 0 \tag{11.24}$$

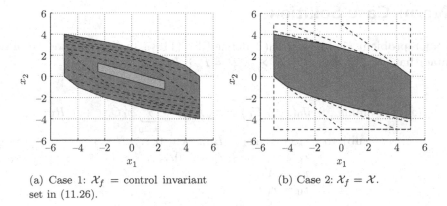

(a) Case 1: $\mathcal{X}_f$ = control invariant
set in (11.26).

(b) Case 2: $\mathcal{X}_f = \mathcal{X}$.

Figure 11.1 Example 11.1. Propagation of the feasible sets $\mathcal{X}_i$ to arrive at
$\mathcal{C}_\infty$ (shaded dark).

and the state constraints

$$\begin{bmatrix} -5 \\ -5 \end{bmatrix} \leq x(k) \leq \begin{bmatrix} 5 \\ 5 \end{bmatrix} \text{ for all } k \geq 0. \tag{11.25}$$

We compute the feasible sets $\mathcal{X}_i$ and plot them in Figure 11.1 in two cases.

Case 1. $\mathcal{X}_f$ is the control invariant set

$$\begin{bmatrix} -0.32132 & -0.94697 \\ 0.32132 & 0.94697 \\ 1 & 0 \\ -1 & 0 \end{bmatrix} x \leq \begin{bmatrix} 0.3806 \\ 0.3806 \\ 2.5 \\ 2.5 \end{bmatrix}. \tag{11.26}$$

After six iterations the sets $\mathcal{X}_i$ converge to the following $\mathcal{K}_\infty(\mathcal{X}_f)$

$$\begin{bmatrix} -0.44721 & -0.89443 \\ -0.24254 & -0.97014 \\ -0.31623 & -0.94868 \\ 0.24254 & 0.97014 \\ 0.31623 & 0.94868 \\ 0.44721 & 0.89443 \\ 1 & 0 \\ -1 & 0 \\ 0.70711 & 0.70711 \\ -0.70711 & -0.70711 \end{bmatrix} x \leq \begin{bmatrix} 2.6833 \\ 2.6679 \\ 2.5298 \\ 2.6679 \\ 2.5298 \\ 2.6833 \\ 5 \\ 5 \\ 3.5355 \\ 3.5355 \end{bmatrix}. \tag{11.27}$$

Note that in this case $\mathcal{C}_\infty = \mathcal{K}_\infty(\mathcal{X}_f)$ and the determinedness index is six.

Case 2. $\mathcal{X}_f = \mathcal{X}$. After six iterations the sets $\mathcal{X}_i$ converge to $\mathcal{K}_\infty(\mathcal{X}_f)$ in (11.27).

11.3 2-Norm Case Solution

Consider problem (11.9) with $J_0(\cdot)$ defined by (11.8). In this chapter we always assume that $Q = Q' \succeq 0$, $R = R' \succ 0$, $P = P' \succeq 0$.

$$
\begin{aligned}
J_0^*(x(0)) = \min_{U_0} \quad & J_0(x(0), U_0) = x_N' P x_N + \sum_{k=0}^{N-1} x_k' Q x_k + u_k' R u_k \\
\text{subj. to} \quad & x_{k+1} = A x_k + B u_k, \ k = 0, \ldots, N-1 \\
& x_k \in \mathcal{X}, \ u_k \in \mathcal{U}, \ k = 0, \ldots, N-1 \\
& x_N \in \mathcal{X}_f \\
& x_0 = x(0).
\end{aligned} \tag{11.28}
$$

11.3.1 Solution via QP

As shown in Section 8.2, problem (11.28) can be rewritten as

$$
\begin{aligned}
J_0^*(x(0)) = \min_{U_0} \quad & J_0(x(0), U_0) = U_0' H U_0 + 2 x'(0) F U_0 + x'(0) Y x(0) \\
= \min_{U_0} \quad & J_0(x(0), U_0) = [U_0' \ x'(0)] \begin{bmatrix} H & F' \\ F & Y \end{bmatrix} [U_0' \ x(0)']' \\
\text{subj. to} \quad & G_0 U_0 \leq w_0 + E_0 x(0),
\end{aligned} \tag{11.29}
$$

with G_0, w_0 and E_0 defined in (11.18) for $i = 0$ and H, F, Y defined in (8.8). As $J_0(x(0), U_0) \geq 0$ by definition it follows that $\begin{bmatrix} H & F' \\ F & Y \end{bmatrix} \succeq 0$.

For a given vector $x(0)$ the optimal input sequence U_0^* solving problem (11.29) can be computed by using a Quadratic Program (QP) solver (see Section 2.3 for QP definition and properties and Chapter 3 for fast numerical methods for solving QPs).

To obtain the problem (11.29) we have eliminated the state variables and equality constraints $x_{k+1} = A x_k + B u_k$ by successive substitution so that we are left with $u_0, \ldots, u_{N-1}$ as the only decision variables and $x(0)$ as a parameter vector. In general, it might be more efficient to solve a QP problem with equality and inequality constraints so that sparsity can be exploited. To this aim we can define the variable $\tilde{z}$ as

$$
\tilde{z} = \begin{bmatrix} x_1' & \cdots & x_N' & u_0' & \cdots & u_{N-1}' \end{bmatrix}'
$$

and rewrite problem (11.28) as

$$
\begin{aligned}
J_0^*(x(0)) = \min_{\tilde{z}} \quad & [\tilde{z}' \ x(0)'] \begin{bmatrix} \bar{H} & 0 \\ 0 & \bar{Q} \end{bmatrix} [\tilde{z}' \ x(0)']' \\
\text{subj. to} \quad & G_{0,\text{eq}} \tilde{z} = E_{0,\text{eq}} x(0) \\
& G_{0,\text{in}} \tilde{z} \leq w_{0,\text{in}} + E_{0,\text{in}} x(0).
\end{aligned} \tag{11.30}
$$

For a given vector $x(0)$ the optimal input sequence U_0^* solving problem (11.30) can be computed by using a Quadratic Program (QP) solver.

To obtain problem (11.30) we have rewritten the equalities from system dynamics $x_{k+1} = Ax_k + Bu_k$ as $G_{0,\mathrm{eq}}\tilde{z} = E_{0,\mathrm{eq}}x(0)$ where

$$G_{0,\mathrm{eq}} = \begin{bmatrix} I & & & & & -B & & & \\ -A & I & & & & & -B & & \\ & -A & I & & & & & -B & \\ & & \ddots & \ddots & & & & & \ddots \\ & & & -A & I & & & & & -B \end{bmatrix}, \quad E_{0,\mathrm{eq}} = \begin{bmatrix} A \\ 0 \\ \vdots \\ 0 \end{bmatrix},$$

and rewritten state and input constraints as $G_{0,\mathrm{in}}\tilde{z} \le w_{0,\mathrm{in}} + E_{0,\mathrm{in}}x(0)$ where

$$G_{0,\mathrm{in}} = \left[\begin{array}{ccccc|ccccc} 0 & & & & & 0 & & & \\ A_x & 0 & & & & 0 & & & \\ & A_x & & & & & 0 & & \\ & & \ddots & & & & & \ddots & \\ & & & A_x & & & & & 0 \\ & & & & A_f & & & & & 0 \\ \hline 0 & & & & & A_u & & & \\ & 0 & & & & & A_u & & \\ & & \ddots & & & & & \ddots & \\ & & & 0 & & & & & A_u & \\ & & & & 0 & & & & & A_u \end{array}\right], \quad w_{0,\mathrm{in}} = \begin{bmatrix} b_x \\ b_x \\ \vdots \\ b_x \\ b_f \\ b_u \\ b_u \\ \vdots \\ b_u \\ b_u \end{bmatrix},$$

$$E_{0,\mathrm{in}} = \begin{bmatrix} -A_x' & 0 & \cdots & 0 \end{bmatrix}',$$

and constructed the cost matrix $\bar{H}$ as

$$\bar{H} = \left[\begin{array}{cccc|cccc} Q & & & & & & & \\ & \ddots & & & & & & \\ & & Q & & & & & \\ & & & P & & & & \\ \hline & & & & \bar{R} & & & \\ & & & & & \ddots & & \\ & & & & & & R \end{array}\right].$$

11.3.2   State Feedback Solution via Batch Approach

As shown in Section 8.2, problem (11.28) can be rewritten as

$$J_0^*(x(0)) = \min_{U_0} \quad J_0(x(0), U_0) = [U_0' \ x'(0)] \begin{bmatrix} H & F' \\ F & Y \end{bmatrix} [U_0' \ x(0)']'$$

$$\text{subj. to} \quad G_0 U_0 \le w_0 + E_0 x(0),$$

(11.31)

with G_0, w_0 and E_0 defined in (11.18) for $i = 0$ and H, F, Y defined in (8.8). As $J_0(x(0), U_0) \ge 0$ by definition it follows that $\begin{bmatrix} H & F' \\ F & Y \end{bmatrix} \succeq 0$. Note that the optimizer U_0^* is independent of the term involving Y in (11.31).

We view $x(0)$ as a vector of parameters and our goal is to solve (11.31) for all values of $x(0) \in \mathcal{X}_0$ and to make this dependence *explicit*. The computation of the set $\mathcal{X}_0$ of initial states for which problem (11.31) is feasible was discussed in Section 11.2.

Before proceeding further, it is convenient to define

$$z = U_0 + H^{-1}F'x(0), \tag{11.32}$$

$z \in \mathbb{R}^s$, remove $x(0)'Yx(0)$ and to transform (11.31) to obtain the equivalent problem

$$\hat{J}^*(x(0)) = \min_z \quad z'Hz$$
$$\text{subj. to} \quad G_0 z \le w_0 + S_0 x(0), \tag{11.33}$$

where $S_0 = E_0 + G_0 H^{-1} F'$, and $\hat{J}^*(x(0)) = J_0^*(x(0)) - x(0)'(Y - FH^{-1}F')x(0)$. In the transformed problem the parameter vector $x(0)$ appears only on the right-hand side of the constraints.

Problem (11.33) is a multiparametric quadratic program that can be solved by using the algorithm described in Section 6.3.1. Once the multiparametric problem (11.33) has been solved, the solution $U_0^* = U_0^*(x(0))$ of CFTOC (11.28) and therefore $u^*(0) = u^*(x(0))$ is available explicitly as a function of the initial state $x(0)$ for all $x(0) \in \mathcal{X}_0$.

Theorem 6.7 states that the solution $z^*(x(0))$ of the mp-QP problem (11.33) is a continuous and piecewise affine function on polyhedra of $x(0)$. Clearly the same properties are inherited by the controller. The following corollaries of Theorem 6.7 establish the analytical properties of the optimal control law and of the value function.

Corollary 11.1   *The control law $u^*(0) = f_0(x(0))$, $f_0 : \mathbb{R}^n \to \mathbb{R}^m$, obtained as a solution of the CFTOC (11.28) is continuous and piecewise affine on polyhedra*

$$f_0(x) = F_0^j x + g_0^j \quad \text{if} \quad x \in CR_0^j, \quad j = 1, \dots, N_0^r, \tag{11.34}$$

where the polyhedral sets $CR_0^j = \{x \in \mathbb{R}^n \ : \ H_0^j x \le K_0^j\}$, $j = 1, \dots, N_0^r$ are a partition of the feasible polyhedron $\mathcal{X}_0$.

Proof: From (11.32) $U_0^*(x(0)) = z^*(x(0)) - H^{-1}F'x(0)$. From Theorem 6.7 we know that $z^*(x(0))$, solution of (11.33), is PPWA and continuous. As $U_0^*(x(0))$ is a linear combination of a linear function and a PPWA function, it is PPWA. As $U_0^*(x(0))$ is a linear combination of two continuous functions it is continuous. In particular, these properties hold for the first component $u^*(0)$ of U_0^*. ∎

Remark 11.5   Note that, as discussed in Remark 6.8, the critical regions defined in (6.4) are in general sets that are neither closed nor open. In Corollary 11.1 the polyhedron CR_0^i describes the closure of a critical region. The function $f_0(x)$ is continuous and therefore it is simpler to use a polyhedral partition rather than a strict polyhedral partition.

Corollary 11.2   *The value function $J_0^*(x(0))$ obtained as solution of the CFTOC (11.28) is convex and piecewise quadratic on polyhedra. Moreover, if the mp-QP problem (11.33) is not degenerate, then the value function $J_0^*(x(0))$ is $C^{(1)}$.*

Proof: By Theorem 6.7 $\hat{J}^*(x(0))$ is a convex function of $x(0)$. As $\left[\begin{smallmatrix} H & F' \\ F & Y \end{smallmatrix}\right] \succeq 0$, its Schur complement[1] $Y - FH^{-1}F' \succeq 0$, and therefore $J_0^*(x(0)) = \hat{J}^*(x(0)) + x(0)'(Y - FH^{-1}F')x(0)$ is a convex function, because it is the sum of convex functions. If the mp-QP problem (11.33) is not degenerate, then Theorem 6.9 implies that $\hat{J}^*(x(0))$ is a $C^{(1)}$ function of $x(0)$ and therefore $J_0^*(x(0))$ is a $C^{(1)}$ function of $x(0)$. The results of Corollary 11.1 imply that $J_0^*(x(0))$ is piecewise quadratic. ∎

Remark 11.6   The relation between the design parameters of the optimal control problem (11.28) and the degeneracy of the mp-QP problem (11.33) is complex, in general.

The solution of the multiparametric problem (11.33) provides the state feedback solution $u^*(k) = f_k(x(k))$ of CFTOC (11.28) for $k = 0$ and it also provides the open-loop optimal control $u^*(k)$ as function of the initial state, i.e., $u^*(k) = u^*(k, x(0))$. The state feedback PPWA optimal controllers $u^*(k) = f_k(x(k))$ with $f_k : \mathcal{X}_k \mapsto \mathcal{U}$ for $k = 1, \ldots, N$ are computed in the following way. Consider the same CFTOC (11.28) over the shortened time-horizon $[i, N]$

$$\min_{U_i} \quad x_N'Px_N + \sum_{k=i}^{N-1} x_k'Qx_k + u_k'Ru_k$$

$$\text{subj. to} \quad x_{k+1} = Ax_k + Bu_k, \ k = i, \ldots, N-1 \tag{11.35}$$
$$x_k \subset \mathcal{X}, \ u_k \in \mathcal{U}, \ k = i, \ldots, N-1$$
$$x_N \in \mathcal{X}_f$$
$$x_i = x(i),$$

where $U_i = [u_i', \ldots, u_{N-1}']$. As defined in (11.16) and discussed in Section 11.2, $\mathcal{X}_i \subseteq \mathbb{R}^n$ is the set of initial states $x(i)$ for which the optimal control problem (11.35) is feasible. We denote by U_i^* the optimizer of the optimal control problem (11.35).

Problem (11.35) can be translated into the mp-QP

$$\min \quad U_i'H_iU_i + 2x'(i)F_iU_i + x'(i)Y_ix(i)$$

$$\text{subj. to} \quad G_iU_i \leq w_i + E_ix(i), \tag{11.36}$$

where $H_i = H_i' \succ 0$, F_i, Y_i are appropriately defined for each i and G_i, w_i, E_i are defined in (11.18). The first component of the multiparametric solution of (11.36) has the form

$$u_i^*(x(i)) = f_i(x(i)), \ \forall x(i) \in \mathcal{X}_i, \tag{11.37}$$

where the control law $f_i : \mathbb{R}^n \to \mathbb{R}^m$, is continuous and PPWA

$$f_i(x) = F_i^j x + g_i^j \quad \text{if} \quad x \in CR_i^j, \ j = 1, \ldots, N_i^r, \tag{11.38}$$

---

[1] Let $X = \left[\begin{smallmatrix} A & B' \\ B & C \end{smallmatrix}\right]$ and $A \succ 0$. Then $X \succeq 0$ if and only if the Schur complement $S = C - BA^{-1}B' \succeq 0$.

and where the polyhedral sets $CR_i^j = \{x \in \mathbb{R}^n \; : \; H_i^j x \leq K_i^j\}$, $j = 1, \ldots, N_i^r$ are a partition of the feasible polyhedron $\mathcal{X}_i$. Therefore the feedback solution $u^*(k) = f_k(x(k))$, $k = 0, \ldots, N-1$ of the CFTOC (11.28) is obtained by solving N mp-QP problems of decreasing size. The following corollary summarizes the final result.

Corollary 11.3   *The state feedback control law* $u^*(k) = f_k(x(k))$, $f_k \; : \; \mathcal{X}_k \subseteq \mathbb{R}^n \to \mathcal{U} \subseteq \mathbb{R}^m$, *obtained as a solution of the CFTOC (11.28) and* $k = 0, \ldots, N-1$ *is time-varying, continuous and piecewise affine on polyhedra*

$$f_k(x) = F_k^j x + g_k^j \quad if \quad x \in CR_k^j, \quad j = 1, \ldots, N_k^r, \tag{11.39}$$

where the polyhedral sets $CR_k^j = \{x \in \mathbb{R}^n \; : \; H_k^j x \leq K_k^j\}$, $j = 1, \ldots, N_k^r$ *are a partition of the feasible polyhedron* $\mathcal{X}_k$.

11.3.3   State Feedback Solution via Recursive Approach

Consider the dynamic programming formulation of the CFTOC (11.28)

$$
\begin{aligned}
J_j^*(x_j) = \min_{u_j} \quad & x_j' Q x_j + u_j' R u_j + J_{j+1}^*(A x_j + B u_j) \\
\text{subj. to} \quad & x_j \in \mathcal{X}, \; u_j \in \mathcal{U}, \\
& A x_j + B u_j \in \mathcal{X}_{j+1}
\end{aligned}
\tag{11.40}
$$

for $j = 0, \ldots, N-1$, with boundary conditions

$$J_N^*(x_N) = x_N' P x_N \tag{11.41}$$

$$\mathcal{X}_N = \mathcal{X}_f, \tag{11.42}$$

where $\mathcal{X}_j$ denotes the set of states x for which the CFTOC (11.28) is feasible at time j (as defined in (11.16)). Note that according to Corollary 11.2, $J_{j+1}^*(A x_j + B u_j)$ is piecewise quadratic for $j < N-1$. Therefore (11.40) is not simply an mp-QP and, contrary to the unconstrained case (Section 8.2), the computational advantage of the iterative over the batch approach is not obvious. Nevertheless an algorithm was developed and can be found in Section 17.6.

11.3.4   Infinite Horizon Problem

Assume $Q \succ 0$, $R \succ 0$ and that the constraint sets $\mathcal{X}$ and $\mathcal{U}$ contain the origin in their interior.[2] Consider the following infinite-horizon linear quadratic regulation problem with constraints (CLQR)

$$
\begin{aligned}
J_\infty^*(x(0)) = \min_{u_0, u_1, \ldots} \quad & \sum_{k=0}^{\infty} x_k' Q x_k + u_k' R u_k \\
\text{subj. to} \quad & x_{k+1} = A x_k + B u_k, \; k = 0, \ldots, \infty \\
& x_k \in \mathcal{X}, \; u_k \in \mathcal{U}, \; k = 0, \ldots, \infty \\
& x_0 = x(0)
\end{aligned}
\tag{11.43}
$$

---

[2] As in the unconstrained case, the assumption $Q \succ 0$ can be relaxed by requiring that $(Q^{1/2}, A)$ is observable (Section 8.5).

and the set (see Remark 7.1 for notation)

$$\mathcal{X}_\infty = \{x(0) \in \mathbb{R}^n \ : \ \text{Problem (11.43) is feasible and } J_\infty^*(x(0)) < +\infty\}. \quad (11.44)$$

Because $Q \succ 0$, $R \succ 0$ *any* optimizer u_k^* of problem (11.43) *must* converge to the origin ($u_k^* \to 0$) and so must the state trajectory resulting from the application of u_k^* ($x_k^* \to 0$). Thus the origin $x = 0$, $u = 0$ must lie in the interior of the constraint set $(\mathcal{X}, \mathcal{U})$ (if the origin were not contained in the constraint set then $J_\infty^*(x(0))$ would be infinite). For this reason, the set $\mathcal{X}_\infty$ in (11.44) is the maximal stabilizable set $\mathcal{K}_\infty(\mathcal{O})$ of system (11.1) subject to the constraints (11.2) with $\mathcal{O}$ being the origin (Definition 10.13).

If the initial state $x_0 = x(0)$ is sufficiently close to the origin, then the constraints will never become active and the solution of problem (11.43) will yield the same control input as the *unconstrained* LQR (8.33). More formally we can define a corresponding invariant set around the origin.

Definition 11.1 (Maximal LQR Invariant Set $\mathcal{O}_\infty^{\mathbf{LQR}}$) *Consider the system* $x(k+1) = Ax(k) + Bu(k)$. $\mathcal{O}_\infty^{LQR} \subseteq \mathbb{R}^n$ *denotes the maximal positively invariant set for the autonomous constrained linear system:*

$$x(k+1) = (A + BF_\infty)x(k), \ x(k) \in \mathcal{X}, \ u(k) \in \mathcal{U}, \ \forall\, k \geq 0,$$

where $u(k) = F_\infty x(k)$ *is the unconstrained LQR control law (8.33) obtained from the solution of the ARE (8.32).*

Therefore, from the previous discussion, there is some finite time $\bar{N}(x_0)$, depending on the initial state x_0, at which the state enters $\mathcal{O}_\infty^{LQR}$ and after which the system evolves in an unconstrained manner ($x_k^* \in \mathcal{X}$, $u_k^* \in \mathcal{U}$, $\forall k > \bar{N}$). This consideration allows us to split problem (11.43) into two parts by using the dynamic programming principle, one up to time $k = \bar{N}$ where the constraints may be active and one for longer times $k > \bar{N}$ where there are no constraints.

$$\begin{aligned}
J_\infty^*(x(0)) = \min_{u_0,u_1,\ldots} \ & \sum_{k=0}^{\bar{N}-1} x_k' Q x_k + u_k' R u_k + J_{\bar{N}\to\infty}^*(x_{\bar{N}}) \\
\text{subj. to} \quad & x_k \in \mathcal{X}, \ u_k \in \mathcal{U}, \ k = 0,\ldots,\bar{N}-1 \\
& x_{k+1} = Ax_k + Bu_k, \ k \geq 0 \\
& x_0 = x(0),
\end{aligned} \quad (11.45)$$

where

$$\begin{aligned}
J_{\bar{N}\to\infty}^*(x_{\bar{N}}) = \min_{u_{\bar{N}},u_{\bar{N}+1},\ldots} \ & \sum_{k=\bar{N}}^{\infty} x_k' Q x_k + u_k' R u_k \\
\text{subj. to} \quad & x_{k+1} = Ax_k + Bu_k, \ k \geq \bar{N} \\
= \ & x_{\bar{N}}' P_\infty x_{\bar{N}}.
\end{aligned} \quad (11.46)$$

This key insight due to Sznaier and Damborg [272] is formulated precisely in the following.

Theorem 11.3 (Equality of Finite and Infinite Optimal Control, [260])
For any given initial state $x(0)$, the solution to (11.45, 11.46) is equal to the infinite time solution of (11.43), if the terminal state $x_{\bar{N}}$ of (11.45) lies in the positive invariant set $\mathcal{O}_\infty^{LQR}$ and no terminal set constraint is applied in (11.45), i.e., the state 'voluntarily' enters the set $\mathcal{O}_\infty^{LQR}$ after $\bar{N}$ steps.

Theorem 11.3 suggests that we can obtain the infinite horizon constrained linear quadratic regulator CLQR by solving the finite horizon problem for a horizon of $\bar{N}$ with a terminal weight of $P = P_\infty$ and *no* terminal constraint. The critical question of how to determine $\bar{N}(x_0)$ or at least an upper bound was studied by several researchers. Chmielewski and Manousiouthakis [85] presented an approach that provides a conservative estimate N_{est} of the finite horizon $\bar{N}(x_0)$ for all x_0 belonging to a compact set of initial conditions $\mathcal{S} \subseteq \mathcal{X}_\infty = \mathcal{K}_\infty(\mathbf{0})$ ($N_{\text{est}} \geq \bar{N}_{\mathcal{S}}(x_0)$, $\forall x_0 \in \mathcal{S}$). They solve a single, finite dimensional, convex program to obtain N_{est}. Their estimate can be used to compute the PWA solution of (11.45) for a particular set $\mathcal{S}$.

Alternatively, the quadratic program with horizon N_{est} can be solved to determine u_0^*, $u_1^*, \ldots, u_{N_{\text{est}}}^*$ for a particular $x(0) \in \mathcal{S}$. For a given initial state $x(0)$, rather then a set $\mathcal{S}$, Scokaert and Rawlings [260] presented an algorithm that attempts to identify $\bar{N}(x(0))$ iteratively. In summary, we can state the following Theorem.

Theorem 11.4 (Explicit solution of CLQR) *Assume that (A, B) is a stabilizable pair and $(Q^{1/2}, A)$ is an observable pair, $R \succ 0$. The state feedback solution to the CLQR problem (11.43) in a compact set of the initial conditions $\mathcal{S} \subseteq \mathcal{X}_\infty = \mathcal{K}_\infty(\mathbf{0})$ is time-invariant, continuous and piecewise affine on polyhedra*

$$u^*(k) = f_\infty(x(k)), \quad f_\infty(x) = F^j x + g^j \quad \text{if} \quad x \in CR_\infty^j, \quad j = 1, \ldots, N_\infty^r, \quad (11.47)$$

where the polyhedral sets $CR_\infty^j = \{x \in \mathbb{R}^n : H^j x \leq K^j\}$, $j = 1, \ldots, N_\infty^r$ are a finite partition of the feasible compact polyhedron $\mathcal{S} \subseteq \mathcal{X}_\infty$.

As argued previously, the complexity of the solution manifested by the number of polyhedral regions depends on the chosen horizon. As the various discussed techniques yield an N_{est} that may be too large by orders of magnitude this is not a viable proposition. An efficient algorithm for computing the PPWA solution to the CLQR problem is presented next.

11.3.5 CLQR Algorithm

In this section we will sketch an efficient algorithm to compute the PWA solution to the CLQR problem in (11.43) for a given set $\mathcal{S}$ of initial conditions. Details are available in [132, 133]. As a side product, the algorithm also computes $\bar{N}_{\mathcal{S}}$, the shortest horizon $\bar{N}$ for which the problem (11.45), (11.46) is equivalent to the infinite horizon problem (11.43).

The idea is as follows. For the CFTOC problem (11.28) with a horizon N with no terminal constraint ($\mathcal{X}_f = \mathbb{R}^n$) and terminal cost $P = P_\infty$, where P_∞ is the

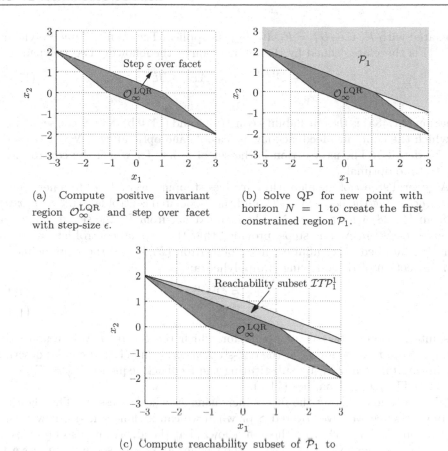

(a)   Compute   positive   invariant region $\mathcal{O}_\infty^{\text{LQR}}$ and step over facet with step-size ϵ.

(b) Solve QP for new point with horizon $N = 1$ to create the first constrained region $\mathcal{P}_1$.

(c) Compute reachability subset of $\mathcal{P}_1$ to obtain $\mathcal{ITP}_1^1$.

Figure 11.2 CLQR Algorithm. Region Exploration.

solution to the ARE (8.32), we solve an mp-QP and obtain the PWA control law. From Theorem 11.3 we can conclude that for all states which enter the invariant set $\mathcal{O}_\infty^{\text{LQR}}$ introduced in Definition 11.1 with the computed control law in N steps, the infinite-horizon problem has been solved. For these states, which we can identify via a reachability analysis, the computed feedback law is infinite-horizon optimal.

In more detail, we start the procedure by computing the Maximal LQR Invariant Set $\mathcal{O}_\infty^{\text{LQR}}$ introduced in Definition 11.1, the polyhedron $\mathcal{P}_0 = \mathcal{O}_\infty^{\text{LQR}} = \{x \in \mathbb{R}^n : H_0 x \le K_0\}$. Figure 11.2(a) depicts $\mathcal{O}_\infty^{\text{LQR}}$. Then, the algorithm finds a point $\bar{x}$ by stepping over a facet of $\mathcal{O}_\infty^{\text{LQR}}$ with a small step ϵ, as described in [19]. If (11.28) is feasible for horizon $N = 1$ (terminal set constraint $\mathcal{X}_f = \mathbb{R}^n$, terminal cost $P = P_\infty$ and $x(0) = \bar{x}$), the active constraints will define the neighboring polyhedron $\mathcal{P}_1 = \{x \in \mathbb{R}^n : H_1 x \le K_1\}$ ($\bar{x} \in \mathcal{P}_1$, see Figure 11.2(b)) [44]. By Theorem 11.3, the finite time optimal solution computed above equals the infinite time optimal solution if $x_1 \in \mathcal{O}_\infty^{\text{LQR}}$. Therefore we extract from $\mathcal{P}_1$ the set of points that will enter $\mathcal{O}_\infty^{\text{LQR}}$ in $N = 1$ time-steps, provided that the optimal control law associated with $\mathcal{P}_1$ (i.e., $U_1^* = F_1 x(0) + g_1$) is applied. The Infinite Time Polyhedron $(\mathcal{ITP}_1^1)$ is therefore defined by the intersection of the following two polyhedra:

$$x_1 \in \mathcal{O}_\infty^{\text{LQR}}, \ x_1 = Ax_0 + BU_1^*, \tag{11.48a}$$

$$x_0 \in \mathcal{P}_1. \tag{11.48b}$$

Equation (11.48a) is the reachability constraint and (11.48b) defines the set of states for which the computed feedback law is feasible and optimal over $N = 1$ steps (see [44] for details). The intersection is the set of points for which the control law is infinite time optimal.

A general step r of the algorithm involves stepping over a facet to a new point $\bar{x}$ and determining the polyhedron $\mathcal{P}_r$ and the associated control law ($U_N^* = F_r x(0) + g_r$) from (11.28) with horizon N. Then we extract from $\mathcal{P}_r$ the set of points that will enter $\mathcal{O}_\infty^{\text{LQR}}$ in N time-steps, provided that the optimal control law associated with $\mathcal{P}_r$ is applied. The Infinite Time Polyhedron $(\mathcal{ITP}_r^N)$ is therefore defined by the intersection of the following two polyhedra:

$$x_N \in \mathcal{O}_\infty^{\text{LQR}} \tag{11.49a}$$

$$x_0 \in \mathcal{P}_r. \tag{11.49b}$$

This intersection is the set of points for which the control law is infinite time optimal. Note that, as for x_1 in the one-step case, x_N in (11.49a) can be described as a linear function of x_0 by substituting the feedback sequence $U_N^* = F_r x_0 + g_r$ into the LTI system dynamics (11.1).

We continue exploring the facets increasing N when necessary. The algorithm terminates when we have covered $\mathcal{S}$ or when we can no longer find a new feasible polyhedron $\mathcal{P}_r$. The following theorem shows that the algorithm also provides the horizon $\bar{N}_\mathcal{S}$ for compact sets. Exact knowledge of $\bar{N}_\mathcal{S}$ can serve to improve the performance of a wide array of algorithms presented in the literature.

Theorem 11.5 (Exact Computation of $\bar{N}_\mathcal{S}$, [132, 133]) *If we explore any given compact set $\mathcal{S}$ with the proposed algorithm, the largest resulting horizon is equal to $\bar{N}_\mathcal{S}$, i.e.,*

$$\bar{N}_\mathcal{S} = \max_{\substack{\mathcal{ITP}_r^N \\ r=0,\dots,R}} N.$$

Often the proposed algorithm is more efficient than standard multiparametric solvers, even if finite horizon optimal controllers are sought. The initial polyhedral representation $\mathcal{P}_r$ contains redundant constraints which need to be removed in order to obtain a minimal representation of the controller region. The intersection with the reachability constraint, as proposed here, can simplify this constraint removal.

11.3.6   Examples

Example 11.2 Consider the double integrator (11.23). We want to compute the state feedback optimal controller that solves problem (11.28) with $N = 6$, $Q = \begin{bmatrix} 1 & 0 \\ 0 & 1 \end{bmatrix}$, $R = 0.1$, P is equal to the solution of the Riccati equation (8.32), $\mathcal{X}_f = \mathbb{R}^2$. The input constraints are

$$-1 \leq u(k) \leq 1, \ k = 0,\dots,5 \tag{11.50}$$

and the state constraints

$$\begin{bmatrix} -10 \\ -10 \end{bmatrix} \leq x(k) \leq \begin{bmatrix} 10 \\ 10 \end{bmatrix}, \ k = 0, \ldots, 5. \tag{11.51}$$

This task is addressed as shown in Section (11.3.2). The feedback optimal solution $u^*(0), \ldots, u^*(5)$ is computed by solving six mp-QP problems and the corresponding polyhedral partitions of the state space are depicted in Figure 11.3. Only the last two optimal control moves are reported below:

$$u^*(5) = \begin{cases} \begin{bmatrix} -0.58 & -1.55 \end{bmatrix} x(5) & \text{if } \begin{bmatrix} -0.35 & -0.94 \\ 0.35 & 0.94 \\ 1.00 & 0.00 \\ -1.00 & 0.00 \end{bmatrix} x(5) \leq \begin{bmatrix} 0.61 \\ 0.61 \\ 10.00 \\ 10.00 \end{bmatrix} & \text{(Region \#1)} \\[3em] 1.00 & \text{if } \begin{bmatrix} 0.35 & 0.94 \\ 0.00 & -1.00 \\ -0.71 & -0.71 \\ 1.00 & 0.00 \\ -1.00 & 0.00 \end{bmatrix} x(5) \leq \begin{bmatrix} -0.61 \\ 10.00 \\ 7.07 \\ 10.00 \\ 10.00 \end{bmatrix} & \text{(Region \#2)} \\[3em] -1.00 & \text{if } \begin{bmatrix} -0.35 & -0.94 \\ 1.00 & 0.00 \\ 0.00 & 1.00 \\ -1.00 & 0.00 \\ 0.71 & 0.71 \end{bmatrix} x(5) \leq \begin{bmatrix} -0.61 \\ 10.00 \\ 10.00 \\ 10.00 \\ 7.07 \end{bmatrix} & \text{(Region \#3)} \end{cases}$$

$$u^*(4) = \begin{cases} \begin{bmatrix} -0.58 & -1.55 \end{bmatrix} x(4) & \text{if } \begin{bmatrix} -0.35 & -0.94 \\ 0.35 & 0.94 \\ -0.77 & -0.64 \\ 0.77 & 0.64 \end{bmatrix} x(4) \leq \begin{bmatrix} 0.61 \\ 0.61 \\ 2.43 \\ 2.43 \end{bmatrix} & \text{(Region \#1)} \\[3em] 1.00 & \text{if } \begin{bmatrix} 0.29 & 0.96 \\ 0.00 & -1.00 \\ -0.71 & -0.71 \\ -0.45 & -0.89 \\ 1.00 & 0.00 \\ -1.00 & 0.00 \end{bmatrix} x(4) \leq \begin{bmatrix} -0.98 \\ 10.00 \\ 7.07 \\ 4.92 \\ 10.00 \\ 10.00 \end{bmatrix} & \text{(Region \#2)} \\[3em] 1.00 & \text{if } \begin{bmatrix} 0.29 & 0.96 \\ 0.35 & 0.94 \\ 0.00 & -1.00 \\ -0.71 & -0.71 \\ -0.45 & -0.89 \\ 1.00 & 0.00 \\ -1.00 & 0.00 \end{bmatrix} x(4) \leq \begin{bmatrix} -0.37 \\ -0.61 \\ 10.00 \\ 7.07 \\ 4.92 \\ 10.00 \\ 10.00 \end{bmatrix} & \text{(Region \#3)} \\[3em] -1.00 & \text{if } \begin{bmatrix} -0.29 & -0.96 \\ 1.00 & 0.00 \\ 0.00 & 1.00 \\ -1.00 & 0.00 \\ 0.71 & 0.71 \\ 0.45 & 0.89 \end{bmatrix} x(4) \leq \begin{bmatrix} -0.98 \\ 10.00 \\ 10.00 \\ 10.00 \\ 7.07 \\ 4.92 \end{bmatrix} & \text{(Region \#4)} \\[3em] 1.00 & \text{if } \begin{bmatrix} -0.29 & -0.96 \\ -0.35 & -0.94 \\ 1.00 & 0.00 \\ 0.00 & 1.00 \\ -1.00 & 0.00 \\ 0.71 & 0.71 \\ 0.45 & 0.89 \end{bmatrix} x(4) \leq \begin{bmatrix} -0.37 \\ -0.61 \\ 10.00 \\ 10.00 \\ 10.00 \\ 7.07 \\ 4.92 \end{bmatrix} & \text{(Region \#5)} \\[3em] \begin{bmatrix} -0.44 & -1.43 \end{bmatrix} x(4) - 0.46 & \text{if } \begin{bmatrix} -0.29 & -0.96 \\ 0.29 & 0.96 \\ -0.77 & -0.64 \\ 1.00 & 0.00 \end{bmatrix} x(4) \leq \begin{bmatrix} 0.98 \\ 0.37 \\ -2.43 \\ 10.00 \end{bmatrix} & \text{(Region \#6)} \\[3em] \begin{bmatrix} -0.44 & -1.43 \end{bmatrix} x(4) + 0.46 & \text{if } \begin{bmatrix} -0.29 & -0.96 \\ 0.29 & 0.96 \\ 0.77 & 0.64 \\ -1.00 & 0.00 \end{bmatrix} x(4) \leq \begin{bmatrix} 0.37 \\ 0.98 \\ -2.43 \\ 10.00 \end{bmatrix} & \text{(Region \#7)} \end{cases}$$

Note that by increasing the horizon N, the control law changes only far away from the origin. This must be expected from the results of Section 11.3.5. The control law does not change anymore with increasing N in the set where the CFTOC law becomes equal to the constrained infinite-horizon linear quadratic regulator (CLQR) problem. This set gets larger as N increases [85, 260].

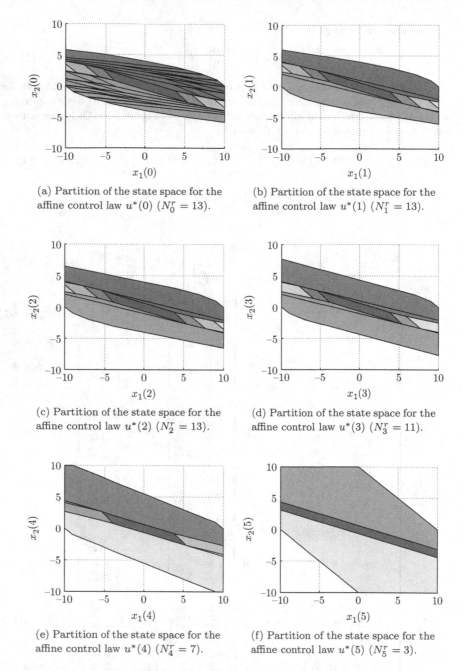

(a) Partition of the state space for the affine control law $u^*(0)$ ($N_0^r = 13$).

(b) Partition of the state space for the affine control law $u^*(1)$ ($N_1^r = 13$).

(c) Partition of the state space for the affine control law $u^*(2)$ ($N_2^r = 13$).

(d) Partition of the state space for the affine control law $u^*(3)$ ($N_3^r = 11$).

(e) Partition of the state space for the affine control law $u^*(4)$ ($N_4^r = 7$).

(f) Partition of the state space for the affine control law $u^*(5)$ ($N_5^r = 3$).

Figure 11.3 Example 11.2. Double Integrator, 2-norm objective function, horizon $N = 6$. Partition of the state space for the time-varying optimal control law. Polyhedra with the same control law were merged.

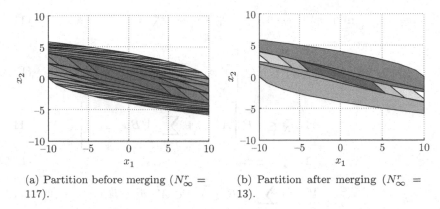

(a) Partition before merging ($N^r_\infty =$ 117).

(b) Partition after merging ($N^r_\infty =$ 13).

Figure 11.4 Example 11.3. Double Integrator, 2-norm objective function, horizon $N = \infty$. Partition of the state space for the time invariant optimal control law.

Example 11.3 The infinite time CLQR (11.43) was determined for Example 11.2 by using the approach presented in Section 11.3.4. The resulting $\bar{N}_S$ is 12. The state space is divided into 117 polyhedral regions and is depicted in Figure 11.4(a). In Figure 11.4(b) the same control law is represented where polyhedra with the same affine control law were merged.

11.4 1-Norm and ∞-Norm Case Solution

Next, we consider problem (11.9) with $J_0(\cdot)$ defined by (11.7) with $p = 1$ or $p = \infty$. In the following section we will concentrate on the ∞-norm, the results can be extended easily to cost functions with 1-norm or mixed $1/\infty$ norms.

$$J_0^*(x(0)) = \min_{U_0} \quad J_0(x(0), U_0) = \|Px_N\|_p + \sum_{k=0}^{N-1} \|Qx_k\|_p + \|Ru_k\|_p$$

$$
\begin{aligned}
\text{subj. to} \quad & x_{k+1} = Ax_k + Bu_k, \; k = 0, \dots, N-1 \\
& x_k \in \mathcal{X}, \; u_k \in \mathcal{U}, \; k = 0, \dots, N-1 \\
& x_N \in \mathcal{X}_f \\
& x_0 = x(0).
\end{aligned}
\tag{11.52}
$$

11.4.1 Solution via LP

The optimal control problem (11.52) with $p = \infty$ can be rewritten as a linear program by using the approach presented in Section 9.2. Therefore, problem (11.52) can be reformulated as the following LP problem

$$\min_{z_0} \quad \varepsilon_0^x + \cdots + \varepsilon_N^x + \varepsilon_0^u + \cdots + \varepsilon_{N-1}^u \tag{11.53a}$$

$$\text{subj. to} \quad -\mathbf{1}_n \varepsilon_k^x \leq \pm Q \left[A^k x_0 + \sum_{j=0}^{k-1} A^j B u_{k-1-j} \right], \tag{11.53b}$$

$$-\mathbf{1}_r \varepsilon_N^x \leq \pm P \left[A^N x_0 + \sum_{j=0}^{N-1} A^j B u_{N-1-j} \right], \tag{11.53c}$$

$$-\mathbf{1}_m \varepsilon_k^u \leq \pm R u_k, \tag{11.53d}$$

$$A^k x_0 + \sum_{j=0}^{k-1} A^j B u_{k-1-j} \in \mathcal{X}, \; u_k \in \mathcal{U}, \tag{11.53e}$$

$$A^N x_0 + \sum_{j=0}^{N-1} A^j B u_{N-1-j} \in \mathcal{X}_f, \tag{11.53f}$$

$$k = 0, \ldots, N-1$$

$$x_0 = x(0), \tag{11.53g}$$

where constraints (11.53b)–(11.53f) are componentwise, and $\pm$ means that the constraint appears once with each sign.

Problem (11.53) can be rewritten in the more compact form

$$\begin{aligned} \min_{z_0} \quad & c_0' z_0 \\ \text{subj. to} \quad & \bar{G}_0 z_0 \leq \bar{w}_0 + \bar{S}_0 x(0), \end{aligned} \tag{11.54}$$

where $z_0 = \{\varepsilon_0^x, \ldots, \varepsilon_N^x, \varepsilon_0^u, \ldots, \varepsilon_{N-1}^u, u_0', \ldots, u_{N-1}'\} \in \mathbb{R}^s$, $s = (m+1)N + N + 1$ and

$$\bar{G}_0 = \begin{bmatrix} G_\varepsilon^x & G_\varepsilon^u & G_\varepsilon^c \\ 0 & 0 & G_0 \end{bmatrix}, \quad \bar{S}_0 = \begin{bmatrix} S_\varepsilon \\ E_0 \end{bmatrix}, \quad \bar{w}_0 = \begin{bmatrix} w_\varepsilon \\ w_0 \end{bmatrix}, \tag{11.55}$$

where $[G_\varepsilon^x, \; G_\varepsilon^u, \; G_\varepsilon^c]$ is the block partition of the matrix G_ε into the three parts corresponding to the variables ε_i^x, ε_i^u and u_i, respectively. The vector c_0 and the submatrices G_ε, w_ε, S_ε associated with the constraints (11.53b)–(11.53d) are defined in (9.10). The matrices G_0, w_0 and E_0 are defined in (11.18) for $i = 0$.

For a given vector $x(0)$ the optimal input sequence U_0^* solving problem (11.54) can be computed by using a Linear Program (LP) solver (see Section 2.2 for LP definition and properties and Chapter 3 for fast numerical methods for solving LPs).

To obtain the problem (11.54) we have eliminated the state variables and equality constraints $x_{k+1} = A x_k + B u_k$ by successive substitution so that we are left with $u_0, \ldots, u_{N-1}$ and the slack variables ϵ as the only decision variables, and $x(0)$ as a parameter vector. As in the 2-norm case, it might be more efficient to solve an LP problem with equality and inequality constraints so that sparsity can be exploited. We omit the details and refer the reader to the construction of the QP problem without substitution in Section 11.3.1.

11.4.2   State Feedback Solution via Batch Approach

As shown in the previous section, problem (11.52) can be rewritten in the compact form

$$\min_{z_0} \quad c_0' z_0$$
$$\text{subj. to} \quad \bar{G}_0 z_0 \leq \bar{w}_0 + \bar{S}_0 x(0), \tag{11.56}$$

where $z_0 = \{\varepsilon_0^x, \dots, \varepsilon_N^x, \varepsilon_0^u, \dots, \varepsilon_{N-1}^u, u_0', \dots, u_{N-1}'\} \in \mathbb{R}^s$, $s = (m+1)N + N + 1$. As in the 2-norm case, by treating $x(0)$ as a vector of parameters, problem (11.56) becomes a *multiparametric linear program* (mp-LP) that can be solved as described in Section 6.2. Once the multiparametric problem (11.56) has been solved, the explicit solution $z_0^*(x(0))$ of (11.56) is available as a piecewise affine function of $x(0)$, and the optimal control law $u^*(0)$ is also available explicitly, as the optimal input $u^*(0)$ consists simply of m components of $z_0^*(x(0))$

$$u^*(0) = [0 \ \dots 0 \ I_m \ 0 \ \dots \ 0] z_0^*(x(0)). \tag{11.57}$$

Theorem 6.5 states that there always exists a continuous PPWA solution $z_0^*(x)$ of the mp-LP problem (11.56). Clearly the same properties are inherited by the controller. The following corollaries of Theorem 6.5 summarize the analytical properties of the optimal control law and of the value function.

Corollary 11.4   *There exists a control law $u^*(0) = f_0(x(0))$, $f_0 : \mathbb{R}^n \to \mathbb{R}^m$, obtained as a solution of the CFTOC (11.52) with $p = 1$ or $p = \infty$, which is continuous and PPWA*

$$f_0(x) = F_0^j x + g_0^j \quad \text{if} \quad x \in CR_0^j, \ j = 1, \dots, N_0^r, \tag{11.58}$$

where the polyhedral sets $CR_0^j = \{H_0^j x \leq k_0^j\}$, $j = 1, \dots, N_0^r$, are a partition of the feasible set $\mathcal{X}_0$.

Corollary 11.5   *The value function $J^*(x)$ obtained as a solution of the CFTOC (11.52) is convex and PPWA.*

Remark 11.7   Note that if the optimizer of problem (11.52) is unique for all $x(0) \in \mathcal{X}_0$, then Corollary 11.4 reads: "The control law $u^*(0) = f_0(x(0))$, $f_0 : \mathbb{R}^n \to \mathbb{R}^m$, obtained as a solution of the CFTOC (11.52) with $p = 1$ or $p = \infty$, **is** continuous and PPWA,…." From the results of Section 6.2 we know that in case of multiple optimizers for some $x(0) \in \mathcal{X}_0$, a continuous control law of the form (11.58) can always be computed.

The multiparametric solution of (11.56) provides the open-loop optimal sequence $u^*(0), \dots, u^*(N-1)$ as an affine function of the initial state $x(0)$. The state feedback PPWA optimal controllers $u^*(k) = f_k(x(k))$ with $f_k : \mathcal{X}_k \mapsto \mathcal{U}$ for

$k = 1, \ldots, N$ are computed in the following way. Consider the same CFTOC (11.52) over the shortened time horizon $[i, N]$

$$\min_{U_i} \quad \|Px_N\|_p + \sum_{k=i}^{N-1} \|Qx_k\|_p + \|Ru_k\|_p$$

$$\text{subj. to} \quad \begin{aligned} &x_{k+1} = Ax_k + Bu_k, \ k = i, \ldots, N-1 \\ &x_k \in \mathcal{X}, \ u_k \in \mathcal{U}, \ k = i, \ldots, N-1 \\ &x_N \in \mathcal{X}_f \\ &x_i = x(i), \end{aligned} \tag{11.59}$$

where $U_i = [u_i', \ldots, u_{N-1}']$ and $p = 1$ or $p = \infty$. As defined in (11.16) and discussed in Section 11.2, $\mathcal{X}_i \subseteq \mathbb{R}^n$ is the set of initial states $x(i)$ for which the optimal control problem (11.59) is feasible. We denote by U_i^* one of the optimizers of the optimal control problem (11.59).

Problem (11.59) can be translated into the mp-LP

$$\begin{aligned} &\min_{z_i} && c_i' z_i \\ &\text{subj. to} && \bar{G}_i z_i \leq \bar{w}_i + \bar{S}_i x(i), \end{aligned} \tag{11.60}$$

where $z_i = \{\varepsilon_i^x, \ldots, \varepsilon_N^x, \varepsilon_i^u, \ldots, \varepsilon_{N-1}^u, u_i', \ldots, u_{N-1}'\}$ and c_i, $\bar{G}_i$, $\bar{S}_i$, $\bar{w}_i$, are appropriately defined for each i. The component u_i^* of the multiparametric solution of (11.60) has the form

$$u_i^*(x(i)) = f_i(x(i)), \ \forall x(i) \in \mathcal{X}_i, \tag{11.61}$$

where the control law $f_i : \mathbb{R}^n \to \mathbb{R}^m$, is continuous and PPWA

$$f_i(x) = F_i^j x + g_i^j \quad \text{if} \quad x \in CR_i^j, \ j = 1, \ldots, N_i^r \tag{11.62}$$

and where the polyhedral sets $CR_i^j = \{x \in \mathbb{R}^n : H_i^j x \leq K_i^j\}$, $j = 1, \ldots, N_i^r$ are a partition of the feasible polyhedron $\mathcal{X}_i$. Therefore the feedback solution $u^*(k) = f_k(x(k))$, $k = 0, \ldots, N-1$ of the CFTOC (11.52) with $p = 1$ or $p = \infty$ is obtained by solving N mp-LP problems of decreasing size. The following corollary summarizes the final result.

Corollary 11.6   *There exists a state feedback control law $u^*(k) = f_k(x(k))$, $f_k : \mathcal{X}_k \subseteq \mathbb{R}^n \to \mathcal{U} \subseteq \mathbb{R}^m$, solution of the CFTOC (11.52) for $p = 1$ or $p = \infty$ and $k = 0, \ldots, N-1$ which is time-varying, continuous and piecewise affine on polyhedra*

$$f_k(x) = F_k^j x + g_k^j \quad \text{if} \quad x \in CR_k^j, \ j = 1, \ldots, N_k^r, \tag{11.63}$$

where the polyhedral sets $CR_k^j = \{x \in \mathbb{R}^n : H_k^j x \leq K_k^j\}$, $j = 1, \ldots, N_k^r$ are a partition of the feasible polyhedron $\mathcal{X}_k$.

11.4.3 State Feedback Solution via Recursive Approach

Consider the dynamic programming formulation of (11.52) with $J_0(\cdot)$ defined by (11.7) with $p = 1$ or $p = \infty$

$$
\begin{aligned}
J_j^*(x_j) = \min_{u_j} \quad & \|Qx_j\|_p + \|Ru_j\|_p + J_{j+1}^*(Ax_j + Bu_j) \\
\text{subj. to } \quad & x_j \in \mathcal{X}, \; u_j \in \mathcal{U} \\
& Ax_j + Bu_j \in \mathcal{X}_{j+1},
\end{aligned}
\tag{11.64}
$$

for $j = 0, \ldots, N-1$, with boundary conditions

$$
J_N^*(x_N) = \|Px_N\|_p
\tag{11.65}
$$

$$
\mathcal{X}_N = \mathcal{X}_f.
\tag{11.66}
$$

Unlike for the 2-norm case the dynamic program (11.64)–(11.66) can be solved as explained in the next theorem.

Theorem 11.6 *The state feedback piecewise affine solution (11.63) of the CFTOC (11.52) for $p = 1$ or $p = \infty$ is obtained by solving the optimization problem (11.64)–(11.66) via N mp-LPs.*

Proof: Consider the first step $j = N - 1$ of dynamic programming (11.64)–(11.66)

$$
\begin{aligned}
J_{N-1}^*(x_{N-1}) = \min_{u_{N-1}} \quad & \|Qx_{N-1}\|_p + \|Ru_{N-1}\|_p + J_N^*(Ax_{N-1} + Bu_{N-1}) \\
\text{subj. to } \quad & x_{N-1} \in \mathcal{X}, \; u_{N-1} \in \mathcal{U} \\
& Ax_{N-1} + Bu_{N-1} \in \mathcal{X}_f.
\end{aligned}
\tag{11.67}
$$

$J_{N-1}^*(x_{N-1})$, $u_{N-1}^*(x_{N-1})$ and $\mathcal{X}_{N-1}$ are computable via the mp-LP:

$$
\begin{aligned}
J_{N-1}^*(x_{N-1}) = \min_{\mu, u_{N-1}} \quad & \mu \\
\text{subj. to } \quad & \mu \geq \|Qx_{N-1}\|_p + \|Ru_{N-1}\|_p + \|P(Ax_{N-1} + Bu_{N-1})\|_p \\
& x_{N-1} \in \mathcal{X}, \; u_{N-1} \in \mathcal{U} \\
& Ax_{N-1} + Bu_{N-1} \in \mathcal{X}_f.
\end{aligned}
\tag{11.68}
$$

The constraint $\mu \geq \|Qx_{N-1}\|_p + \|Ru_{N-1}\|_p + \|P(Ax_{N-1} + Bu_{N-1})\|_p$ in (11.68) is converted into a set of linear constraints as discussed in Remark 2.1 of Section 2.2.5. For instance, if $p = \infty$ we follow the approach of Section 9.3 and rewrite the constraint as

$$
\begin{aligned}
\mu &\geq \varepsilon_{N-1}^x + \varepsilon_{N-1}^u + \varepsilon_N^x \\
-\mathbf{1}_n \varepsilon_{N-1}^x &\leq \pm Qx_{N-1} \\
-\mathbf{1}_m \varepsilon_{N-1}^u &\leq \pm Ru_{N-1} \\
-\mathbf{1}_{r_N} \varepsilon_N^x &\leq \pm P_N [Ax_{N-1} + Bu_{N-1}]
\end{aligned}
\tag{11.69}
$$

By Theorem 6.5, J_{N-1}^* is a convex and piecewise affine function of x_{N-1}, the corresponding optimizer u_{N-1}^* is piecewise affine and continuous, and the feasible set $\mathcal{X}_{N-1}$ is a polyhedron. Without any loss of generality we assume J_{N-1}^* to be described as follows: $J_{N-1}^*(x_{N-1}) = \max_{i=1,\ldots,n_{N-1}} \{c_i' x_{N-1} + d_i\}$ (see Section 2.2.5 for convex PPWA functions representation) where n_{N-1} is the number of affine components comprising the value function J_{N-1}^*. At step $j = N - 2$ of dynamic programming (11.64)–(11.66) we have

$$
\begin{aligned}
J_{N-2}^*(x_{N-2}) = \min_{u_{N-2}} \quad & \|Qx_{N-2}\|_p + \|Ru_{N-2}\|_p + J_{N-1}^*(Ax_{N-2} + Bu_{N-2}) \\
\text{subj. to} \quad & x_{N-2} \in \mathcal{X}, \ u_{N-2} \in \mathcal{U} \\
& Ax_{N-2} + Bu_{N-2} \in \mathcal{X}_{N-1}.
\end{aligned}
\tag{11.70}
$$

Since $J_{N-1}^*(x)$ is a convex and piecewise affine function of x, the problem (11.70) can be recast as the following mp-LP (see Section 2.2.5 for details)

$$
\begin{aligned}
J_{N-2}^*(x_{N-2}) = \min_{\mu, u_{N-2}} \quad & \mu \\
\text{subj. to} \quad & \mu \geq \|Qx_{N-2}\|_p + \|Ru_{N-2}\|_p + c_i(Ax_{N-2} + Bu_{N-2}) + d_i \\
& i = 1, \ldots, n_{N-1} \\
& x_{N-2} \in \mathcal{X}, \ u_{N-2} \in \mathcal{U} \\
& Ax_{N-2} + Bu_{N-2} \in \mathcal{X}_{N-1}.
\end{aligned}
\tag{11.71}
$$

$J_{N-2}^*(x_{N-2})$, $u_{N-2}^*(x_{N-2})$ and $\mathcal{X}_{N-2}$ are computed by solving the mp-LP (11.71). By Theorem 6.5, J_{N-2}^* is a convex and piecewise affine function of x_{N-2}, the corresponding optimizer u_{N-2}^* is piecewise affine and continuous, and the feasible set $\mathcal{X}_{N-2}$ is a convex polyhedron.

The convexity and piecewise linearity of J_j^* and the polyhedra representation of $\mathcal{X}_j$ still hold for $j = N - 3, \ldots, 0$ and the procedure can be iterated backwards in time, proving the theorem.   ∎

Consider the state feedback piecewise affine solution (11.63) of the CFTOC (11.52) for $p = 1$ or $p = \infty$ and assume we are interested only in the optimal controller at time 0. In this case, by using duality arguments we can solve the equations (11.64)–(11.66) by using vertex enumerations and one mp-LP. This is proven in the next theorem.

Theorem 11.7 *The state feedback piecewise affine solution (11.63) at time $k = 0$ of the CFTOC (11.52) for $p = 1$ or $p = \infty$ is obtained by solving the optimization problem (11.64)–(11.66) via one mp-LP.*

Proof: Consider the first step $j = N - 1$ of dynamic programming (11.64)–(11.66)

$$
\begin{aligned}
J_{N-1}^*(x_{N-1}) = \min_{u_{N-1}} \quad & \|Qx_{N-1}\|_p + \|Ru_{N-1}\|_p + J_N^*(Ax_{N-1} + Bu_{N-1}) \\
\text{subj. to} \quad & x_{N-1} \in \mathcal{X}, \ u_{N-1} \in \mathcal{U} \\
& Ax_{N-1} + Bu_{N-1} \in \mathcal{X}_f
\end{aligned}
\tag{11.72}
$$

and the corresponding mp-LP:

$$
\begin{aligned}
J_{N-1}^*(x_{N-1}) = \min_{\mu, u_{N-1}} \quad & \mu \\
\text{subj. to} \quad & \mu \geq \|Qx_{N-1}\|_p + \|Ru_{N-1}\|_p + \|P(Ax_{N-1} + Bu_{N-1})\|_p \\
& x_{N-1} \in \mathcal{X}, \ u_{N-1} \in \mathcal{U} \\
& Ax_{N-1} + Bu_{N-1} \in \mathcal{X}_f.
\end{aligned}
\tag{11.73}
$$

By Theorem 6.5, J_{N-1}^* is a convex and piecewise affine function of x_{N-1}, and the feasible set $\mathcal{X}_{N-1}$ is a polyhedron. J_{N-1}^* and $\mathcal{X}_{N-1}$ are computed without explicitly solving the mp-LP (11.73). Rewrite problem (11.73) in the more compact form

$$
\begin{aligned}
\min_{z_{N-1}} \quad & c_{N-1}' z_{N-1} \\
\text{subj. to} \quad & \bar{G}_{N-1} z_{N-1} \leq \bar{w}_{N-1} + \bar{S}_{N-1} x_{N-1},
\end{aligned}
\tag{11.74}
$$

where z_{N-1} collects the optimization variables μ, u_{N-1} and the auxiliary variables need to transform the constraint $\mu \geq \|Qx_{N-1}\|_p + \|Ru_{N-1}\|_p + \|P(Ax_{N-1} + Bu_{N-1})\|_p$ into a set of linear constraints. Consider the LP dual of (11.74)

$$\begin{array}{ll} \max_v & -(\bar{w}_{N-1} + \bar{S}_{N-1}x_{N-1})'v \\ \text{subj. to} & \bar{G}'_{N-1}v = -c_{N-1} \\ & v \geq 0. \end{array} \qquad (11.75)$$

Consider the dual feasibility polyheron $\mathcal{P}_d = \{v \geq 0 \ : \ \bar{G}'_{N-1}v = -c_{N-1}\}$. Let $\{V_1, \ldots, V_k\}$ be the vertices of $\mathcal{P}_d$ and $\{y_1, \ldots, y_e\}$ be the rays of $\mathcal{P}_d$. Since we have a zero duality gap, we have that

$$J^*_{N-1}(x_{N-1}) = \max_{i=1,\ldots,k} \{-(\bar{w}_{N-1} + \bar{S}_{N-1}x_{N-1})'V_i\}$$

i.e.,

$$J^*_{N-1}(x_{N-1}) = \max_{i=1,\ldots,k} \{-(V_i'\bar{S}_{N-1})x_{N-1} - \bar{w}'_{N-1}V_i\}.$$

Recall that if the dual of a mp-LP is unbounded, then the primal is infeasible (Theorem 6.3). For this reason the feasible set $\mathcal{X}_{N-1}$ is obtained by requiring that the cost of (11.75) does not increase in the direction of the rays:

$$\mathcal{X}_{N-1} = \{x_{N-1} \ : \ -(\bar{w}_{N-1} + \bar{S}_{N-1}x_{N-1})'y_i < 0, \ \forall\, i - 1, \ldots, e\}$$

with $J^*_{N-1}(x_{N-1})$ and $\mathcal{X}_{N-1}$ available, we can iterate the procedure backwards in time for steps $N-2, N-3, \ldots, 1$. At step 0 one mp-LP will be required in order to compute $u^*_0(x(0))$. ∎

11.4.4 Example

Example 11.4 Consider the double integrator system (11.23). We want to compute the state feedback optimal controller that solves (11.52) with $p = \infty$, $N = 6$, $P = \begin{bmatrix} 1 & 0 \\ 0 & 1 \end{bmatrix}$, $Q = \begin{bmatrix} 1 & 0 \\ 0 & 1 \end{bmatrix}$, $R = 0.8$, subject to the input constraints

$$\mathcal{U} = \{u \in \mathbb{R} \ : \ -1 \leq u \leq 1\} \qquad (11.76)$$

and the state constraints

$$\mathcal{X} = \left\{ x \in \mathbb{R}^2 \ : \ \begin{bmatrix} -10 \\ -10 \end{bmatrix} \leq x \leq \begin{bmatrix} 10 \\ 10 \end{bmatrix} \right\} \qquad (11.77)$$

and $\mathcal{X}_f = \mathcal{X}$. The optimal feedback solution $u^*(0), \ldots, u^*(5)$ was computed by solving six mp-LP problems and the corresponding polyhedral partitions of the state space are depicted in Figure 11.5, where polyhedra with the same control law were merged. Only the last optimal control move is reported below:

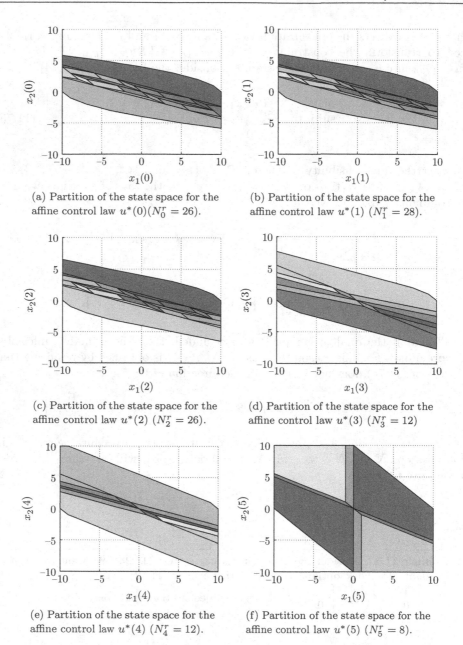

(a) Partition of the state space for the affine control law $u^*(0)$ ($N_0^r = 26$).

(b) Partition of the state space for the affine control law $u^*(1)$ ($N_1^r = 28$).

(c) Partition of the state space for the affine control law $u^*(2)$ ($N_2^r = 26$).

(d) Partition of the state space for the affine control law $u^*(3)$ ($N_3^r = 12$)

(e) Partition of the state space for the affine control law $u^*(4)$ ($N_4^r = 12$).

(f) Partition of the state space for the affine control law $u^*(5)$ ($N_5^r = 8$).

Figure 11.5 Example 11.4. Double Integrator, ∞-norm objective function, horizon $N = 6$. Partition of the state space for the time-varying optimal control law. Polyhedra with the same control law were merged.

$$
u^*(5) = \begin{cases}
0 & \text{if } \begin{bmatrix} 0.45 & 0.89 \\ 1.00 & 0.00 \\ -0.71 & -0.71 \\ -1.00 & -0.00 \end{bmatrix} x(5) \le \begin{bmatrix} 0.00 \\ 0.00 \\ 7.07 \\ 10.00 \end{bmatrix} & \text{(Region \#1)} \\[24pt]
0 & \text{if } \begin{bmatrix} -0.45 & -0.89 \\ -1.00 & 0.00 \\ 0.71 & 0.71 \\ 1.00 & -0.00 \end{bmatrix} x(5) \le \begin{bmatrix} 0.00 \\ 0.00 \\ 7.07 \\ 10.00 \end{bmatrix} & \text{(Region \#2)} \\[24pt]
\begin{bmatrix} -1.00 & -2.00 \end{bmatrix} x(5) & \text{if } \begin{bmatrix} -0.45 & -0.89 \\ 0.45 & 0.89 \\ 0.71 & 0.71 \\ -1.00 & -0.00 \end{bmatrix} x(5) \le \begin{bmatrix} 0.00 \\ 0.45 \\ 0.00 \\ 10.00 \end{bmatrix} & \text{(Region \#3)} \\[24pt]
\begin{bmatrix} -1.00 & -2.00 \end{bmatrix} x(5) & \text{if } \begin{bmatrix} -0.45 & -0.89 \\ 0.45 & 0.89 \\ -0.71 & -0.71 \\ 1.00 & -0.00 \end{bmatrix} x(5) \le \begin{bmatrix} 0.45 \\ 0.00 \\ 0.00 \\ 10.00 \end{bmatrix} & \text{(Region \#4)} \\[24pt]
\begin{bmatrix} 1.00 & 0.00 \end{bmatrix} x(5) & \text{if } \begin{bmatrix} -0.71 & -0.71 \\ -1.00 & 0.00 \\ 1.00 & 0.00 \\ -0.00 & 1.00 \end{bmatrix} x(5) \le \begin{bmatrix} 0.00 \\ 1.00 \\ 0.00 \\ 10.00 \end{bmatrix} & \text{(Region \#5)} \\[24pt]
\begin{bmatrix} 1.00 & 0.00 \end{bmatrix} x(5) & \text{if } \begin{bmatrix} 0.71 & 0.71 \\ -1.00 & 0.00 \\ 1.00 & 0.00 \\ -0.00 & -1.00 \end{bmatrix} x(5) \le \begin{bmatrix} 0.00 \\ 0.00 \\ 1.00 \\ 10.00 \end{bmatrix} & \text{(Region \#6)} \\[24pt]
1.00 & \text{if } \begin{bmatrix} 0.45 & 0.89 \\ -1.00 & 0.00 \\ -0.00 & -1.00 \\ 1.00 & -0.00 \end{bmatrix} x(5) \le \begin{bmatrix} -0.45 \\ -1.00 \\ 10.00 \\ 10.00 \end{bmatrix} & \text{(Region \#7)} \\[24pt]
-1.00 & \text{if } \begin{bmatrix} -0.45 & -0.89 \\ 1.00 & 0.00 \\ -1.00 & -0.00 \\ -0.00 & 1.00 \end{bmatrix} x(5) \le \begin{bmatrix} -0.45 \\ -1.00 \\ 10.00 \\ 10.00 \end{bmatrix} & \text{(Region \#8)}
\end{cases}
$$

$$\tag{11.78}$$

Note that the controller (11.78) is piecewise linear around the origin. In fact, the origin belongs to multiple regions (1 to 6). Note that the number N_i^r of regions is not always increasing with decreasing i ($N_5^r = 8$, $N_4^r = 12$, $N_3^r = 12$, $N_2^r = 26$, $N_1^r = 28$, $N_0^r = 26$). This is due to the merging procedure, before merging we have $N_5^r = 12$, $N_4^r = 22$, $N_3^r = 40$, $N_2^r = 72$, $N_1^r = 108$, $N_0^r = 152$.

11.4.5 Infinite-Time Solution

Assume that Q and R have full column rank and that the constraint sets $\mathcal{X}$ and $\mathcal{U}$ contain the origin in their interior. Consider the following infinite-horizon problem with constraints

$$
\begin{aligned}
J_\infty^*(x(0)) = \min_{u_0, u_1, \ldots} \quad & \sum_{k=0}^{\infty} \|Qx_k\|_p + \|Ru_k\|_p \\
\text{subj. to} \quad & x_{k+1} = Ax_k + Bu_k, \ k = 0, \ldots, \infty \\
& x_k \in \mathcal{X}, \ u_k \in \mathcal{U}, \ k = 0, \ldots, \infty \\
& x_0 = x(0)
\end{aligned}
\tag{11.79}
$$

and the set

$$
\mathcal{X}_\infty = \{x(0) \in \mathbb{R}^n : \text{ Problem (11.79) is feasible and } J_\infty^*(x(0)) < +\infty\}. \tag{11.80}
$$

Because Q and R have full column rank, *any* optimizer u_k^* of problem (11.43) *must* converge to the origin ($u_k^* \to 0$) and so must the state trajectory resulting from the application of u_k^* ($x_k^* \to 0$). Thus the origin $x = 0, u = 0$ must lie in the interior of the constraint set $(\mathcal{X}, \mathcal{U})$. (If the origin were not contained in the constraint set then $J_\infty^*(x(0))$ would be infinite.) Furthermore, if the initial state $x_0 = x(0)$ is sufficiently close to the origin, then the state and input constraints will never become active and the solution of problem (11.43) will yield the *unconstrained* optimal controller (9.31).

The discussion for the solution of the infinite horizon constrained linear quadratic regulator (Section 11.3.4) by means of the batch approach can be repeated here with one precaution. Since the unconstrained optimal controller (if it exists) is PPWA the computation of the Maximal Invariant Set for the autonomous constrained piecewise linear system is more involved and requires algorithms which will be presented later in Chapter 17.

Differently from the 2-norm case, here the use of dynamic programming for computing the infinite horizon solution is a viable alternative to the batch approach. Convergence conditions for the dynamic programming strategy and convergence guarantees for the resulting possibly discontinuous closed-loop system are given in [87]. A computationally efficient algorithm to obtain the infinite time optimal solution, based on a dynamic programming exploration strategy with an mp-LP solver and basic polyhedral manipulations, is also presented in [87].

Example 11.5 We consider the double integrator system (11.23) from Example 11.4 with $N = \infty$.

The partition of the state space for the time invariant optimal control law is shown in Figure 11.6(a) and consists of 202 polyhedral regions. In Figure 11.6(b) the same control law is represented where polyhedra with the same affine control law were merged.

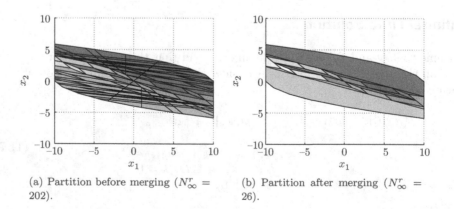

(a) Partition before merging ($N_\infty^r = 202$).

(b) Partition after merging ($N_\infty^r = 26$).

Figure 11.6 Example 11.5. Double Integrator, ∞-norm objective function, horizon $N = \infty$. Partition of the state space for the time invariant optimal control law.

11.5 State Feedback Solution, Minimum-Time Control

In this section we consider the solution of minimum-time optimal control problems

$$
\begin{aligned}
J_0^*(x(0)) = \min_{U_0, N} \quad & N \\
\text{subj. to} \quad & x_{k+1} = A x_k + B u_k, \ k = 0, \ldots, N-1 \\
& x_k \in \mathcal{X}, \ u_k \in \mathcal{U}, \ k = 0, \ldots, N-1 \\
& x_N \in \mathcal{X}_f \\
& x_0 = x(0),
\end{aligned}
\tag{11.81}
$$

where $\mathcal{X}_f \subset \mathbb{R}^n$ is a terminal target set to be reached in minimum time.

We can find the controller that brings the states into $\mathcal{X}_f$ in one time step by solving the following multiparametric program

$$
\begin{aligned}
\min_{u_0} \quad & c(x_0, u_0) \\
\text{subj. to} \quad & x_1 = A x_0 + B u_0 \\
& x_0 \in \mathcal{X}, \ u_0 \in \mathcal{U} \\
& x_1 \in \mathcal{X}_f,
\end{aligned}
\tag{11.82}
$$

where $c(x_0, u_0)$ is any convex quadratic function. Let us assume that the solution of the multiparametric program generates R^1 regions $\{\mathcal{P}_r^1\}_{r=1}^{R^1}$ with the affine control law $u_0 = F_r^1 x + g_r^1$ in each region r. By construction we have

$$
\mathcal{X}_0 = \mathcal{K}_1(\mathcal{X}_f).
$$

Continuing to set up simple multiparametric programs bring the states into $\mathcal{X}_f$ in $2, 3, \ldots$ steps, we have for step j

$$
\begin{aligned}
\min_{u_0} \quad & c(x_0, u_0) \\
\text{subj. to} \quad & x_1 = A x_0 + B u_0 \\
& x_0 \in \mathcal{X}, \ u_0 \in \mathcal{U} \\
& x_1 \in \mathcal{K}_{j-1}(\mathcal{X}_f),
\end{aligned}
\tag{11.83}
$$

which yields R^j regions $\{\mathcal{P}_r^j\}_{r=1}^{R^j}$ with the affine control law $u_0 = F_r^j x + g_r^j$ in each region r. By construction we have

$$
\mathcal{X}_0 = \mathcal{K}_j(\mathcal{X}_f).
$$

Thus to obtain $\mathcal{K}_1(\mathcal{X}_f), \ldots, \mathcal{K}_N(\mathcal{X}_f)$ we need to solve N multiparametric programs with a prediction horizon of 1. Since the overall complexity of a multiparametric program is exponential in N, this scheme can be exploited to yield controllers of lower complexity than the optimal control schemes introduced in the previous sections.

Since N multiparametric programs have been solved, the controller regions overlap in general. In order to achieve minimum time behavior, the feedback law associated with the region computed for the smallest number of steps c, is selected for any given state x.

Algorithm 11.1   *On-line computation of minimum-time control input*

Input: State measurement x, N controller partitions solution to (11.83) for $j = 1, \ldots, N$

Output: Minimum time control action $u(x)$

 Find controller partition $c_{\min} = \min_{c \in \{0, \ldots, N\}} c$, s.t. $x \in \mathcal{K}_c(\mathcal{X}_f)$

 Find controller region r, such that $x \in \mathcal{P}_r^{c_{\min}}$ and compute $u = F_r^{c_{\min}} x + g_r^{c_{\min}}$

 Return u

Note that the region identification for this type of controller partition is much more efficient than simply checking all the regions. The two steps of "finding a controller partition" and "finding a controller region" in Algorithm 11.1 correspond to two levels of a search tree, where the search is first performed over the feasible sets $\mathcal{K}_c(\mathcal{X}_f)$ and then over the controller partition $\{\mathcal{P}_r^c\}_{r=1}^{R^c}$. Furthermore, one may discard all regions $\mathcal{P}_r^i$ which are completely covered by previously computed controllers (i.e., $\mathcal{P}_r^i \subseteq \bigcup_{j \in \{1, \ldots, i-1\}} \mathcal{K}_j(\mathcal{X}_f)$) since they are not time optimal.

Example 11.6   Consider again the double integrator from Example 11.2. The Minimum-Time Controller is computed that steers the system to the Maximal LQR Invariant Set $\mathcal{O}_\infty^{\mathrm{LQR}}$ in the minimum number of time steps N. The Algorithm terminated after 11 iterations, covering the Maximal Controllable Set $\mathcal{K}_\infty(\mathcal{O}_\infty^{\mathrm{LQR}})$. The resulting controller is defined over 33 regions. The regions are depicted in Figure 11.7(a). The Maximal LQR Invariant Set $\mathcal{O}_\infty^{\mathrm{LQR}}$ is the central shaded region.

The control law on this partition is depicted in Figure 11.7(b). Note that, in general, the minimum-time control law is not continuous as can be seen in Figure 11.7(b).

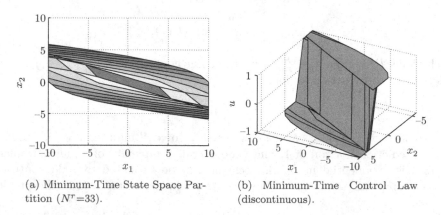

(a) Minimum-Time State Space Partition (N^r=33).

(b) Minimum-Time Control Law (discontinuous).

Figure 11.7 Example 11.6. Double integrator, minimum-time objective function. Partition of the state space and control law. Maximal LQR Invariant Set $\mathcal{O}_\infty^{\mathrm{LQR}}$ is the central shaded region in (a).

11.6   Comparison of the Design Approaches and Controllers

For the design, the storage and the on-line execution time of the control law we are interested in the *controller complexity*. All the controllers we discussed in this chapter are PPWA. Their storage and execution times are closely related to the number of polyhedral regions, which is related to the number of constraints in the mp-LP or mp-QP.

When the control objective is expressed in terms of the 1- or ∞- norm, it is translated into a set of constraints in the mp-LP. When the control objective is expressed in terms of the 2-norm such a translation is not necessary. Thus controllers minimizing the 1- or ∞- norm do generally involve more regions than those minimizing the 2-norm and therefore tend to be more complex.

In mp-LPs degeneracies are common and have to be taken care of in the respective algorithms. Strictly convex mp-QPs for which the Linear Independent Constraint Qualification holds are comparatively well behaved with unique solutions and full-dimensional polyhedral regions. Only for the 1-norm and ∞-norm objective, however, we can use for the controller design an efficient Dynamic Programming scheme involving a sequence of mp-LPs.

As we let the horizon N go to infinity, the number of regions stays finite for the 2-norm objective, for the 1- and ∞-norm nothing can be said in general. In the examples, we have usually observed a finite number, but cases can be constructed where the number can be proven to be infinite.

As we will argue in the following chapter, infinite horizon controllers based on the 2-norm render the closed-loop system exponentially stable, while controllers based on the 1- or ∞-norm render the closed-loop system stable, only.

Among the controllers proposed in this chapter, the minimum-time controller is usually the least complex involving the smallest number of regions. Minimum time controllers are often considered to be too aggressive for practical use. The controller here is different, however. It is minimum time only until it reaches the terminal set $\mathcal{X}_f$. Inside the terminal set a different unconstrained control law can be used. Overall the operation of the minimum-time controller is observed to be very similar to that of the other infinite time controllers in this chapter.

12

Receding Horizon Control

In this chapter we review the basics of Receding Horizon Control (RHC). In the first part we discuss the stability and the feasibility of RHC and we provide guidelines for choosing the terminal weight so that closed-loop stability is achieved.

The second part of the chapter focuses on the RHC implementation. Since RHC requires at each sampling time to solve an open-loop constrained finite time optimal control problem as a function of the current state, the results of the previous chapters imply two possible approaches for RHC implementation.

In the first approach a mathematical program is solved at each time step for the current initial state. In the second approach the explicit piecewise affine feedback policy (that provides the optimal control for all states) is precomputed off-line. This reduces the on-line computation of the RHC law to a function evaluation, thus avoiding the on-line solution of a quadratic or linear program. This technique is attractive for a wide range of practical problems where the computational complexity of on-line optimization is prohibitive. It also provides insight into the structure underlying optimization-based controllers, describing the behavior of the RHC controller in different regions of the state space. Moreover, for applications where safety is crucial, the correctness of a piecewise affine control law is easier to verify than that of a mathematical program solver.

12.1 RHC Idea

In the previous chapter we discussed the solution of constrained finite time and infinite time optimal control problems for linear systems. An infinite horizon suboptimal controller can be designed by repeatedly solving finite time optimal control problems in a receding horizon fashion as described next. At each sampling time, starting at the current state, an open-loop optimal control problem is solved over a finite horizon (top diagram in Figure 12.1). The computed optimal manipulated input signal is applied to the process only during the following sampling interval $[t, t+1]$. At the next time step $t+1$ a new optimal control problem based on new measurements of the state is solved over a shifted horizon (bottom

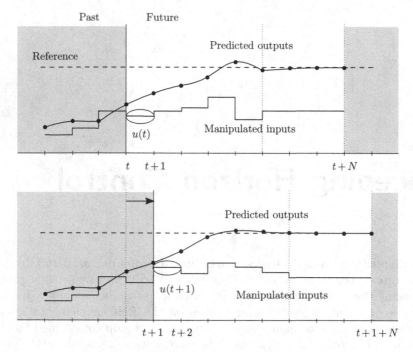

Figure 12.1 Receding Horizon Idea.

diagram in Figure 12.1). The resulting controller is referred to as a Receding Horizon Controller (RHC). A receding horizon controller where the finite time optimal control law is computed by solving an optimization problem on-line is usually referred to as *Model Predictive Control* (MPC).

12.2 RHC Implementation

Consider the problem of regulating to the origin the discrete-time linear time-invariant system

$$x(t+1) = Ax(t) + Bu(t), \tag{12.1}$$

where $x(t) \in \mathbb{R}^n$, $u(t) \in \mathbb{R}^m$ are the state and input vectors, respectively, subject to the constraints

$$x(t) \in \mathcal{X}, \ u(t) \in \mathcal{U}, \ \forall t \geq 0, \tag{12.2}$$

where the sets $\mathcal{X} \subseteq \mathbb{R}^n$ and $\mathcal{U} \subseteq \mathbb{R}^m$ are polyhedra. Receding Horizon Control (RHC) approaches such a constrained regulation problem in the following way. Assume that a full measurement or estimate of the state $x(t)$ is available at the current time t. Then the finite time optimal control problem

$$J_t^*(x(t)) = \min_{U_{t \to t+N|t}} \quad J_t(x(t), U_{t \to t+N|t}) = p(x_{t+N|t}) + \sum_{k=0}^{N-1} q\left(x_{t+k|t}, u_{t+k|t}\right)$$

$$\text{subj. to} \quad x_{t+k+1|t} = Ax_{t+k|t} + Bu_{t+k|t}, \ k = 0, \dots, N-1$$
$$x_{t+k|t} \in \mathcal{X}, \ u_{t+k|t} \in \mathcal{U}, \ k = 0, \dots, N-1$$
$$x_{t+N|t} \in \mathcal{X}_f$$
$$x_{t|t} = x(t)$$

$$(12.3)$$

is solved at time t, where $U_{t \to t+N|t} = \{u_{t|t}, \dots, u_{t+N-1|t}\}$ and where $x_{t+k|t}$ denotes the state vector at time $t + k$ predicted at time t obtained by starting from the current state $x_{t|t} = x(t)$ and applying to the system model

$$x_{t+k+1|t} = Ax_{t+k|t} + Bu_{t+k|t}$$

the input sequence $u_{t|t}, \dots, u_{t+k-1|t}$. Often the symbol $x_{t+k|t}$ is read as "the state x at time $t + k$ predicted at time t." Similarly, $u_{t+k|t}$ is read as "the input u at time $t + k$ computed at time t." For instance, $x_{3|1}$ represents the predicted state at time 3 when the prediction is done at time $t = 1$ starting from the current state $x(1)$. It is different, in general, from $x_{3|2}$ which is the predicted state at time 3 when the prediction is done at time $t = 2$ starting from the current state $x(2)$.

Let $U_{t \to t+N|t}^* = \{u_{t|t}^*, \dots, u_{t+N-1|t}^*\}$ be the optimal solution of (12.3) at time t and $J_t^*(x(t))$ the corresponding value function. Then, the first element of $U_{t \to t+N|t}^*$ is applied to system (12.1)

$$u(t) = u_{t|t}^*(x(t)). \tag{12.4}$$

The optimization problem (12.3) is repeated at time $t + 1$, based on the new state $x_{t+1|t+1} = x(t + 1)$, yielding a *moving* or *receding horizon* control strategy.

Let $f_t : \mathbb{R}^n \to \mathbb{R}^m$ denote the *receding horizon* control law that associates the optimal input $u_{t|t}^*$ to the current state $x(t)$, $f_t(x(t)) = u_{t|t}^*(x(t))$. Then the closed-loop system obtained by controlling (12.1) with the RHC (12.3)–(12.4) is

$$x(k + 1) = Ax(k) + Bf_k(x(k)) = f_{cl}(x(k), k), \ k \geq 0. \tag{12.5}$$

Note that the notation used in this chapter is slightly different from the one used in Chapter 11. Because of the receding horizon strategy, there is the need to distinguish between the input $u^*(t + k)$ applied to the plant at time $t + k$, and optimizer $u_{t+k|t}^*$ of the problem (12.3) at time $t + k$ obtained by solving (12.3) at time t with $x_{t|t} = x(t)$.

Consider problem (12.3). As the system, the constraints and the cost function are time-invariant, the solution to problem (12.3) is a time-invariant function of the initial state $x(t)$. Therefore, in order to simplify the notation, we can set $t = 0$ in (12.3) and remove the term "$|0$" since it is now redundant and rewrite (12.3) as

$$J_0^*(x(t)) = \min_{U_0} \quad J_0(x(t), U_0) = p(x_N) + \sum_{k=0}^{N-1} q(x_k, u_k)$$

$$\text{subj. to} \quad x_{k+1} = Ax_k + Bu_k, \ k = 0, \dots, N-1$$
$$x_k \in \mathcal{X}, \ u_k \in \mathcal{U}, \ k = 0, \dots, N-1$$
$$x_N \in \mathcal{X}_f$$
$$x_0 = x(t), \tag{12.6}$$

where $U_0 = \{u_0, \ldots, u_{N-1}\}$ and the notation in Remark 7.1 applies. Similarly as in previous chapters, we will focus on two classes of cost functions. If the 1-norm or ∞-norm is used in (12.6), then we set $p(x_N) = \|Px_N\|_p$ and $q(x_k, u_k) = \|Qx_k\|_p + \|Ru_k\|_p$ with $p = 1$ or $p = \infty$ and P, Q, R full column rank matrices. The cost function is rewritten as

$$J_0(x(0), U_0) = \|Px_N\|_p + \sum_{k=0}^{N-1} \|Qx_k\|_p + \|Ru_k\|_p. \tag{12.7}$$

If the squared Euclidian norm is used in (12.6), then we set $p(x_N) = x_N' P x_N$ and $q(x_k, u_k) = x_k' Q x_k + u_k' R u_k$ with $P \succeq 0$, $Q \succeq 0$ and $R \succ 0$. The cost function is rewritten as

$$J_0(x(0), U_0) = x_N' P x_N + \sum_{k=0}^{N-1} x_k' Q x_k + u_k' R u_k. \tag{12.8}$$

The control law (12.4)

$$u(t) = f_0(x(t)) = u_0^*(x(t)) \tag{12.9}$$

and closed-loop system (12.5)

$$x(k+1) = Ax(k) + Bf_0(x(k)) = f_{cl}(x(k)), \ k \geq 0 \tag{12.10}$$

are time-invariant as well.

Note that the notation in (12.6) does not allow us to distinguish at which time step a certain state prediction or optimizer is computed and is valid for time-invariant problems only. Nevertheless, we will prefer the RHC notation in (12.6) to the one in (12.3) in order to simplify the exposition.

Compare problem (12.6) and the CFTOC (11.9). The *only* difference is that problem (12.6) is solved for $x_0 = x(t)$, $t \geq 0$ rather than for $x_0 = x(0)$. For this reason we can make use of all the results of the previous chapter. In particular, $\mathcal{X}_0$ denotes the set of feasible states $x(t)$ for problem (12.6) as defined and studied in Section 11.2. Recall from Section 11.2 that $\mathcal{X}_0$ is a polyhedron.

From the above explanations it is clear that a fixed prediction horizon is shifted or *recedes* over time, hence its name, receding horizon control. The procedure of this *on-line* optimal control technique is summarized in the following algorithm.

Algorithm 12.1   *On-line receding horizon control*

Input: State $x(t)$ at time instant t
Output: Receding horizon control input $u(x(t))$

> **Obtain** $U_0^*(x(t))$ by solving the optimization problem (12.6)
> **If** 'problem infeasible' **Then** stop
> **Return** the first element u_0^* of U_0^*

Example 12.1   Consider the double integrator system (11.23) rewritten below:

$$x(t+1) = \begin{bmatrix} 1 & 1 \\ 0 & 1 \end{bmatrix} x(t) + \begin{bmatrix} 0 \\ 1 \end{bmatrix} u(t) \tag{12.11}$$

The aim is to compute the receding horizon controller that solves the optimization problem (12.6) with $p(x_N) = x_N' P x_N$, $q(x_k, u_k) = x_k' Q x_k + u_k' R u_k$ $N = 3, P = Q = \begin{bmatrix} 1 & 0 \\ 0 & 1 \end{bmatrix}$, $R = 10$, $\mathcal{X}_f = \mathbb{R}^2$ subject to the input constraints

$$-0.5 \leq u(k) \leq 0.5, \; k = 0, \ldots, 3 \tag{12.12}$$

and the state constraints

$$\begin{bmatrix} -5 \\ -5 \end{bmatrix} \leq x(k) \leq \begin{bmatrix} 5 \\ 5 \end{bmatrix}, \; k = 0, \ldots, 3. \tag{12.13}$$

The QP problem associated with the RHC has the form (11.31) with

$$H = \begin{bmatrix} 13.50 & -10.00 & -0.50 \\ -10.00 & 22.00 & -10.00 \\ -0.50 & -10.00 & 31.50 \end{bmatrix}, \; F = \begin{bmatrix} -10.50 & 10.00 & -0.50 \\ -20.50 & 10.00 & 9.50 \end{bmatrix}, \; Y = \begin{bmatrix} 14.50 & 23.50 \\ 23.50 & 54.50 \end{bmatrix} \tag{12.14}$$

and

$$G_0 = \begin{bmatrix} 0.50 & -1.00 & 0.50 \\ -0.50 & 1.00 & -0.50 \\ -0.50 & 0.00 & 0.50 \\ -0.50 & 0.00 & -0.50 \\ 0.50 & 0.00 & -0.50 \\ 0.50 & 0.00 & 0.50 \\ -1.00 & 0.00 & 0.00 \\ 0.00 & -1.00 & 0.00 \\ 1.00 & 0.00 & 0.00 \\ 0.00 & 1.00 & 0.00 \\ 0.00 & 0.00 & -1.00 \\ 0.00 & 0.00 & 1.00 \\ 0.00 & 0.00 & 0.00 \\ -0.50 & 0.00 & 0.50 \\ 0.00 & 0.00 & 0.00 \\ 0.50 & 0.00 & -0.50 \\ -0.50 & 0.00 & 0.50 \\ 0.50 & 0.00 & -0.50 \\ 0.00 & 0.00 & 0.00 \\ 0.00 & 0.00 & 0.00 \\ 0.00 & 0.00 & 0.00 \\ 0.00 & 0.00 & 0.00 \end{bmatrix}, \; E_0 = \begin{bmatrix} 0.50 & 0.50 \\ 0.50 & 0.50 \\ 0.50 & 0.50 \\ -0.50 & -0.50 \\ -0.50 & -0.50 \\ 0.50 & 0.50 \\ 0.00 & 0.00 \\ 0.00 & 0.00 \\ 0.00 & 0.00 \\ 0.00 & 0.00 \\ 0.00 & 0.00 \\ 0.00 & 0.00 \\ 1.00 & 1.00 \\ -0.50 & -0.50 \\ -1.00 & -1.00 \\ 0.50 & 0.50 \\ -0.50 & -1.50 \\ 0.50 & 1.50 \\ 1.00 & 0.00 \\ 0.00 & 1.00 \\ -1.00 & 0.00 \\ 0.00 & -1.00 \end{bmatrix}, \; w_0 = \begin{bmatrix} 0.50 \\ 0.50 \\ 5.00 \\ 5.00 \\ 5.00 \\ 5.00 \\ 5.00 \\ 5.00 \\ 5.00 \\ 5.00 \\ 0.50 \\ 0.50 \\ 5.00 \\ 5.00 \\ 5.00 \\ 0.50 \\ 0.50 \\ 5.00 \\ 5.00 \\ 5.00 \\ 5.00 \end{bmatrix} \tag{12.15}$$

The RHC (12.6)–(12.9) algorithm becomes Algorithm 12.2. We refer to the solution U_0^* of the QP (11.31) as $[U_0^*, \text{Flag}] = \text{QP}(H, 2F'x(0), G_0, w_0 + E_0 x(0))$ where "Flag" indicates if the QP was found to be feasible or not.

Algorithm 12.2   *QP-based on-line receding horizon control*

Input: State $x(t)$ at time instant t

Output: Receding horizon control input $u(x(t))$

    Compute $\tilde{F} = 2F'x(t)$ and $\tilde{w}_0 = w_0 + E_0 x(t)$

    Obtain $U_0^*(x(t))$ by solving the optimization problem $[U_0^*, \text{Flag}] = \text{QP}(H, \tilde{F}, G_0, \tilde{w}_0)$

    If Flag = infeasible **Then** stop

    Return the first element u_0^* of U_0^*

Figure 12.2 shows two closed-loop trajectories starting at state $x(0) = [-4.5, 2]$ and $x(0) = [-4.5, 3]$. The trajectory starting from $x(0) = [-4.5, 2]$ converges to the origin and satisfies input and state constraints. The trajectory starting from $x(0) = [-4.5, 3]$ stops at $x(2) = [1, 2]$ because of infeasibility. At each time step, the open-loop predictions are depicted with dashed lines. This shows that the closed-loop

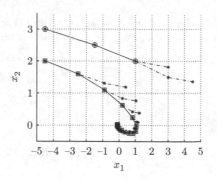

Figure 12.2 Example 12.1. Closed-loop trajectories realized (solid) and predicted (dashed) for two initial states x(0)=[−4.5,2] (*boxes*) and x(0)=[−4.5,3] (*circles*).

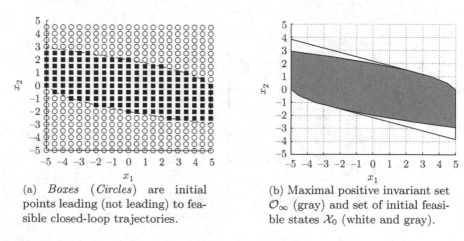

(a) *Boxes* (*Circles*) are initial points leading (not leading) to feasible closed-loop trajectories.

(b) Maximal positive invariant set $\mathcal{O}_\infty$ (gray) and set of initial feasible states $\mathcal{X}_0$ (white and gray).

Figure 12.3 Example 12.1. Double integrator with RHC.

trajectories are different from the open-loop predicted trajectories because of the receding horizon nature of the controller.

In Figure 12.3(a) the feasible state space was gridded and each point of the grid was marked with a square if the RHC law (12.6)–(12.9) generates feasible closed-loop trajectories and with a circle if it does not. The set of all initial conditions generating feasible closed-loop trajectories is the maximal positive invariant set $\mathcal{O}_\infty$ of the autonomous system (12.10). We remark that this set is different from the set $\mathcal{X}_0$ of feasible initial conditions for the QP problem (11.31) with matrices (12.15). Both sets $\mathcal{O}_\infty$ and $\mathcal{X}_0$ are depicted in Figure 12.3(b). The computation of f_0 is discussed later in this chapter. Because of the nonlinear nature of f_0, the computation of $\mathcal{O}_\infty$ for the system (12.10) is not an easy task. Therefore, we will show how to choose a terminal invariant set $\mathcal{X}_f$ such that $\mathcal{O}_\infty = \mathcal{X}_0$ is guaranteed automatically.

Note that a feasible closed-loop trajectory does not necessarily converge to the origin. Feasibility, convergence and stability of RHC are discussed in detail in the next sections. Before that we want to illustrate these issues through another example.

Example 12.2   Consider the unstable system

$$x(t+1) = \begin{bmatrix} 2 & 1 \\ 0 & 0.5 \end{bmatrix} x(t) + \begin{bmatrix} 1 \\ 0 \end{bmatrix} u(t) \tag{12.16}$$

with the input constraints

$$-1 \leq u(k) \leq 1, \ k = 0, \dots, N-1 \tag{12.17}$$

and the state constraints

$$\begin{bmatrix} -10 \\ -10 \end{bmatrix} \leq x(k) \leq \begin{bmatrix} 10 \\ 10 \end{bmatrix}, \ k = 0, \dots, N-1. \tag{12.18}$$

In the following, we study the receding horizon control problem (12.6) with $p(x_N) = x_N' P x_N$, $q(x_k, u_k) = x_k' Q x_k + u_k' R u_k$ for different horizons N and weights R. We set $Q = I$ and omit both the terminal set constraint and the terminal weight, i.e., $\mathcal{X}_f = \mathbb{R}^2$, $P = 0$.

Figure 12.4 shows closed-loop trajectories for receding horizon control loops that were obtained with the following parameter settings

- *Setting 1: $N = 2, R = 10$*
- *Setting 2: $N = 3, R = 2$*
- *Setting 3: $N = 4, R = 1$*

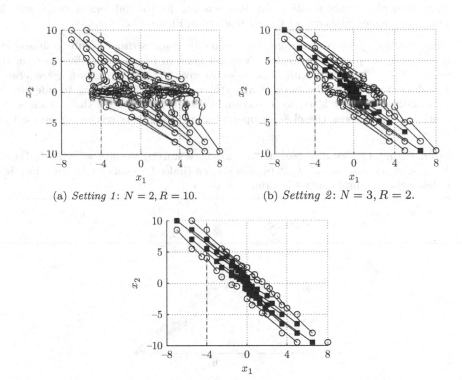

(a) *Setting 1: $N = 2, R = 10$.*   (b) *Setting 2: $N = 3, R = 2$.*

(c) *Setting 3: $N = 4, R = 1$.*

Figure 12.4 Example 12.2. Closed-loop trajectories for different settings of horizon N and weight R. *Boxes* (*Circles*) are initial points leading (not leading) to feasible closed-loop trajectories.

For *Setting 1* (Figure 12.4(a)) there is evidently no initial state that can be steered to the origin. Indeed, it turns out, that *all* nonzero initial states $x(0) \in \mathbb{R}^2$ diverge from the origin and eventually become infeasible. In contrast, *Setting 2* leads to a receding horizon controller, that manages to get some of the initial states converge to the origin, as seen in Figure 12.4(b). Finally, Figure 12.4(c) shows that *Setting 3* can expand the set of those initial states that can be brought to the origin.

Note the behavior of particular initial states:

1. Closed-loop trajectories starting at state $x(0) = [-4, 7]$ behave differently depending on the chosen setting. Both *Setting 1* and *Setting 2* cannot bring this state to the origin, but the controller with *Setting 3* succeeds.

2. There are initial states, e.g., $x(0) = [-4, 8.5]$, that always lead to infeasible trajectories independent of the chosen settings. It turns out, that *no* setting can be found that brings those states to the origin.

These results illustrate that the choice of parameters for receding horizon control influences the behavior of the resulting closed-loop trajectories in a complex manner. A better understanding of the effect of parameter changes can be gained from an inspection of maximal positive invariant sets $\mathcal{O}_\infty$ for the different settings, and the maximal control invariant set $\mathcal{C}_\infty$ as depicted in Figure 12.5.

The maximal positive invariant set stemming from *Setting 1* only contains the origin ($\mathcal{O}_\infty = \{0\}$) which explains why all nonzero initial states diverge from the origin. For *Setting 2* the maximal positive invariant set has grown considerably, but does not contain the initial state $x(0) = [-4, 7]$, thus leading to infeasibility eventually. *Setting 3* leads to a maximal positive invariant set that contains this state and thus keeps the closed-loop trajectory inside this set for all future time steps.

From Figure 12.5 we also see that a trajectory starting at $x(0) = [-4, 8.5]$ cannot be kept inside any bounded set by *any* setting (indeed, by any controller) since it is outside the maximal control invariant set $\mathcal{C}_\infty$.

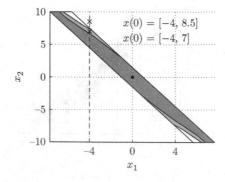

Figure 12.5 Example 12.2. Maximal positive invariant sets $\mathcal{O}_\infty$ for different parameter settings: *Setting 1* (origin), *Setting 2* (dark-gray) and *Setting 3* (gray and dark-gray). Also depicted is the maximal control invariant set $\mathcal{C}_\infty$ (white and gray and dark-gray).

12.3 RHC Main Issues

If we solve the receding horizon problem for the special case of an infinite horizon (setting $N = \infty$ in (12.6) as we did for LQR in Section 8.5 and CLQR in Section 11.3.4) then it is almost immediate that the closed-loop system with this controller has some nice properties. Most importantly, the differences between the open-loop predicted and the actual closed-loop trajectories observed in Example 12.1 disappear. As a consequence, if the optimization problem is feasible, then the closed-loop trajectories will be feasible for all times. If the optimization problem has a finite solution, then in closed-loop the states and inputs will converge to the origin asymptotically.

In RHC, when we solve the optimization problem over a finite horizon repeatedly at each time step, we hope that the controller resulting from this "short-sighted" strategy will lead to a closed-loop behavior that mimics that of the infinite horizon controller. The examples in the last section indicated that at least two problems may occur. First of all, the controller may lead us into a situation where after a few steps the finite horizon optimal control problem that we need to solve at each time step is infeasible, i.e., that there does not exist a sequence of control inputs for which the constraints are obeyed. Second, even if the feasibility problem does not occur, the generated control inputs may not lead to trajectories that converge to the origin, i.e., that the closed-loop system is asymptotically stable.

In general, stability and feasibility are not ensured by the RHC law (12.6)–(12.9). In principle, we could analyze the RHC law for feasibility, stability and convergence but this is difficult as the examples in the last section illustrated. Therefore, conditions will be derived on how the terminal weight P and the terminal constraint set $\mathcal{X}_f$ should be chosen such that closed-loop stability and feasibility are ensured.

12.3.1 Feasibility of RHC

The examples in the last section illustrate that feasibility at the initial time $x(0) \in \mathcal{X}_0$ does not necessarily imply feasibility for all future times. It is desirable to design an RHC such that feasibility for all future times is guaranteed, a property we refer to as *persistent feasibility*.

We would like to gain some insight when persistent feasibility occurs and how it is affected by the formulation of the control problem and the choice of the controller parameters. Let us first recall the various sets introduced in Section 10.2 and Section 11.2 and how they are influenced.

$\mathcal{C}_\infty$: The maximal control invariant set $\mathcal{C}_\infty$ is only affected by the sets $\mathcal{X}$ and $\mathcal{U}$, the constraints on states and inputs. It is the largest set over which we can expect *any* controller to work.

$\mathcal{X}_0$: A control input U_0 can only be found, i.e., the control problem is feasible, if $x(0) \in \mathcal{X}_0$. The set $\mathcal{X}_0$ depends on $\mathcal{X}$ and $\mathcal{U}$, on the controller horizon N and on the controller terminal set $\mathcal{X}_f$. It does not depend on the objective function and it has generally no relation with $\mathcal{C}_\infty$ (it can be larger, smaller, etc.).

$\mathcal{O}_\infty$: The maximal positive invariant set for the closed-loop system depends on the controller and as such on all parameters affecting the controller, i.e., $\mathcal{X}$, $\mathcal{U}$, N, $\mathcal{X}_f$ and the objective function with its parameters P, Q and R. Clearly $\mathcal{O}_\infty \subseteq \mathcal{X}_0$ because if it were not there would be points in $\mathcal{O}_\infty$ for which the control problem is not feasible. Because of invariance, the closed-loop is persistently feasible for all states $x(0) \in \mathcal{O}_\infty$. Clearly, $\mathcal{O}_\infty \subseteq \mathcal{C}_\infty$.

We can now state necessary and sufficient conditions guaranteeing persistent feasibility by means of invariant set theory.

Lemma 12.1   *Let $\mathcal{O}_\infty$ be the maximal positive invariant set for the closed-loop system $x(k+1) = f_{cl}(x(k))$ in (12.10) with constraints (12.2). The RHC problem is persistently feasible if and only if $\mathcal{X}_0 = \mathcal{O}_\infty$.*

Proof: For the RHC problem to be persistently feasible $\mathcal{X}_0$ must be positive invariant for the closed-loop system. We argued above that $\mathcal{O}_\infty \subseteq \mathcal{X}_0$. As the positive invariant set $\mathcal{X}_0$ cannot be larger than the maximal positive invariant set $\mathcal{O}_\infty$, it follows that $\mathcal{X}_0 = \mathcal{O}_\infty$.   ∎

As $\mathcal{X}_0$ does not depend on the controller parameters P, Q and R but $\mathcal{O}_\infty$ does, the requirement $\mathcal{X}_0 = \mathcal{O}_\infty$ for persistent feasibility shows that, in general, only some P, Q and R are allowed. The parameters P, Q and R affect the performance. The complex effect they have on persistent feasibility makes their choice extremely difficult for the design engineer. In the following we will remedy this undesirable situation. We will make use of the following important *sufficient* condition for persistent feasibility.

Lemma 12.2   *Consider the RHC law (12.6)–(12.9) with $N \geq 1$. If $\mathcal{X}_1$ is a control invariant set for system (12.1)–(12.2) then the RHC is persistently feasible. Also, $\mathcal{O}_\infty = \mathcal{X}_0$ is independent of P, Q and R.*

Proof: If $\mathcal{X}_1$ is control invariant then, by definition, $\mathcal{X}_1 \subseteq \mathrm{Pre}(\mathcal{X}_1)$. Also recall that $\mathrm{Pre}(\mathcal{X}_1) = \mathcal{X}_0$ from the properties of the feasible sets in equation (11.20) (note that $\mathrm{Pre}(\mathcal{X}_1) \cap \mathcal{X} = \mathrm{Pre}(\mathcal{X}_1)$ from control invariance). Pick some $x \in \mathcal{X}_0$ and some feasible control u for that x and define $x^+ = Ax + Bu \in \mathcal{X}_1$. Then $x^+ \in \mathcal{X}_1 \subseteq \mathrm{Pre}(\mathcal{X}_1) = \mathcal{X}_0$. As u was arbitrary (as long as it is feasible) $x^+ \in \mathcal{X}_0$ for all feasible u. As $\mathcal{X}_0$ is positive invariant, $\mathcal{X}_0 = \mathcal{O}_\infty$ from Lemma 12.1. As $\mathcal{X}_0$ is positive invariant for all feasible u, $\mathcal{O}_\infty$ does not depend on P, Q and R.   ∎

Note that in the proof of Lemma 12.2, persistent feasibility does not depend on the input u as long as it is feasible. For this reason, sometimes in the literature this property is referred to "persistently feasible for all feasible u."

We can use Lemma 12.2 in the following manner. For $N = 1$, $\mathcal{X}_1 = \mathcal{X}_f$. If we choose the terminal set to be control invariant then $\mathcal{X}_0 = \mathcal{O}_\infty$ and RHC will be persistently feasible independent of chosen control objectives and parameters. Thus the designer can choose the parameters to affect performance without affecting persistent feasibility. A control horizon of $N = 1$ is often too restrictive, but we can easily extend Lemma 12.2.

Theorem 12.1   *Consider the RHC law (12.6)–(12.9) with $N \geq 1$. If $\mathcal{X}_f$ is a control invariant set for system (12.1)–(12.2) then the RHC is persistently feasible.*

Proof: If $\mathcal{X}_f$ is control invariant, then $\mathcal{X}_{N-1}, \mathcal{X}_{N-2}, \ldots, \mathcal{X}_1$ are control invariant and Lemma 12.2 establishes persistent feasibility for all feasible u. ∎

Corollary 12.1 *Consider the RHC law (12.6)–(12.9) with $N \geq 1$. If there exists $i \in [1, N]$ such that $\mathcal{X}_i$ is a control invariant set for system (12.1)–(12.2), then the RHC is persistently feasible for all cost functions.*

Proof: Follows directly from the proof of Theorem 12.1. ∎

Recall that Theorem 11.2 together with Remark 11.4 define the properties of the set $\mathcal{X}_0$ as N varies. Therefore, Theorem 12.1 and Corollary 12.1 provide also guidelines on the choice of the horizon N for guaranteeing persistent feasibility for all feasible u. For instance, if the RHC problem (12.6) for $N = \bar{N}$ yields a control invariant set $\mathcal{X}_0$, then from Theorem 11.2 the RHC law (12.6)–(12.9) with $N = \bar{N} + 1$ is persistently feasible for all feasible u. Moreover, from Corollary 12.1 the RHC law (12.6)–(12.9) with $N \geq \bar{N} + 1$ is persistently feasible for all feasible u.

Corollary 12.2 *Consider the RHC problem (12.6)–(12.9). If N is greater than the determinedness index $\bar{N}$ of $\mathcal{K}_\infty(\mathcal{X}_f)$ for system (12.1)–(12.2), then the RHC is persistently feasible.*

Proof: The feasible set $\mathcal{X}_i$ for $i = 1, \ldots, N - 1$ is equal to the $(N - i)$-step controllable set $\mathcal{X}_i = \mathcal{K}_{N-i}(\mathcal{X}_f)$. If the maximal controllable set is finitely determined then $\mathcal{X}_i = \mathcal{K}_\infty(\mathcal{X}_f)$ for $i \leq N - \bar{N}$. Note that $\mathcal{K}_\infty(\mathcal{X}_f)$ is control invariant. Then persistent feasibility for all feasible u follows from Corollary 12.1. ∎

Persistent feasibility does not guarantee that the closed-loop trajectories converge towards the desired equilibrium point. From Theorem 12.1 it is clear that one can only guarantee that $x(k) \in \mathcal{X}_1$ for all $k > 0$ if $x(0) \in \mathcal{X}_0$.

One of the most popular approaches to guarantee persistent feasibility and stability of the RHC law (12.6)–(12.9) makes use of a control invariant terminal set $\mathcal{X}_f$ and a terminal cost P which drives the closed-loop optimal trajectories towards the origin. A detailed discussion follows in the next section.

12.3.2 Stability of RHC

In this section we will derive the main stability result for RHC. Our objective is to find a Lyapunov function for the closed-loop system. We will show next that if the terminal cost and constraint are appropriately chosen, then the value function $J_0^*(\cdot)$ is a Lyapunov function.

Theorem 12.2 *Consider system (12.1)–(12.2), the RHC law (12.6)–(12.9) and the closed-loop system (12.10). Assume that*

(A0) The stage cost $q(x, u)$ and terminal cost $p(x)$ are continuous and positive definite functions.

(A1) The sets $\mathcal{X}$, $\mathcal{X}_f$ and $\mathcal{U}$ contain the origin in their interior and are closed.

(A2) $\mathcal{X}_f$ *is control invariant,* $\mathcal{X}_f \subseteq \mathcal{X}$.

(A3) $\displaystyle\min_{v \in \mathcal{U}, \ Ax+Bv \in \mathcal{X}_f}(-p(x) + q(x,v) + p(Ax + Bv)) \leq 0, \ \forall x \in \mathcal{X}_f$.

Then, the origin of the closed-loop system (12.10) is asymptotically stable with domain of attraction $\mathcal{X}_0$.

Proof: From hypothesis (A2), Theorem 12.1 and Lemma 12.1, we conclude that $\mathcal{X}_0 = \mathcal{O}_\infty$ is a positive invariant set for the closed-loop system (12.10) for any choice of the cost function. Thus persistent feasibility for any feasible input is guaranteed in $\mathcal{X}_0$.

Next, we prove convergence and stability. We establish that the function $J_0^*(\cdot)$ in (12.6) is a Lyapunov function for the closed-loop system. Because the cost J_0, the system and the constraints are time-invariant we can study the properties of J_0^* between step $k = 0$ and step $k + 1 = 1$.

Consider problem (12.6) at time $t = 0$. Let $x(0) \in \mathcal{X}_0$ and let $U_0^* = \{u_0^*, \ldots, u_{N-1}^*\}$ be the optimizer of problem (12.6) and $\mathbf{x}_0 = \{x(0), x_1, \ldots, x_N\}$ be the corresponding optimal state trajectory. After the implementation of u_0^* we obtain $x(1) = x_1 = Ax(0) + Bu_0^*$. Consider now problem (12.6) for $t = 1$. We will construct an upper bound on $J_0^*(x(1))$. Consider the sequence $\tilde{U}_1 = \{u_1^*, \ldots, u_{N-1}^*, v\}$ and the corresponding state trajectory resulting from the initial state $x(1)$, $\tilde{\mathbf{x}}_1 = \{x_1, \ldots, x_N, Ax_N + Bv\}$. Because $x_N \in \mathcal{X}_f$ and (A2) there exists a feasible v such that $x_{N+1} = Ax_N + Bv \in \mathcal{X}_f$ and with this v the sequence $\tilde{U}_1 = \{u_1^*, \ldots, u_{N-1}^*, v\}$ is feasible. Because $\tilde{U}_1$ is not optimal $J_0(x(1), \tilde{U}_1)$ is an upper bound on $J_0^*(x(1))$.

Since the trajectories generated by U_0^* and $\tilde{U}_1$ overlap, except for the first and last sampling intervals, it is immediate to show that

$$\begin{aligned} J_0^*(x(1)) \leq J_0(x(1), \tilde{U}_1) = \ &J_0^*(x(0)) - q(x_0, u_0^*) - p(x_N) \\ &+ (q(x_N, v) + p(Ax_N + Bv)). \end{aligned} \tag{12.19}$$

Let $x = x_0 = x(0)$ and $u = u_0^*$. Under assumption (A3) equation (12.19) becomes

$$J_0^*(Ax + Bu) - J_0^*(x) \leq -q(x, u), \ \forall x \in \mathcal{X}_0. \tag{12.20}$$

Equation (12.20) and the hypothesis (A0) on the stage cost $q(\cdot)$ ensure that $J_0^*(x)$ strictly decreases along the state trajectories of the closed-loop system (12.10) for any $x \in \mathcal{X}_0$, $x \neq 0$. In addition to the fact that $J_0^*(x)$ decreases, $J_0^*(x)$ is lower-bounded by zero and since the state trajectories generated by the closed-loop system (12.10) starting from any $x(0) \in \mathcal{X}_0$ lie in $\mathcal{X}_0$ for all $k \geq 0$, equation (12.20) is sufficient to ensure that the state of the closed-loop system converges to zero as $k \to 0$ if the initial state lies in $\mathcal{X}_0$. We have proven (i).

In order to prove stability via Theorem 7.2 we have to establish that $J_0^*(x)$ is a Lyapunov function. Positivity holds by the hypothesis (A0), decrease follows from (12.20). For continuity at the origin we will show that $J_0^*(x) \leq p(x), \ \forall x \in \mathcal{X}_f$ and as $p(x)$ is continuous at the origin (by hypothesis (A0)) $J_0^*(x)$ must be continuous as well. From assumption (A2), $\mathcal{X}_f$ is control invariant and thus for any $x \in \mathcal{X}_f$ there exists a feasible input sequence $\{u_0, \ldots, u_{N-1}\}$ for problem (12.6) starting from the initial state $x_0 = x$ whose corresponding state trajectory is

$\{x_0, x_1, \ldots, x_N\}$ stays in $\mathcal{X}_f$, i.e., $x_i \in \mathcal{X}_f \; \forall \; i = 0, \ldots, N$. Among all the aforementioned input sequences $\{u_0, \ldots, u_{N-1}\}$ we focus on the one where u_i satisfies assumption (A3) for all $i = 0, \ldots, N-1$. Such a sequence provides an upper bound on the function J_0^*:

$$J_0^*(x_0) \leq \left(\sum_{i=0}^{N-1} q(x_i, u_i) \right) + p(x_N), \; x_i \in \mathcal{X}_f, \; i = 0, \ldots, N, \qquad (12.21)$$

which can be rewritten as

$$J_0^*(x_0) \leq \left(\sum_{i=0}^{N-1} q(x_i, u_i) \right) + p(x_N)$$
$$= p(x_0) + \left(\sum_{i=0}^{N-1} q(x_i, u_i) + p(x_{i+1}) - p(x_i) \right) \; x_i \in \mathcal{X}_f, \; i = 0, \ldots, N, \tag{12.22}$$

which from assumption (A3) yields

$$J_0^*(x) \leq p(x), \; \forall x \in \mathcal{X}_f. \qquad (12.23)$$

In conclusion, there exist a finite time in which any $x \in \mathcal{X}_0$ is steered to a level set of $J_0^*(x)$ contained in $\mathcal{X}_f$ after which convergence to and stability of the origin follows. ∎

Remark 12.1 The assumption on the positive definiteness of the stage cost $q(\cdot)$ in Theorem 12.2 can be relaxed as in standard optimal control. For instance, for the 2-norm based cost function (12.8), one can allow $Q \succeq 0$ with $(Q^{\frac{1}{2}}, A)$ observable.

Remark 12.2 The procedure outlined in Theorem 12.2 is, in general, conservative because it requires the introduction of an artificial terminal set $\mathcal{X}_f$ to guarantee persistent feasibility and a terminal cost to guarantee stability. Requiring $x_N \in \mathcal{X}_f$ usually decreases the size of the region of attraction $\mathcal{X}_0 = \mathcal{O}_\infty$. Also the performance may be negatively affected.

Remark 12.3 A function $p(x)$ satisfying assumption (A3) of Theorem 12.2 is often called control Lyapunov function.

The hypothesis (A2) of Theorem 12.2 is required for guaranteeing persistent feasibility as discussed in Section 12.3.1. In some part of the literature the constraint $\mathcal{X}_f$ is not used. However, in this literature the terminal region constraint $\mathcal{X}_f$ is implicit. In fact, it is typically required that the horizon N is sufficiently large to ensure feasibility of the RHC (12.6)–(12.9) at all time instants t. Technically this means that N has to be greater than the determinedness index $\bar{N}$ of system (12.1)–(12.2) which by Corollary 12.2 guarantees persistent feasibility for all inputs. We refer the reader to Section 12.3.1 for more details on feasibility.

Next we will show a few simple choices for P and $\mathcal{X}_f$ satisfying the hypothesis (A2) and (A3) of Theorem 12.2.

Stability, 2-Norm Case

Consider system (12.1)–(12.2), the RHC law (12.6)–(12.9), the cost function (12.8) and the closed-loop system (12.10). A simple choice for $\mathcal{X}_f$ is the maximal positive invariant set (see Section 10.1) for the closed-loop system $x(k+1) = (A+BF_\infty)x(k)$ where F_∞ is the associated unconstrained infinite time optimal controller (8.33). With this choice the assumption (A3) in Theorem 12.2 becomes

$$x'(A'(P - PB(B'PB + R)^{-1}BP)A + Q - P)x \leq 0, \ \forall x \in \mathcal{X}_f, \qquad (12.24)$$

which is satisfied as an equality if P is chosen as the solution P_∞ of the Algebraic Riccati Equation (8.32) for system (12.1).

In general, instead of F_∞ we can choose any controller F which stabilizes $A + BF$. With $v = Fx$ the assumption (A3) in Theorem 12.2 becomes

$$-P + (Q + F'RF) + (A + BF)'P(A + BF) \leq 0. \qquad (12.25)$$

It is satisfied as an equality if we choose P as a solution of the corresponding Lyapunov equation.

We learned in Section 7.5.2 that P satisfying (12.24) or (12.25) as equalities expresses the infinite horizon cost

$$J_\infty^*(x_0) = x_0'Px_0 = \sum_{k=0}^{\infty} x_k'Qx_k + u_k'Ru_k. \qquad (12.26)$$

In summary, we conclude that the closed-loop system (12.10) is stable if there exist a controller F which stabilizes the unconstrained system inside the controlled invariant terminal region $\mathcal{X}_f$ and if the infinite horizon cost $x'Px$ incurred with this controller is used in the cost function (12.8). Thus, the two terms in the objective (12.8) reflect the infinite horizon cost, one the initial finite horizon cost when the controller is constrained and the second one the cost for the infinite tail of the trajectory incurred after the system enters $\mathcal{X}_f$ and the controller is unconstrained.

If the open loop system (12.1) is asymptotically stable, then we may even select $F = 0$. Note that depending on the choice of the controller the controlled invariant terminal region $\mathcal{X}_f$ changes.

For any of the discussed choices for F and $\mathcal{X}_f$ stability implies exponential stability. The argument is simple. As the system is closed-loop stable it enters the terminal region in finite time. If $\mathcal{X}_f$ is chosen as suggested, the closed-loop system is unconstrained after entering $\mathcal{X}_f$. For an unconstrained linear system the convergence to the origin is exponential.

Stability, 1-Norm and ∞-Norm Case

Consider system (12.1)–(12.2), the RHC law (12.6)–(12.9), the cost function (12.7) and the closed-loop system (12.10). Let $p = 1$ or $p = \infty$. If system (12.1) is asymptotically stable, then $\mathcal{X}_f$ can be chosen as the positively invariant set of the autonomous system $x(k+1) = Ax(k)$ subject to the state constraints $x \in \mathcal{X}$. Therefore in $\mathcal{X}_f$ the input $\mathbf{0}$ is feasible and the assumption (A3) in Theorem 12.2 becomes

$$-\|Px\|_p + \|PAx\|_p + \|Qx\|_p \le 0, \forall x \in \mathcal{X}_f, \tag{12.27}$$

which is the corresponding Lyapunov inequality for the 1-norm and ∞-norm case (7.55) whose solution has been discussed in Section 7.5.3.

In general, if the unconstrained optimal controller (9.31) exists it is PPWA. In this case the computation of the maximal invariant set $\mathcal{X}_f$ for the closed-loop PWA system

$$x(k+1) = (A+F^i)x(k) \quad \text{if} \quad H^i x \le 0, \; i = 1, \dots, N^r \tag{12.28}$$

is more involved. However if such $\mathcal{X}_f$ can be computed it can be used as terminal constraint in Theorem 12.2. With this choice the assumption (A3) in Theorem 12.2 is satisfied by the infinite time unconstrained optimal cost matrix P_∞ in (9.32).

12.4 State Feedback Solution of RHC, 2-Norm Case

The state feedback receding horizon controller (12.9) with cost (12.8) for system (12.1) is

$$u(t) = f_0^*(x(t)), \tag{12.29}$$

where $f_0^*(x_0) : \mathbb{R}^n \to \mathbb{R}^m$ is the piecewise affine solution to the CFTOC (12.6) and is obtained as explained in Section 11.3.

We remark that the implicit form (12.6) and the explicit form (12.29) describe the same function, and therefore the stability, feasibility, and performance properties mentioned in the previous sections are automatically inherited by the piecewise affine control law (12.29). Clearly, the explicit form (12.29) has the advantage of being easier to implement, and provides insight into the type of controller action in different regions CR_i of the state space.

Example 12.3 Consider the double integrator system (12.11) subject to the input constraints

$$-1 \le u(k) \le 1 \tag{12.30}$$

and the state constraints

$$-10 \le x(k) \le 10. \tag{12.31}$$

We want to regulate the system to the origin by using the RHC problem (12.6)–(12.9) with cost (12.8), $Q = \begin{bmatrix} 1 & 0 \\ 0 & 1 \end{bmatrix}$, $R = 0.01$, and $P = P_\infty$ where P_∞ solves the algebraic Riccati equation (8.32). We consider three cases:

Case 1. $N = 2$, $\mathcal{X}_f = 0$,
Case 2. $N = 2$, $\mathcal{X}_f$ is the positively invariant set of the closed-loop system $x(k+1) = (A + BF_\infty)$ where F_∞ is the infinite time unconstrained optimal controller (8.33).
Case 3. No terminal state constraints: $\mathcal{X}_f = \mathbb{R}^2$ and $N = 6 = $ determinedness index+1.

From the results presented in this chapter, all three cases guarantee persistent feasibility for all cost functions and asymptotic stability of the origin with region of attraction $\mathcal{X}_0$ (with $\mathcal{X}_0$ different for each case). Next, we will detail the matrices of the quadratic program for the on-line solution as well as the explicit solution for the three cases.

Case 1. $\mathcal{X}_f = 0$. The mp-QP problem associated with the RHC has the form (11.31) with

$$H = \begin{bmatrix} 19.08 & 8.55 \\ 8.55 & 5.31 \end{bmatrix}, \quad F = \begin{bmatrix} -10.57 & -5.29 \\ -10.59 & -5.29 \end{bmatrix}, \quad Y = \begin{bmatrix} 10.31 & 9.33 \\ 9.33 & 10.37 \end{bmatrix} \tag{12.32}$$

and

$$G_0 = \begin{bmatrix} 0.00 & -1.00 \\ 0.00 & 1.00 \\ 0.00 & 0.00 \\ -1.00 & 0.00 \\ 0.00 & 0.00 \\ 1.00 & 0.00 \\ -1.00 & 0.00 \\ -1.00 & -1.00 \\ 1.00 & 0.00 \\ 1.00 & 1.00 \\ 0.00 & 0.00 \\ -1.00 & 0.00 \\ 0.00 & 0.00 \\ 1.00 & 0.00 \\ -1.00 & 0.00 \\ 1.00 & 0.00 \\ 0.00 & 0.00 \\ 0.00 & 0.00 \\ 0.00 & 0.00 \\ 0.00 & 0.00 \\ 0.00 & 0.00 \\ 0.00 & 0.00 \\ 0.00 & 0.00 \\ 0.00 & 0.00 \\ 1.00 & 0.00 \\ 1.00 & 1.00 \\ -1.00 & 0.00 \\ -1.00 & -1.00 \end{bmatrix}, \quad E_0 = \begin{bmatrix} 0.00 & 0.00 \\ 0.00 & 0.00 \\ 1.00 & 1.00 \\ -1.00 & -1.00 \\ -1.00 & -1.00 \\ 1.00 & 1.00 \\ 0.00 & 0.00 \\ -1.00 & -1.00 \\ 0.00 & 0.00 \\ 1.00 & 1.00 \\ 1.00 & 1.00 \\ -1.00 & -1.00 \\ -1.00 & -1.00 \\ 1.00 & 1.00 \\ -1.00 & -2.00 \\ 1.00 & 2.00 \\ 1.00 & 1.00 \\ 0.00 & 1.00 \\ -1.00 & 0.00 \\ 0.00 & -1.00 \\ 1.00 & 0.00 \\ 0.00 & 1.00 \\ -1.00 & 0.00 \\ 0.00 & -1.00 \\ 0.00 & 0.00 \\ 1.00 & 1.00 \\ 0.00 & 0.00 \\ -1.00 & -1.00 \end{bmatrix}, \quad w_0 = \begin{bmatrix} 1.00 \\ 1.00 \\ 10.00 \\ 10.00 \\ 10.00 \\ 10.00 \\ 10.00 \\ 10.00 \\ 10.00 \\ 10.00 \\ 5.00 \\ 5.00 \\ 5.00 \\ 5.00 \\ 1.00 \\ 1.00 \\ 10.00 \\ 10.00 \\ 10.00 \\ 10.00 \\ 5.00 \\ 5.00 \\ 5.00 \\ 5.00 \\ 0.00 \\ 0.00 \\ 0.00 \\ 0.00 \end{bmatrix} \tag{12.33}$$

The corresponding polyhedral partition of the state space is depicted in Figure 12.6(a).

The RHC law is:

$$u = \begin{cases} [-0.61 \ -1.61]\, x & \text{if } \begin{bmatrix} 0.70 & 0.71 \\ -0.70 & -0.71 \\ -0.70 & -0.71 \\ 0.70 & 0.71 \end{bmatrix} x \le \begin{bmatrix} 0.00 \\ 0.00 \\ 0.00 \\ 0.00 \end{bmatrix} & \text{(Region \#1)} \\[4ex] [-1.00 \ -2.00]\, x & \text{if } \begin{bmatrix} -0.71 & -0.71 \\ -0.70 & -0.71 \\ -0.45 & -0.89 \\ 0.45 & 0.89 \\ 0.71 & 0.71 \\ -0.70 & -0.71 \end{bmatrix} x \le \begin{bmatrix} 0.00 \\ -0.00 \\ 0.45 \\ 0.45 \\ 0.71 \\ -0.00 \end{bmatrix} & \text{(Region \#2)} \\[4ex] [-1.00 \ -2.00]\, x & \text{if } \begin{bmatrix} 0.45 & 0.89 \\ -0.70 & -0.71 \\ 0.71 & 0.71 \end{bmatrix} x \le \begin{bmatrix} 0.45 \\ -0.00 \\ -0.00 \end{bmatrix} & \text{(Region \#3)} \\[4ex] [-0.72 \ -1.72]\, x & \text{if } \begin{bmatrix} 0.39 & 0.92 \\ 0.70 & 0.71 \\ -0.70 & -0.71 \\ 0.70 & 0.71 \end{bmatrix} x \le \begin{bmatrix} 0.54 \\ 0.00 \\ 0.00 \\ -0.00 \end{bmatrix} & \text{(Region \#4)} \\[4ex] [-1.00 \ -2.00]\, x & \text{if } \begin{bmatrix} 0.45 & 0.89 \\ -0.71 & -0.71 \\ 0.70 & 0.71 \\ -0.45 & -0.89 \\ 0.71 & 0.71 \\ 0.70 & 0.71 \end{bmatrix} x \le \begin{bmatrix} 0.45 \\ 0.71 \\ -0.00 \\ 0.45 \\ 0.00 \\ -0.00 \end{bmatrix} & \text{(Region \#5)} \\[4ex] [-1.00 \ -2.00]\, x & \text{if } \begin{bmatrix} -0.45 & -0.89 \\ -0.71 & -0.71 \\ 0.70 & 0.71 \end{bmatrix} x \le \begin{bmatrix} 0.45 \\ -0.00 \\ -0.00 \end{bmatrix} & \text{(Region \#6)} \\[4ex] [-0.72 \ -1.72]\, x & \text{if } \begin{bmatrix} -0.39 & -0.92 \\ 0.70 & 0.71 \\ -0.70 & -0.71 \\ -0.70 & -0.71 \end{bmatrix} x \le \begin{bmatrix} 0.54 \\ 0.00 \\ 0.00 \\ -0.00 \end{bmatrix} & \text{(Region \#7)} \end{cases}$$

The union of the regions depicted in Figure 12.6(a) is $\mathcal{X}_0$. From Theorem 12.2, $\mathcal{X}_0$ is also the domain of attraction of the RHC law.

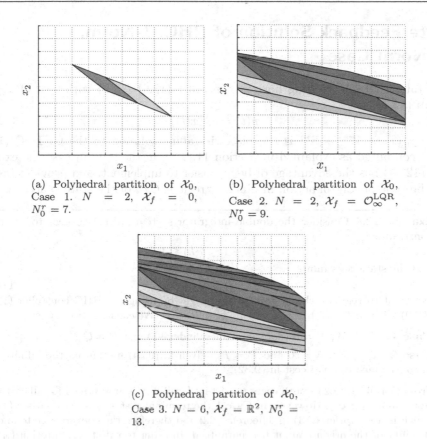

(a) Polyhedral partition of $\mathcal{X}_0$, Case 1. $N = 2$, $\mathcal{X}_f = 0$, $N_0^r = 7$.

(b) Polyhedral partition of $\mathcal{X}_0$, Case 2. $N = 2$, $\mathcal{X}_f = \mathcal{O}_\infty^{\mathrm{LQR}}$, $N_0^r = 9$.

(c) Polyhedral partition of $\mathcal{X}_0$, Case 3. $N = 6$, $\mathcal{X}_f = \mathbb{R}^2$, $N_0^r = 13$.

Figure 12.6 Example 12.3. Double integrator. RHC with 2-norm. Region of attraction $\mathcal{X}_0$ for different horizons N and terminal regions $\mathcal{X}_f$.

Case 2. $\mathcal{X}_f$ positively invariant set. The set $\mathcal{X}_f$ is

$$\mathcal{X}_f = \left\{ x \in \mathbb{R}^2 : \begin{bmatrix} -0.35617 & -0.93442 \\ 0.35617 & 0.93442 \\ 0.71286 & 0.70131 \\ -0.71286 & -0.70131 \end{bmatrix} x \leq \begin{bmatrix} 0.58043 \\ 0.58043 \\ 1.9049 \\ 1.9049 \end{bmatrix} \right\} \qquad (12.34)$$

The corresponding polyhedral partition of the state space is depicted in Figure 12.6(b). The union of the regions depicted in Figure 12.6(b) is $\mathcal{X}_0$. Note that from Theorem 12.2 the set $\mathcal{X}_0$ is also the domain of attraction of the RHC law.

Case 3. $\mathcal{X}_f = \mathbb{R}^n$, $N = 6$. The corresponding polyhedral partition of the state space is depicted in Figure 12.6(c).

Comparing the feasibility regions $\mathcal{X}_0$ in Figure 12.6 we notice that in Case 2 we obtain a larger region than in Case 1 and that in Case 3 we obtain a feasibility region larger than Case 1 and Case 2. This can be easily explained from the theory presented in this and the previous chapter. In particular we have seen that if a control invariant set is chosen as terminal constraint $\mathcal{X}_f$, the size of the feasibility region increases with the number of control moves (increase from Case 2 to Case 3) (Remark 11.3). Actually in Case 3, $\mathcal{X}_0 = \mathcal{K}_\infty(\mathcal{X}_f)$ with $\mathcal{X}_f = \mathbb{R}^2$, the maximal controllable set. Also, the size of the feasibility region increases with the size of the target set (increase from Case 1 to Case 2).

12.5 State Feedback Solution of RHC, 1-Norm, ∞-Norm Case

The state feedback receding horizon controller (12.6)–(12.9) with cost (12.7) for system (12.1) is

$$u(t) = f_0^*(x(t)) \tag{12.35}$$

where $f_0^*(x_0) : \mathbb{R}^n \to \mathbb{R}^m$ is the piecewise affine solution to the CFTOC (12.6) and is computed as explained in Section 11.4. As in the 2-norm case the explicit form (12.35) has the advantage of being easier to implement, and provides insight into the type of control action in different regions CR_i of the state space.

Example 12.4 Consider the double integrator system (12.11) subject to the input constraints

$$-1 \le u(k) \le 1 \tag{12.36}$$

and the state constraints

$$-5 \le x(k) \le 5. \tag{12.37}$$

We want to regulate the system to the origin by using the RHC controller (12.6)–(12.9) with cost (12.7), $p = \infty$, $Q = \begin{bmatrix} 1 & 0 \\ 0 & 1 \end{bmatrix}$, $R = 20$. We consider two cases:

Case 1. $\mathcal{X}_f = \mathbb{R}^n$, $N = 6$ (determinedness index+1) and $P = Q$

Case 2. $\mathcal{X}_f = \mathbb{R}^n$, $N = 6$ and $P = P_\infty$ given in (9.34) measuring the infinite time unconstrained optimal cost in (9.32).

From Corollary 12.2 in both cases persistent feasibility is guaranteed for all cost functions and $\mathcal{X}_0 = \mathcal{C}_\infty$. However, in Case 1 the terminal cost P does not satisfy (12.27) which is assumption (A3) in Theorem 12.2 and therefore the convergence to and the stability of the origin cannot be guaranteed. In order to satisfy assumption (A3) in Theorem 12.2, in Case 2 we select the terminal cost to be equal to the infinite time unconstrained optimal cost computed in Example 9.1.

Next we will detail the explicit solutions for the two cases.

Case 1. The LP problem associated with the RHC has the form (11.56) with $\bar{G}_0 \in \mathbb{R}^{124 \times 18}$, $\bar{S}_0 \in \mathbb{R}^{124 \times 2}$ and $c' = [0_6 \ 1_{12}]$. The corresponding polyhedral partition of the state space is depicted in Figure 12.7(a). The RHC law is:

$$u = \begin{cases}
0 & \text{if } \begin{bmatrix} 0.16 & 0.99 \\ -0.16 & -0.99 \\ -1.00 & 0.00 \\ 1.00 & 0.00 \end{bmatrix} x \le \begin{bmatrix} 0.82 \\ 0.82 \\ 5.00 \\ 5.00 \end{bmatrix} & \text{(Region \#1)} \\[4em]
\begin{bmatrix} -0.29 & -1.71 \end{bmatrix} x + 1.43 & \text{if } \begin{bmatrix} 1.00 & 0.00 \\ -0.16 & -0.99 \\ -1.00 & 0.00 \\ 0.16 & 0.99 \end{bmatrix} x \le \begin{bmatrix} 5.00 \\ -0.82 \\ 5.00 \\ 1.40 \end{bmatrix} & \text{(Region \#2)} \\[4em]
-1.00 & \text{if } \begin{bmatrix} -0.16 & -0.99 \\ 1.00 & 0.00 \\ 0.71 & 0.71 \\ -1.00 & 0.00 \\ 0.20 & 0.98 \\ 0.16 & 0.99 \\ 0.24 & 0.97 \\ 0.45 & 0.89 \\ 0.32 & 0.95 \end{bmatrix} x \le \begin{bmatrix} -1.40 \\ 5.00 \\ 4.24 \\ 5.00 \\ 2.94 \\ 3.04 \\ 2.91 \\ 3.35 \\ 3.00 \end{bmatrix} & \text{(Region \#3)} \\[6em]
\begin{bmatrix} -0.29 & -1.71 \end{bmatrix} x - 1.43 & \text{if } \begin{bmatrix} -1.00 & 0.00 \\ 0.16 & 0.99 \\ 1.00 & 0.00 \\ -0.16 & -0.99 \end{bmatrix} x \le \begin{bmatrix} 5.00 \\ -0.82 \\ 5.00 \\ 1.40 \end{bmatrix} & \text{(Region \#4)} \\[4em]
1.00 & \text{if } \begin{bmatrix} -0.32 & -0.95 \\ -0.24 & -0.97 \\ -0.20 & -0.98 \\ -0.16 & -0.99 \\ -1.00 & 0.00 \\ 0.16 & 0.99 \\ -0.71 & -0.71 \\ -0.45 & -0.89 \\ 1.00 & 0.00 \end{bmatrix} x \le \begin{bmatrix} 3.00 \\ 2.91 \\ 2.94 \\ 3.04 \\ 5.00 \\ -1.40 \\ 4.24 \\ 3.35 \\ 5.00 \end{bmatrix} & \text{(Region \#5)}
\end{cases}$$

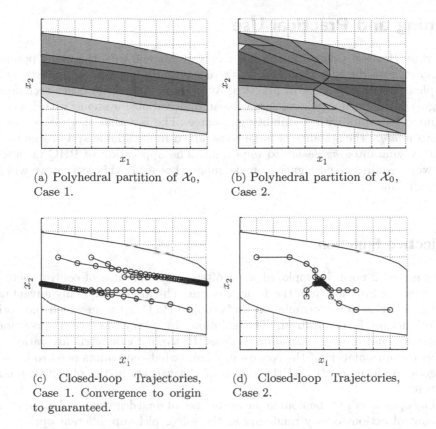

(a) Polyhedral partition of $\mathcal{X}_0$, Case 1.

(b) Polyhedral partition of $\mathcal{X}_0$, Case 2.

(c) Closed-loop Trajectories, Case 1. Convergence to origin to guaranteed.

(d) Closed-loop Trajectories, Case 2.

Figure 12.7 Example 12.4. Double integrator. RHC with ∞-norm cost function, behavior for different terminal weights.

The union of the regions depicted in Figure 12.7(a) is $\mathcal{X}_0$ and is shown in white in Figure 12.7(c). Since N is equal to the determinedness index plus one, $\mathcal{X}_0$ is a positive invariant set for the closed-loop system and thus persistent feasibility is guaranteed for all $x(0) \in \mathcal{X}_0$. However, it can be noticed from Figure 12.7(c) that convergence to the origin is not guaranteed. Starting from the initial conditions $[-4,2]$, $[-2, 2]$, $[0, 0.5]$, $[4,-2]$, $[-1,-1]$ and $[2,-0.5]$, the closed-loop system converges to either $[-5, 0]$ or $[5, 0]$.

Case 2. The LP problem associated with the RHC has the form (11.56) with $\bar{G}_0 \in \mathbb{R}^{174 \times 18}$, $\bar{S}_0 \in \mathbb{R}^{174 \times 2}$ and $c' = [\mathbf{0}_6 \ \mathbf{1}_{12}]$. The RHC law is defined over 21 regions and the corresponding polyhedral partition of the state space is depicted in Figure 12.7(b).

The union of the regions depicted in Figure 12.7(b) is $\mathcal{X}_0$ and is shown in white in Figure 12.7(d). Since N is equal to the determinedness index plus one, $\mathcal{X}_0$ is a positive invariant set for the closed-loop system and thus persistent feasibility is guaranteed for all $x(0) \in \mathcal{X}_0$. Convergence to the origin is also guaranteed by the choice of P as shown by the closed-loop trajectories in Figure 12.7(d) starting from the same initial conditions as in Case 1.

12.6   Tuning and Practical Use

Receding Horizon Control in the different variants and with the many parameter options as introduced in this chapter is a very powerful control technique to address complex control problems in practice. At present there is no other technique to design controllers for general large linear multivariable systems with input and output constraints with a stability guarantee. The fact that there is no effective tool to *analyze* the stability of large constrained multivariable systems makes this stability guarantee *by design* so important. The application of RHC in practice, however, requires the designer to make many choices. In this section we will offer some guidance.

Objective Function

The squared 2-norm is employed more often as an indicator of control quality in the objective function than the 1- or ∞-norm. The former has many advantages.

The 1- or ∞-norm formulation leads to an LP (11.56) from which the optimal control action is determined. The solution lies always at the intersection of constraints and changes *discontinuously* as the tuning parameters are varied. This makes the formulation of the control problem (what constraints need to be added for good performance?) and the choice of the weights in the objective function difficult.

The optimizer may be nonunique in the case of dual degeneracy. This may cause the control action to vary randomly as the solver picks up different optimizers in succeeding time steps.

The construction of the control invariant target set $\mathcal{X}_f$ and of a suitable terminal cost $p(x_N)$ needed in Theorem 12.2 is more difficult for the 1- or ∞-norm formulation because the unconstrained controller is PPWA (Section 12.3.2).

If the optimal control problem is stated in an *ad hoc* manner by trial and error rather than on the basis of Theorem 12.2 undesirable response characteristics may result. The controller may cease to take any action and the system may get stuck. This happens if any control action leads to a short-term increase in the objective function, which can occur, for example, for a system with inverse response characteristics.

Finally, the 1- or ∞-norm formulation involves many more constraints than the 2-norm formulation. In general, this will lead to a larger number of regions of the explicit control law. These regions do not need to be stored, however, but only the value function and the controller, and the search for the region containing the present state can be executed very efficiently (Chapter 14).

Despite the discussed deficiencies it may be advantageous to use the 1- or ∞-norm in cases when the control objective is not just an indirect tool to achieve a control specification but reflects the economics of the process. For example, the costs of electrical power depend sometimes on peak power that can be captured with the ∞-norm. However, even in these cases, quadratic terms are often added to the objective for "regularization."

Design via Theorem 12.2

First, we need to choose the horizon length N and the control invariant target set $\mathcal{X}_f$. Then we can vary the parameters Q and R freely to affect the control performance in the same spirit as we do for designing an LQR. Stability is assured as long as we adjust P according to the outlined procedures when changing Q and R.

The longer the horizon N, the larger the maximal controllable set $\mathcal{K}_N(\mathcal{X}_f)$ over which the closed-loop system is guaranteed to be able to operate (this is true as long as the horizon is smaller than the determinedness index of $\mathcal{K}_\infty(\mathcal{X}_f)$). On the other hand, with the control horizon increases the on-line computational effort or the complexity of the explicit controller determined off-line.

We need to keep in mind that the terminal set $\mathcal{X}_f$ is introduced artificially for the sole purpose of leading to a *sufficient* condition for persistent feasibility. We want it to be large so that it does not compromise closed-loop performance. The larger $\mathcal{X}_f$, the larger $\mathcal{K}_N(\mathcal{X}_f)$. Though it is simplest to choose $\mathcal{X}_f = 0$, it is undesirable unless N is chosen large. Ideally $\mathcal{X}_f$ should be the maximal control invariant set achieved with the unconstrained controller.

Specifically, we first design the unconstrained optimal controller as suggested at the end of Section 12.3.2. From this construction we also obtain the terminal cost satisfying condition (A3) of Theorem 12.2 to use in the RHC design. Then we determine the maximal positive invariant set for the closed-loop system with the unconstrained controller and use this set as $\mathcal{X}_f$. This set is usually difficult to compute for systems of large dimension with the algorithms we introduced in Section (10.2).

Note that for stable systems without state constraints $\mathcal{K}_\infty(\mathcal{X}_f) = \mathbb{R}^n$ always, i.e., the choice of $\mathcal{X}_f$ is less critical. For unstable systems $\mathcal{K}_\infty(\mathcal{X}_f)$ is the region over which the system can operate stably in the presence of input constraints and is of eminent practical importance.

State/Output Constraints

State constraints arise from practical restrictions on the allowed operating range of the system. Thus, contrary to input constraints, they are rarely "hard." They can lead to complications in the controller implementation, however. As it can never be excluded that the state of the real system moves outside the constraint range chosen for the controller design, special provisions must be made (patches in the algorithm) to move the state back into the range. This is difficult and these types of patches are exactly what one wanted to avoid by choosing MPC in the first place.

Thus, typically, state constraints are "softened" in the MPC formulation. For example,

$$x \le x_{\max}$$

is approximated by

$$x \le x_{\max} + \epsilon, \ \epsilon \ge 0,$$

and a term $l(\epsilon)$ is added to the objective function to penalize violations of the constraint. This formulation may, however, lead to a violation of the constraint even when a feasible input exists that avoids it but is not optimal for the new objective function with the penalty term added. Let us analyze how to choose $l(\cdot)$ such that this does not occur by using the exact penalty method [173].

As an example, take

$$J^* = \min_z \quad f(z) \\ \text{subj. to} \quad g(z) \le 0, \tag{12.38}$$

where $g(z)$ is assumed to be scalar for simplicity and $f(z)$ to be strictly convex. Let us soften the constraints and add the penalty as suggested

$$p(\epsilon) = \min_z \quad f(z) + l(\epsilon) \\ \text{subj. to} \quad g(z) \le \epsilon, \tag{12.39}$$

where $\epsilon \ge 0$. We want to choose the penalty $l(\epsilon)$ so that by minimizing the augmented objective we recover the solution to the original problem if it exists:

$$p(0) = J^*$$

and

$$\arg \min_{\epsilon \ge 0} p(\epsilon) = 0$$

which are equivalent to

$$p(0) = J^* \tag{12.40}$$

and

$$p(\epsilon) > p(0), \ \forall \epsilon > 0. \tag{12.41}$$

For (12.40) we require $l(0) = 0$. To construct $l(\epsilon)$ to satisfy (12.41) we assume that strong duality holds and u^* exists so that

$$J^* = \min_z (f(z) + u^* g(z)), \tag{12.42}$$

where u^* is an optimal dual variable. As the optimizer z^* of (12.38) satisfies $g(z^*) \le 0$ we can add a redundant constraint without affecting the solution

$$J^* = \min_z \quad (f(z) + u^* g(z)) \\ \text{subj. to} \quad g(z) \le \epsilon \qquad \forall \epsilon \ge 0. \tag{12.43}$$

Then we can state the bounds

$$J^* \le \min_z \quad (f(z) + u^* \epsilon) \\ \text{subj. to} \quad g(z) \le \epsilon \quad < \quad p(\epsilon) = \min_z \quad (f(z) + u\epsilon) \\ \text{subj. to} \quad g(z) \le \epsilon \quad \forall \epsilon > 0, \ u > u^*. \tag{12.44}$$

Thus $l(\epsilon) = u\epsilon$ with $u > u^* \ge 0$ is a possible penalty term satisfying the requirements (12.40)–(12.41).

Because of smoothness

$$l(\epsilon) = u\epsilon + v\epsilon^2, \ u > u^*, v > 0 \tag{12.45}$$

is preferable. On the other hand, note that $l(\epsilon) = v\epsilon^2$ does not satisfy an inequality of the form (12.44). Therefore, it should not be used as it can lead to optimizers z^* in (12.39) which violate $g(z) \leq 0$ even if a feasible optimizer to the original problem exists.

These ideas can be extended to multiple constraints $g_j(z) \leq 0$, $j = 1, \ldots, r$ via the penalty term

$$l(\epsilon) = u \sum_{j=1}^{r} \epsilon_j + v \sum_{j=1}^{r} \epsilon_j^2, \tag{12.46}$$

where

$$u > \max_{j \in \{1, \ldots, r\}} u_j^*, \ v \geq 0. \tag{12.47}$$

Formulations also exist where the necessary constraint violations are following a prescribed order so that less important constraints are violated first [174].

Time-varying references, constraints, disturbances and system parameters

The standard RHC formulation (12.6) can be easily extended to include these features. Known disturbances are simply included in the prediction model. If the system states are not to return to the origin, but some output y is to follow some trajectory r, then appropriate penalties of the error $e = y - r$ are included in the control objective. How to do this and to achieve offset-free tracking is described in Section 12.7. Theorem 12.2 can be used to design the controller in these cases.

If the constraints are time-varying then $\mathcal{X}$ and $\mathcal{U}$ become time-varying. For example, the constraints may be shaped like a funnel tightening towards the end of the horizon.

If the underlying system is nonlinear, one often uses a locally linearized model for the prediction and updates it at each time step. Note that the results of Theorem 12.2 do not apply when the system model and/or the constraints are time-varying.

In all these cases, the optimization problem to be solved on line, (11.31) or (11.56), does not change in structure but some of the defining matrices will now change at each time step, which will increase the necessary on-line computational effort somewhat.

If the controller is to be computed explicitly off-line, then all the varying parameters (disturbances, constraints, references), which we will denote by θ, become *parameters* in the multiparametric optimization problem and the resulting controller becomes an explicit function of them: $u(t) = F(x(t), \theta)$. As we have learned, the complexity of the solution of mp-QP and mp-LP problems depends primarily on the number of constraints. If the number of parameters affects the number of constraints then this may only be possible for a relatively small number of parameters. Thus, the possibilities to take into account time-varying control problem features in explicit MPC are rather limited. Time-varying models cannot be handled at all.

Multiple Horizons and Move-Blocking

Unfortunately, as we started to discuss in the previous paragraphs, in challenging applications, the design procedure implied by Theorem 12.2 may not be strictly applicable. Then the closed-loop behavior with the RHC has to be analyzed by other means, in the worst case through extensive simulation studies. In principle, this offers us also more freedom in the problem formulation and the choice of the parameters, for example, in order to reduce the computational complexity. The basic RHC formulation (12.6) may be modified as follows:

$$\min_{U_0} \quad p(x_{N_y}) + \sum_{k=0}^{N_y-1} q(x_k, u_k)$$

$$\text{subj. to} \quad x_{k+1} = Ax_k + Bu_k, \ k = 0, \ldots, N_y - 1 \qquad (12.48)$$
$$x_k \in \mathcal{X}, k = 0, \ldots, N_c$$
$$u_k \in \mathcal{U}, \ k = 0, \ldots, N_u$$
$$u_k = Kx_k, \ N_u < k < N_y$$

where K is some feedback gain, N_y, N_u, N_c are the prediction, input, and state constraint horizons, respectively, with $N_u \leq N_y$ and $N_c \leq N_y$. This formulation reduces the number of constraints and as a consequence makes the long horizon prediction used in the optimization less accurate as it is not forced to obey all the constraints. As this approximation affects only the states far in the future, it is hoped that it will not influence significantly the present control action.

Generalized Predictive Control (GPC) in its most common formulation [89] has multiple horizons as an inherent feature but does not include constraints. Experience and theoretical analysis [56] have shown that it is very difficult to choose all these horizons that affect not only performance but even stability in a nonintuitive fashion. Thus, for problems where constraints are not important and the adaptive features of GPC are not needed, it is much preferable to resort to the well established LQR and LQG controllers for which a wealth of stability, performance and robustness results have been established.

Another more effective way to reduce the computational effort is *move-blocking* where the manipulated variables are assumed to be fixed over time intervals in the future thus reducing the degrees of freedom in the optimization problem. By choosing the blocking strategies carefully, the available RHC stability results remain applicable [72].

12.7 Offset-Free Reference Tracking

This section describes how the RHC problem has to be formulated to track constant references without offset under model mismatch. We distinguish between the number p of measured outputs, the number r of outputs which one desires to track (called "tracked outputs"), and the number n_d of disturbances. First, we summarize the conditions that need to be satisfied to obtain offset-free RHC by using the arguments of the internal model principle. Then, we provide a simple proof of zero steady-state offset when $r \leq p = n_d$. Extensive treatment of reference tracking for RHC can be found in in [13, 218, 225, 226, 198]. Consider the discrete-time time-invariant system

$$\begin{cases} x_m(t+1) = f(x_m(t), u(t)) \\ y_m(t) = g(x_m(t)) \\ z(t) = Hy_m(t). \end{cases} \tag{12.49}$$

In (12.49), $x_m(t) \in \mathbb{R}^n$, $u(t) \in \mathbb{R}^m$ and $y_m(t) \in \mathbb{R}^p$ are the state, input, measured output vector, respectively. The controlled variables $z(t) \in \mathbb{R}^r$ are a linear combination of the measured variables for which offset-free behavior is sought. Without any loss of generality we assume H to have full row rank.

The objective is to design an RHC based on the linear system model (12.1) of (12.49) in order to have $z(t)$ track $r(t)$, where $r(t) \in \mathbb{R}^r$ is the reference signal, which we assume to converge to a constant, i.e., $r(t) \to r_\infty$ as $t \to \infty$. We require zero steady-state tracking error, i.e., $(z(t) - r(t)) \to 0$ for $t \to \infty$.

The Observer Design

The plant model (12.1) is augmented with a disturbance model in order to capture the mismatch between (12.49) and (12.1) in steady state. Several disturbance models have been presented in the literature [13, 205, 193, 226, 225, 294]. Here we follow [226] and use the form:

$$\begin{cases} x(t+1) = Ax(t) + Bu(t) + B_d d(t) \\ d(t+1) = d(t) \\ y(t) = Cx(t) + C_d d(t) \end{cases} \tag{12.50}$$

with $d(t) \in \mathbb{R}^{n_d}$. With abuse of notation we have used the same symbols for state and outputs of system (12.1) and system (12.50). Later we will focus on specific versions of the model (12.50).

The observer estimates both states and disturbances based on this augmented model. Conditions for the observability of (12.50) are given in the following theorem.

Theorem 12.3   *[215, 216, 226, 13] The augmented system (12.50) is observable if and only if (C, A) is observable and*

$$\begin{bmatrix} A - I & B_d \\ C & C_d \end{bmatrix} \tag{12.51}$$

has full column rank.

Proof: From the Hautus observability condition system (12.50) is observable iff

$$\begin{bmatrix} A' - \lambda I & 0 & C' \\ B_d' & I - \lambda I & C_d' \end{bmatrix} \text{ has full row rank } \forall \lambda. \tag{12.52}$$

Again from the Hautus condition, the first set of rows is linearly independent iff (C, A) is observable. The second set of rows is linearly independent from the first n rows except possibly for $\lambda = 1$. Thus, for the augmented system the Hautus condition needs to be checked for $\lambda = 1$ only, where it becomes (12.51). ∎

Remark 12.4 Note that for condition (12.51) to be satisfied the number of disturbances in d needs to be smaller or equal to the number of available measurements in y, $n_d \leqslant p$. Condition (12.51) can be nicely interpreted. It requires that the model of the disturbance effect on the output $d \to y$ must not have a zero at $(1, 0)$. Alternatively we can look at the steady state of system (12.50)

$$
\begin{bmatrix} A - I & B_d \\ C & C_d \end{bmatrix} \begin{bmatrix} x_\infty \\ d_\infty \end{bmatrix} = \begin{bmatrix} 0 \\ y_\infty \end{bmatrix}, \tag{12.53}
$$

where we have denoted the steady state values with a subscript ∞ and have omitted the forcing term u for simplicity. We note that from the observability condition (12.51) for system (12.50) Equation (12.53) is required to have a unique solution, which means, that we must be able to deduce a unique value for the disturbance d_∞ from a measurement of y_∞ in steady state.

The following corollary follows directly from Theorem 12.3.

Corollary 12.3 *The augmented system (12.50) with $n_d = p$ and $C_d = I$ is observable if and only if (C, A) is observable and*

$$
det \begin{bmatrix} A - I & B_d \\ C & I \end{bmatrix} = det(A - I - B_d C) \neq 0. \tag{12.54}
$$

Remark 12.5 We note here how the observability requirement restricts the choice of the disturbance model. If the plant has no integrators, then $det\,(A - I) \neq 0$ and we can choose $B_d = 0$. If the plant has integrators then B_d has to be chosen specifically to make $det\,(A - I - B_d C) \neq 0$.

The state and disturbance estimator is designed based on the augmented model as follows:

$$
\begin{bmatrix} \hat{x}(t+1) \\ \hat{d}(t+1) \end{bmatrix} = \begin{bmatrix} A & B_d \\ 0 & I \end{bmatrix} \begin{bmatrix} \hat{x}(t) \\ \hat{d}(t) \end{bmatrix} + \begin{bmatrix} B \\ 0 \end{bmatrix} u(t) + \begin{bmatrix} L_x \\ L_d \end{bmatrix} (-y_m(t) + C\hat{x}(t) + C_d\hat{d}(t)), \tag{12.55}
$$

where L_x and L_d are chosen so that the estimator is stable. We remark that the results below are independent of the choice of the method for computing L_x and L_d. We then have the following property.

Lemma 12.3 *Suppose the observer (12.55) is stable. Then, $rank(L_d) = n_d$.*

Proof: From (12.55) it follows

$$
\begin{bmatrix} \hat{x}(t+1) \\ \hat{d}(t+1) \end{bmatrix} = \begin{bmatrix} A + L_x C & B_d + L_x C_d \\ L_d C & I + L_d C_d \end{bmatrix} \begin{bmatrix} \hat{x}(t) \\ \hat{d}(t) \end{bmatrix} + \begin{bmatrix} B \\ 0 \end{bmatrix} u(t) - \begin{bmatrix} L_x \\ L_d \end{bmatrix} y_m(t). \tag{12.56}
$$

By stability, the observer has no poles at $(1, 0)$ and therefore

$$\det \left(\begin{bmatrix} A - I + L_x C & B_d + L_x C_d \\ L_d C & L_d C_d \end{bmatrix} \right) \neq 0. \tag{12.57}$$

For (12.57) to hold, the last n_d rows of the matrix have to be of full row rank. A necessary condition is that L_d has full row rank. ∎

In the rest of this section we will focus on the case $n_d = p$.

Lemma 12.4 *Suppose the observer (12.55) is stable. Choose $n_d = p$. The steady state of the observer (12.55) satisfies:*

$$\begin{bmatrix} A - I & B \\ C & 0 \end{bmatrix} \begin{bmatrix} \hat{x}_\infty \\ u_\infty \end{bmatrix} = \begin{bmatrix} -B_d \hat{d}_\infty \\ y_{m,\infty} - C_d \hat{d}_\infty \end{bmatrix}, \tag{12.58}$$

where $y_{m,\infty}$ and u_∞ are the steady state measured output and input of the system (12.49), $\hat{x}_\infty$ and $\hat{d}_\infty$ are state and disturbance estimates from the observer (12.55) at steady state, respectively.

Proof: From (12.55) we note that the disturbance estimate $\hat{d}$ converges only if $L_d(-y_{m,\infty} + C\hat{x}_\infty + C_d \hat{d}_\infty) = 0$. As L_d is square by assumption and nonsingular by Lemma 12.3 this implies that at steady state, the observer estimates (12.55) satisfy

$$-y_{m,\infty} + C\hat{x}_\infty + C_d \hat{d}_\infty = 0. \tag{12.59}$$

Equation (12.58) follows directly from (12.59) and (12.55). ∎

The MPC Design

Denote by $z_\infty = H y_{m,\infty}$ and r_∞ the tracked measured outputs and their references at steady state, respectively. For offset-free tracking at steady state we want $z_\infty = r_\infty$. The observer condition (12.58) suggests that at steady state the MPC should satisfy

$$\begin{bmatrix} A - I & B \\ HC & 0 \end{bmatrix} \begin{bmatrix} x_\infty \\ u_\infty \end{bmatrix} = \begin{bmatrix} -B_d \hat{d}_\infty \\ r_\infty - H C_d \hat{d}_\infty \end{bmatrix}, \tag{12.60}$$

where x_∞ is the MPC state at steady state. For x_∞ and u_∞ to exist for any $\hat{d}_\infty$ and r_∞ the matrix $\begin{bmatrix} A - I & B \\ HC & 0 \end{bmatrix}$ must be of full row rank which implies $m \geq r$.

The MPC is designed as follows

$$\min_{U_0} \quad (x_N - \bar{x}_t)' P (x_N - \bar{x}_t) + \sum_{k=0}^{N-1} (x_k - \bar{x}_t)' Q (x_k - \bar{x}_t) + (u_k - \bar{u}_t)' R (u_k - \bar{u}_t)$$

$$\begin{aligned} \text{subj. to} \quad & x_{k+1} = A x_k + B u_k + B_d d_k, \ k = 0, \ldots, N \\ & x_k \in \mathcal{X}, \ u_k \in \mathcal{U}, \ k = 0, \ldots, N-1 \\ & x_N \in \mathcal{X}_f \\ & d_{k+1} = d_k, \ k = 0, \ldots, N-1 \\ & x_0 = \hat{x}(t) \\ & d_0 = \hat{d}(t), \end{aligned} \tag{12.61}$$

with the targets $\bar{u}_t$ and $\bar{x}_t$ given by

$$\begin{bmatrix} A - I & B \\ HC & 0 \end{bmatrix} \begin{bmatrix} \bar{x}_t \\ \bar{u}_t \end{bmatrix} = \begin{bmatrix} -B_d \hat{d}(t) \\ r(t) - HC_d \hat{d}(t) \end{bmatrix} \tag{12.62}$$

and where $Q \succeq 0$, $R \succ 0$, and $P \succ 0$.

Let $U^*(t) = \{u_0^*, \ldots, u_{N-1}^*\}$ be the optimal solution of (12.61)–(12.62) at time t. Then, the first sample of $U^*(t)$ is applied to system (12.49)

$$u(t) = u_0^*. \tag{12.63}$$

Denote by $c_0(\hat{x}(t), \hat{d}(t), r(t)) = u_0^*(\hat{x}(t), \hat{d}(t), r(t))$ the control law when the estimated state and disturbance are $\hat{x}(t)$ and $\hat{d}(t)$, respectively. Then the closed-loop system obtained by controlling (12.49) with the MPC (12.61)–(12.62)–(12.63) and the observer (12.55) is:

$$x(t+1) = f(x(t), c_0(\hat{x}(t), \hat{d}(t), r(t)))$$
$$\hat{x}(t+1) = (A + L_x C)\hat{x}(t) + (B_d + L_x C_d)\hat{d}(t) + Bc_0(\hat{x}(t), \hat{d}(t), r(t)) - L_x y_m(t)$$
$$\hat{d}(t+1) = L_d C \hat{x}(t) + (I + L_d C_d)\hat{d}(t) - L_d y_m(t). \tag{12.64}$$

Often in practice, one desires to track all measured outputs with zero offset. Choosing $n_d = p = r$ is thus a natural choice. Such a zero-offset property continues to hold if only a subset of the measured outputs are to be tracked, i.e., $n_d = p > r$. Next we provide a very simple proof for offset-free control when $n_d = p$.

Theorem 12.4   *Consider the case $n_d = p$. Assume that for $r(t) \to r_\infty$ as $t \to \infty$, the MPC problem (12.61)–(12.62) is feasible for all $t \in \mathbb{N}_+$, unconstrained for $t \geq j$ with $j \in \mathbb{N}_+$ and the closed-loop system (12.64) converges to $\hat{x}_\infty$, $\hat{d}_\infty$, $y_{m,\infty}$, i.e., $\hat{x}(t) \to \hat{x}_\infty$, $\hat{d}(t) \to \hat{d}_\infty$, $y_m(t) \to y_{m,\infty}$ as $t \to \infty$. Then $z(t) = Hy_m(t) \to r_\infty$ as $t \to \infty$.*

Proof: Consider the MPC problem (12.61)–(12.62). At steady state $u(t) \to u_\infty = c_0(\hat{x}_\infty, \hat{d}_\infty, r_\infty)$, $\bar{x}_t \to \bar{x}_\infty$ and $\bar{u}_t \to \bar{u}_\infty$. Note that the steady state controller input u_∞ (computed and implemented) might be different from the steady state target input $\bar{u}_\infty$.

The asymptotic values $\hat{x}_\infty$, $\bar{x}_\infty$, u_∞ and $\bar{u}_\infty$ satisfy the observer conditions (12.58)

$$\begin{bmatrix} A - I & B \\ C & 0 \end{bmatrix} \begin{bmatrix} \hat{x}_\infty \\ u_\infty \end{bmatrix} = \begin{bmatrix} -B_d \hat{d}_\infty \\ y_{m,\infty} - C_d \hat{d}_\infty \end{bmatrix} \tag{12.65}$$

and the controller requirement (12.62)

$$\begin{bmatrix} A - I & B \\ HC & 0 \end{bmatrix} \begin{bmatrix} \bar{x}_\infty \\ \bar{u}_\infty \end{bmatrix} = \begin{bmatrix} -B_d \hat{d}_\infty \\ r_\infty - HC_d \hat{d}_\infty \end{bmatrix}. \tag{12.66}$$

Define $\delta x = \hat{x}_\infty - \bar{x}_\infty$, $\delta u = u_\infty - \bar{u}_\infty$ and the offset $\varepsilon = z_\infty - r_\infty$. Notice that the steady state target values $\bar{x}_\infty$ and $\bar{u}_\infty$ are both functions of r_∞ and $\hat{d}_\infty$ as given by (12.66). Left multiplying the second row of (12.65) by H and subtracting (12.66) from the result, we obtain

$$(A - I)\delta x + B\delta u = 0$$
$$HC\delta x = \varepsilon. \tag{12.67}$$

Next we prove that $\delta x = 0$ and thus $\varepsilon = 0$.

Consider the MPC problem (12.61)–(12.62) and the following change of variables $\delta x_k = x_k - \bar{x}_t$, $\delta u_k = u_k - \bar{u}_t$. Notice that $Hy_k - r(t) = HCx_k + HC_dd_k - r(t) = HC\delta x_k + HC\bar{x}_t + HC_dd_k - r(t) = HC\delta x_k$ from condition (12.62) with $\hat{d}(t) = d_k$. Similarly, one can show that $\delta x_{k+1} = A\delta x_k + B\delta u_k$. Then, around the origin where all constraints are inactive the MPC problem (12.61) becomes:

$$\min_{\delta u_0, \ldots, \delta u_{N-1}} \quad \delta x_N' P \delta x_N + \sum_{k=0}^{N-1} \delta x_k' Q \delta x_k + \delta u_k' R \delta u_k$$

$$\text{subj. to} \qquad \delta x_{k+1} = A\delta x_k + B\delta u_k, \quad 0 \le k \le N \tag{12.68}$$
$$\delta x_0 = \delta x(t),$$
$$\delta x(t) - \hat{x}(t) - \bar{x}_t.$$

Denote by K_{MPC} the unconstrained MPC controller (12.68), i.e., $\delta u_0^* = K_{MPC}\delta x(t)$. At steady state $\delta u_0^* \to u_\infty - \bar{u}_\infty = \delta u$ and $\delta x(t) \to \hat{x}_\infty - \bar{x}_\infty = \delta x$. Therefore, at steady state, $\delta u = K_{MPC}\delta x$. From (12.67)

$$(A - I + BK_{MPC})\delta x = 0. \tag{12.69}$$

By assumption the unconstrained system with the MPC controller converges. Thus K_{MPC} is a stabilizing control law, which implies that $(A - I + BK_{MPC})$ is nonsingular and hence $\delta x = 0$. ∎

Remark 12.6   Theorem 12.4 was proven in [226] by using a different approach.

Remark 12.7   Theorem 12.4 can be extended to prove local Lyapunov stability of the closed-loop system (12.64) under standard regularity assumptions on the state update function f in (12.64) [204].

Remark 12.8   The proof of Theorem 12.4 assumes only that the models used for the control design (12.1) and the observer design (12.50) are identical in steady state in the sense that they give rise to the same relation $z = z(u, d, r)$. It does not make any assumptions about the behavior of the real plant (12.49), i.e., the model-plant mismatch, with the exception that the closed-loop system (12.64) must converge to a fixed point. The models used in the controller and the observer could even be different as long as they satisfy the same steady state relation.

Remark 12.9   If condition (12.62) does not specify $\bar{x}_t$ and $\bar{u}_t$ uniquely, it is customary to determine $\bar{x}_t$ and $\bar{u}_t$ through an optimization problem, for example, minimizing the magnitude of $\bar{u}_t$ subject to the constraint (12.62) [226].

Remark 12.10   Note that in order to achieve no offset we augmented the model of the plant with as many disturbances (and integrators) as we have measurements ($n_d = p$) (cf. Equation (12.56)). Our design procedure requires the addition of p integrators even if we wish to control only a subset of $r < p$ measured variables. This is actually not necessary as we suspect from basic system theory. The design procedure for the case $n_d = r < p$ is, however, more involved [198].

If the squared 2-norm in the objective function of (12.61) is replaced with a 1- or ∞-norm ($\|P(x_N - \bar{x}_t)\|_p + \sum_{k=0}^{N-1} \|Q(x_k - \bar{x}_t)\|_p + \|R(u_k - \bar{u}_t)\|_p$, where $p = 1$ or $p = \infty$), then our results continue to hold. In particular, Theorem 12.4 continues to hold. The unconstrained MPC controlled K_{MPC} in (12.68) will be piecewise linear around the origin [44]. In particular, around the origin, $\delta u^*(t) = \delta u_0^* = K_{MPC}(\delta x(t))$ is a continuous piecewise linear function of the state variation δx:

$$K_{MPC}(\delta x) = F^j \delta x \quad \text{if} \quad H^j \delta x \leq K^j, \; j = 1, \ldots, N^r, \qquad (12.70)$$

where H^j and K^j in equation (12.70) are the matrices describing the j-th polyhedron $CR^j = \{\delta x \in \mathbb{R}^n \; : \; H^j \delta x \leq K^j\}$ inside which the feedback optimal control law $\delta u^*(t)$ has the linear form $F^j \delta x(k)$. The polyhedra CR^j, $j = 1, \ldots, N^r$ are a partition of the set of feasible states of problem (12.61) and they all contain the origin.

Explicit Controller

Examining (12.61), (12.62) we note that the control law depends on $\hat{x}(t)$, $\hat{d}(t)$ and $r(t)$. Thus in order to achieve offset free tracking of r outputs out of p measurements we had to add the $p + r$ "parameters" $\hat{d}(t)$ and $r(t)$ to the usual parameters $\hat{x}(t)$.

There are more involved RHC design techniques to obtain offset-free control for models with $n_d < p$ and in particular, with minimum order disturbance models $n_d = r$. The total size of the parameter vector can thus be reduced to $n + 2r$. This is significant only if a small subset of the plant outputs are to be controlled. A greater reduction of parameters can be achieved by the following method. By Corollary 12.3, we are allowed to choose $B_d = 0$ in the disturbance model if the plant has no integrators. Recall the target conditions 12.62 with $B_d = 0$

$$\begin{bmatrix} A - I & B \\ HC & 0 \end{bmatrix} \begin{bmatrix} \bar{x}_t \\ \bar{u}_t \end{bmatrix} = \begin{bmatrix} 0 \\ r(t) - HC_d \hat{d}(t) \end{bmatrix}. \qquad (12.71)$$

Clearly, any solution to (12.71) can be parameterized by $r(t) - HC_d \hat{d}(t)$. The explicit control law is written as $u(t) = c_0(\hat{x}(t), r(t) - HC_d \hat{d}(t))$, with only $n + r$ parameters. Since the observer is unconstrained, complexity is much less of an issue. Hence, a full disturbance model with $n_d = p$ can be chosen to yield offset-free control.

Remark 12.11   The choice of $B_d = 0$ might be limiting in practice. In [14], the authors have shown that for a wide range of systems, if $B_d = 0$ and a Kalman filter is chosen as observer, then the closed-loop system might suffer a dramatic performance deterioration.

Delta Input (δu) Formulation

In the δu formulation, the MPC scheme uses the following linear time-invariant system model of (12.49):

$$\begin{cases} x(t+1) = Ax(t) + Bu(t) \\ \quad u(t) = u(t-1) + \delta u(t) \\ \quad y(t) = Cx(t). \end{cases} \tag{12.72}$$

System (12.72) is controllable if (A, B) is controllable. The δu formulation often arises naturally in practice when the actuator is subject to uncertainty, e.g., the exact gain is unknown or is subject to drift. In these cases, it can be advantageous to consider changes in the control value as input to the plant. The absolute control value is estimated by the observer, which is expressed as follows

$$\begin{bmatrix} \hat{x}(t+1) \\ \hat{u}(t+1) \end{bmatrix} = \begin{bmatrix} A & B \\ 0 & I \end{bmatrix} \begin{bmatrix} \hat{x}(t) \\ \hat{u}(t) \end{bmatrix} + \begin{bmatrix} B \\ I \end{bmatrix} \delta u(t) + \begin{bmatrix} L_x \\ L_u \end{bmatrix} (-y_m(t) + C\hat{x}(t)). \tag{12.73}$$

The MPC problem is readily modified

$$\begin{aligned} \min_{\delta u_0, \dots, \delta u_{N-1}} & \quad \|y_k - r_k\|_Q^2 + \|\delta u_k\|_R^2 \\ \text{subj. to} & \quad x_{k+1} = Ax_k + Bu_k, \ k \geq 0 \\ & \quad y_k = Cx_k \ k \geq 0 \\ & \quad x_k \in \mathcal{X}, \ u_k \in \mathcal{U}, \ k = 0, \dots, N-1 \\ & \quad x_N \in \mathcal{X}_f \\ & \quad u_k = u_{k-1} + \delta u_k, \ k \geq 0 \\ & \quad u_{-1} = \hat{u}(t) \\ & \quad x_0 = \hat{x}(t). \end{aligned} \tag{12.74}$$

The control input applied to the system is

$$u(t) = \delta u_0^* + u(t-1). \tag{12.75}$$

The input estimate $\hat{u}(t)$ is not necessarily equal to the actual input $u(t)$. This scheme inherently achieves offset-free control, there is no need to add a disturbance model. To see this, we first note that $\delta u_0^* = 0$ in steady-state. Hence our analysis applies as the δu formulation is equivalent to a disturbance model in steady-state. This is due to the fact that any plant/model mismatch is lumped into $\hat{u}(t)$. Indeed this approach is equivalent to an input disturbance model ($B_d = B$, $C_d = 0$). If in (12.74) the measured $u(t)$ were substituted for its estimate, i.e., $u_{-1} = u(t-1)$, then the algorithm would show offset.

In this formulation the computation of a target input $\bar{u}_t$ and state $\bar{x}_t$ is not required. A disadvantage of the formulation is that it is not applicable when there is an excess of manipulated variables u compared to measured variables y, since detectability of the augmented system (12.72) is then lost.

Minimum-Time Controller

In minimum-time control, the cost function minimizes the predicted number of steps necessary to reach a target region, usually the invariant set associated with the unconstrained LQR controller [167]. This scheme can reduce the on-line computation time significantly, especially for explicit controllers (Section 11.5). While minimum-time MPC is computed and implemented differently from standard MPC controllers, there is no difference between the two control schemes at steady-state. In particular, one can choose the target region to be the unconstrained region of (12.61)–(12.62). When the state and disturbance estimates and reference are within this region, the control law is switched to (12.61)–(12.62). The analysis and methods presented in this section therefore apply directly.

12.8   Literature Review

Although the basic idea of receding horizon control can be found in the theoretical work of Propoi [239] in 1963 it did not gain much attention until the mid-1970s, when Richalet and coauthors [248, 249] introduced their "Model Predictive Heuristic Control (MPHC)". Independently, in 1969 Charles Cutler proposed the concept to his PhD advisor Dr. Huang at the University of Houston. In 1973 Cutler implemented it successfully in the Shell Refinery in New Orleans, Louisiana [90].

Several years later Cutler and Ramaker [91] described this predictive control algorithm named Dynamic Matrix Control (DMC) in the literature. It has been hugely successful in the petro-chemical industry. A vast variety of different methodologies with different names followed, such as Quadratic Dynamic Matrix Control (QDMC), Adaptive Predictive Control (APC), Generalized Predictive Control (GPC), Sequential Open Loop Optimization (SOLO), and others.

While the mentioned algorithms are seemingly different, they all share the same structural features: a model of the plant, the receding horizon idea, and an optimization procedure to obtain the control action by optimizing the system's predicted evolution.

Some of the first industrial MPC algorithms like IDCOM [249] and DMC [91] were developed for constrained MPC with quadratic performance indices. However, in those algorithms input and output constraints were treated in an indirect ad-hoc fashion. Only later, algorithms like QDMC [118] overcame this limitation by employing quadratic programming to solve constrained MPC problems with quadratic performance indices. During the same period, the use of linear programming was studied by Gutman and coauthors [137, 138, 140].

An extensive theoretical effort was devoted to analyze receding horizon control schemes, provide conditions for guaranteeing feasibility and closed-loop stability, and highlight the relations between MPC and the linear quadratic regulator [204, 203]. Theorem 12.2 in this book is the main result on feasibility and stability of MPC and was adopted from these publications.

The idea behind Theorem 12.2 dates back to Keerthi and Gilbert [168], the first researchers to propose specific choices for the terminal cost P and the terminal constraint $\mathcal{X}_f$, namely $\mathcal{X}_f = 0$ and $P = 0$. Under these assumptions Keerthi and Gilbert prove the stability for general nonlinear performance functions and nonlinear models. Their work has been followed by many other stability conditions for RHC including those in [168, 34, 35, 154, 84]. If properly analyzed all these results are based on the same concepts as Theorem 12.2.

In the past fifty years, the richness of theoretical and computational issues surrounding MPC has generated a large number of research studies that appeared in conference proceedings, archival journals and research monographs. The preceding paragraphs are not intended to summarize this vast literature. We refer the interested reader to some books on Model Predictive Control [76, 197, 247] that appeared in the last decade.

For complex constrained multivariable control problems, model predictive control has long been the accepted standard in the process industries [240, 241]. The results reported in this book have greatly reduced the on-line computational effort and have opened up this methodology to other application areas where hardware speed and costs are dominant — unlike in process control.

13

Approximate Receding Horizon Control

Contributed by Prof. Colin N. Jones
Ecole Polytechnique Fédérale de Lausanne
colin.jones@epfl.ch

This chapter explains the construction of approximate explicit control laws of desired complexity that provide certificates of recursive feasibility and stability. The methods introduced all have the following properties:

1. Operations are only on the problem data, and the methods do not require the computation of the explicit control law.

2. Any convex parametric problem can be approximated, not just those for which the explicit solution can be computed.

3. The desired level of suboptimality or the desired complexity (in terms of online storage and/or FLOPS) can be specified.

We begin by assuming that a controller with the desired properties has already been defined via one of the predictive control methods introduced in Chapter 12, but that solving the requisite optimization problem online, or evaluating the explicit solution offline, is not possible due to computational limitations. The chapter begins by listing the specific assumptions on this optimal control law, which is to be approximated. Then, the key theorem is introduced, which provides sufficient conditions under which an approximate control law will be stabilizing and persistently feasible. Section 13.2 describes an abstract algorithm that will provide these properties, and Section 13.3 reviews a number of specific implementations of this algorithm that have been proposed in the literature.

13.1   Stability of Approximate Receding Horizon Control

Recall problem (12.6)

$$J_0^*(x(t)) = \min_{U_0} \quad J_0\Big(x(t), U_0\Big) = p(x_N) + \sum_{k=0}^{N-1} q(x_k, u_k)$$

$$\text{subj. to} \quad x_{k+1} = Ax_k + Bu_k, \ k = 0, \ldots, N-1$$
$$x_k \in \mathcal{X}, \ u_k \in \mathcal{U}, \ k = 0, \ldots, N-1 \qquad (13.1)$$
$$x_N \in \mathcal{X}_f$$
$$x_0 = x(t).$$

The cost function is either

$$J_0(x(0), U_0) = \|Px_N\|_p + \sum_{k=0}^{N-1} \|Qx_k\|_p + \|Ru_k\|_p \qquad (13.2)$$

with $p = 1$ or $p = \infty$, or

$$J_0(x(0), U_0) = x_N' P x_N + \sum_{k=0}^{N-1} x_k' Q x_k + u_k' R u_k. \qquad (13.3)$$

Denote a suboptimal feasible solution of (13.1) by $\tilde{U}_0 = \{\tilde{u}_0, \ldots, \tilde{u}_{N-1}\}$. The corresponding suboptimal MPC law is

$$\tilde{u}(t) = \tilde{f}_0(x(t)) = \tilde{u}_0(x(t)) \qquad (13.4)$$

and the closed-loop system

$$x(k+1) = Ax(k) + B\tilde{f}_0(x(k)) = \tilde{f}_{cl}(x(k)), \ k \geq 0. \qquad (13.5)$$

In the following we will prove that the stability of receding horizon control is preserved when the suboptimal control law $\tilde{u}_0(x(t))$ is used as long as the level of suboptimality is less than the stage cost, i.e.,

$$J_0(x, \tilde{U}_0(x)) < J_0^*(x) + q(x, 0).$$

This is stated more precisely in the following Theorem 13.1.

Theorem 13.1   *Assume*

(A0)   The stage cost $q(x, u)$ and terminal cost $p(x)$ are continuous and positive definite functions.

(A1)   The sets $\mathcal{X}$, $\mathcal{X}_f$ and $\mathcal{U}$ contain the origin in their interior and are closed.

(A2) $\mathcal{X}_f$ is control invariant, $\mathcal{X}_f \subseteq \mathcal{X}$.

(A3) $\displaystyle\min_{v \in \mathcal{U}, \ Ax+Bv \in \mathcal{X}_f} (-p(x) + q(x, v) + p(Ax + Bv)) \leq 0, \ \forall x \in \mathcal{X}_f.$

(A4) The suboptimal $\tilde{U}_0(x)$ satisfies $J_0(x, \tilde{U}_0(x)) \leq J_0^*(x) + \gamma(x)$, $\forall x \in \mathcal{X}_0$, $\gamma \succeq 0$.

(A5) $\gamma(x) - q(x, 0) \prec 0$, $\forall x \in \mathcal{X}_0$, $x \neq 0$.

Then, the origin of the closed-loop system (13.5) is asymptotically stable with domain of attraction $\mathcal{X}_0$.

Proof: The proof follows the proof of Theorem 12.2. We focus only on the arguments for convergence, which are different for the suboptimal MPC controller. We establish that the function $J_0^*(\cdot)$ in (13.1) strictly decreases for the closed-loop system. Because the cost J_0, the system and the constraints are time invariant we can study the properties of J_0^* between step $k = 0$ and step $k + 1 = 1$.

Consider problem (13.1) at time $t = 0$. Let $x(0) \in \mathcal{X}_0$ and let $\tilde{U}_0 = \{\tilde{u}_0, \ldots, \tilde{u}_{N-1}\}$ be a feasible suboptimal solution to problem (13.1) and $\mathbf{x}_0 = \{x(0), x_1, \ldots, x_N\}$ be the corresponding state trajectory. We denote by $\tilde{J}_0(x(0))$ the cost at $x(0)$ when the feasible $\tilde{U}_0$ is applied, $\tilde{J}_0(x(0)) = J_0(x(0), \tilde{U}_0)$.

After the implementation of $\tilde{u}_0$ we obtain $x(1) = x_1 = Ax(0) + B\tilde{u}_0$. Consider now problem (13.1) for $t = 1$. We will construct an upper bound on $J_0^*(x(1))$. Consider the sequence $\tilde{U}_1 = \{\tilde{u}_1, \ldots, \tilde{u}_{N-1}, v\}$ and the corresponding state trajectory resulting from the initial state $x(1)$, $\tilde{\mathbf{x}}_1 = \{x_1, \ldots, x_N, Ax_N + Bv\}$. Because $x_N \in \mathcal{X}_f$ and (A2) there exists a feasible v such that $x_{N+1} = Ax_N + Bv \in \mathcal{X}_f$ and with this v the sequence $\tilde{U}_1 = \{\tilde{u}_1, \ldots, \tilde{u}_{N-1}, v\}$ is feasible. Because $\tilde{U}_1$ is not optimal $J_0(x(1), \tilde{U}_1)$ is an upper bound on $J_0^*(x(1))$.

Since the trajectories generated by $\tilde{U}_0$ and $\tilde{U}_1$ overlap, except for the first and last sampling intervals, it is immediate to show that

$$J_0^*(x(1)) \leq J_0(x(1), \tilde{U}_1) = \tilde{J}_0(x(0)) - q(x_0, \tilde{u}_0) - p(x_N)$$
$$+ (q(x_N, v) + p(Ax_N + Bv)). \quad (13.6)$$

Let $x = x_0 = x(0)$ and $u = \tilde{u}_0$. Under assumption (A3) Equation (13.6) becomes

$$J_0^*(Ax + Bu) \leq \tilde{J}_0(x) - q(x, u), \ \forall x \in \mathcal{X}_0. \quad (13.7)$$

Under assumption (A4) Equation (13.7) becomes

$$J_0^*(Ax + Bu) - J_0^*(x) \leq \gamma(x) - q(x, u) \leq \gamma(x) - q(x, 0), \ \forall x \in \mathcal{X}_0. \quad (13.8)$$

Equation (13.8), the hypothesis (A0) on the matrices Q and R and the hypothesis (A5) ensure that $J_0^*(x)$ strictly decreases along the state trajectories of the closed-loop system (13.5) for any $x \in \mathcal{X}_0$, $x \neq 0$. ∎

Theorem 13.1 has introduced sufficient conditions for stability for the approximate solution $\tilde{U}_0(x)$ based on its level of suboptimality specified by the assumptions (A4) and (A5). In the following section we describe how these conditions can be met by sampling the optimal control law for different values of x and then using a technique called barycentric interpolation.

To simplify the discussion we will adopt the notation for the general multiparametric programming problem from Chapter 5. We consider the multiparametric program

$$J^*(x) = \inf_z \ J(z, x)$$
$$\text{subj. to} \quad (z, x) \in \mathcal{C}, \quad (13.9)$$

where
$$C = \{z, x : \ g(z, x) \leq 0\}, \tag{13.10}$$

$z \in \mathcal{Z} \subseteq \mathbb{R}^s$ is the optimization vector, $x \in \mathcal{X} \subseteq \mathbb{R}^n$ is the parameter vector, $J : \mathbb{R}^s \times \mathbb{R}^n \to \mathbb{R}$ is the cost function and $g : \mathbb{R}^s \times \mathbb{R}^n \to \mathbb{R}^{n_g}$ are the constraints. We assume throughout this chapter that the set C is convex, and that the function $J(z, x)$ is simultaneously convex in z and x. Note that this implies that the optimization problem (13.9) is convex for each fixed value of the parameter x. For a particular value of the parameter x, we denote the set of feasible points as

$$R(x) = \{z \in \mathcal{Z} : \ g(z, x) \leq 0\}, \tag{13.11}$$

and define the feasible set as the set of parameters for which (13.9) has a solution

$$\mathcal{K}^* = \{x \in \mathcal{X} : \ R(x) \neq \emptyset\}, \tag{13.12}$$

We denote by $J^*(x)$ the real-valued function that expresses the dependence of the minimum value of the objective function over $\mathcal{K}^*$ on x

$$J^*(x) = \inf_z \{J(z, x) : \ z \in R(x)\}, \tag{13.13}$$

and by $Z^*(x)$ the point-to-set map which assigns the (possibly empty) set of optimizers $z^* \in 2^{\mathcal{Z}}$ to a parameter $x \in \mathcal{X}$

$$Z^*(x) = \{z \in R(x) : \ J(z, x) = J^*(x)\}. \tag{13.14}$$

We make the following assumptions:

Convexity The function $J^*(x)$ is continuous and convex.

Uniqueness The set of optimizers $Z^*(x)$ is a singleton for each value of the parameter. The function $z^*(x) = Z^*(x)$ will be called *optimizer function*.

Continuity The function $z^*(x)$ is continuous.

Note that the solutions $J^*(x)$, $Z^*(x)$ and $z^*(x)$ of problem (13.1) satisfy these conditions. We emphasize, however, that the approximate methods introduced in the following are applicable also to more general MPC problems as long as these conditions of convexity, uniqueness and continuity are met.

For the approximation methods introduced below convexity of the function $\gamma(x) \succeq 0$ (Theorem 13.1, assumptions (A4)–(A5)) bounding the level of suboptimality is necessary. It is not required, however, for the closed-loop stability with the approximate control law.

13.2 Barycentric Interpolation

Schemes have been proposed in the literature that compute an approximate solution to various problems of the form (13.9) based on interpolating the optimizer $z^*(v_i)$ at a number of samples of the parameter $V = \{v_i\} \subset \mathcal{K}^*$. The goal of interpolated control is to define a function $\tilde{z}(x)$ over the convex hull of the sampled points conv(V), which has the required properties laid out in Theorem 13.1.

Section 13.2.1 will introduce the notion of *barycentric interpolation*, and will demonstrate that such an approach will lead to an approximate function $\tilde{z}(x)$, which is feasible

$$\tilde{z}(x) \in R(x), \forall x \in \text{conv}(V). \tag{13.15}$$

Then Section 13.2.2 introduces a convex optimization problem, whose optimal value, if positive, verifies a specified level of suboptimality

$$J(\tilde{z}(x), x) \leq J^*(x) + \gamma(x), \ \forall x \in \text{conv}(V). \tag{13.16}$$

By Theorem 13.1 the stability of the resulting suboptimal interpolated control law follows.

13.2.1 Feasibility

The class of interpolating functions $\tilde{z}(x)$ satisfying (13.15) is called the class of *barycentric functions*.

Definition 13.1 (Barycentric function) *Let $V = conv(v_1, \ldots, v_n) \subset \mathbb{R}^n$ be a polytope. The set of functions $w_v(x)$, $v \in extreme(V)$ is called* barycentric *if three conditions hold for all $x \in V$*

$$w_v(x) \geq 0 \quad \text{positivity} \tag{13.17a}$$

$$\sum_{v \in extreme(V)} w_v(x) - 1 \quad \text{partition of unity} \tag{13.17b}$$

$$\sum_{v \in extreme(V)} v w_v(x) = x \quad \text{linear precision} \tag{13.17c}$$

Many methods have been proposed in the literature to generate barycentric functions, several of which are detailed in Section 13.3, when we introduce constructive methods for generating approximate controllers. Figure 13.1 illustrates several standard methods of barycentric interpolation.

Given a set of samples V, and a set of barycentric interpolating functions $\{w_v(x)\}$, we can now define an interpolated approximation $\tilde{z}(x)$:

$$\tilde{z}(x) = \sum_{v \in \text{extreme}(V)} z^*(v) w_v(x) \tag{13.18}$$

The key statement that can be made about functions defined via barycentric interpolation is that the function values lie within the convex hull of the function values at the sample points. This property is illustrated in Figure 13.2, and formally defined in the following lemma.

Lemma 13.1 (Convex Hull Property of Barycentric Interpolation) *If $V = \{v_0, \ldots, v_m\} \subset \mathcal{K}^*$, $z^*(v_i)$ is the optimizer of (13.9) for the parameter v_i, and $\{w_v(x)\}$ is a set of barycentric interpolating functions, then*

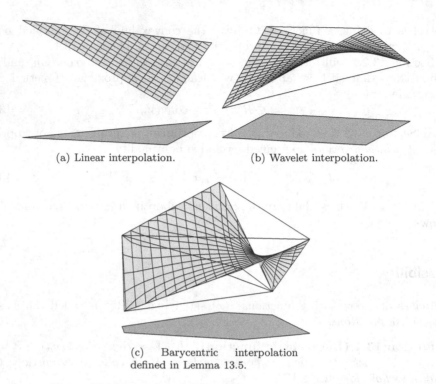

(a) Linear interpolation.   (b) Wavelet interpolation.

(c)   Barycentric   interpolation
defined in Lemma 13.5.

Figure 13.1 Examples of different barycentric interpolations.

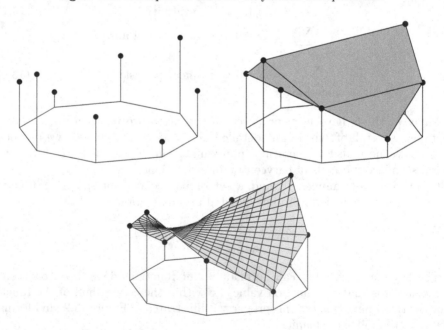

Figure 13.2 Illustration of the Convex-Hull Property of Barycentric Interpolation. The first image (a) shows a number of samples of a function, (b) the convex hull of these sample points and (c) a barycentric interpolation that lies within this convex hull.

$$\tilde{z}(x) = \sum_{v \in V} z^*(v) w_v(x) \in R(x) \text{ for all } x \in conv(V).$$

Proof: The statement holds if for each $x \in conv(V)$ there exists a set of positive multipliers $\lambda_0, \ldots, \lambda_m$ such that

$$\begin{pmatrix} x \\ \tilde{z}(x) \end{pmatrix} = \sum_{i=0}^{m} \lambda_i \begin{pmatrix} v_i \\ z^*(v_i) \end{pmatrix}$$

and $\sum_{i=0}^{m} \lambda_i = 1$. The properties of barycentric functions (13.17) clearly satisfy this requirement. ∎

13.2.2   Suboptimality

Next we investigate the level of suboptimality obtained for a barycentric interpolation. First we demonstrate that the value of the function $J(z, x)$ for the approximate solution $\tilde{z}(x)$ is upper-bounded by the barycentric interpolation of its optimal value taken at the sample points. Then we introduce a convex optimization problem, whose solution upper-bounds the approximation error.

Lemma 13.2   *If $\tilde{z}(x)$ is a barycentric interpolation as defined in (13.18), then the following condition holds:*

$$J^*(x) \le J(\tilde{z}(x), x) \le \sum_{v \in V} w_v(x) J^*(v) \text{ for all } x \in conv(V).$$

Proof: The lower bound follows from the fact that $\tilde{z}(x)$ is feasible for the optimization problem for all $x \in V$ (Lemma 13.1).

The upper bound follows from convexity of the function $J(z, x)$. Consider a fixed value of the parameter $x \in conv(V)$:

$$\begin{aligned} J(\tilde{z}(x), x) &= J\left(\sum_{v \in V} w_v(x) z^*(v), \sum_{v \in V} w_v(x) v \right) \\ &\le \sum_{v \in V} w_v(x) J(z^*(v), v) \\ &= \sum_{v \in V} w_v(x) J^*(v). \end{aligned}$$

The inequality in the above sequence follows because for a fixed value of the parameter x, the barycentric interpolating functions $\{w_v\}$ are all positive, and sum to one. ∎

The previous lemma demonstrates that due to the convexity of the value function $J(z, x)$, the interpolation of the optimal value function at the sample points provides an upper bound of the value function evaluated at the interpolated approximation $\tilde{z}(x)$ (Figure 13.3). We can use this fact to formulate a convex optimization problem, whose solution will provide a certificate for (13.16).

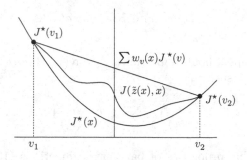

Figure 13.3 Illustration of the relationship between the optimal value function $J^*(x)$, the value function evaluated for the sub-optimal function $\tilde{z}(x)$ and the convex hull of the optimal value at the sample points.

Theorem 13.2   *If $\delta_V \geq 0$, then $J\left(\tilde{z}(x), x\right) \leq J^*(x) + \gamma(x)$ for all $x \in conv(V)$, where*

$$\delta_V = \min_{x, z, \lambda_v} \ J(z, x) + \gamma(x) - \sum_{v \in V} \lambda_v J^*(v) \tag{13.19}$$

$$s.t. \ x = \sum_{v \in V} \lambda_v v, \ \lambda_v \geq 0, \ \sum_{v \in V} \lambda_v = 1$$

$$(x, z) \in C$$

Proof: We see that the suboptimality bound is met within the convex hull of V if and only if the following value is nonnegative:

$$\min_{\lambda_v, x \in \text{conv}(V)} \ J^*(x) + \gamma(x) - J\left(\tilde{z}(x), x\right) \tag{13.20}$$

Lemma 13.2 provides a lower-bound

$$(13.20) \ \geq \ \min_{x \in \text{conv}(V)} \ J^*(x) + \gamma(x) - \sum_{v \in V} w_v(x) J^*(v)$$

and we obtain a further relaxation / lower bound by optimizing over all possible barycentric functions

$$(13.20) \geq \min_{x \in \text{conv}(V)} \ J^*(x) + \gamma(x) - \sum_{v \in V} \lambda_v J^*(v)$$

$$s.t. \quad x = \sum_{v \in V} \lambda_v v, \ \lambda_v \geq 0, \ \sum_{v \in V} \lambda_v = 1.$$

Recalling that $J^*(x) = \min_z \ J(z, x)$ s.t. $(x, z) \in C$ provides the desired result. ∎

In this section we have demonstrated that if the approximate solution $\tilde{z}(x)$ is generated via barycentric interpolation, it will satisfy (13.15) and that the condition for suboptimality (13.16) can be verified by solving a convex optimization problem. In the next section, we will introduce several constructive techniques to incrementally build approximate barycentric interpolating functions.

13.2.3 Constructive Algorithm

In this section, we introduce a simple high-level algorithm to incrementally construct a barycentric function that satisfies conditions (13.15)–(13.16) and therefore can be used in Theorem 13.1 to generate a stabilizing control law. The idea is simple: begin with a set of sample points and solve the optimization problem in Theorem 13.2 to test if the stability condition is met. If it is not, then sample an additional set of points, subdivide the resulting region into subregions and repeat the procedure for each such generated region. In the following we formalize this idea, and demonstrate that this procedure will terminate.

Algorithm 13.1 *Approximate Parametric Programming with Stability Guarantee*

Input: Parametric Optimization Problem (13.9). Initial set of samples V
Output: Sample points $V \subset \mathcal{K}^*$ and set of subsets $\mathcal{R} = \{R_j, \ j = 1, \ldots\}$, $R_j \subseteq V$ such that $\{\mathrm{conv}(R_j), \ j = 1, \ldots\}$ forms a partition of $\mathrm{conv}(V)$ and $\delta_{R_j} \geq 0$ for all j

    Let $\mathcal{R} \leftarrow V$
    Repeat
        $R^* \leftarrow \mathrm{argmin}_{R_j \in \mathcal{R}} \ \delta_{R_j}$ (from problem (13.19))
        If $\delta_{R^*} < 0$ **Then**
            $V \leftarrow V \cup \{x_{R^*}\}$, where x_{R^*} is the optimizer of (13.19) for region R^*
            $\mathcal{R} \leftarrow \mathrm{partition}(V, \mathcal{R})$
    Until $\delta_{R^*} \geq 0$
    Return $\mathcal{R}$

Algorithm 13.1 outlines at a high level a method for constructing a polytopic partition (Definition 4.6) of a set of initial samples, where each polytope in the partition satisfies the suboptimality condition of Theorem 13.1, and therefore any barycentrically interpolated control law defined over the partition will be stabilizing. One can see that in each iteration of the algorithm, the region R^* with the worst fit measured in terms of δ_{R^*} is computed, and the parameter x_{R^*} at which this maximum error occurs is added to the samples. The sample points V are then updated to include this new point, and the set $\mathcal{R}$ is recomputed. This set $\mathcal{R}$ defines a partition splitting $\mathrm{conv}(V)$ into a set of polytopes $\{\mathrm{conv}(R_j), \ j = 1, \ldots\}$. Once this key function "partition" is defined, this algorithm can be directly implemented.

Several methods have been proposed in the literature that roughly follow the scheme of Algorithm 13.1 [37, 159, 3, 156]. Many of these approaches differ in the details by adding caching steps to minimize computation of δ_R, add multiple points at a time, or take an incremental repartitioning approach. All of these procedures can significantly improve the computation speed of the algorithm, and should therefore be incorporated into any practical approach.

13.3 Partitioning and Interpolation Methods

This section outlines three approaches to partitioning the set of samples and the associated barycentric interpolation methods. While many proposals have been put forward in the literature, we focus here on three techniques: inner polytopic approximation (triangulation), outer polytopic approximation (convex hull) and second-order interpolants (hierarchical grids). For each partitioning approach, we present an appropriate barycentric interpolation method.

13.3.1   Triangulation

The simplest approach to generate a barycentric function is to first triangulate the sampled points V, and then define a piece-wise affine control law over the triangulation. The resulting interpolation of the optimal value function at the sampled points will result in a convex piecewise affine *inner* approximation of the optimal value function, and the approximate control law will be a piecewise affine function.

Definition 13.2 (Triangulation)   *A triangulation of a finite set of points $V \subset \mathbb{R}^n$ is a finite collection $T_V = \{R_0, \ldots, R_L\}$ such that*

- $R_i = conv(V_i)$ *is an $n-$dimensional simplex for some $V_i \subset V$*

- $conv(V) = \cup R_i$ *and* $\operatorname{int} R_i \cap \operatorname{int} R_j = \emptyset$ *for all $i \neq j$*

- *If $i \neq j$ then there is a common (possibly empty) face F of the boundaries of S_i and S_j such that $S_i \cap S_j = F$.*

There are various triangulations possible; for example, the recursive triangulation developed in [232, 37] has the strong property of generating a simple hierarchy that can significantly speed online evaluation of the resulting control law. The Delaunay triangulation [109], which has the nice property of minimizing the number of skinny triangles, or those with small angles is a common choice for which incremental update algorithms are well-studied and readily available (i.e., computation of $T_{V \cup \{v\}}$ given T_V). A particularly suitable Delaunay triangulation can be defined by using the optimal cost function as a weighting term, which causes the resulting triangulation to closely match the optimal partition [159].

Given a discrete set of states V, we can now define an interpolated control law $\tilde{z}(x; T_V) : \mathbb{R}^n \mapsto \mathbb{R}^m$.

Definition 13.3   *If $V \subset \mathcal{X} \subset \mathbb{R}^n$ is a finite set and $T = \{conv(V_1), \ldots, conv(V_L)\}$ is a triangulation of V, then the* interpolated function $\tilde{z}(x; T) : conv(V) \mapsto \mathbb{R}^m$ *is*

$$\tilde{z}(x; T) = \sum_{v \in V_j} z^*(v) \lambda_v \ , \ if \ x \in conv(V_j) \tag{13.21}$$

where $\lambda_v \geq 0$, $\sum \lambda_v = 1$ and $x = \sum_{v \in V_j} v \lambda_v$.

For each simplical region $conv(V_j)$ in the triangulation, the set of multipliers λ is unique and can be found by solving the system of linear equations

$$\lambda = \begin{bmatrix} v_1 & \cdots & v_{n+1} \\ 1 & \cdots & 1 \end{bmatrix}^{-1} \begin{pmatrix} x \\ 1 \end{pmatrix}$$

where $v_i \in V_j$.

The function $\tilde{z}(x; T)$ is piecewise affine, and given by:

$$
\tilde{z}(x; T) = \begin{bmatrix} z^*(v_1) \\ \vdots \\ z^*(v_{n+1}) \end{bmatrix}' \begin{bmatrix} v_1 & \cdots & v_{n+1} \\ 1 & \cdots & 1 \end{bmatrix}^{-1} \begin{pmatrix} x \\ 1 \end{pmatrix} \quad \text{if } x \in V_j = \mathrm{conv}(\{v_1, \dots, v_{n+1}\}).
$$

The simplical nature of each region immediately demonstrates that the interpolation 13.21 is barycentric.

13.3.2 Outer Polyhedral Approximation

The previous section introduced an approach that generates an inner polytopic approximation of the optimal value function, which results in a triangulation of the state space. This section introduces an outer polytopic approximation method that will result in a general polytopic partition of the space, which is often significantly less conservative than a triangulation, i.e., for the same number of samples it provides a tighter approximation of the value function.

Implicit Double Description Algorithm: Computing an Outer Polyhedral Approximation of a Convex Set

We begin by introducing an algorithm, called the implicit double description method [162], which computes an outer polyhedral approximation in an incremental fashion that is greedy-optimal in terms of the Hausdorff metric defined below.

The epigraph of the optimal value function plus the specified desired approximation error is defined as:

$$
\mathrm{epi}\, J^{*,\gamma} = \{(x, J) \ : \ J \geq J^*(x) + \gamma(x)\}.
$$

The epigraph of a convex function is a convex set. Our goal is to compute a polyhedron P which is an inner approximation of the epigraph of the optimal

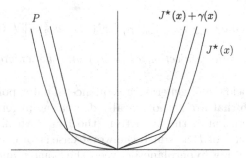

Figure 13.4 Polyhedron P constructed to be an outer approximation of epi $J^{*,\gamma}$ and an inner approximation of epi J^*.

value function and an outer approximation of the optimal value function plus the specified desired approximation error (Figure 13.4):

$$\text{epi } J^* \supseteq P \supseteq \text{epi } J^{*,\gamma}.$$

Once such an approximation is in hand, we will then use it to compute a polytopic partition of $\mathcal{K}^*$ over which the approximate control law will be defined.

The quality of approximation is measured in terms of the Hausdorff distance, which measures the distance between two sets X and Y as:

$$d_H(X,Y) = \max\left\{\sup_{x \in X} \inf_{y \in Y} \|x - y\|_2, \ \sup_{y \in Y} \inf_{x \in X} \|x - y\|_2\right\}.$$

Our goal is to compute the Hausdorff distance between our polytopic approximation, and epi $J^{*,\gamma}$ by solving a convex optimization problem. We first recall the definition of the projection of a point v onto a convex set S

$$\text{proj}_S v = \arg \min_{x \in S} \|v - x\|_2^2$$

The following lemma states that we can compute the projection of a point onto the epigraph of the optimal value function by solving a convex optimization problem.

Lemma 13.3   *The projection of (v, J_v) onto the epigraph* epi $J^{*,\gamma}$ *is the optimizer of the following problem, (x^*, J^*):*

$$\min_{z,x} \|x - v\|_2^2 + \|J(z,x) + \gamma(x) - J_v\|_2^2$$

$$s.t. \ (x, z) \in C.$$

We can now use the fact that the maximizer of a convex function over a convex set will always occur at an extreme point of the set to write the Hausdorff distance as the maximum of the projection of the vertices of the polyhedron P onto the epigraph, which can be done by testing a finite number of points, since a polyhedron has a finite number of vertices.

Lemma 13.4   *If P is a polyhedron, and $P \supseteq$ epi $J^{*,\gamma}$ is a convex set, then the Hausdorff distance is:*

$$d_H(P, \text{epi } J^{*,\gamma}) = max_{(v,J_v) \in extreme(P)} \|x^*(v) - v\|_2^2 + \|J^*(v) + \gamma(x) - J_v\|_2^2$$

where $(x^*(v), J^*(v))$ is the projection of (v, J_v) onto the epigraph.

Algorithm 13.2 below will generate a sequence of outer polyhedral approximations of the epigraph that are monotonically decreasing in terms of the Hausdorff distance. In each iteration of the algorithm, the Hausdorff distance is computed, and the extreme point v of P at which the worst-case error occurs is removed from P by adding a seperating hyperplane between the point v and the epigraph. The addition of a seperating hyperplane ensures that the updated polyhedron remains an outer approximation, and since the worst-case point is removed from the set,

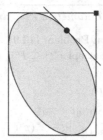

Figure 13.5 Example of one iteration of the implicit double-description algorithm. Square: Point $v_{\max}$ at which the worst-case error occurs. Circle: Point $z^*(v_{\max})$; Projection of $v_{\max}$ onto the set. Line: Separating hyperplane.

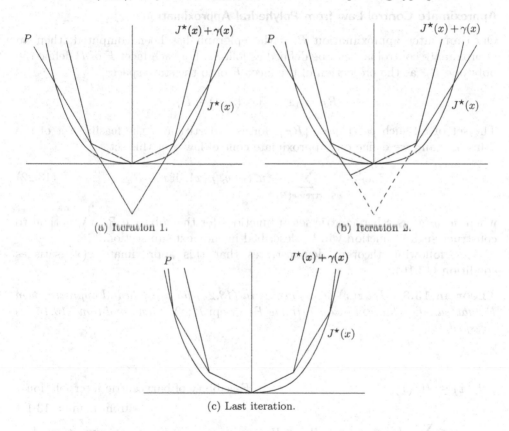

(a) Iteration 1.

(b) Iteration 2.

(c) Last iteration.

Figure 13.6 Several steps of the Double Description algorithm with the termination criterion extreme$(P) \subset$ epi $J^*(x)$ satisfied in the last step.

we have that the approximation error is monotonically decreasing in the Hausdorff distance. The algorithm terminates when all of the vertices of P are contained in the epigraph of J^*. Figure 13.5 illustrates the process of a single cycle of the implicit double description algorithm. Figure 13.6 shows a progression of steps until the termination criterion extreme$(P) \subset$ epi $J^*(x)$ is satisfied.

Algorithm 13.2   *Approximate Parametric Programming with Stability Guarantee*

Input: Parametric Optimization Problem (13.9). Initial polyhedron $P \supset \operatorname{epi} J^{*,\gamma}$
Output: Polyhedron P such that $\operatorname{epi} J^{*,\gamma} \subseteq P \subseteq \operatorname{epi} J^*$

    Repeat
        ForEach $v \in \operatorname{extreme}(P)$
            $z^*(v) = \min_{z \in \operatorname{epi} J^{*,\gamma}} \|v - z\|_2^2$
        $v_{\max} \leftarrow \operatorname{argmax}_{v \in \operatorname{extreme}(P)} \|v - z^*(v)\|$
        $P \leftarrow P \cap \{z \ : \ [z^*(v_{\max}) - v]' \, (z - z^*(v_{\max})) \leq 0\}$
    Until $\operatorname{extreme}(P) \subset \operatorname{epi} J^*(x)$

Approximate Control Law from Polyhedral Approximation

Once an outer approximation P of the epigraph has been computed, then an approximate control law can be defined as follows. For each facet F of P define the polytope R_F as the projection of the facet F onto the state space

$$R_F = \{x \ : \ \exists J, \ (x, J) \in F\}.$$

The set of all such polytopes, $\{R_F\}$, forms a partition of the feasible set of the controller and we define our approximate control law over this set:

$$\tilde{z}(x) = \sum_{v \in \operatorname{extreme}(R_F)} u^*(v) w_{v,F}(x), \text{ if } v \in R_F \tag{13.22}$$

where $w_{v,F}(x)$ is a barycentric set of functions for the polytope R_F. A method to construct such a function will be described in the next sub-section.

The following theorem demonstrates that this approximate $\tilde{z}(x)$ satisfies condition (13.16).

Theorem 13.3   *Let $\tilde{z}(x)$ be defined as in (13.22) for a polyhedral approximation P that satisfies the condition $\operatorname{extreme}(P) \subset \operatorname{epi} J^*(x)$, then condition (13.16) is satisfied.*

    Proof:

$$J^*(x) \leq J(\tilde{z}(x), x) \qquad\qquad \text{Feasibility of barycentric interpolation}$$
$$\text{from Lemma 13.1}$$

$$\leq \sum_{v \in \operatorname{extreme}(R_F)} J^*(v) w_{v,F}(x), \text{ if } v \in R_F \qquad\qquad \text{Convexity of } J^*$$

$$\leq \sum_{(v,J_v) \in \operatorname{extreme}(F)} J_v w_{v,F}(x), \text{ if } v \in R_F \qquad\qquad P \subseteq \operatorname{epi} J^*$$

$$\leq \min_{(x,J) \in P} J \qquad\qquad (v, J_v) \text{ are all on a hyperplane}$$

$$\leq J^*(x) + \gamma(x) \qquad\qquad P \supseteq \operatorname{epi} J^{*,\gamma}$$

∎

Barycentric Functions for Polytopes

Our goal is now to define an easily computable barycentric function for each polytope R_F. If the polytope R_F is a simplex, then the barycentric function is unique, linear and trivially computed and so we focus on the nonsimplical case. In [285] a very elegant method of computing a barycentric function for arbitrary polytopes was proposed that can be put to use here.

Lemma 13.5 (Barycentric coordinates for polytopes [285]) *Let $S = conv(V) \subset \mathbb{R}^d$ be a polytope and for each simple vertex v of S, let $b_v(x)$ be the function*

$$b_v(x) = \frac{\alpha_v}{\|v - x\|_2}$$

where α_v is the area of the polytope $\{y : [V - \mathbf{1}x']y \leq 0, \ (v-x)'y = 1\}$; i.e., the area of the facet of the polar dual of $S - \{x\}$ corresponding to the vertex $v - x$. The function $w_v(x) = b_v(x)/\sum_v b_v(x)$ is barycentric over the polytope S.

The areas of the facets of the polar duals α_v can be precomputed offline and stored. If there are $d + 1$ facets incident with the vertex v (i.e., v is simplical), then the area of the polar facet is $\det([a_0 \cdots a_{d+1}])$, where $\{a_0, \ldots, a_{d+1}\}$ are the normals of the incident facets. If the vertex is not simplical, then the area can be easily computed by perturbing the incident facets [285]. Such computation is straightforward because both the vertices and halfspaces of each region are available due to the double-description representation.

13.3.3 Second-Order Interpolants

This section summarizes the hierarchical grid-based approach proposed in [270]. The main idea is to partition the space into a grid, which is then extremely fast to evaluate online. We first introduce an interpolation method based on second-order interpolants and demonstrate that it is barycentric. We then discuss a sparse hierarchical gridding structure, which allows for extremely sparse storage of the data and rapid evaluation of the function.

Our goal is to define a set of basis functions, one for each sampled vertex of the grid, whose sum is barycentric within each hypercube of the grid. We begin by defining the one-dimensional scaling function (hat function), with support on $[-1, 1]$ as:

$$\phi(x) = \begin{cases} 1 - |x| & \text{if } x \in [-1, 1] \\ 0 & \text{otherwise.} \end{cases}$$

For a particular grid point v, with grid width of w, we can extend the hat function by scaling and shifting:

$$\phi_{v,w}(x) = \phi\left(\frac{x - v}{w}\right).$$

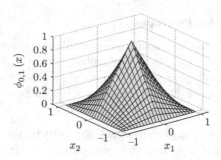

Figure 13.7 Plot of the second order interpolant in 2D centered at zero, with a width of one.

We extend the function to d-dimensions by simply taking the product:

$$\phi_{v,w}(x) = \prod_{i=1}^{d} \phi_{v_i,w}(x_i),$$

where v_i and x_i are the ith components of the vectors v and x respectively.

The resulting basis function $\phi_{\mathbf{0},1}(x)$ is illustrated in two dimensions in Figure 13.7.

We now define our interpolated approximation as:

$$\tilde{z}(x) = \sum_{v \in G} z^*(v)\phi_{v,w}(x)$$

where w is the width of the grid, and G is the set of grid vertices.

Lemma 13.6   *[270] Let R be the hypercube defined by*

$$R = \frac{w}{2}\left(\mathbb{B}_\infty \oplus \mathbf{1}\right) \oplus v$$

where $\mathbb{B}_\infty = \{x \ : \ \|x\|_\infty \leq 1\}$ is the unit infinity-norm ball. The set of functions $\{\phi_{v,w}(x)\}$ is barycentric within the set R.

Lemma 13.6 states that each hypercube in the grid is in fact barycentric. In order to apply Algorithm 13.1 we need a method of repartitioning the current sample points. We do this via a hierarchical approach, in which each hypercube of the grid is tested for compliance with condition (13.16) and, if it fails, is subdivided into 2^d more hypercubes. The result is a hierarchical grid, that is very easy to evaluate online.

What has been discussed so far is termed a "nodal" representation of the approximate function. It is also possible to define what is termed a "hierarchical multi-scale" representation, which, at each level of the regridding procedure, approximates only the residual, rather than the original function. This approach generally results in an extremely sparse representation of the data, and significantly ameliorates the downsides of sampling on a grid. See [270] for more details.

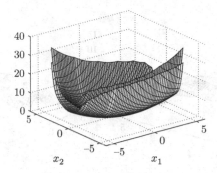

Figure 13.8 Example 13.1. Optimal cost $J^*(x)$ and approximation stability bound $J^*(x) + \gamma(x)$ above it.

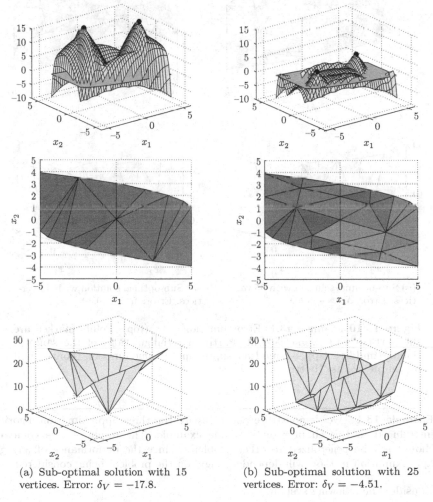

(a) Sub-optimal solution with 15 vertices. Error: $\delta_V = -17.8$.

(b) Sub-optimal solution with 25 vertices. Error: $\delta_V = -4.51$.

Figure 13.9 Example 13.1. Error function $-\delta_V$ (top). Points marked are the vertices with worst-case fit. Partition (triangulation) of the sampled points (middle). Upper-bound on approximate value function (convex hull of sample points) (bottom).

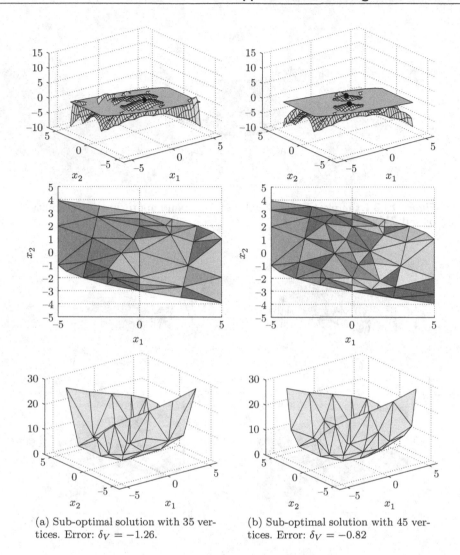

(a) Sub-optimal solution with 35 vertices. Error: $\delta_V = -1.26$.

(b) Sub-optimal solution with 45 vertices. Error: $\delta_V = -0.82$

Figure 13.10 Example 13.1. Error function $-\delta_V$ (top). Points marked are
the vertices with worst-case fit. Partition (triangulation) of the sampled
points (middle). Upper-bound on approximate value function (convex hull
of sample points) (bottom).

Example 13.1 We demonstrate and compare the three approximation algorithms
introduced in this chapter on a simple example. The result of this comparison
should not be generalized to other problems, since the performance of any given
approximation method will depend strongly on the problem being considered.

Consider the following system:

$$x(k+1) = \begin{bmatrix} 1 & 1 \\ 0 & 1 \end{bmatrix} x(k) + \begin{bmatrix} 1 \\ 0.5 \end{bmatrix} u(k)$$

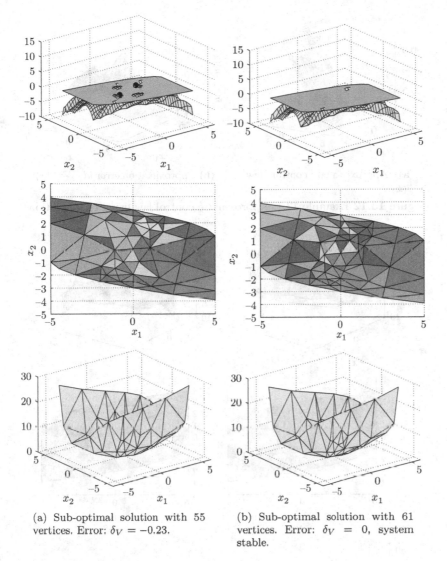

(a) Sub-optimal solution with 55 vertices. Error: $\delta_V = -0.23$.

(b) Sub-optimal solution with 61 vertices. Error: $\delta_V = 0$, system stable.

Figure 13.11 Example 13.1. Error function $-\delta_V$ (top). Points marked are the vertices with worst-case fit. Partition (triangulation) of the sampled points (middle). Upper-bound on approximate value function (convex hull of sample points) (bottom).

with state and input constraints:

$$x \in \mathcal{X} = \{x \mid \|x\|_\infty \leq 5\} \qquad u \in \mathcal{U} = \{u \mid \|u\|_\infty \leq 1\}.$$

We approximate the control law resulting for a standard quadratic-cost MPC problem with a horizon of 10 and weighting matrices taken to be

$$Q = \begin{bmatrix} 0.1 & 0 \\ 0 & 2 \end{bmatrix}, \; R = 1.$$

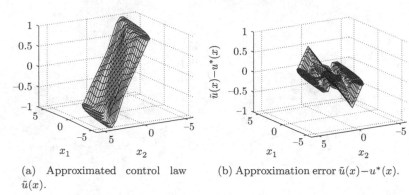

(a) Approximated control law $\tilde{u}(x)$.

(b) Approximation error $\tilde{u}(x) - u^*(x)$.

Figure 13.12 Example 13.1. Approximate control law using triangulation.

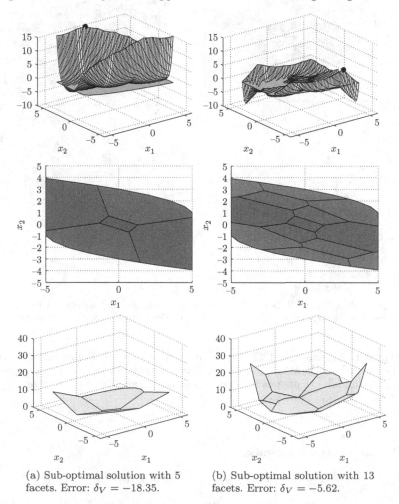

(a) Sub-optimal solution with 5 facets. Error: $\delta_V = -18.35$.

(b) Sub-optimal solution with 13 facets. Error: $\delta_V = -5.62$.

Figure 13.13 Example 13.1. Error function $-\delta_V$ (top). Points marked are the vertices with worst-case fit. Partition (polyhedral approximation) of the sampled points (middle). Upper-bound on approximate value function (convex hull of sample points) (bottom).

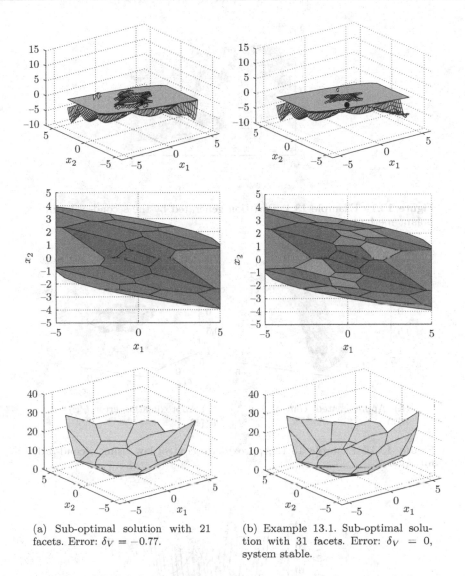

(a) Sub-optimal solution with 21 facets. Error: $\delta_V = -0.77$.

(b) Example 13.1. Sub-optimal solution with 31 facets. Error: $\delta_V = 0$, system stable.

Figure 13.14 Example 13.1. Error function $-\delta_V$ (top). Points marked are the vertices with worst-case fit. Partition (polyhedral approximation) of the sampled points (middle). Upper-bound on approximate value function (convex hull of sample points) (bottom).

The MPC problem formulation includes a terminal set and terminal weight, which are computed as discussed in Chapter 12 as the maximum invariant set for the system when controlled with the optimal LQR control law.

The optimal value function J^* for this problem can be seen in Figure 13.8, along with the approximation target $J^*(x) + \gamma(x)$ for $\gamma(x) = x'Qx$.

Figures 13.9–13.11 shows the evolution of the approximation algorithm using a triangulation approach, and Figure 13.12 shows the resulting approximate control law

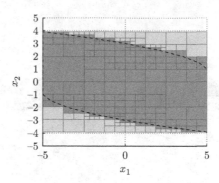

Figure 13.15 Example 13.1. Partition generated by the second-order inter-polant method.

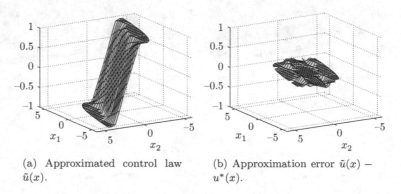

(a) Approximated control law $\tilde{u}(x)$.

(b) Approximation error $\tilde{u}(x) - u^*(x)$.

Figure 13.16 Example 13.1. Approximate control law using second-order interpolants.

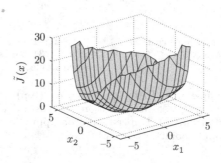

Figure 13.17 Example 13.1. Approximate value function $\tilde{J}(x)$ using second-order interpolants.

and the approximation error. The approximation error seems quite large but this is all what is needed when the only objective is to retain stability rather than performance. Figures 13.13–13.14 show the same evolution, but for the double description/outer approximation approach. One can see that far fewer sample points are required to reach a stability certificate.

Table 13.1 Example 13.1. Comparison of the approximation methods introduced in Section 13.3. The indicated Online FLOPS for triangulation are based on linear search.

	Data numbers	Online FLOPS	Approx error $\|\tilde{u}(x) - u^*(x)\|$ (avg)	(max)
Triangulation	120	1,227	3.3%	24.6%
Polyhedral approx	720	752	2.5%	22.6%
Second-order interpolants	187	26	1.1%	8.3%

The example has also been approximated using the second-order interpolant method until the stability certification was achieved. Figure 13.15 shows the resulting nodal partition of the state space. Note that since it would take an infinite number of hypercubes to approximate a nonorthogonal polytope, a soft-constrained version of the problem is taken in the domain outside the feasible set (although the stability certificate is valid only within the feasible set). One can see the approximate control law, the resulting error and the resulting approximated value function in Figures 13.16 and 13.17, respectively.

Table 13.1 shows the required online data storage and computation time for the various methods introduced in this chapter for the studied example, as well as the average and worst-case approximation errors (based on samples). One can see that for this small example, the second-order interpolants are clearly the best solution. However, this should not be taken as a clear indicator, or a recommendation for the method beyond the others, since this would require a thorough computational comparison of the various methods (or their variants in the literature) on a wide range of practical problems.

14

On-Line Control Computation

In previous chapters we have shown how to compute the solution to constrained finite time optimal control problems. Two approaches can be used. The first approach consists of solving a mathematical program for a given initial state. The second approach employs multiparametric programming to compute the solution as an explicit piecewise affine function of the initial state. This implies that for constrained linear systems, a *Receding Horizon Control* (RHC) policy requires either the on-line solution of a quadratic or linear program, or the evaluation of a piecewise affine on polyhedra function.

In this chapter we focus on efficient on-line methods for the computation of RHC control laws. The main drawback of explicit optimal control laws is that the number of polyhedral regions can grow exponentially with the number of constraints in the optimal control problem. Section 14.1 presents efficient on-line methods for the evaluation of explicit piecewise affine control laws.

If the on-line solution of a quadratic or linear program is preferred, Sections 14.2 and 14.3 briefly discuss how to improve the efficiency of a mathematical programming solver by exploiting the structure of the RHC control problem.

14.1 Storage and On-Line Evaluation of the PWA Control Law

In Chapter 11 we have shown how to compute the solution to the constrained finite time optimal control (CFTOC) problem as an explicit piecewise affine function of the initial state. Such a function is computed off-line by using a multiparametric program solver, which divides the state space into polyhedral regions, and for each region determines the linear gain and offset which produces the optimal control action.

This method reveals its effectiveness when applied to *Receding Horizon Control* (RHC). Having a precomputed solution as an explicit piecewise affine on polyhedra (PPWA) function of the state vector reduces the on-line computation of the RHC control law to a function evaluation, therefore avoiding the on-line solution of a quadratic or linear program.

The main drawback of such an explicit optimal control law is that the number of polyhedral regions can grow dramatically with the number of constraints in the optimal control problem. In this chapter we focus on efficient on-line methods for the evaluation of such a piecewise affine control law.

The simplest way to implement the piecewise affine feedback laws is to store the polyhedral cells $\{H^i x \leq K^i\}$, perform on-line a search through them to locate the one which contains $x(t)$ and then look up the corresponding feedback gain (F^i, g^i) (note that this procedure can be easily parallelized). This chapter presents implementation techniques which avoid the storage and the evaluation of the polyhedra and can significantly reduce the on-line storage demands and computational complexity of RHC. They exploit the properties of the value function and the piecewise affine optimal control law of the constrained finite time optimal control problem.

Let the explicit optimal control law be:

$$u^*(x) = F^i x + g^i, \quad \forall x \in \mathcal{P}_i, \quad i = 1, \dots, N^r, \tag{14.1}$$

where $F^i \in \mathbb{R}^{m \times n}$, $g^i \in \mathbb{R}^m$, and $\mathcal{P}_i = \{x \in \mathbb{R}^n : H^i x \leq K^i, H^i \in \mathbb{R}^{N_c^i \times n}, K^i \in \mathbb{R}^{N_c^i}\}$, $i = 1, \dots, N^r$ is a polyhedral partition of $\mathcal{X}$. In the following H_j^i denotes the j-th row of the matrix H^i, K_j^i denotes the j-th element of the vector K^i and N_c^i is the number of constraints defining the i-th polyhedron $\mathcal{P}_i$. The on-line implementation of the control law (14.1) is simply executed according to the following steps:

Algorithm 14.1

Input: State $x(t)$ at time instant t
Output: Receding horizon control input $u(x(t))$

    Search for the j-th polyhedron that contains $x(t)$, $(H^j x(t) \leq K^j)$
    $u(t) \leftarrow F^j x(t) + g^j$
    Return $u(t)$, the j-th control law evaluated at $x(t)$

In Algorithm 14.1, step (2) is critical and it is the only step whose efficiency can be improved. A simple implementation of step (2) would require searching for the polyhedral region that contains the state $x(t)$ as in the following algorithm.

Algorithm 14.2

Input: State $x(t)$ at time instant t and polyhedral partition $\{\mathcal{P}_i\}_{i=1}^{N^r}$ of the control law (14.1)
Output: Index j of the polyhedron $\mathcal{P}_j$ in the control law (14.1) containing $x(t)$

    $i \leftarrow 1$, notfound$\leftarrow 1$
    While $i \leq N^r$ and notfound
        $j \leftarrow 0$, stillfeasible$\leftarrow 1$

> **While** $j \leq N_c^i$ and stillfeasible=1
>> **If** $H_j^i x(t) > K_j^i$ **Then** stillfeasible← 0
>> **Else** $j \leftarrow j + 1$
> **End**
> **If** stillfeasible=1 **Then** notfound← 0 **Else** $i \leftarrow i + 1$
End
Return j

Algorithm 14.2 requires the storage of all polyhedra $\mathcal{P}_i$, i.e., $(n + 1)N_C$ real numbers (n numbers for each row of the matrix H^i plus one number for the corresponding element in the matrix K^i), $N_C = \sum_{i=1}^{N^r} N_c^i$, and in the worst case (the state is contained in the last region of the list) it will give a solution after nN_C multiplications, $(n - 1)N_C$ sums and N_C comparisons.

In this section, by using the properties of the value function, we show how Algorithm 14.2 can be replaced by more efficient algorithms that *avoid storing the polyhedral regions* $\mathcal{P}_i$, $i = 1, \ldots, N^r$, therefore reducing significantly the storage demand and the computational complexity.

In the following we will distinguish between optimal control based on LP and optimal control based on QP.

14.1.1   Efficient Implementation, 1-Norm, ∞-Norm Case

From Corollary 11.5, the value function $J^*(x)$ corresponding to the solution of the CFTOC problem (12.48) with 1, ∞-norm is convex and PWA:

$$J^*(x) = T^{i'}x + V^i, \quad \forall x \in \mathcal{P}_i, \ i = 1, \ldots, N^r. \qquad (14.2)$$

By exploiting the convexity of the value function the storage of the polyhedral regions $\mathcal{P}_i$ can be avoided. From the equivalence of the representations of PWA convex functions (see Section 2.2.5), the function $J^*(x)$ in equation (14.2) can be represented alternatively as

$$J^*(x) = \max\left\{T^{i'}x + V^i, \ i = 1, \ldots, N^r\right\} \text{ for } x \in \mathcal{X} = \cup_{i=1}^{N^r}\mathcal{P}_i. \qquad (14.3)$$

Thus, the polyhedral region P_j containing x can be identified simply by searching for the maximum number in the list $\{T^{i'}x + V^i\}_{i=1}^{N^r}$:

$$x \in \mathcal{P}_j \Leftrightarrow T^{j'}x + V^j = \max\left\{T^{i'}x + V^i, \ i = 1, \ldots, N^r\right\}. \qquad (14.4)$$

Therefore, instead of searching for the polyhedron j that contains the point x via Algorithm 14.2, we can just store the value function and identify region j by searching for the maximum in the list of numbers composed of the single affine function $T^{i'}x + V^i$ evaluated at x (see Figure 14.1):

Table 14.1 Complexity comparison of Algorithm 14.2 and Algorithm 14.3

	Algorithm 14.2	Algorithm 14.3
Storage demand (real numbers)	$(n+1)N_C$	$(n+1)N^r$
Number of flops (worst case)	$2nN_C$	$2nN^r$

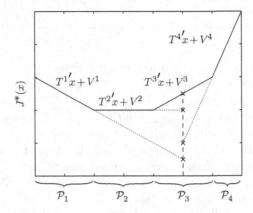

Figure 14.1 Example for Algorithm 14.3 in one dimension: For a given point $x \in \mathcal{P}_3$ $(x = 5)$ we have $J^*(x) = \max(T^1 x + V^1, \ldots, T^4 x + V^4)$.

Algorithm 14.3

Input: State $x(t)$ at time instant t and value function 14.2, $T^{i'} x + V^i$, $i = 1, \ldots, N^r$
Output: Index j of the polyhedron $\mathcal{P}_j$ containing $x(t)$ in the control law (14.1)

    Compute the list $\mathcal{L} = \{n_i = T^{i'} x + V^i, \ i = 1, \ldots, N^r\}$
    Find j such that $n_j = \max_{n_i \in \mathcal{L}} n_i$
    Return j

The search Algorithm 14.3 requires the storage of $(n+1)N^r$ real numbers and will give a solution after nN^r multiplications, $(n-1)N^r$ sums, and $N^r - 1$ comparisons. In Table 14.1 we compare the complexity of Algorithm 14.3 against Algorithm 14.2 in terms of storage demand and number of flops. Algorithm 14.3 will outperform Algorithm 14.2 since typically $N_C \gg N^r$.

14.1.2 Efficient Implementation, 2-Norm Case

Consider the state feedback solution (11.15) of the CFTOC problem (11.9) with $p = 2$. Theorem 11.2 states that the value function $J^*(x)$ is convex and piecewise quadratic on polyhedra and the simple Algorithm 14.3 described in the previous subsection cannot be used here. Instead, a different approach is described below. It uses a surrogate of the value function to uniquely characterize the polyhedral partition of the optimal control law.

We will first establish the following general result: given a general polyhedral partition of the state space, we can locate where the state lies (i.e., in which polyhedron) by using a search procedure based on the information provided by an "appropriate" PWA continuous function defined over the same polyhedral partition. We will refer to such "appropriate" PWA function as a *PWA descriptor function*. First we outline the properties of the PWA descriptor function and then we describe the search procedure itself.

Let $\{\mathcal{P}_i\}_{i=1}^{N^r}$ be the polyhedral partition obtained by solving the mp-QP (11.31) and denote by $C_i = \{j : \mathcal{P}_j \text{ is a neighbor of } \mathcal{P}_i, j = 1, \ldots, N^r, j \neq i\}$ the list of all neighboring polyhedra of $\mathcal{P}_i$. The list C_i has N_c^i elements and we denote by $C_i(k)$ its k-th element.

Definition 14.1 (PWA descriptor function)   *A continuous real-valued PWA function*

$$f(x) = f_i(x) = {A^i}'x + B^i, \ if \ x \in \mathcal{P}_i \tag{14.5}$$

is called a descriptor function if

$$A^i \neq A^j, \ \forall j \subset C_i, \ i = 1, \ldots, N_r. \tag{14.6}$$

Theorem 14.1   *Let $f(x)$ be a PWA descriptor function on the polyhedral partition $\{\mathcal{P}_i\}_{i=1}^{N_r}$.*

Let $O^i(x) \in \mathbb{R}^{N_c^i}$ be a vector associated with region $\mathcal{P}_i$, and let the j-th element of $O^i(x)$ be defined as

$$O_j^i(x) = \begin{cases} +1 & f_i(x) \geqslant f_{C_i(j)}(x) \\ -1 & f_i(x) < f_{C_i(j)}(x) \end{cases} \tag{14.7}$$

Then $O^i(x)$ has the following properties:

(i) $O^i(x) = S^i = const, \ \forall x \in \mathcal{P}_i, \ i = 1, \ldots, N_r.$

(ii) $O^i(x) \neq S^i, \ \forall x \notin \mathcal{P}_i, \ i = 1, \ldots, N_r.$

  Proof: Let $\mathcal{F} = \mathcal{P}_i \cap \mathcal{P}_{C_i(j)}$ be the common facet of $\mathcal{P}_i$ and $\mathcal{P}_{C_i(j)}$. Define the linear function

$$g_j^i(x) = f_i(x) - f_{C_i(j)}(x). \tag{14.8}$$

From the continuity of $f(x)$ it follows that $g_j^i(x) = 0, \forall x \in \mathcal{F}$. As $\mathcal{P}_i$ and $\mathcal{P}_{C_i(j)}$ are disjoint convex polyhedra and $A^i \neq A^{C_i(j)}$ it follows that $g_j^i(\xi_i) > 0$ (or $g_j^i(\xi_i) < 0$, but not both) for any interior point ξ_i of $\mathcal{P}_i$. Similarly for any interior point $\xi_{C_i(j)}$ of $\mathcal{P}_{C_i(j)}$ we have $g_j^i(\xi_{C_i(j)}) < 0$ (or $g_i^j(\xi_i) > 0$, but not both). Consequently, $g_j^i(x) = 0$ is the separating hyperplane between $\mathcal{P}_i$ and $\mathcal{P}_{C_i(j)}$.

(i) Because $g_j^i(x) = 0$ is a separating hyperplane, the function $g_j^i(x)$ does not change its sign for all $x \in \mathcal{P}_i$, i.e., $O_j^i(x) = s_j^i, \forall x \in \mathcal{P}_i$ with $s_j^i = +1$ or $s_j^i = -1$. The same reasoning can be applied to all neighbors of $\mathcal{P}_i$ to get the vector $S^i = \{s_j^i\} \in \mathbb{R}^{N_c^i}$.

(ii) $\forall x \notin \mathcal{P}_i, \ \exists j \in C_i$ such that $H_j^i x > K_j^i$. Since $g_j^i(x) = 0$ is a separating hyperplane $O_j^i(x) = -s_j^i$. ∎

Equivalently, Theorem 14.1 states that

$$x \in \mathcal{P}_i \Leftrightarrow O^i(x) = S^i, \tag{14.9}$$

which means that the function $O^i(x)$ and the vector S^i uniquely characterize $\mathcal{P}_i$. Therefore, to check on-line if the polyhedral region i contains the state x it is sufficient to compute the binary vector $O^i(x)$ and compare it with S^i. Vectors S^i are calculated off-line for all $i = 1, \ldots, N^r$, by comparing the values of $f_i(x)$ and $f_{C_i(j)}(x)$ for $j = 1, \ldots, N_c^i$, for a point x belonging to $\mathcal{P}_i$, for instance, the Chebyshev center of $\mathcal{P}_i$.

Remark 14.1   Theorem 14.1 does not hold if one stores the closures of critical regions $\mathcal{P}_i$ (see Remark 6.8). In this case one needs to modify Theorem 14.1 as follows

(i) $O^i(x) = S^i = const, \quad \forall x \in \text{int}(\mathcal{P}_i), \quad i = 1, \ldots, N_r,$

(ii) $O^i(x) \neq S^i, \quad \forall x \notin (\text{int}(\mathcal{P}_i) \bigcup \partial \mathcal{P}_i), \quad i = 1, \ldots, N_r.$

The results in the remainder of this section can be easily extended to this case.

In Figure 14.2 a one-dimensional example illustrates the procedure with $N^r = 4$ regions. The list of neighboring regions C_i and the vector S^i can be constructed by simply looking at the figure: $C_1 = \{2\}$, $C_2 = \{1, 3\}$, $C_3 = \{2, 4\}$, $C_4 = \{3\}$, $S^1 = -1$, $S^2 = [-1 \ 1]'$, $S^3 = [1 \ -1]'$, $S^4 = -1$. The point $x = 4$ is in region 2 and we have $O^2(x) = [-1 \ 1]' = S^2$, while $O^3(x) = [-1 \ -1]' \neq S^3$, $O^1(x) = 1 \neq S^1$, $O^4(x) = 1 \neq S^4$. The failure of a match $O^i(x) = S^i$ provides information on a good search direction(s). The solution can be found by searching in the direction where a constraint is violated, i.e., one should check the neighboring region $\mathcal{P}_j$ for which $O_j^i(x) \neq s_j^i$.

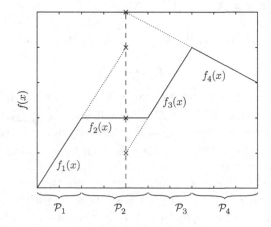

Figure 14.2 Example for Algorithm 14.4 in one dimension: For a given point $x \in \mathcal{P}_2$ ($x = 4$) we have $O_2(x) = [-1 \ 1]' = S_2$, while $O_1(x) = 1 \neq S_1 = -1$, $O_3(x) = [-1 \ -1]' \neq S_3 = [1 \ -1]'$, $O_4(x) = 1 \neq S_4 = -1$.

The overall procedure is composed of two parts:

1. *(off-line)* Construction of the PWA function $f(x)$ in (14.5) satisfying (14.6) and computation of the list of neighbors C_i and the vector S^i

2. *(on-line)* Execution of the following algorithm

Algorithm 14.4

Input: State $x(t)$ at time instant t, list of neighboring regions C_i and the vectors S^i
Output: Index i of the polyhedron $\mathcal{P}_i$ containing $x(t)$ in the control law (14.1)

    $i \leftarrow 1, \mathcal{I} = \{1, \ldots, N_r\}$, notfound$\leftarrow 1$
    While notfound and $\mathcal{I} \neq \emptyset$
        $\mathcal{I} \leftarrow \mathcal{I} \setminus i$
        Compute $O^i(x)$
        If $O^i(x) = S^i$ **Then** notfound$\leftarrow 0$
        Else $i \leftarrow C_i(q)$, where $O_q^i(x) \neq s_q^i$
    End
    Return i

Algorithm 14.4 does not require the storage of the polyhedra $\mathcal{P}_i$, but only the storage of one linear function $f_i(x)$ per polyhedron, i.e., $N^r(n+1)$ real numbers and the list of neighbors C_i which requires N_C integers. In the worst case, Algorithm 14.4 terminates after $N^r n$ multiplications, $N^r(n-1)$ sums and N_C comparisons.

In Table 14.2 we compare the complexity of Algorithm 14.4 against the standard Algorithm 14.2 in terms of storage demand and number of flops.

Remark 14.2   Note that the computation of $O^i(x)$ in Algorithm 14.4 requires the evaluation of N_c^i linear functions, but the overall computation never exceeds N^r linear function evaluations. Consequently, Algorithm 14.4 will outperform Algorithm 14.2, since typically $N_C \gg N^r$.

Now that we have shown how to locate the polyhedron in which the state lies by using a PWA descriptor function, we need a procedure for the construction of such a function.

The image of the descriptor function is the set of real numbers $\mathbb{R}$. In the following we will show how the descriptor function can be generated from a vector-valued function $m : \mathbb{R}^n \to \mathbb{R}^s$. This general result will be used in the following subsections.

Table 14.2 Complexity comparison of Algorithm 14.2 and Algorithm 14.4

	Algorithm 14.2	Algorithm 14.4
Storage demand (real numbers)	$(n+1)N_C$	$(n+1)N^r$
Number of flops (worst case)	$2nN_C$	$(2n-1)N^r + N_C$

Definition 14.2 (Vector-valued PWA descriptor function) *A continuous vector-valued PWA function*

$$m(x) = \bar{A}^i x + \bar{B}^i, \ \ if \ x \in \mathcal{P}_i, \tag{14.10}$$

is called a vector-valued PWA descriptor function if

$$\bar{A}^i \neq \bar{A}^j, \ \forall j \in C_i, \ i = 1, \dots, N_r, \tag{14.11}$$

where $\bar{A}^i \in \mathbb{R}^{s \times n}$, $\bar{B}^i \in \mathbb{R}^s$.

Theorem 14.2 *Given a vector-valued PWA descriptor function $m(x)$ defined over a polyhedral partition $\{\mathcal{P}_i\}_{i=1}^{N_r}$ it is possible to construct a PWA descriptor function $f(x)$ over the same polyhedral partition.*

Proof: Let $\mathcal{N}_{i,j}$ be the null-space of $(\bar{A}^i - \bar{A}^j)'$. Since by the definition $\bar{A}^i - \bar{A}^j \neq \mathbf{0}$ it follows that $\mathcal{N}_{i,j}$ is not full-dimensional, i.e., $\mathcal{N}_{i,j} \subseteq \mathbb{R}^{s-1}$. Consequently, it is always possible to find a vector $w \in \mathbb{R}^s$ such that $w(\bar{A}^i - \bar{A}^j) \neq \mathbf{0}$ holds for all $i = 1, \dots, N_r$ and $\forall j \in C_i$. Clearly, $f(x) = w'm(x)$ is then a valid PWA descriptor function. ∎

As shown in the proof of Theorem 14.2, once we have a vector-valued PWA descriptor function, practically any randomly chosen vector $w \in \mathbb{R}^s$ is likely to be satisfactory for the construction of PWA descriptor function. But, from a numerical point of view, we would like to obtain a w that is as far away as possible from the null-spaces $\mathcal{N}_{i,j}$. We show one algorithm for finding such a vector w.

For a given vector-valued PWA descriptor function we form a set of vectors $a_k \in \mathbb{R}^s$, $\|a_k\| = 1$, $k = 1, \dots, N_C/2$, by choosing and normalizing one (and only one) nonzero column from each matrix $(\bar{A}^i - \bar{A}^j)$, $\forall j \in C_i$, $i = 1, \dots, N_r$. The vector $w \in \mathbb{R}^s$ satisfying the set of equations $w'a_k \neq 0$, $k = 1, \dots, N_C/2$, can then be constructed by using the following algorithm. Note that the index k goes to $N_C/2$ since the term $(\bar{A}^j - \bar{A}^i)$ is the same as $(\bar{A}^i - \bar{A}^j)$ and thus there is no need to consider it twice.

Algorithm 14.5

Input: Vectors $a_i \in \mathbb{R}^s$, $i = 1, \dots, N$
Output: The vector $w \in \mathbb{R}^s$ satisfying the set of equations $w'a_i \neq 0$, $i = 1, \dots, N$

  $w \leftarrow [1, \dots, 1]'$, $R \leftarrow 1$
  While $k \leq N_C/2$
    $d \leftarrow w'a_k$
    If $0 \leq d \leq R$ **Then** $w \leftarrow w + \frac{1}{2}(R - d)a_k$, $R \leftarrow \frac{1}{2}(R + d)$
    If $-R \leq d < 0$ **Then** $w \leftarrow w - \frac{1}{2}(R + d)a_k$, $R \leftarrow \frac{1}{2}(R - d)$
  End
  Return w

Algorithm 14.5 is based on a construction of a sequence of balls $\mathcal{B} = \{x \ : \ x = w + r, \ \|r\|_2 \leq R\}$. As depicted in Figure 14.3, Algorithm 14.5 starts with the initial

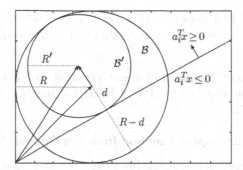

Figure 14.3 Illustration for Algorithm 14.5 in two dimensions.

ball of radius $R = 1$, centered at $w = [1, \ldots, 1]'$. Iteratively one hyperplane $a'_k x = 0$ at the time is introduced and the largest ball $\mathcal{B}' \subseteq \mathcal{B}$ that does not intersect this hyperplane is designed. The center w of the final ball is the vector w we wanted to construct, while R gives some information about the degree of nonorthogonality: $|w'a_k| \geq R, \forall k$.

In the following subsections we will show that the gradient of the value function, and the optimizer, are vector-valued PWA descriptor functions and therefore we can use Algorithm 14.5 for the construction of the PWA descriptor function.

Generating a PWA Descriptor Function from the Value Function

Let $J^*(x)$ be the convex and piecewise quadratic (CPWQ) value function obtained as a solution of the CFTOC (11.9) problem for $p = 2$:

$$J^*(x) = q_i(x) = x'Q_i x + T_i' x + V_i, \quad \text{if } x \in \mathcal{P}_i, \ i = 1, \ldots, N^r. \tag{14.12}$$

In Section 6.3.4 we have proven that for nondegenerate problems the value function $J^*(x)$ is a $C^{(1)}$ function. We can obtain a vector-valued PWA descriptor function by differentiating $J^*(x)$. We first need to introduce the following theorem.

Theorem 14.3 ([21]) *Assume that the CFTOC (11.9) problem leads to a non-degenerate mp-QP (11.33). Consider the value function $J^*(x)$ in (14.12) and let CR_i, CR_j be the closure of two neighboring critical regions corresponding to the set of active constraints A_i and A_j, respectively, then*

$$Q_i - Q_j \preceq 0 \text{ or } Q_i - Q_j \succeq 0 \text{ and } Q_i \neq Q_j \tag{14.13}$$

and

$$Q_i - Q_j \preceq 0 \text{ iff } A_i \subset A_j. \tag{14.14}$$

Theorem 14.4 *Consider the value function $J^*(x)$ in (14.12) and assume that the CFTOC (11.9) problem leads to a nondegenerate mp-QP (11.33). Then the gradient $m(x) = \nabla J^*(x)$, is a vector-valued PWA descriptor function.*

Proof: From Theorem 6.9 we see that $m(x)$ is a continuous vector-valued PWA function, while from equation (14.12) we get

$$m(x) = \nabla J^*(x) = 2Q_i x + T_i. \tag{14.15}$$

Since from Theorem 14.3 we know that $Q_i \neq Q_j$ for all neighboring polyhedra, it follows that $m(x)$ satisfies all conditions for a vector-valued PWA descriptor function.　　　　　　　　　　　　　　　　　　　　　　　　　　■

Combining results of Theorem 14.4 and Theorem 14.2 it follows that by using Algorithm 14.5 we can construct a PWA descriptor function from the gradient of the value function $J^*(x)$.

Generating a PWA Descriptor Function from the Optimal Inputs

Another way to construct a descriptor function $f(x)$ emerges naturally if we look at the properties of the optimizer $U_0^*(x)$ corresponding to the state feedback solution of the CFTOC problem (11.9). From Theorem 6.7 it follows that the optimizer $U_0^*(x)$ is continuous in x and piecewise affine on polyhedra:

$$U_0^*(x) = l_i(x) = F^i x + g^i, \text{ if } x \in \mathcal{P}_i, \ i = 1, \dots, N^r, \tag{14.16}$$

where $F^i \in \mathbb{R}^{s \times n}$ and $g^i \in \mathbb{R}^s$. We will assume that $\mathcal{P}_i$ are critical regions (as defined in Section 6.1.2). Before going further we need the following lemma.

Lemma 14.1　*Consider the state feedback solution (14.16) of the CFTOC problem (11.9) and assume that the CFTOC (11.9) leads to a nondegenerate mp-QP (11.33). Let $\mathcal{P}_i$, $\mathcal{P}_j$ be two neighboring polyhedra, then $F^i \neq F^j$.*

Proof: The proof is a simple consequence of Theorem 14.3. As in Theorem 14.3, without loss of generality we can assume that the set of active constraints $\mathcal{A}_i$ associated with the critical region $\mathcal{P}_i$ is empty, i.e., $\mathcal{A}_i = \emptyset$. Suppose that the optimizer is the same for both polyhedra, i.e., $[F^i \ g^i] = [F^j \ g^j]$. Then, the cost functions $q_i(x)$ and $q_j(x)$ are also equal. From the proof of Theorem 14.3 this implies that $\mathcal{P}_i = \mathcal{P}_j$, which is a contradiction. Thus we have $[F^i \ g^i] \neq [F^j \ g^j]$. Note that $F^i = F^j$ cannot happen since, from the continuity of $U_0^*(x)$, this would imply $g^i = g^j$. Consequently we have $F^i \neq F^j$.　　　　　■

From Lemma 14.1 and Theorem 14.2 it follows that an appropriate PWA descriptor function $f(x)$ can be calculated from the gradient of the optimizer $U^*(x)$ by using Algorithm 14.5.

Remark 14.3　Note that even if we are implementing a receding horizon control strategy, the construction of the PWA descriptor function is based on the full optimization vector $U^*(x)$ and the corresponding matrices F^i and g^i.

Remark 14.4　In some cases the use of the optimal control profile $U^*(x)$ for the construction of descriptor function $f(x)$ can be extremely simple. If there is a row r, $r \leq m$ (m is the dimension of u) for which $(F^i)_r \neq (F^j)_r$, $\forall i = 1 \dots, N^r$, $\forall j \in C_i$, it is enough to set $A^{i'} = (F^i)_r$ and $B^i = (g^i)_r$, where $(F^i)_r$ and $(g^i)_r$ denote the i-th row of the matrices F^i and g^i, respectively. The row r has to be the same for all $i = 1 \dots, N^r$. In this way we *avoid* the storage of the descriptor function altogether since it is equal to one component of the control law, which is stored anyway.

14.1.3 Example

As an example, we compare the performance of Algorithm 14.2, 14.3 and 14.4 on CFTOC problem for the discrete-time system

$$
\begin{cases}
x(t+1) = \begin{bmatrix} 4 & -1.5 & 0.5 & -0.25 \\ 4 & 0 & 0 & 0 \\ 0 & 2 & 0 & 0 \\ 0 & 0 & 0.5 & 0 \end{bmatrix} x(t) + \begin{bmatrix} 0.5 \\ 0 \\ 0 \\ 0 \end{bmatrix} u(t) \\
y(t) = \begin{bmatrix} 0.083 & 0.22 & 0.11 & 0.02 \end{bmatrix} x(t)
\end{cases}
\tag{14.17}
$$

resulting from the linear system

$$
y = \frac{1}{s^4} u
\tag{14.18}
$$

sampled at $T_s = 1$, subject to the input constraint

$$
-1 \le u(t) \le 1
\tag{14.19}
$$

and the output constraint

$$
-10 \le y(t) \le 10.
\tag{14.20}
$$

CFTOC based on LP

To regulate (14.17), we design a receding horizon controller based on the optimization problem (12.6), (12.7) where $p = \infty$, $N = 2$, $Q = \mathrm{diag}\{5, 10, 10, 10\}$, $R = 0.8$, $P = 0$, $\mathcal{X}_f = \mathbb{R}^4$. The PPWA solution of the mp-LP problem comprises 136 regions. In Table 14.3 we report the comparison between the complexity of Algorithm 14.2 and Algorithm 14.3 for this example.

The average on-line evaluation effort of the PPWA solution for a set of 1000 random points in the state space is 2259 flops (Algorithm 14.2), and 1088 flops (Algorithm 14.3). We note that the solution using MATLAB® LP solver (function linprog.m with interior point algorithm and LargeScale set to "off") takes $25,459$ flops on average.

CFTOC based on QP

To regulate (14.17), we design a receding horizon controller based on the optimization problem (12.6), (12.7)) where $N = 7$, $Q = I$, $R = 0.01$, $P = 0$, $\mathcal{X}_f = \mathbb{R}^4$. The PPWA solution of the mp-QP problem comprises 213 regions. We obtained a descriptor function from the value function and for this example the choice of

Table 14.3 Complexity comparison of Algorithm 14.2 and Algorithm 14.3 for the example in Section 14.1.3

	Algorithm 14.2	Algorithm 14.3
Storage demand (real numbers)	5690	680
Number of flops (worst case)	9104	1088
Number of flops (average for 1000 random points)	2259	1088

Table 14.4 Complexity comparison of Algorithm 14.2 and Algorithm 14.4 for the example in Section 14.1.3

	Algorithm 14.2	Algorithm 14.4
Storage demand (real numbers)	9740	1065
Number of flops (worst case)	15584	3439
Number of flops (average for 1000 random points)	2114	175

$w = [1\ 0\ 0\ 0]'$ is satisfactory. In Table 14.4 we report the comparison between the complexity of Algorithm 14.2 and Algorithm 14.4 for this example.

The average computation effort of the PPWA solution for a set of 1000 random points in the state space is 2114 flops (Algorithm 14.2), and 175 flops (Algorithm 14.4). The solution of the corresponding quadratic program with MATLAB® QP solver (function quadprog.m and LargeScale set to "off") takes $25,221$ flops on average.

14.1.4 Literature Review

The problem considered in this chapter has been approached by several other researchers. For instance, in [275] the authors propose to organize the controller gains of the PWA control law on a balanced search tree. By doing so, the search for the region containing the current state has on average a logarithmic computation complexity although it is less efficient in terms of memory requirements. At the expense of the optimality a similar computational complexity can be achieved with the approximative point location algorithm described in [160].

The comparison of the proposed algorithms with other efficient solvers in the literature, e.g., [21, 55, 256, 105, 209, 212, 284, 61, 25] requires the simultaneous analysis of several issues such as speed of computation, storage demand and real time code verifiability.

14.2 Gradient Projection Methods Applied to MPC

In this section, we investigate the gradient projection methods from Section 3.3.1 for the iterative solution of MPC problems with a quadratic cost function given as

$$J_0^*(x(0)) = \min \quad x_N' P x_N + \sum_{k=0}^{N-1} x_k' Q x_k + u_k' R u_k$$

$$\begin{aligned}
\text{subj. to} \quad & x_{k+1} = A x_k + B u_k, \ k = 0, \ldots, N-1 \\
& x_k \in \mathcal{X}, \ u_k \in \mathcal{U}, \ k = 0, \ldots, N-1 \\
& x_N \in \mathcal{X}_f \\
& x_0 = x(0).
\end{aligned} \qquad (14.21)$$

Note that 1- and ∞-norm cost functions (see Section 11.1) cannot be handled by the gradient (projection) methods discussed in this book since they require a smooth objective function.

In Section 14.2.1 we first investigate a special case of the general setup in (14.21) which is input-constrained MPC. In this case, we impose constraints on the control inputs only, i.e., for the state sets we have $\mathcal{X} = \mathcal{X}_f = \mathbb{R}^n$. If there are no state constraints, the MPC problem can be formulated with the sequence of inputs $U_0 = [u_0', \ldots, u_{N-1}']'$ as the optimization vector (so-called *condensing*) where at the same time the feasible set $\mathcal{U}^N = \mathcal{U} \times \ldots \times \mathcal{U}$ remains "simple" if the input set $\mathcal{U}$ was simple (recall from Section 3.3.1 that a "simple" set is a convex set with a projection operator that can be evaluated efficiently). This allows one to solve the MPC problem right in the primal domain and thus to achieve linear convergence with the gradient projection methods of this book.

In the case of input and state constraints, the feasible set of problem (14.21) does not allow for an easy-to-evaluate projection operator in general, neither in condensed form nor with the state sequence as an additional optimization vector (so-called *sparse formulation*). However, gradient methods can still be applied to the sparse formulation of the MPC problem if a dual approach, as discussed in Section 3.3.1, is applied. However, only sublinear convergence can be achieved in the dual domain. Section 14.2.2 will summarize problem formulation, important aspects regarding the computation of the dual gradient and optimal step size selection.

14.2.1   Input-Constrained MPC

In order to apply the classic or the fast gradient projection method from Section 3.3.1, we first eliminate the states from the MPC problem (14.21) and rewrite it in condensed form (see (8.8)) as

$$J_0^*(x(0)) = x'(0)Yx(0) + \min_{U_0} \quad U_0'HU_0 + 2x'(0)FU_0$$
$$\text{subj. to} \quad U_0 \in \mathcal{U}^N,$$

which resembles the general problem setup for gradient methods given by (3.23). Since the objective function is twice continuously differentiable, L-smoothness and strong convexity follow from Lemma (3.1) and Theorem (3.2), respectively. In fact, the Lipschitz constant L and the strong convexity parameter μ are given by

$$L = 2\,\lambda_{\max}(H), \quad \mu = 2\,\lambda_{\min}(H),$$

and are both independent of the initial state $x(0)$. Thus, no computationally expensive eigenvalue computations are necessary during runtime if the Hessian H stays constant.

From the definition of H in (8.8) it follows that it is positive definite whenever the penalty matrix R is positive definite. In this case, the constant μ is positive and both the classic gradient method given by Algorithm 3.6 and the fast gradient projection method in Algorithm 3.7 converge linearly.

The projection operator for the feasible set $\mathcal{U}^N$ has to be computed in every iteration of any gradient projection method. In the MPC context, this operation translates to

$$\pi_{\mathcal{U}^N}\left(Y_i - \frac{2}{L}\big(HY_i + F'x(0)\big)\right), \tag{14.22}$$

where Y_i is the previous iterate in case of the classic gradient projection method or the previous secondary iterate in case of the fast gradient projection method. Since the feasible set in input-constrained MPC is the direct product of N sets, evaluation of the projection operator in (14.22) can be separated into N evaluations of the projection operator $\pi_{\mathcal{U}}$, i.e., if we let $U_f = Y_i - \frac{2}{L}\left(HY_i + F'x(0)\right)$, then

$$\pi_{\mathcal{U}^N}(U_f) = \left[\pi_{\mathcal{U}}(u_{f,1})', \ldots, \pi_{\mathcal{U}}(u_{f,N})'\right]',$$

with $U_f = \left[u'_{f,1}, \ldots, u'_{f,N}\right]'$ and $u_{f,k} \in \mathbb{R}^m$, $k = 1, \ldots, N$.

14.2.2   Input- and State-Constrained MPC

In the input- and state-constrained case, gradient projection methods cannot be applied in the primal domain, since the projection operator for the feasible set

$$\left\{u_k \in \mathcal{U}, x_k \in \mathcal{X}, x_N \in \mathcal{X}_f \mid x_{k+1} = Ax_k + Bu_k, \, x_0 = x(0), \, k = 0 \ldots N-1\right\}$$

is nontrivial in general, even if the convex sets $\mathcal{U}$, $\mathcal{X}$ and $\mathcal{X}_f$ allow efficient projections individually. The inherent complication comes from the intersection of these sets with the affine set, which is determined by the state update equation.

As an alternative, the input- and state-constrained MPC problem can be solved in the dual domain by first relaxing the state update equation and then solving for the associated optimal dual multiplier (or Lagrange multiplier) vector using gradient methods. This approach was thoroughly discussed in Section 3.3.1 for a generic convex objective function and a feasible set that can be written as the intersection of an affine set and a simple set (see (3.27)). If we rewrite the MPC problem in (14.21) analogously to the generic setup of Section 3.3.1 with the optimization vector $z = \left[x'_0, x'_1, \ldots, x'_N, u'_0, \ldots, u'_{N-1}\right]'$, we obtain

$$\begin{aligned}
J_0^*(x(0)) = \min \quad & z'Mz \\
\text{subj. to} \quad & Gz = Ex(0) \\
& z \in \mathcal{K},
\end{aligned} \qquad (14.23)$$

where the problem data are defined as

$$M = \text{blockdiag}\{Q, \ldots, Q, P, R, \cdots, R\},$$

$$G = -\begin{bmatrix}
-I_n & 0 & \cdots & \cdots & 0 & 0 & \cdots & \cdots & 0 \\
A & -I_n & 0 & \cdots & 0 & B & 0 & \cdots & 0 \\
0 & \ddots & \ddots & \ddots & \vdots & 0 & B & \ddots & \vdots \\
\vdots & \ddots & A & -I_n & 0 & \vdots & \ddots & \ddots & 0 \\
0 & \cdots & 0 & A & -I_n & 0 & \cdots & 0 & B
\end{bmatrix},$$

$$E = \left[I_n, 0_n, \ldots, 0_n\right]',$$

$$\mathcal{K} = \mathcal{X} \times \ldots \times \mathcal{X} \times \mathcal{X}_f \times \mathcal{U} \times \ldots \mathcal{U}.$$

The dual problem is an unconstrained maximization problem with the dual gradient given as $Gz^*(\nu) - Ex(0)$, where

$$z^*(\nu) = \arg \min_{z \in \mathcal{K}} z'Mz + \nu'(Gz - Ex(0)). \qquad (14.24)$$

As a prerequisite for the above statement, the minimizer $z^*(\nu)$ must be unique for every dual multiplier vector ν (see Section 3.3.1). A sufficient condition for uniqueness of the minimizer is a strongly convex objective function in (14.24). In view of the definition of matrix M this is the case whenever the penalty matrices Q, P and R are positive definite, which is not a restrictive assumption in an application.

The so-called *inner problem* (14.24) needs to be solved in every outer iteration of the classic or the fast gradient method in order to determine the dual gradient. This can be accomplished in one of the following ways.

- *Exact analytic solution*: This is possible if all penalty matrices are positive diagonal matrices and all sets $\mathcal{X}$, $\mathcal{X}_f$ and $\mathcal{U}$ are boxes. In practice, this is a frequent setup.

- *Exact multiparametric solution*: By the block-diagonal structure of matrix M and the structure of set $\mathcal{K}$, the inner problem can be separated into $2N+1$ subproblems. If all sets $\mathcal{X}$, $\mathcal{X}_f$ and $\mathcal{U}$ are polyhedral, each subproblem can be pre-solved by means of multiparametric programming with the gradient of the linear objective term as the parameter.

- *Approximate iterative solution*: This is the most general approach for the solution of the inner problem as it only requires simple sets $\mathcal{X}$, $\mathcal{X}_f$ and $\mathcal{U}$ (not necessarily polyhedral). Each of the $2N+1$ subproblems can then be solved approximately by a gradient projection method. Note that convergence issues of the outer gradient method might arise if the accuracy of the approximate inner solution is too low.

Finally, we note that under mild assumptions, the tight, i.e., smallest possible, Lipschitz constant of the dual gradient is given by $L_d = \frac{1}{2}\|GM^{-1/2}\|^2$ which, same as in the input-constrained case, is independent of the initial state $x(0)$. The Lipschitz constant determines the step size of the outer gradient method and being able to compute a tight value ensures the best possible (sublinear) convergence (see Section 3.3.1).

14.3 Interior Point Method Applied to MPC

All presented interior-point methods can be applied to linear model predictive control problems with convex constraints. To obtain an efficient method with low solve times, it is crucial to exploit the sparsity structure of the resulting quadratic program (or any other form of problem P-IPM (3.30)) when computing the search directions. Therefore, when using interior point solvers, the MPC problem is formulated as a sparse QP (sometimes with a different ordering of states and inputs in the optimization vector). As a result, general sparse routines for solving linear systems with KKT-structure can be applied to the Newton systems (3.41) and (3.45), or specific linear system solvers can be used that exploit the MPC-specific problem structure. A good starting point for details on this topic are the references [284, 246, 98, 201, 110, 171, 170].

15

Constrained Robust Optimal Control

In the previous chapters we assumed perfect model knowledge for the plant to be controlled. This chapter focuses on the case when the model is uncertain. We focus on discrete-time uncertain linear systems with linear constraints on inputs and states. Uncertainty is modeled as additive disturbance and/or parametric uncertainty in the state space matrices. We first introduce the concepts of robust control design with "open-loop prediction" and "closed-loop prediction." Then we discuss how to robustly enforce constraints satisfaction, and present several objective functions one might be interested in minimizing. For each robust control design we discuss the conservatism and show how to obtain the optimal control action by using convex or nonconvex programming.

For robust controllers which can be implemented by using linear and quadratic programming, we show that the robust optimal control law is a continuous and piecewise affine function of the state vector. The robust state feedback controller can be computed by means of multiparametric linear or quadratic programming. Thus, when the optimal control law is implemented in a moving horizon scheme, the on-line computation of the resulting controller requires simply the evaluation of a piecewise affine function.

15.1 Problem Formulation

Consider the linear uncertain system

$$x(t+1) = A(w^p(t))x(t) + B(w^p(t))u(t) + Ew^a(t) \qquad (15.1)$$

where $x(t) \in \mathbb{R}^n$ and $u(t) \in \mathbb{R}^m$ are the state and input vectors, respectively, subject to the constraints

$$x(t) \in \mathcal{X}, \ u(t) \in \mathcal{U}, \ \forall t \geq 0. \qquad (15.2)$$

The sets $\mathcal{X} \subseteq \mathbb{R}^n$ and $\mathcal{U} \subseteq \mathbb{R}^m$ are polytopes. Vectors $w^a(t) \in \mathbb{R}^{n_a}$ and $w^p(t) \in \mathbb{R}^{n_p}$ are unknown additive disturbances and parametric uncertainties, respectively. The disturbance vector is

$$w(t) = [w^a(t)', \ w^p(t)']' \in \mathcal{W} \subset \mathbb{R}^{n_w} \tag{15.3}$$

with $n_w = n_a + n_p$. We assume that bounds on $w^a(t)$ and $w^p(t)$ are known. In particular $\mathcal{W} = \mathcal{W}^a \times \mathcal{W}^p$ with $w^a(t) \in \mathcal{W}^a$ and $w^p(t) \in \mathcal{W}^p$, where $\mathcal{W}^a \subset \mathbb{R}^{n_a}$ and $\mathcal{W}^p \subset \mathbb{R}^{n_p}$ are given polytopes. We also assume that $A(\cdot)$, $B(\cdot)$ are affine functions of w^p

$$A(w^p) = A^0 + \sum_{i=1}^{n_p} A^i w_c^{p,i}, \ B(w^p) = B^0 + \sum_{i=1}^{n_p} B^i w_c^{p,i} \tag{15.4}$$

where $A^i \in \mathbb{R}^{n \times n}$ and $B^i \in \mathbb{R}^{n \times m}$ are given matrices for $i = 0 \ldots, n_p$ and $w_c^{p,i}$ is the i-th component of the vector w^p, i.e., $w^p = [w_c^{p,1}, \ldots, w_c^{p,n_p}]$.

Robust finite time optimal control problems can be formulated by optimizing over "Open-Loop" policies or "Closed-Loop" policies. We will discuss the two cases next.

Open-Loop Predictions

Define the cost function as

$$J_0(x(0), U_0) = \mathcal{J}_\mathcal{W} \left[p(x_N) + \sum_{k=0}^{N-1} q(x_k, u_k) \right] \tag{15.5}$$

where the operation $\mathcal{J}_\mathcal{W}$ evaluates the cost $p(x_N) + \sum_{k=0}^{N-1} q(x_k, u_k)$ for the set of uncertainties $\{w_0, \ldots, w_{N-1}\} \in \mathcal{W} \times \cdots \times \mathcal{W}$ and the input sequence U_0. The following options are typically considered.

- **Nominal Cost**
  In the simplest case, the objective function is evaluated for a single disturbance profile $\bar{w}_0^a, \ldots, \bar{w}_{N-1}^a, \bar{w}_0^p, \ldots, \bar{w}_{N-1}^p$:

$$\mathcal{J}_\mathcal{W} = \left[p(x_N) + \sum_{k=0}^{N-1} q(x_k, u_k) \right]$$
$$\text{where} \begin{cases} x_{k+1} = A(w_k^p)x_k + B(w_k^p)u_k + Ew_k^a \\ x_0 = x(0) \\ w_k^a = \bar{w}_k^a, w_k^p = \bar{w}_k^p \\ k = 0, \ldots, N - 1. \end{cases} \tag{15.6}$$

- **Expected Cost**
  Another option is to consider the probability density function $f(w)$ of the disturbance w:

$$\text{Probability}\,[w(t) \in \mathcal{W}] = 1 = \int_{w \in \mathcal{W}} f(w)dw.$$

Then the expected value of a function $g(w)$ of the disturbance w is defined as:

$$\mathbf{E}_w[g(w)] = \int_{w \in \mathcal{W}} g(w)f(w)dw.$$

In this case we consider the expected cost over the admissible disturbance set.

$$\mathcal{J}_\mathcal{W} = \mathbf{E}_{w_0,\ldots,w_{N-1}} \left[p(x_N) + \sum_{k=0}^{N-1} q(x_k, u_k) \right]$$

$$\text{where} \begin{cases} x_{k+1} = A(w_k^p)x_k + B(w_k^p)u_k + Ew_k^a \\ x_0 = x(0) \\ w_k^a \in \mathcal{W}^a, w_k^p \in \mathcal{W}^p \\ k = 0, \ldots, N-1. \end{cases} \qquad (15.7)$$

- **Worst Case Cost**
  Finally the worst case cost may be of interest

$$\mathcal{J}_\mathcal{W} = \max_{w_0,\ldots,w_{N-1}} \left[p(x_N) + \sum_{k=0}^{N-1} q(x_k, u_k) \right]$$

$$\text{where} \begin{cases} x_{k+1} = A(w_k^p)x_k + B(w_k^p)u_k + Ew_k^a \\ x_0 = x(0) \\ w_k^a \in \mathcal{W}^a, w_k^p \in \mathcal{W}^p \\ k = 0, \ldots, N-1. \end{cases} \qquad (15.8)$$

If the 1-norm or ∞-norm is used in the cost function of problems (15.6), (15.7) and (15.8), then we set

$$p(x_N) = \|Px_N\|_p, \ q(x_k, u_k) = \|Qx_k\|_p + \|Ru_k\|_p \qquad (15.9)$$

with $p = 1$ or $p = \infty$. If the squared Euclidian norm is used in the cost function of problems (15.6), (15.7) and (15.8), then we set

$$p(x_N) = x_N'Px_N, \ q(x_k, u_k) = x_k'Qx_k + u_k'Ru_k. \qquad (15.10)$$

Note that in (15.6), (15.7) and (15.8) x_k denotes the state vector at time k obtained by starting from the state $x_0 = x(0)$ and applying to the system model

$$x_{k+1} = A(w_k^p)x_k + B(w_k^p)u_k + Ew_k^a$$

the input sequence $u_0, \ldots, u_{k-1}$ and the disturbance sequences $\mathbf{w}^a = \{w_0^a, \ldots, w_{N-1}^a\}$, $\mathbf{w}^p = \{w_0^p, \ldots, w_{N-1}^p\}$.

Consider the optimal control problem

$$J_0^*(x(0)) = \min_{U_0} \ J_0(x(0), U_0)$$

$$\text{subj. to} \begin{cases} x_{k+1} = A(w_k^p)x_k + B(w_k^p)u_k + Ew_k^a \\ x_k \in \mathcal{X}, \ u_k \in \mathcal{U} \\ k = 0, \ldots, N-1 \\ x_N \in \mathcal{X}_f \\ x_0 = x(0) \end{cases} \left. \begin{array}{l} \forall w_k^a \in \mathcal{W}^a, \forall w_k^p \in \mathcal{W}^p \\ \forall k = 0, \ldots, N-1. \end{array} \right.$$

$$(15.11)$$

where $\mathcal{X}_f \subseteq \mathbb{R}^n$ is a terminal polyhedral region. In (15.11) N is the time horizon and $U_0 = [u_0', \ldots, u_{N-1}']' \in \mathbb{R}^s$, $s = mN$ the vector of the input sequence. We denote by $U_0^* = \{u_0^*, \ldots, u_{N-1}^*\}$ the optimal solution to (15.11).

We denote with $\mathcal{X}_i^{\mathrm{OL}} \subseteq \mathcal{X}$ the set of states x_i for which the robust optimal control problem (15.8)–(15.11) is feasible, i.e.,

$$\begin{aligned}
\mathcal{X}_i^{\mathrm{OL}} = \{x_i \in \mathcal{X} : \ &\exists (u_i, \ldots, u_{N-1}) \text{ such that } x_k \in \mathcal{X}, \ u_k \in \mathcal{U}, \ k = i, \ldots, N-1, \\
&x_N \in \mathcal{X}_f \ \forall \ w_k^a \in \mathcal{W}^a, \ \forall w_k^p \in \mathcal{W}^p \ k = i, \ldots, N-1, \\
&\text{where } x_{k+1} = A(w_k^p)x_k + B(w_k^p)u_k + Ew_k^a\}.
\end{aligned}$$

$$(15.12)$$

Problem (15.11) minimizes the performance index $\mathcal{J}_{\mathcal{W}}$ subject to the constraint that the input sequence must be feasible *for all* possible disturbance realizations. In other words, the nominal, expected or worst-case performance is minimized with the requirement of constraint fulfillment for all possible realizations of $\mathbf{w}^a$, $\mathbf{w}^p$.

Remark 15.1 Note that we distinguish between the *current* state $x(k)$ of system (15.1) at time k and the variable x_k in the optimization problem (15.11), that is the *predicted* state of system (15.1) at time k obtained by starting from the state $x_0 = x(0)$ and applying to system $x_{k+1} = A(w_k^p)x_k + B(w_k^p)u_k + Ew_k^a$ the input sequence $u_0, \ldots, u_{k-1}$ and the disturbance sequences $w_0^p, \ldots, w_{k-1}^p, \ w_0^a, \ldots, w_{k-1}^a$. Analogously, $u(k)$ is the input applied to system (15.1) at time k while u_k is the k-th optimization variable of the optimization problem (15.11).

The formulation (15.11) is based on an *open-loop* prediction and thus referred to as Constrained Robust Optimal Control with open-loop predictions (CROC-OL). The optimal control problem (15.11) can be viewed as a game played between two players: the controller U and the disturbance W [24, p. 266–272]. Regardless of the cost function in (15.11) the player U tries to counteract *any* feasible disturbance realization with just *one* single sequence $\{u_0, \ldots, u_{N-1}\}$. This prediction model does not consider that at the next time step, the player can measure the state $x(1)$ and "adjust" his input $u(1)$ based on the current measured state. By not considering this fact, the effect of the uncertainty may grow over the prediction horizon and may easily lead to infeasibility of the min problem (15.11). Alternatively, in the closed-loop prediction scheme presented next, the optimization scheme *takes into account* that the disturbance and the controller play one move at a time.

The game is played differently depending on the cost function $\mathcal{J}_{\mathcal{W}}$:

- In problem (15.6), (15.11), among all control actions which robustly satisfy the constraints, the player U chooses the one which minimizes a cost for a "guessed" sequence played by W.

- In problem (15.7), (15.11), among all control actions which robustly satisfy the constraints, the player U chooses the one which minimizes an expected cost over all the sequences, which can be played by W.

- The optimal control problem (15.8) and (15.11) is somehow more involved and can be viewed as the solution of a zero-sum dynamic game. The player U plays first. Given the initial state $x(0)$, U chooses his action over the whole horizon $\{u_0, \ldots, u_{N-1}\}$, reveals his plan to the opponent W, who decides on his actions next $\{w_0^a, w_0^p, \ldots, w_{N-1}^a, w_{N-1}^p\}$ by solving (15.8). By solving (15.8), (15.11), the player U selects the action corresponding to the smallest worst-case cost.

Closed-Loop Predictions

The constrained robust optimal control problem based on closed-loop predictions (CROC-CL) is defined as follows:

$$J_0^*(x(0)) = \min_{\pi_0(\cdot), \ldots, \pi_{N-1}(\cdot)} J_0(x(0), U_0)$$

$$\text{subj. to} \quad \left.\begin{cases} x_{k+1} = A(w_k^p)x_k + B(w_k^p)u_k + Ew_k^a \\ x_k \in \mathcal{X}, \ u_k \in \mathcal{U} \\ u_k - \pi_k(x_k) \\ k = 0, \ldots, N-1 \\ x_N \in \mathcal{X}_f \\ x_0 = x(0) \end{cases}\right\}$$

$$\begin{array}{c} \forall w_k^a \in \mathcal{W}^a, \ \forall w_k^p \in \mathcal{W}^p \\ \forall k = 0, \ldots, N-1. \end{array} \quad (15.13)$$

In (15.13) we allow the same cost functions as in (15.6), (15.7) and (15.8).

In problem (15.13) we look for a set of time-varying feedback policies $\pi_0(\cdot), \ldots, \pi_{N-1}(\cdot)$ which minimizes the performance index $\mathcal{J}_{\mathcal{W}}$ and generates input sequences $\pi_0(x_0), \ldots, \pi_{N-1}(x_{N-1})$ satisfying state and input constraints *for all* possible disturbance realizations.

We can formulate the search for the optimal policies $\pi_0(\cdot), \ldots, \pi_{N-1}(\cdot)$ as a dynamic program.

$$J_j^*(x_j) = \min_{u_j} J_j(x_j, u_j)$$

$$\text{subj. to} \quad \begin{cases} x_j \in \mathcal{X}, \ u_j \in \mathcal{U} \\ A(w_j^p)x_j + B(w_j^p)u_j + Ew_j^a \in \mathcal{X}_{j+1} \end{cases} \ \forall w_j^a \in \mathcal{W}^a, w_j^p \in \mathcal{W}^p$$

$$(15.14)$$

$$J_j(x_j, u_j) = \mathcal{J}_{\mathcal{W}}\left[q(x_j, u_j) + J_{j+1}^*(A(w_j^p)x_j + B(w_j^p)u_j + Ew_j^a)\right], \quad (15.15)$$

for $j = 0, \ldots, N-1$ and with boundary conditions

$$J_N^*(x_N) = p(x_N) \quad (15.16a)$$

$$\mathcal{X}_N = \mathcal{X}_f, \quad (15.16b)$$

where $\mathcal{X}_j$ denotes the set of states x for which (15.14)–(15.16) is feasible

$$\mathcal{X}_j = \{x \in \mathcal{X} : \exists u \in \mathcal{U} \text{ s.t. } A(w^p)x + B(w^p)u + Ew^a \in \mathcal{X}_{j+1} \ \forall w^a \in \mathcal{W}^a, w^p \in \mathcal{W}^p\}, \quad (15.17)$$

and the cost operator $\mathcal{J}_{\mathcal{W}}$ belongs to one of the following classes:

- **Nominal Cost**

$$\mathcal{J}_{\mathcal{W}} = \left[q(x_j, u_j) + J_{j+1}^*(A(\bar{w}_j^p)x_j + B(\bar{w}_j^p)u_j + E\bar{w}_j^a)\right]. \quad (15.18)$$

- **Expected Cost**

$$\mathcal{J}_{\mathcal{W}} = \mathbf{E}_{w_j^a \in \mathcal{W}^a, w_j^p \in \mathcal{W}^p} \left[q(x_j, u_j) + J_{j+1}^*(A(w_j^p)x_j + B(w_j^p)u_j + Ew_j^a) \right].$$
(15.19)

- **Worst Case Cost**

$$\mathcal{J}_{\mathcal{W}} = \max_{w_j^a \in \mathcal{W}^a, w_j^p \in \mathcal{W}^p} \left[q(x_j, u_j) + J_{j+1}^*(A(w_j^p)x_j + B(w_j^p)u_j + Ew_j^a) \right].$$
(15.20)

If the 1-norm or ∞-norm is used in the cost function of problems (15.18), (15.19) and (15.20), then we set

$$p(x_N) = \|Px_N\|_p, \ q(x_k, u_k) = \|Qx_k\|_p + \|Ru_k\|_p$$
(15.21)

with $p = 1$ or $p = \infty$. If the squared Euclidian norm is used then we set

$$p(x_N) = x_N' Px_N, \ q(x_k, u_k) = x_k' Qx_k + u_k' Ru_k.$$
(15.22)

The reason for including constraints (15.14) in the minimization problem and not in the inner problem (15.15) is that in (15.15) w_j^a and w_j^p are free to act regardless of the state constraints. On the other hand in (15.14), the input u_j has the duty to keep the state within the constraints (15.14) for all possible disturbance realization. By solving problem (15.14) at step j of the dynamic program, we obtain the function $u_j(x_j)$, i.e., the policy $\pi_j(\cdot)$.

Again, the optimal control problem (15.14)–(15.15) can be viewed as a game between two players: the controller U and the disturbance W. This time the player U predicts that the game is going to be played as follows. At the generic time j player U observes x_j and responds with $u_j = \pi_j(x_j)$. Player W observes (x_j, u_j) and responds with w_j^a and w_j^p. Therefore, regardless of the cost function in (15.15) the player U counteracts *any* feasible disturbance realization w_j with a feedback controller $\pi_j(x_j)$. This prediction model takes into account that the disturbance and the controller play one move at a time.

The game is played differently depending on the cost function $\mathcal{J}_{\mathcal{W}}$.

- In problem (15.14) and (15.18), among all the feedback policies $u_j = \pi_j(x_j)$ which robustly satisfy the constraints, the player U chooses the one which minimizes a cost for a "guessed" action $\bar{w}_j$ played by W.

- In problem (15.14) and (15.19), among all the feedback policies $u_j = \pi_j(x_j)$ which robustly satisfy the constraints, the player U chooses the one which minimizes the expected cost at time j.

- The optimal control problem (15.14)–(15.20) is somehow more involved and can be viewed as the solution of a zero-sum dynamic game. The player U plays first. At the generic time j player U observes x_j and responds with $u_j = \pi_j(x_j)$. Player W observes (x_j, u_j) and responds with the w_j^a and w_j^p which lead to the worst case cost. The player W will always play the worst case action only if it has knowledge of both x_j and $u_j = \pi_j(x_j)$. In fact, w_j^a and w_j^p in (15.20) are a function of x_j and u_j. If U does *not* reveal his action to player W, then we can only claim that the player W *might* play the worst case action. Problem (15.14)–(15.20) is meaningful in both cases. Robust constraint satisfaction and worst case minimization will always be guaranteed.

Example 15.1   Consider the system

$$x_{k+1} = x_k + u_k + w_k \tag{15.23}$$

where x, u and w are state, input and disturbance, respectively. Let $u_k \in \{-1, 0, 1\}$ and $w_k \in \{-1, 0, 1\}$ be the feasible input and disturbance. Here $\{-1, 0, 1\}$ denotes the set with three elements: -1, 0 and 1. Let $x(0) = 0$ be the initial state. The objective for player U is to play two moves in order to keep the state x_2 at time 2 in the set $[-1, 1]$. If U is able to do so for any possible disturbance, then he will win the game.

The open-loop formulation (15.11) is infeasible. In fact, in open-loop U can choose from nine possible sequences: (0,0), (1,1), (−1,−1), (−1,1) (1,−1), (−1,0), (1,0), (0,1) and (0,−1). For any of those sequences there will always exist a disturbance sequence w_0, w_1 which will bring x_2 outside the feasible set [−1,1].

The closed-loop formulation (15.14)–(15.20) is feasible and has a simple feasible solution: $u_k = -x_k$. In this case system (15.23) becomes $x_{k+1} = w_k$ and $x_2 = w_1$ lies in the feasible set $[-1, 1]$ for all admissible disturbances w_1. The optimal feedback policy might be different from $u_k = -x_k$ and depends on the choice of the objective function.

15.1.1   State Feedback Solutions Summary

In the following sections we will describe how to compute the solution to CROC-OL and CROC-CL problems. In particular we will show that the solution to CROC-OL and CROC-CL problem can be expressed in feedback form where $u^*(k)$ is a continuous piecewise affine function on polyhedra of the state $x(k)$, i.e., $u^*(k) = f_k(x(k))$ where

$$f_k(x) = F_k^i x + g_k^i \quad \text{if} \quad H_k^i x \le K_k^i, \ i = 1, \ldots, N_k^r. \tag{15.24}$$

H_k^i and K_k^i in equation (15.24) are the matrices describing the i-th polyhedron $CR_k^i = \{x \in \mathbb{R}^n : H_k^i x \le K_k^i\}$ inside which the feedback optimal control law $u^*(k)$ at time k has the affine form $F_k^i x + g_k^i$.

The optimal solution is continuous and has the form (15.24) in the following cases.

- Nominal or worst-case performance index based on 1- and ∞-norm,

  - CROC-OL with no uncertainty in the dynamics matrix A (i.e., $A(\cdot) = A$)
  - CROC-CL

- Nominal performance index based on 2-norm,

  - CROC-OL with no uncertainty in the dynamics matrix A (i.e., $A(\cdot) = A$)

If CROC-OL problems are considered, the set of polyhedra CR_k^i, $i = 1, \ldots, N_k^r$ is a *polyhedral partition* of the set of feasible states $\mathcal{X}_k^{\mathrm{OL}}$ (15.12) at time k.

If CROC-CL problems are considered, the set of polyhedra CR_k^i, $i = 1, \ldots, N_k^r$ is a *polyhedral partition* of the set of feasible states $\mathcal{X}_k$ of problem (15.17) at time k.

The difference between the feasible sets $\mathcal{X}_k^{\mathrm{OL}}$ and $\mathcal{X}_k$ associated with *open-loop* prediction and *closed-loop* prediction, respectively, are discussed in the next section.

Remark 15.2   CROC-OL and CROC-CL problems with expected cost performance index are not treated in this book. In general, the calculation of expected costs over polyhedral domains requires approximations via sampling or bounding even for disturbances characterized by simple probability distribution functions (e.g., uniform or Gaussian). Once the cost has been approximated, one can easily modify the approaches described in this chapter to solve CROC-OL and CROC-CL problems with expected cost performance index.

15.2 Feasible Solutions

As in the nominal case (Section 11.2), there are two ways to define and compute the robust feasible sets: the *batch approach* and the *recursive approach*. While for systems without disturbances, both approaches yield the same result, in the robust case the *batch approach* provides the feasible set $\mathcal{X}_i^{\mathrm{OL}}$ of CROC with open-loop predictions and the *recursive approach* provides the feasible set $\mathcal{X}_i$ of CROC with closed-loop predictions. From the discussion in the previous sections, clearly we have $\mathcal{X}_i^{\mathrm{OL}} \subseteq \mathcal{X}_i$. We will detail the *batch approach* and the *recursive approach* next and at the end of the section we will show how to modify the batch approach in order to compute $\mathcal{X}_i$.

Batch Approach: Open-Loop Prediction

Consider the set $\mathcal{X}_i^{\mathrm{OL}}$ (15.12) of feasible states x_i at time i for which (15.11) is feasible, for $i = 0, \ldots, N$, rewritten below

$$\mathcal{X}_i^{\mathrm{OL}} = \{x_i \in \mathcal{X} : \exists (u_i, \ldots, u_{N-1}) \text{ such that } x_k \in \mathcal{X},\ u_k \in \mathcal{U},\ k = i, \ldots, N-1,$$
$$x_N \in \mathcal{X}_f \ \forall\ w_k^a \in \mathcal{W}^a,\ \forall\ w_k^p \in \mathcal{W}^p \ k = i, \ldots, N-1,$$
$$\text{where } x_{k+1} = A(w_k^p)x_k + B(w_k^p)u_k + Ew_k^a\}.$$

Thus for any initial state $x_i \in \mathcal{X}_i^{\mathrm{OL}}$ there exists a feasible sequence of inputs $U_i = [u_i', \ldots, u_{N-1}']$ which keeps the state evolution in the feasible set $\mathcal{X}$ at future time instants $k = i+1, \ldots, N-1$ and forces x_N into $\mathcal{X}_f$ at time N for all feasible disturbance sequences $w_k^a \in \mathcal{W}^a$, $w_k^p \in \mathcal{W}^p$, $k = i, \ldots, N-1$. Clearly $\mathcal{X}_N^{\mathrm{OL}} = \mathcal{X}_f$.

Next we show how to compute $\mathcal{X}_i^{\mathrm{OL}}$ for $i = 0, \ldots, N-1$. Let the state and input constraint sets $\mathcal{X}$, $\mathcal{X}_f$ and $\mathcal{U}$ be the $\mathcal{H}$-polyhedra $A_x x \leq b_x$, $A_f x \leq b_f$, $A_u u \leq b_u$, respectively. Assume that the disturbance sets are represented in terms of their vertices: $\mathcal{W}^a = \text{conv}\{w^{a,1}, \ldots, w^{a,n_{\mathcal{W}^a}}\}$ and $\mathcal{W}^p = \text{conv}\{w^{p,1}, \ldots, w^{p,n_{\mathcal{W}^p}}\}$. Define $U_i = [u'_i, \ldots, u'_{N-1}]$ and the polyhedron $\mathcal{P}_i$ of robustly feasible states and input sequences at time i,

$$\mathcal{P}_i = \{(U_i, x_i) \in \mathbb{R}^{m(N-i)+n} : G_i U_i - E_i x_i \le W_i\}. \tag{15.25}$$

In (15.25) G_i, E_i and W_i are obtained by collecting all the following inequalities:

- Input Constraints
$$A_u u_k \le b_u, \quad k = i, \ldots, N - 1.$$

- State Constraints
$$A_x x_k \le b_x, \quad k = i, \ldots, N-1 \text{ for all } w^a_l \in \mathcal{W}^a, \ w^p_l \in \mathcal{W}^p, \ l = i, \ldots, k-1. \tag{15.26}$$

- Terminal State Constraints
$$A_f x_N \le b_f, \quad \text{for all } w^a_l \in \mathcal{W}^a, \ w^p_l \in \mathcal{W}^p, \ l = i, \ldots, N-1. \tag{15.27}$$

Constraints (15.26)–(15.27) are enforced for all feasible disturbance sequences. In order to do so, we rewrite constraints (15.26)–(15.27) at time k as a function of x_i and the input sequence U_i

$$A_x \left(\Pi_{j=i}^{k-1} A(w^p_j) x_i + \sum_{l=i}^{k-1} \Pi_{j=i}^{l-1} A(w^p_j)(B(w^p_{k-1-l}) u_{k-1-l} + E w^a_{k-1-l}) \right) \le b_x. \tag{15.28}$$

In general, the constraint (15.28) is nonconvex in the disturbance sequences $w^a_i, \ldots, w^a_{k-1}, w^p_i, \ldots, w^p_{k-1}$ because of the product $\Pi_{j=i}^{k-1} A(w^p_j)$.

Assume that there is no uncertainty in the dynamics matrix A, i.e., $A(\cdot) = A$. In this case, since $B(\cdot)$ is an affine function of w^p (cf. Equation (15.4)) and since the composition of a convex constraint with an affine map generates a convex constraint (Section 1.2), we can use Lemma 10.1 to rewrite constraints (15.26) at time k as

$$A_x \left(A^{k-i-1} x_i + \sum_{l=i}^{k-1} A^{l-i-1}(B(w^p_{k-1-l}) u_{k-1-l} + E w^a_{k-1-l}) \right) \le b_x, \tag{15.29}$$

for all $w^a_j \in \{w^{a,i}\}_{i=1}^{n_{\mathcal{W}^a}}, \ w^p_j \in \{w^{p,i}\}_{i=1}^{n_{\mathcal{W}^p}}, \ \forall j = i, \ldots, k-1,$
$$k = i, \ldots, N-1,$$

and imposing the constraints at all the vertices of the sets $\underbrace{\mathcal{W}^a \times \mathcal{W}^a \times \ldots \times \mathcal{W}^a}_{i,\ldots,N-1}$

and $\underbrace{\mathcal{W}^p \times \mathcal{W}^p \times \ldots \times \mathcal{W}^p}_{i,\ldots,N-1}$. Note that the constraints (15.29) are now linear in x_i

and U_i. The same procedure can be repeated for constraint (15.27).

Remark 15.3 When only additive disturbances are present (i.e., $n_p = 0$), vertex enumeration is not required and a set of linear programs can be used to transform the constraints (15.26) into a smaller number of constraints than (15.29) as explained in Lemma 10.2.

Once the matrices G_i, E_i and W_i have been computed, the set $\mathcal{X}_i^{\mathrm{OL}}$ is a polyhedron and can be computed by projecting the polyhedron $\mathcal{P}_i$ in (15.25) on the x_i space.

Recursive Approach: Closed-Loop Prediction

In the *recursive approach* we have

$$\mathcal{X}_i = \{x \in \mathcal{X} : \exists u \in \mathcal{U} \text{ such that } A(w_i^p)x + B(w_i^p)u + Ew_i^a \in \mathcal{X}_{i+1},$$
$$\forall w_i^a \in \mathcal{W}^a, \ \forall w_i^p \in \mathcal{W}^p\}, \ i = 0, \dots, N-1$$
$$\mathcal{X}_N = \mathcal{X}_f. \tag{15.30}$$

The definition of $\mathcal{X}_i$ in (15.30) is recursive and it requires that for any feasible initial state $x_i \in \mathcal{X}_i$ there exists a feasible input u_i which keeps the next state $A(w_i^p)x + B(w_i^p)u + Ew_i^a$ in the feasible set $\mathcal{X}_{i+1}$ for all feasible disturbances $w_i^a \in \mathcal{W}^a$, $w_i^p \in \mathcal{W}^p$.

Initializing $\mathcal{X}_N$ to $\mathcal{X}_f$ and solving (15.30) backward in time yields the feasible set $\mathcal{X}_0$ for the CROC-CL (15.14)–(15.16) which, as shown in Example 15.1, is different from $\mathcal{X}_0^{\mathrm{OL}}$.

Let $\mathcal{X}_i$ be the $\mathcal{H}$-polyhedron $A_{\mathcal{X}_i}x \leq b_{\mathcal{X}_i}$. Then the set $\mathcal{X}_{i-1}$ is the projection of the following polyhedron

$$\begin{bmatrix} A_u \\ 0 \\ A_{\mathcal{X}_i} B(w_i^p) \end{bmatrix} u_{i-1} + \begin{bmatrix} 0 \\ A_x \\ A_{\mathcal{X}_i} A(w_i^p) \end{bmatrix} x_{i-1} \leq \begin{bmatrix} b_u \\ b_x \\ b_{\mathcal{X}_i} - Ew_i^a \end{bmatrix}$$
$$\text{for all} \ \ w_i^a \in \{w^{a,i}\}_{i=1}^{n_{\mathcal{W}^a}},$$
$$\text{for all} \ \ w_i^p \in \{w^{p,i}\}_{i=1}^{n_{\mathcal{W}^p}} \tag{15.31}$$

on the x_{i-1} space. (Note that we have used Lemma 10.1 and imposed state constraints at all the vertices of the sets $\mathcal{W}^a \times \mathcal{W}^p$.)

Remark 15.4   When only additive disturbances are present (i.e., $n_p = 0$), vertex enumeration is not required and a set of linear programs can be used to transform the constraints in (15.30) into a smaller number of constraints than (15.31) as explained in Lemma 10.2.

The backward evolution in time of the feasible sets $\mathcal{X}_i$ enjoys the properties described by Theorems 11.1 and 11.2 for the nominal case. In particular, if a robust controlled invariant set is chosen as terminal constraint $\mathcal{X}_f$, then set $\mathcal{X}_i$ grows as i becomes smaller and stops growing when it becomes the maximal robust stabilizable set.

Theorem 15.1   *Let the terminal constraint set $\mathcal{X}_f$ be a robust control invariant subset of $\mathcal{X}$. Then,*

1. *The feasible set $\mathcal{X}_i$, $i = 0, \ldots, N - 1$ is equal to the $(N - i)$-step robust stabilizable set:*

$$\mathcal{X}_i = \mathcal{K}_{N-i}(\mathcal{X}_f, \mathcal{W}).$$

2. *The feasible set $\mathcal{X}_i$, $i = 0, \ldots, N - 1$ is robust control invariant and contained within the maximal robust control invariant set:*

$$\mathcal{X}_i \subseteq \mathcal{C}_\infty.$$

3. $\mathcal{X}_i \supseteq \mathcal{X}_j$ *if $i < j$, $i = 0, \ldots, N - 1$. The size of the feasible $\mathcal{X}_i$ set stops increasing (with decreasing i) if and only if the maximal robust stabilizable set is finitely determined and $N - i$ is larger than its determinedness index $\bar{N}$, i.e.,*

$$\mathcal{X}_i \supset \mathcal{X}_j \ \text{if} \ N - \bar{N} < i < j < N.$$

Furthermore,

$$\mathcal{X}_i = \mathcal{K}_\infty(\mathcal{X}_f, \mathcal{W}) \ \text{if} \ i \leq N - \bar{N}.$$

Batch Approach: Closed-Loop Prediction

The batch approach can be modified in order to obtain $\mathcal{X}_i$ instead of $\mathcal{X}_i^{\text{OL}}$. The main idea is to augment the number of inputs by allowing one input sequence for each vertex l of the set containing all disturbance sequences over the horizon [0,N-1]. We will explain the approach under the assumption that there is no parametric uncertainty ($w = w^a$). From causality, the input u_j is a function of x_0 and the sequence $w_0, \ldots, w_{j-1}$. Therefore, u_0 will be unique, u_1 a function of the vertices of the disturbance set $\mathcal{W}^a$, u_2 a function of the vertices of the disturbance set $\mathcal{W}^a \times \mathcal{W}^a$, and u_{N-1} a function of $\underbrace{\mathcal{W}^a \times \mathcal{W}^a \times \ldots \times \mathcal{W}^a}_{N-1}$. The generic l-th input sequence can be written as

$$U_l = [u_0, \tilde{u}_1^{j_1}, \tilde{u}_2^{j_2}, \ldots \tilde{u}_{N-1}^{j_{N-1}}]$$

where the lower index denotes time, and the upper index is used to associate a different input for every disturbance realization. For instance, at time 0 the extreme realizations of additive uncertainty are $n_{\mathcal{W}^a}$. At time 1 there is one input for each realization and thus $j_1 \in \{1, \ldots, n_{\mathcal{W}^a}\}$. At time 2 the number of extreme realizations of the disturbance sequence w_0, w_1 is $n_{\mathcal{W}^a}^2$ and therefore $j_2 \in \{1, \ldots, n_{\mathcal{W}^a}^2\}$. By repeating this argument we have $j_k \in \{1, \ldots, n_{\mathcal{W}^a}^k\}$, $k = 1, \ldots, N - 1$. In total we will have $n_{\mathcal{W}^a}^{N-1}$ different control sequences U_l and $m\left(1 + n_{\mathcal{W}^a} + n_{\mathcal{W}^a}^2 + \cdots + n_{\mathcal{W}^a}^{N-1}\right)$ control variables to optimize over.

Figure 15.1 depicts an example where the disturbance set has two vertices w^1 and w^2 ($n_{\mathcal{W}^a} = 2$) and the horizon is $N = 4$. The continuous line represents the occurrence of w^1, while the dashed line represents the occurrence of w^2. Over the horizon there are eight possible scenarios associated to eight extreme disturbance realizations. The state $\tilde{x}_3^j$ denotes the predicted state at time 3 associated to the j-th disturbance scenario.

Since the approach is computationally demanding, we prefer to present the main idea through a very simple example rather than including all the tedious details.

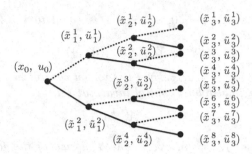

Figure 15.1 Batch Approach with Closed-Loop Prediction. State propagation and associated input variables. The disturbance set has two vertices w^1 and w^2 ($n_{\mathcal{W}^a} = 2$) and the horizon is $N = 4$. The continuous line represents the occurrence of w^1, while the dashed line represents the occurrence of w^2. Over the horizon there are eight possible scenarios associated to eight extreme disturbance realizations. The state $\tilde{x}_3^j$ denotes the predicted state at time 3 associated to the j-th disturbance scenario.

Example 15.2   Consider the system

$$x_{k+1} = x_k + u_k + w_k \tag{15.32}$$

where x, u and w are state, input and disturbance, respectively. Let $u_k \in [-1,1]$ and $w_k \in [-1,1]$ be the feasible input and disturbance. The objective for player U is to play two moves in order to keep the state at time three x_3 in the set $\mathcal{X}_f = [-1,1]$.

- **Batch approach**
  We rewrite the terminal constraint as

$$\begin{aligned} x_3 = x_0 + u_0 + u_1 + u_2 + w_0 + w_1 + w_2 \in [-1,1] \\ \text{for all } \ w_0 \in [-1,1], \ w_1 \in [-1,1], \ w_2 \in [-1,1] \end{aligned} \tag{15.33}$$

which by Lemma 10.1 becomes

$$\begin{aligned} -1 \le x_0 + u_0 + u_1 + u_2 + 3 \le 1 \\ -1 \le x_0 + u_0 + u_1 + u_2 + 1 \le 1 \\ -1 \le x_0 + u_0 + u_1 + u_2 - 1 \le 1 \\ -1 \le x_0 + u_0 + u_1 + u_2 - 3 \le 1 \end{aligned} \tag{15.34}$$

which by removing redundant constraints becomes the constraint

$$2 \le x_0 + u_0 + u_1 + u_2 \le -2 \tag{15.35}$$

which is infeasible and thus $\mathcal{X}_0 = \emptyset$. For the sake of clarity we continue with the procedure as if we did not notice that constraint (15.35) was infeasible.
The set $\mathcal{X}_0^{\text{OL}}$ is the projection on the x_0 space of the polyhedron $\mathcal{P}_0$ of combined feasible input and state constraints

$$\mathcal{P}_0 = \left\{ (u_0, u_1, u_2, x_0) \in \mathbb{R}^4 : \begin{bmatrix} 1 & 0 & 0 \\ -1 & 0 & 0 \\ 0 & 1 & 0 \\ 0 & -1 & 0 \\ 0 & 0 & 1 \\ 0 & 0 & -1 \\ 1 & 1 & 1 \\ -1 & -1 & -1 \end{bmatrix} \begin{bmatrix} u_0 \\ u_1 \\ u_2 \end{bmatrix} + \begin{bmatrix} 0 \\ 0 \\ 0 \\ 0 \\ 0 \\ 0 \\ 1 \\ -1 \end{bmatrix} x_0 \le \begin{bmatrix} 1 \\ 1 \\ 1 \\ 1 \\ 1 \\ 1 \\ -2 \\ -2 \end{bmatrix} \right\}.$$

The projection will provide an empty set since the last two constraints correspond to the infeasible terminal state constraint (15.35).

- **Recursive approach**
  For the recursive approach we have $\mathcal{X}_3 = \mathcal{X}_f = [-1, 1]$. We rewrite the terminal constraint as

$$x_3 = x_2 + u_2 + w_2 \in [-1, 1] \quad \text{for all } w_2 \in [-1, 1] \tag{15.36}$$

which by Lemma 10.1 becomes

$$-1 \le x_2 + u_2 + 1 \le 1$$
$$-1 \le x_2 + u_2 - 1 \le 1 \tag{15.37}$$

which by removing redundant constraints becomes

$$0 \le x_2 + u_2 \le 0.$$

The set $\mathcal{X}_2$ is the projection on the x_2 space of the polyhedron

$$\begin{bmatrix} 1 & 0 \\ -1 & 0 \\ 1 & 1 \\ -1 & -1 \end{bmatrix} \begin{bmatrix} u_2 \\ x_2 \end{bmatrix} \le \begin{bmatrix} 1 \\ 1 \\ 0 \\ 0 \end{bmatrix} \tag{15.38}$$

which yields $\mathcal{X}_2 - [-1, 1]$. Since $\mathcal{X}_2 = \mathcal{X}_3$ one can conclude that $\mathcal{X}_2$ is the maximal controllable robust invariant set and $\mathcal{X}_0 = \mathcal{X}_1 = \mathcal{X}_2 = [-1, 1]$.

- **Batch approach with closed-loop predictions***
  The horizon N is three and therefore we have $2^{(N-1)} = 4$ different control sequences: $\{u_0, \tilde{u}_1^1, \tilde{u}_2^1\}$, $\{u_0, \tilde{u}_1^1, \tilde{u}_2^2\}$, $\{u_0, \tilde{u}_1^2, \tilde{u}_2^3\}$, and $\{u_0, \tilde{u}_1^2, \tilde{u}_2^4\}$, corresponding to the extreme disturbance realizations $\{w_0 = -1, w_1 = -1, w_2\}$, $\{w_0 = -1, w_1 = 1, w_2\}$, $\{w_0 = 1, w_1 = -1, w_2\}$ and $\{w_0 = 1, w_1 = 1, w_2\}$, respectively.

The terminal constraint is thus rewritten as

$$-1 \le x_0 + u_0 + \tilde{u}_1^1 + \tilde{u}_2^1 - 1 - 1 + w_2 \le 1, \quad \forall w_2 \in [-1, 1]$$
$$-1 \le x_0 + u_0 + \tilde{u}_1^1 + \tilde{u}_2^2 - 1 + 1 + w_2 \le 1, \quad \forall w_2 \in [-1, 1]$$
$$-1 \le x_0 + u_0 + \tilde{u}_1^2 + \tilde{u}_2^3 + 1 - 1 + w_2 \le 1, \quad \forall w_2 \in [-1, 1]$$
$$-1 \le x_0 + u_0 + \tilde{u}_1^2 + \tilde{u}_2^4 + 1 + 1 + w_2 \le 1, \quad \forall w_2 \in [-1, 1]$$
$$-1 \le u_0 \le 1$$
$$-1 \le \tilde{u}_1^1 \le 1$$
$$-1 \le \tilde{u}_1^2 \le 1 \tag{15.39}$$
$$-1 \le \tilde{u}_2^1 \le 1$$
$$-1 \le \tilde{u}_2^2 \le 1$$
$$-1 \le \tilde{u}_2^3 \le 1$$
$$-1 \le \tilde{u}_2^4 \le 1$$

which becomes by using Lemma 10.1

$$-1 \leq x_0 + u_0 + \tilde{u}_1^1 + \tilde{u}_2^1 - 1 - 1 + 1 \leq 1$$
$$-1 \leq x_0 + u_0 + \tilde{u}_1^1 + \tilde{u}_2^1 - 1 - 1 - 1 \leq 1$$
$$-1 \leq x_0 + u_0 + \tilde{u}_1^1 + \tilde{u}_2^2 - 1 + 1 + 1 \leq 1$$
$$-1 \leq x_0 + u_0 + \tilde{u}_1^1 + \tilde{u}_2^2 - 1 + 1 - 1 \leq 1$$
$$-1 \leq x_0 + u_0 + \tilde{u}_1^2 + \tilde{u}_2^3 + 1 - 1 + 1 \leq 1$$
$$-1 \leq x_0 + u_0 + \tilde{u}_1^2 + \tilde{u}_2^3 + 1 - 1 - 1 \leq 1$$
$$-1 \leq x_0 + u_0 + \tilde{u}_1^2 + \tilde{u}_2^4 + 1 + 1 + 1 \leq 1$$
$$-1 \leq x_0 + u_0 + \tilde{u}_1^2 + \tilde{u}_2^4 + 1 + 1 - 1 \leq 1 \qquad (15.40)$$
$$-1 \leq u_0 \leq 1$$
$$-1 \leq \tilde{u}_1^1 \leq 1$$
$$-1 \leq \tilde{u}_1^2 \leq 1$$
$$-1 \leq \tilde{u}_2^1 \leq 1$$
$$-1 \leq \tilde{u}_2^2 \leq 1$$
$$-1 \leq \tilde{u}_2^3 \leq 1$$
$$-1 \leq \tilde{u}_2^4 \leq 1.$$

The set $\mathcal{X}_0$ can be obtained by projecting the polyhedron (15.40) in the $(x_0, u_0, \tilde{u}_1^1, \tilde{u}_1^2, \tilde{u}_2^1, \tilde{u}_2^2, \tilde{u}_2^3, \tilde{u}_2^4)$-space on the x_0 space. This yields $\mathcal{X}_0 = [-1, 1]$, the same as in the recursive approach.

15.3 State Feedback Solution, Nominal Cost

Theorem 15.2 *Consider the CROC-OL (15.11) with cost (15.5), (15.6) and (15.9) or (15.10). Assume that the parametric uncertainties are in the B matrix only $(A(w^p) = A)$. Then, there exists a solution $u^*(0) = f_0^{OL}(x(0))$, $f_0 : \mathbb{R}^n \to \mathbb{R}^m$, which is continuous and PPWA*

$$f_0^{OL}(x) = F_0^i x + g_0^i \quad \text{if} \quad x \in CR_0^i, \ i = 1, \dots, N_0^r \qquad (15.41)$$

where the polyhedral sets $CR_0^i = \{H_0^i x \leq k_0^i\}$, $i = 1, \dots, N_0^r$, are a partition of the feasible set $\mathcal{X}_0^{OL}$. Moreover f_0 can be found by solving an mp-LP for cost (15.9) and an mp-QP for cost (15.10). The same result holds for $u^*(k) = f_k^{OL}(x(0))$, $k = 1, \dots, N - 1$.

Proof: The proof can be easily derived from the results in Chapter 11. The only difference is that the constraints have to be first "robustified" as explained in Section 15.2. ∎

Theorem 15.3 *There exists a state feedback control law $u^*(k) = f_k(x(k))$, $f_k : \mathcal{X}_k \subseteq \mathbb{R}^n \to \mathcal{U} \subseteq \mathbb{R}^m$, solution of the CROC-CL (15.14)–(15.16) with cost (15.18), (15.21) and $k = 0, \dots, N - 1$ which is time-varying, continuous and piecewise affine on polyhedra*

$$f_k(x) = F_k^i x + g_k^i \quad \text{if} \quad x \in CR_k^i, \quad i = 1, \dots, N_k^r \qquad (15.42)$$

where the polyhedral sets $CR_k^i = \{x \in \mathbb{R}^n \colon H_k^i x \le K_k^i\}$, $i = 1, \ldots, N_k^r$ *are a partition of the feasible polyhedron* $\mathcal{X}_k$. *Moreover* f_i, $i = 0, \ldots, N-1$ *can be found by solving* N *mp-LPs.*

Proof: The proof can be easily derived from the results in Chapter 11. The only difference is that the constraints have to be first robustified as explained in Section 15.2. ∎

15.4   State Feedback Solution, Worst-Case Cost, 1-Norm and ∞-Norm Case

In this chapter we show how to find a *state feedback* solution to CROC problems when a worst case performance index is used. The following results will be used in the next sections.

Lemma 15.1   *Let* $f : \mathbb{R}^s \times \mathbb{R}^n \times \mathbb{R}^{n_w} \to \mathbb{R}$ *and* $g : \mathbb{R}^s \times \mathbb{R}^n \times \mathbb{R}^{n_w} \to \mathbb{R}^{n_g}$ *be functions of* (z, x, w) *convex in* w *for each* (z, x). *Assume that the variable* w *belongs to the polyhedron* $\mathcal{W}$ *with vertices* $\{\bar{w}_i\}_{i=1}^{N_\mathcal{W}}$. *Then the min-max multiparametric problem*

$$J^*(x) = \min_z \quad \max_{w \in \mathcal{W}} f(z, x, w) \atop \textit{subj. to} \quad g(z, x, w) \le 0 \ \forall w \in \mathcal{W} \tag{15.43}$$

is equivalent to the multiparametric optimization problem

$$J^*(x) = \min_{\mu, z} \quad \mu \atop \begin{aligned} \textit{subj. to} \quad & \mu \ge f(z, x, \bar{w}_i), \ i = 1, \ldots, N_\mathcal{W} \\ & g(z, x, \bar{w}_i) \le 0, \ i = 1, \ldots, N_\mathcal{W}. \end{aligned} \tag{15.44}$$

Proof: Easily follows from the fact that the maximum of a convex function over a convex set is attained at an extreme point of the set, cf. also [259]. ∎

Corollary 15.1   *If* f *is convex and piecewise affine in* (z, x), *i.e.,* $f(z, x, w) = \max_{i=1,\ldots,n_f} \{L_i(w) z + H_i(w) x + K_i(w)\}$ *and* g *is linear in* (z, x) *for all* $w \in \mathcal{W}$, $g(z, x, w) = L_g(w) z + H_g(w) x + K_g(w)$ *(with* $L_g(\cdot)$, $H_g(\cdot)$, $K_g(\cdot)$, $L_i(\cdot)$, $H_i(\cdot)$, $K_i(\cdot)$, $i = 1, \ldots, n_f$, *convex functions), then the min-max multiparametric problem (15.43) is equivalent to the mp-LP problem*

$$J^*(x) = \min_{\mu, z} \quad \mu$$
$$\begin{aligned} \textit{subj. to} \quad & \mu \ge L_j(\bar{w}_i) z + H_j(\bar{w}_i) x + K_j(\bar{w}_i), \ i = 1, \ldots, N_\mathcal{W}, \ j = 1, \ldots, n_f \\ & L_g(\bar{w}_i) z + H_g(\bar{w}_i) x \le -K_g(\bar{w}_i), \ i = 1, \ldots, N_\mathcal{W}. \end{aligned}$$
$$\tag{15.45}$$

Remark 15.5   As discussed in Lemma 10.2, in the case $g(z, x, w) = g_1(z, x) + g_2(w)$, the second constraint in (15.44) can be replaced by $g_1(z, x) \le -\bar{g}$, where $\bar{g} = [\bar{g}_1, \ldots, \bar{g}_{n_g}]'$ is a vector whose i-th component is

$$\bar{g}_i = \max_{w \in \mathcal{W}} g_2^i(w), \tag{15.46}$$

and $g_2^i(w)$ denotes the i-th component of $g_2(w)$. Similarly, if $f(z, x, w) = f_1(z, x) + f_2(w)$, the first constraint in (15.44) can be replaced by $\mu \geq f_1(z, x) + \bar{f}$, where

$$\bar{f}_i = \max_{w \in \mathcal{W}} f_2^i(w). \tag{15.47}$$

Clearly, this has the advantage of reducing the number of constraints in the multiparametric program from $N_{\mathcal{W}} n_g$ to n_g for the second constraint in (15.44) and from $N_{\mathcal{W}} n_f$ to n_f for the first constraint in (15.44).

In the following subsections we show how to solve CROC problems in state feedback form by using multiparametric linear programming.

15.4.1 Batch Approach: Open-Loop Predictions

Theorem 15.4 *Consider the CROC-OL (15.5), (15.8), (15.9), (15.11). Assume that the parametric uncertainties are in the B matrix only ($A(w^p) = A$). Then, there exists a solution $u^*(0) = f_0^{OL}(x(0))$, $f_0^{OL} : \mathbb{R}^n \to \mathbb{R}^m$, which is continuous and PPWA*

$$f_0^{OL}(x) = F_0^i x + g_0^i \quad \text{if} \quad x \in CR_0^i, \ i = 1, \dots, N_0^r \tag{15.48}$$

where the polyhedral sets $CR_0^i = \{H_0^i x \leq k_0^i\}$, $i = 1, \dots, N_0^r$, are a partition of the feasible set $\mathcal{X}_0^{OL}$. Moreover f_0 can be found by solving an mp-LP. The same result holds for $u^*(k) = f_k^{OL}(x(0))$, $k = 1, \dots, N-1$.

Proof: Since $x_k = A^k x_0 + \sum_{k=0}^{k-1} A^i [B(w_{k-1-i}^p) u_{k-1-i} + E w_{k-1-i}^a]$ is a linear function of the disturbances $\mathbf{w}^a = \{w_0^a, \dots, w_{N-1}^a\}$, and $\mathbf{w}^p = \{w_0^p, \dots, w_{N-1}^p\}$ for a fixed input sequence and x_0, the cost function in the maximization problem (15.8) is convex and piecewise affine with respect to the optimization vectors $\mathbf{w}^a$ and $\mathbf{w}^p$ and the parameters U_0, x_0. The constraints in (15.11) are linear in U_0 and x_0, for any $\mathbf{w}^a$ and $\mathbf{w}^p$. Therefore, by Corollary 15.1, problem (15.8)–(15.11) can be solved by specifying the constraints (15.11) only at the vertices of the set $\mathcal{W}^a \times \mathcal{W}^a \times \dots \times \mathcal{W}^a$ and $\mathcal{W}^p \times \mathcal{W}^p \times \dots \times \mathcal{W}^p$ and solving an mp-LP. The theorem follows from the mp-LP properties described in Theorem 6.5. ∎

Remark 15.6 Note that Theorem 15.4 does not hold when parametric uncertainties are present also in the A matrix. In this case the predicted state x_k is a nonlinear function of the vector $\mathbf{w}^p$.

Remark 15.7 In case of CROC-OL with additive disturbances only ($w^p(t) = 0$) the number of constraints in (15.11) can be reduced as explained in Remark 15.5.

15.4.2 Recursive Approach: Closed-Loop Predictions

Theorem 15.5 *There exists a state feedback control law $u^*(k) = f_k(x(k))$, $f_k \colon \mathcal{X}_k \subseteq \mathbb{R}^n \to \mathcal{U} \subseteq \mathbb{R}^m$, solution of the CROC-CL (15.14)–(15.17) with* *cost (15.20), (15.21) and $k = 0, \ldots, N-1$ which is time-varying, continuous and piecewise affine on polyhedra*

$$f_k(x) = F_k^i x + g_k^i \quad if \quad x \in CR_k^i, \quad i = 1, \ldots, N_k^r \tag{15.49}$$

where the polyhedral sets $CR_k^i = \{x \in \mathbb{R}^n \ : \ H_k^i x \leq K_k^i\}$, $i = 1, \ldots, N_k^r$ are a partition of the feasible polyhedron $\mathcal{X}_k$. Moreover f_i, $i = 0, \ldots, N-1$ can be found by solving N mp-LPs.

Proof: Consider the first step $j = N-1$ of dynamic programming applied to the CROC-CL problem (15.14)–(15.20) with cost (15.9)

$$J_{N-1}^*(x_{N-1}) = \min_{u_{N-1}} J_{N-1}(x_{N-1}, u_{N-1}) \tag{15.50}$$

$$\text{subj. to} \quad \begin{cases} x_{N-1} \in \mathcal{X}, \ u_{N-1} \in \mathcal{U} \\ A(w_{N-1}^p)x_{N-1} + B(w_{N-1}^p)u_{N-1} + Ew_{N-1}^a \in \mathcal{X}_f \\ \forall w_{N-1}^a \in \mathcal{W}^a, \ \forall w_{N-1}^p \in \mathcal{W}^p \end{cases}$$

$$\tag{15.51}$$

$$J_{N-1}(x_{N-1}, u_{N-1}) = \max_{w_{N-1}^a \in \mathcal{W}^a, \ w_{N-1}^p \in \mathcal{W}^p} \left\{ \begin{array}{l} \|Qx_{N-1}\|_p + \|Ru_{N-1}\|_p + \\ + \|P(A(w_{N-1}^p)x_{N-1} + \\ + B(w_{N-1}^p)u_{N-1} + Ew_{N-1}^a)\|_p \end{array} \right\}. \tag{15.52}$$

The cost function in the maximization problem (15.52) is piecewise affine and convex with respect to the optimization vector w_{N-1}^a, w_{N-1}^p and the parameters u_{N-1}, x_{N-1}. Moreover, the constraints in the minimization problem (15.51) are linear in (u_{N-1}, x_{N-1}) for all vectors w_{N-1}^a, w_{N-1}^p. Therefore, by Corollary 15.1, $J_{N-1}^*(x_{N-1})$, $u_{N-1}^*(x_{N-1})$ and $\mathcal{X}_{N-1}$ are computable via the mp-LP:

$$J_{N-1}^*(x_{N-1}) = \min_{\mu, u_{N-1}} \mu$$

subj. to $\mu \geq \|Qx_{N-1}\|_p + \|Ru_{N-1}\|_p +$
$$+ \|P(A(\bar{w}_h^p)x_{N-1} + B(\bar{w}_h^p)u_{N-1} + E\bar{w}_i^a)\|_p \tag{15.53a}$$

$$x_{N-1} \in \mathcal{X}, \ u_{N-1} \in \mathcal{U} \tag{15.53b}$$

$$A(\bar{w}_h^p)x_{N-1} + B(\bar{w}_h^p)u_{N-1} + E\bar{w}_i^a \in \mathcal{X}_N \tag{15.53c}$$

$$\forall i = 1, \ldots, n_{\mathcal{W}^a}, \ \forall h = 1, \ldots, n_{\mathcal{W}^p}.$$

where $\{\bar{w}_i^a\}_{i=1}^{n_{\mathcal{W}^a}}$ and $\{\bar{w}_h^p\}_{h=1}^{n_{\mathcal{W}^p}}$ are the vertices of the disturbance sets $\mathcal{W}^a$ and $\mathcal{W}^p$, respectively. By Theorem 6.5, J_{N-1}^* is a convex and piecewise affine function of x_{N-1}, the corresponding optimizer u_{N-1}^* is piecewise affine and continuous, and the feasible set $\mathcal{X}_{N-1}$ is a convex polyhedron. Therefore, the convexity and linearity arguments still hold for $j = N-2, \ldots, 0$ and the procedure can be iterated backwards in time j, proving the theorem. The theorem follows from the mp-LP properties described in Theorem 6.5. ∎

Remark 15.8   Let n_a and n_b be the number of inequalities in (15.53a) and (15.53c), respectively, for any i and h. In case of additive disturbances only ($n_{\mathcal{W}^p} = 0$) the total number of constraints in (15.53a) and (15.53c) for all i and h can be reduced from $(n_a + n_b)n_{\mathcal{W}^a}n_{\mathcal{W}^p}$ to $n_a + n_b$ as shown in Remark 15.5.

Remark 15.9   The closed-loop solution $u^*(k) = f_k(x(k))$ can be also obtained by using the modified batch approach with closed-loop prediction as discussed in Section 15.2. The idea there is to augment the number of free inputs by allowing one sequence $\tilde{u}_0^i, \ldots \tilde{u}_{N-1}^i$ for each vertex i of the disturbance set $\underbrace{\mathcal{W}^a \times \mathcal{W}^a \times \ldots \times \mathcal{W}^a}_{N-1}$. The large number of extreme points of such a set and the resulting large number of inputs and constraints make this approach computationally hard.

Solution to CROC-CL and CROC-OL via mp-MILP

Theorems 15.2–15.5 propose different ways of finding the PPWA solution to constrained robust optimal control by using multiparametric programs. The solution approach presented in this section is more general than the one of Theorems 15.2–15.5 as it does not exploit convexity, so that it may also be used in other contexts, for instance for CROC-CL of hybrid systems, or for CROC of uncertain systems of the type $x(t + 1) = Ax(t) + Bu(t) + Ew^a(t)$ where A and B belong to a finite discrete set of known matrices.

Consider the multiparametric mixed-integer linear program (mp-MILP)

$$J^*(x) = \min_z \quad \{J(z,x) = c'z\}$$
$$\text{subj. to} \quad Gz \leq w + Sx. \tag{15.54}$$

where $z = [z_c, z_d]$, $z_c \in \mathbb{R}^{n_c}$, $z_d \in \{0,1\}^{n_d}$, $s = n_c + n_d$, $z \in \mathbb{R}^s$ is the optimization vector and $x \in \mathbb{R}^n$ is the vector of parameters.

For a given polyhedral set $\mathcal{K} \subseteq \mathbb{R}^n$ of parameters, solving (15.54) amounts to determining the set $\mathcal{K}^* \subseteq \mathcal{K}$ of parameters for which (15.54) is feasible, the value function $J : \mathcal{K}^* \to \mathbb{R}$, and the optimizer function[1] $z^* : \mathcal{K}^* \to \mathbb{R}^s$.

The properties of $J^*(\cdot)$ and $z^*(\cdot)$ have been analyzed in Section 6.4.1 and summarized in Theorem 6.10. Below we state some properties based on these theorems.

Lemma 15.2   *Let $J : \mathbb{R}^s \times \mathbb{R}^n \to \mathbb{R}$ be a continuous piecewise affine (possibly nonconvex) function of (z,x),*

$$J(z,x) = L_i z + H_i x + K_i \text{ for } \begin{bmatrix} z \\ x \end{bmatrix} \in \mathcal{R}_i, \tag{15.55}$$

where $\{\mathcal{R}_i\}_{i=1}^{n_J}$ are polyhedral sets with disjoint interiors, $\mathcal{R} = \bigcup_{i=1}^{n_J} \mathcal{R}_i$ is a (possibly nonconvex) polyhedral set and L_i, H_i and K_i are matrices of suitable dimensions. Then the multiparametric optimization problem

---

[1] In case of multiple solutions, we define $z^*(x)$ as one of the optimizers.

$$J^*(x) = \min_z \qquad J(z,x)$$
$$subj.\ to\ \ Cz \le c + Sx \qquad\qquad (15.56)$$

is an mp-MILP.

Proof: By following the approach of [42] to transform piecewise affine functions into a set of mixed-integer linear inequalities, introduce the auxiliary binary optimization variables $\delta_i \in \{0,1\}$, defined as

$$[\delta_i = 1] \leftrightarrow \left[\left[\begin{smallmatrix} z \\ x \end{smallmatrix} \right] \in \mathcal{R}_i \right], \qquad\qquad (15.57)$$

where δ_i, $i = 1, \ldots, n_J$, satisfy the exclusive-or condition $\sum_{i=1}^{n_J} \delta_i = 1$, and set

$$J(z,x) = \sum_{i=1}^{n_J} q_i \qquad\qquad (15.58)$$

$$q_i = [L_i z + H_i x + K_i]\delta_i \qquad\qquad (15.59)$$

where q_i are auxiliary continuous optimization vectors. By transforming (15.57)–(15.59) into mixed-integer linear inequalities [42], it is easy to rewrite (15.56) as a multiparametric MILP. ■

Next we present two theorems which describe how to use mp-MILP to solve CROC-OL and CROC-CL problems.

Theorem 15.6   *By solving two mp-MILPs, the solution $U_0^*(x_0)$ to the CROC-OL (15.8), (15.9), (15.11) with additive disturbances only ($n_p = 0$) can be computed in explicit piecewise affine form (15.48).*

Proof: The objective function in the maximization problem (15.8) is convex and piecewise affine with respect to the optimization vector $\mathbf{w}^a = \{w_0^a, \ldots, w_{N-1}^a\}$ and the parameters $U = \{u_0, \ldots, u_{N-1}\}$, x_0. By Lemma 15.2, it can be solved via mp-MILP. By Theorem 6.10, the resulting value function J is a piecewise affine function of U and x_0 and the constraints in (15.11) are a linear function of the disturbance $\mathbf{w}^a$ for any given U and x_0. Then, by Lemma 15.2 and Corollary 15.1 the minimization problem is again solvable via mp-MILP, and the optimizer $U^* = \{u_0^*, \ldots, u_{N-1}^*\}$ is a piecewise affine function of x_0. ■

Theorem 15.7   *By solving $2N$ mp-MILPs, the solution of the CROC-CL (15.14)–(15.17) problem with cost (15.20), (15.21) and additive disturbances only ($n_p = 0$) can be obtained in state feedback piecewise affine form (15.49).*

Proof: Consider the first step $j = N - 1$ of the dynamic programming solution to CROC-CL. The cost-to-go at step $N - 1$ is $p(x_N)$ from (15.16). Therefore, the cost function in the maximization problem is piecewise affine with respect to both the optimization vector w_{N-1}^a and the parameters u_{N-1}, x_{N-1}. By Lemma 15.2, the worst case cost function (15.20) can be computed via mp-MILP, and, from Theorem 6.10 we know that $J_{N-1}(u_{N-1}, x_{N-1})$ is a piecewise affine function. Then, since constraints (15.51) are linear with respect to w_{N-1}^a for each u_{N-1}, x_{N-1}, we can apply Corollary 15.1 and Remark 15.5 by solving LPs of the form (10.54).

Then, by Lemma 15.2, the optimal cost $J_{N-1}^*(x_{N-1})$ of the minimization problem (15.14) is again computable via mp-MILP, and by Theorem 6.10 it is a piecewise affine function of x_{N-1}. By virtue of Theorem 6.10, $\mathcal{X}_{N-1}$ is a (possible nonconvex) polyhedral set and therefore the above maximization and minimization procedures can be iterated to compute the solution (15.49) to the CROC-CL problem. ∎

15.5 **Parametrizations of the Control Policies**

From the previous sections it is clear that CROC-OL (15.5)–(15.11) is conservative since we are optimizing an open-loop control sequence that has to cope with all possible future disturbance realizations, without taking future measurements into account. The CROC-CL (15.14)–(15.16) formulation overcomes this issue but it can quickly lead to a problem so large that it becomes intractable.

This section presents an alternative approach which introduces feedback in the system and, in some cases, can be more efficient than CROC-CL. The idea is to parameterize the control sequence in the state vector and optimize over these parameters. The approach is described next for systems with additive uncertainties.

Consider the worst case cost function as

$$J_0(x(0), U_0) = \max_{w_0^a, \dots, w_{N-1}^a} p(x_N) + \sum_{k=0}^{N-1} q(x_k, u_k)$$
$$\text{subj. to} \quad \begin{cases} x_{k+1} = Ax_k + Bu_k + Ew_k^a \\ w_k^a \in \mathcal{W}^a, \\ k = 0, \dots, N-1 \end{cases} \tag{15.60}$$

where N is the time horizon and $U_0 = [u_0', \dots, u_{N-1}']' \in \mathbb{R}^s$, $s = mN$ the vector of the input sequence. Consider the robust optimal control problem

$$J_0^*(x_0) = \min_{U_0} J_0(x_0, U_0) \tag{15.61}$$

$$\text{subj. to} \quad \left. \begin{cases} x_k \in \mathcal{X}, \ u_k \in \mathcal{U} \\ x_{k+1} = Ax_k + Bu_k + Ew_k^a \\ x_N \in \mathcal{X}_f \\ k = 0, \dots, N-1 \end{cases} \right\} \begin{array}{l} \forall w_k^a \in \mathcal{W}^a \\ \forall k = 0, \dots, N-1 \end{array} \tag{15.62}$$

Consider the parametrization of the control sequence

$$u_k = \sum_{i=0}^{k} L_{k,i} x_i + g_i, \quad k \in \{0, \dots, N-1\} \tag{15.63}$$

with the compact notation:

$$U_0 = Lx + g,$$

where $x = [x_0', x_1', \dots, x_N']'$ and

$$L = \begin{bmatrix} L_{0,0} & 0 & \cdots & 0 \\ \vdots & \ddots & \ddots & \vdots \\ L_{N-1,0} & \cdots & L_{N-1,N-1} & 0 \end{bmatrix}, \quad g = \begin{bmatrix} g_0 \\ \vdots \\ g_{N-1} \end{bmatrix} \tag{15.64}$$

where $L \in \mathbb{R}^{mN \times nN}$ and $g \in \mathbb{R}^{mN}$ are unknown feedback control gain and offset, respectively. With the parametrization (15.63) the robust control problem (15.60)–(15.62) becomes

$$J_0^{Lg}(x(0), L, g) = \max_{w_0^a, \dots, w_{N-1}^a} p(x_N) + \sum_{k=0}^{N-1} q(x_k, u_k)$$

$$\text{subj. to} \begin{cases} x_{k+1} = Ax_k + Bu_k + Ew_k^a \\ w_k^a \in \mathcal{W}^a, \\ u_k = \sum_{i=0}^{k} L_{k,i} x_i + g_i \\ k = 0, \dots, N-1 \end{cases} \tag{15.65}$$

$$J_0^{Lg^*}(x_0) = \min_{L,g} J_0^{Lg}(x_0, L, g) \tag{15.66}$$

$$\text{subj. to} \begin{cases} x_k \in \mathcal{X}, \ u_k \in \mathcal{U} \\ x_{k+1} = Ax_k + Bu_k + Ew_k^a \\ u_k = \sum_{i=0}^{k} L_{k,i} x_i + g_i \\ x_N \in \mathcal{X}_f \\ k = 0, \dots, N-1 \end{cases} \left. \begin{array}{l} \forall w_k^a \subset \mathcal{W}^a \\ \forall k = 0, \dots, N-1. \end{array} \right. \tag{15.67}$$

We denote with $\mathcal{X}_0^{Lg} \subset \mathcal{X}$ the set of states x_0 for which the robust optimal control problem (15.66)–(15.67) is feasible, i.e.,

$$\mathcal{X}_0^{Lg} = \left\{ x_0 \in \mathbb{R}^n : \mathcal{P}_0^{Lg}(x_0) \neq \emptyset \right\}$$

$$\mathcal{P}_0^{Lg}(x_0) = \Big\{ L, g : x_k \in \mathcal{X}, \ u_k \in \mathcal{U}, \ k = 0, \dots, N-1, \ x_N \in \mathcal{X}_f$$

$$\forall \ w_k^a \in \mathcal{W}^a \ k = 0, \dots, N-1,$$

$$\text{where } x_{k+1} = Ax_k + Bu_k + Ew_k^a, \ u_k = \sum_{i=0}^{k} L_{k,i} x_i + g_i \Big\}. \tag{15.68}$$

Problem (15.65) looks for the worst value of the performance index $J_0^{Lg}(x_0, L, g)$ and the corresponding worst sequences $\mathbf{w}^{a*}$ as a function of x_0 and the controller gain L and offset g.

Problem (15.66)–(15.67) minimizes (over L and g) the worst performance subject to the constraint that the input sequence $U_0 = Lx + g$ must be feasible *for all* possible disturbance realizations. Notice that formulation (15.65)–(15.67) is based on a *closed-loop* prediction. Unfortunately the set $\mathcal{P}_0^{Lg}(x_0)$ is nonconvex, in general [194]. Therefore, finding L and g for a given x_0 may be difficult.

Consider now the parametrization of the control sequence in past disturbances

$$u_k = \sum_{i=0}^{k-1} M_{k,i} w_i^a + v_i, \quad k \in \{0, \dots, N-1\}, \tag{15.69}$$

which can be compactly written as:

$$U_0 = M\mathbf{w}^a + v$$

where

$$
M = \begin{bmatrix} 0 & \cdots & & \cdots & 0 \\ M_{1,0} & 0 & & \cdots & 0 \\ \vdots & \ddots & \ddots & & \vdots \\ M_{N-1,0} & \cdots & M_{N-1,N-2} & & 0 \end{bmatrix}, \quad v = \begin{bmatrix} v_0 \\ \vdots \\ \vdots \\ v_{N-1} \end{bmatrix}. \tag{15.70}
$$

Notice that since

$$Ew_k^a = x_{k+1} - Ax_k - Bu_k, \quad k \in \{0, \ldots, N-1\}.$$

the parametrization (15.69) can also be interpreted as a parametrization in the states. The advantages of using (15.69) are explained next.

With the parametrization (15.69) the robust control problem (15.60)–(15.62) becomes

$$
J_0^{Mv}(x_0, M, v) = \max_{w_0^a, \ldots, w_{N-1}^a} p(x_N) + \sum_{k=0}^{N-1} q(x_k, u_k)
$$

$$
\text{subj. to} \begin{cases} x_{k+1} = Ax_k + Bu_k + Ew_k^a \\ w_k^a \in \mathcal{W}^a, \\ u_k = \sum_{i=0}^{k-1} M_{k,i} w_i^a + v_i \\ k = 0, \ldots, N-1 \end{cases} \tag{15.71}
$$

$$
J_0^{Mv*}(x_0) = \min_{M,v} J_0^{Mv}(x_0, M, v) \tag{15.72}
$$

$$
\text{subj. to} \begin{cases} x_k \in \mathcal{X}, \; u_k \in \mathcal{U} \\ x_{k+1} = Ax_k + Bu_k + Ew_k^a \\ u_k = \sum_{i=0}^{k-1} M_{k,i} w_i^a + v_i \\ x_N \in \mathcal{X}_f \\ k = 0, \ldots, N-1 \end{cases} \quad \begin{array}{l} \forall w_k^a \in \mathcal{W}^a \\ \forall k = 0, \ldots, N-1. \end{array}
$$

$$\tag{15.73}$$

The problem (15.71)–(15.73) is now convex in the controller parameters M and v. As solution we obtain $u^*(0) = f_0(x(0)) = v_0(x(0))$.

We denote with $\mathcal{X}_0^{Mv} \subseteq \mathcal{X}$ the set of states x_0 for which the robust optimal control problem (15.72)–(15.73) is feasible, i.e.,

$$\mathcal{X}_0^{Mv} = \left\{ x_0 \in \mathbb{R}^n : \; \mathcal{P}_0^{Mv}(x_0) \neq \emptyset \right\}$$

$$
\mathcal{P}_0^{Mv}(x_0) = \Big\{ M, v : x_k \in \mathcal{X}, \; u_k \in \mathcal{U}, \; k = 0, \ldots, N-1, \; x_N \in \mathcal{X}_f
$$

$$
\forall \, w_k^a \in \mathcal{W}^a \; k = 0, \ldots, N-1, \; \text{where } x_{k+1} = Ax_k + Bu_k + Ew_k^a,
$$

$$
u_k = \sum_{i=0}^{k-1} M_{k,i} w_i^a + v_i \Big\}.
$$

$$\tag{15.74}$$

The following result has been proven in [130], the convexity property has also appeared in [194][Section 7.4].

Theorem 15.8 *Consider the control parameterizations (15.63), (15.69) and the corresponding feasible sets $\mathcal{X}_0^{Lg}$ in (15.68) and $\mathcal{X}_0^{Mv}$ in (15.74). Then,*

$$\mathcal{X}_0^{Lg} = \mathcal{X}_0^{Mv}$$

and $\mathcal{P}_0^{Mv}(x_0)$ is convex in M and v.

Note that in general $\mathcal{X}_0^{Mv}$ and $J_0^{Mv^*}(x_0)$ are different from the corresponding CROC-CL solutions $\mathcal{X}_0$ and $J_0^*(x_0)$. In particular $\mathcal{X}_0^{Mv} \subseteq \mathcal{X}_0$ and $J_0^{Mv^*}(x_0) \geq J_0^*(x_0)$.

Next we solve the problem in Example 15.2 by using the approach discussed in this section. The example will shed some light on how to compute the disturbance feedback gains by means of convex optimization.

Example 15.3 Consider the system

$$x_{k+1} = x_k + u_k + w_k \tag{15.75}$$

where x, u and w are state, input and disturbance, respectively. Let $u_k \in [-1, 1]$ and $w_k \in [-1, 1]$ be the feasible input and disturbance. The objective for player U is to play two moves in order to keep the state x_3 in the set $\mathcal{X}_f = [-1, 1]$.
Rewrite the terminal constraint as

$$\begin{aligned}
&x_3 = x_0 + u_0 + u_1 + u_2 + w_0 + w_1 + w_2 \subset [-1, 1] \\
&\forall\ w_0 \in [-1, 1],\ w_1 \in [-1, 1],\ w_2 \in [-1, 1].
\end{aligned} \tag{15.76}$$

The control inputs u are parameterized in past disturbances as in equation (15.69)

$$\begin{aligned}
u_0 &= v_0 \\
u_1 &= v_1 + M_{1,0} w_0 \\
u_2 &= v_2 + M_{2,0} w_0 + M_{2,1} w_1.
\end{aligned} \tag{15.77}$$

Input constraints and terminal constraint are rewritten as

$$\begin{aligned}
&x_0 + v_0 + v_1 + v_2 + (1 + M_{1,0} + M_{2,0}) w_0 + (1 + M_{2,1}) w_1 + w_2 \in [-1, 1] \\
&u_0 = v_0 \in [-1, 1] \\
&u_1 = v_1 + M_{1,0} w_0 \in [-1, 1] \\
&u_2 = v_2 + M_{2,0} w_0 + M_{2,1} w_1 \in [-1, 1] \\
&\forall\ w_0 \in [-1, 1],\ w_1 \in [-1, 1],\ w_2 \in [-1, 1].
\end{aligned} \tag{15.78}$$

There are two approaches to convert the infinite number of constraints (15.78) into a set of finite number of constraint. The first approach has been already presented in this chapter and resorts to disturbance vertex enumeration. The second approach uses duality. We will present both approaches next.

- Enumeration method

  Constraint (15.78) are linear in w_0, w_1 and w_2 for fixed v_0, v_1, v_2, $M_{1,0}$, $M_{2,0}$, $M_{2,1}$. Therefore the worst case will occur at the extremes of the admissible disturbance realizations, i.e., $\{(1, 1, 1), (1, 1, -1), (-1, 1, 1), (-1, 1, -1), (1, -1, 1),$

$(1, -1, -1)$, $(-1, -1, 1)$, $(-1, -1, -1)\}$. Therefore the constraints (15.78) can be rewritten as:

$$
\begin{aligned}
&x_0 + v_0 + v_1 + v_2 + (1 + M_{1,0} + M_{2,0}) + (1 + M_{2,1}) + 1 \in [-1, 1] \\
&x_0 + v_0 + v_1 + v_2 + (1 + M_{1,0} + M_{2,0}) + (1 + M_{2,1}) - 1 \in [-1, 1] \\
&x_0 + v_0 + v_1 + v_2 - (1 + M_{1,0} + M_{2,0}) + (1 + M_{2,1}) + 1 \in [-1, 1] \\
&x_0 + v_0 + v_1 + v_2 - (1 + M_{1,0} + M_{2,0}) + (1 + M_{2,1}) - 1 \in [-1, 1] \\
&x_0 + v_0 + v_1 + v_2 + (1 + M_{1,0} + M_{2,0}) - (1 + M_{2,1}) + 1 \in [-1, 1] \\
&x_0 + v_0 + v_1 + v_2 + (1 + M_{1,0} + M_{2,0}) - (1 + M_{2,1}) - 1 \in [-1, 1] \\
&x_0 + v_0 + v_1 + v_2 - (1 + M_{1,0} + M_{2,0}) - (1 + M_{2,1}) + 1 \in [-1, 1] \\
&x_0 + v_0 + v_1 + v_2 - (1 + M_{1,0} + M_{2,0}) - (1 + M_{2,1}) - 1 \in [-1, 1] \qquad (15.79) \\
&u_0 = v_0 \in [-1, 1] \\
&u_1 = v_1 + M_{1,0} \in [-1, 1] \\
&u_1 = v_1 - M_{1,0} \in [-1, 1] \\
&u_2 = v_2 + M_{2,0} + M_{2,1} \in [-1, 1] \\
&u_2 = v_2 + M_{2,0} - M_{2,1} \in [-1, 1] \\
&u_2 = v_2 - M_{2,0} + M_{2,1} \in [-1, 1] \\
&u_2 = v_2 - M_{2,0} - M_{2,1} \in [-1, 1].
\end{aligned}
$$

A feasible solution to (15.79) can be obtained by solving a linear optimization problem for a fixed initial condition x_0. If we set $x_0 = 0.8$, a feasible solution to (15.79) is

$$
\begin{aligned}
v_0 = -0.8, \; v_1 = 0, \; v_2 = 0 \\
M_{1,0} = -1, \; M_{2,0} = 0, \; M_{2,1} = -1
\end{aligned} \qquad (15.80)
$$

which provides the controller

$$
\begin{aligned}
u_0 &= -0.8 \\
u_1 &= -w_0 \\
u_2 &= -w_1
\end{aligned} \qquad (15.81)
$$

which corresponds to

$$
\begin{aligned}
u_0 &= -0.8 \\
u_1 &= -x_1 \\
u_2 &= -x_2.
\end{aligned} \qquad (15.82)
$$

- Duality method

Duality can be used to avoid vertex enumeration. Consider the first constraint in (15.78):

$$
\begin{aligned}
&x_0 + v_0 + v_1 + v_2 + (1 + M_{1,0} + M_{2,0})w_0 + (1 + M_{2,1})w_1 + w_2 \le 1, \\
&\forall \, w_0 \in [-1, 1], \; \forall w_1 \in [-1, 1], \; \forall w_2 \in [-1, 1].
\end{aligned} \qquad (15.83)
$$

We want to replace it with the most stringent constraint, i.e.,

$$
x_0 + v_0 + v_1 + v_2 + J^*(M_{1,0}, M_{2,0}, M_{2,1}) \le 1 \qquad (15.84)
$$

where

$$
\begin{aligned}
J^*(M_{1,0}, M_{2,0}, M_{2,1}) = \max_{w_0, w_1, w_2} \; &(1 + M_{1,0} + M_{2,0})w_0 + (1 + M_{2,1})w_1 + w_2 \\
\text{subj. to } \; &w_0 \in [-1, 1], w_1 \in [-1, 1], w_2 \in [-1, 1].
\end{aligned} \qquad (15.85)
$$

For fixed $M_{1,0}, M_{2,0}, M_{2,1}$ the optimization problem in (15.85) is a linear program and can be replaced by its dual

$$
x_0 + v_0 + v_1 + v_2 + d^*(M_{1,0}, M_{2,0}, M_{2,1}) \le 1 \qquad (15.86)
$$

where

$$d^*(M_{1,0}, M_{2,0}, M_{2,1}) = \min_{\lambda_i^u, \lambda_i^l} \lambda_0^u + \lambda_1^u + \lambda_2^u + \lambda_0^l + \lambda_1^l + \lambda_2^l$$
$$\text{subject to } 1 + M_{1,0} + M_{2,0} + \lambda_0^l - \lambda_0^u = 0$$
$$1 + M_{2,1} + \lambda_1^l - \lambda_1^u = 0 \tag{15.87}$$
$$1 + \lambda_2^l - \lambda_2^u = 0$$
$$\lambda_i^u \geq 0, \; \lambda_i^l \geq 0, \; i = 0, 1, 2$$

where λ_i^u and λ_i^l are the dual variables corresponding to the upper- and lower-bounds of w_i, respectively. From strong duality we know that for a given v_0, v_1, v_2, $M_{1,0}$, $M_{2,0}$, $M_{2,1}$ problem (15.84)–(15.85) is feasible if and only if there exists a set of λ_i^u and λ_i^l which together with v_0, v_1, v_2, $M_{1,0}$, $M_{2,0}$, $M_{2,1}$ solves the system of equalities and inequalities

$$x_0 + v_0 + v_1 + v_2 + \lambda_0^u + \lambda_1^u + \lambda_2^u + \lambda_0^l + \lambda_1^l + \lambda_2^l \leq 1$$
$$1 + M_{1,0} + M_{2,0} + \lambda_0^l - \lambda_0^u = 0$$
$$1 + M_{2,1} + \lambda_1^l - \lambda_1^u = 0 \tag{15.88}$$
$$1 + \lambda_2^l - \lambda_2^u = 0$$
$$\lambda_i^u \geq 0, \; \lambda_i^l \geq 0, \; i = 0, 1, 2.$$

We have transformed an infinite dimensional set of constraints in (15.83) into a finite dimensional set of linear constraints (15.88). By repeating the same procedure for every constraint in (15.78) we obtain a set of linear constraints. In particular,

$$x_0 + v_0 + v_1 + v_2 + (1 + M_{1,0} + M_{2,0})w_0 + (1 + M_{2,1})w_1 + w_2 \in [-1, 1]$$
$$\forall \, w_0 \in [-1, 1], \; w_1 \in [-1, 1], \; w_2 \in [-1, 1]$$

is transformed into

$$x_0 + v_0 + v_1 + v_2 + \lambda_0^u + \lambda_1^u + \lambda_2^u + \lambda_0^l + \lambda_1^l + \lambda_2^l \leq 1$$
$$-x_0 - v_0 - v_1 - v_2 + \mu_0^u + \mu_1^u + \mu_2^u + \mu_0^l + \mu_1^l + \mu_2^l \leq 1$$
$$1 + M_{1,0} + M_{2,0} + \lambda_0^l - \lambda_0^u = 0$$
$$1 + M_{2,1} + \lambda_1^l - \lambda_1^u = 0$$
$$1 + \lambda_2^l - \lambda_2^u = 0 \tag{15.89}$$
$$-1 - M_{1,0} - M_{2,0} + \mu_0^l - \mu_0^u = 0$$
$$-1 - M_{2,1} + \mu_1^l - \mu_1^u = 0$$
$$-1 - \mu_2^l - \mu_2^u = 0$$
$$\lambda_i^u, \; \lambda_i^l, \; \mu_i^u, \; \mu_i^l \geq 0, \; i = 0, 1, 2,$$

and

$$u_1 = v_1 + M_{1,0}w_0 \in [-1, 1]$$
$$\forall \, w_0 \in [-1, 1], \; w_1 \in [-1, 1], \; w_2 \in [-1, 1]$$

is transformed into

$$v_1 + \nu_0^u + \nu_0^l \leq 1$$
$$-v_1 + \kappa_0^u + \kappa_0^l \leq 1$$
$$M_{1,0} + \nu_1^l - \nu_1^u = 0 \tag{15.90}$$
$$-M_{1,0} + \kappa_1^l - \kappa_1^u = 0$$
$$\kappa_0^u, \; \kappa_0^l, \; \nu_0^u, \; \nu_0^l \geq 0,$$

and

$$u_2 = v_2 + M_{2,0}w_0 + M_{2,1}w_1 \in [-1, 1]$$
$$\forall \, w_0 \in [-1, 1], \; w_1 \in [-1, 1], \; w_2 \in [-1, 1]$$

is transformed into

$$
\begin{aligned}
v_2 + \rho_0^u + \rho_1^u + \rho_0^l + \rho_1^l &\leq 1 \\
-v_2 + \pi_0^u + \pi_1^u + \pi_0^l + \pi_1^l &\leq 1 \\
M_{2,0} + \rho_0^l - \rho_0^u &= 0 \\
M_{2,1} + \rho_1^l - \rho_1^u &= 0 \\
-M_{2,0} + \pi_0^l - \pi_0^u &= 0 \\
-M_{2,1} + \pi_1^l - \pi_1^u &= 0 \\
\rho_i^u,\ \rho_i^l,\ \pi_i^u,\ \pi_i^l &\geq 0,\ i = 0,1
\end{aligned}
\tag{15.91}
$$

and the constraint

$$
u_0 = v_0 \in [-1, 1]
\tag{15.92}
$$

remains unchanged. The solution to the set of linear of equalities and inequalities (15.89)–(15.92) for a given x_0 provides the solution to our problem.

15.6 Example

Example 15.4 Consider the problem of robustly regulating to the origin the system

$$
x(t+1) = \begin{bmatrix} 1 & 0.1 \\ -0.2 & 1 \end{bmatrix} x(t) + \begin{bmatrix} -0.5 \\ 1 \end{bmatrix} u(t) + w^a(t)
$$

subject to the input constraints

$$
\mathcal{U} = \{u \in \mathbb{R} \ : \ -3 \leq u \leq 1\}
$$

and the state constraints

$$
\mathcal{X} = \{x \in \mathbb{R}^2 \ : \ -10 \leq x \leq 10,\ k = 0, \dots, 3\}.
$$

The two-dimensional disturbance w^a is restricted to the set $\mathcal{W}^a = \{v : \|w^a\|_\infty \leq 2\}$. We use the cost function

$$
\|Px_N\|_\infty + \sum_{k=0}^{N-1} (\|Qx_k\|_\infty + |Ru_k|)
$$

with $N = 3$, $P = Q = \begin{bmatrix} 1 & 1 \\ 0 & 1 \end{bmatrix}$, $R = 1.8$ and we set $\mathcal{X}_f = \mathcal{X}$.

We compute three robust controllers *CROC-OL*, *CROC-CL* and *CROC* with disturbance feedback parametrization.

CROC-OL. The min-max state feedback control law $u^*(k) = f_k^{\mathrm{OL}}(x(0))$, $k = 1, \dots, N-1$ is obtained by solving the CROC-OL (15.8), (15.9), (15.11). The polyhedral partition corresponding to $u^*(0)$ consists of 64 regions and it is depicted in Figure 15.2(a).

CROC-CL. The min-max state feedback control law $u^*(k) = f_k^{\mathrm{CL}}(x(k))$, $k = 1, \dots, N-1$ obtained by solving problem (15.14)–(15.16), (15.20), (15.21) using the approach of Theorem 15.5. The resulting polyhedral partition for $k = 0$ consists of 14 regions and is depicted in Figure 15.2(b).

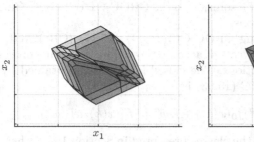
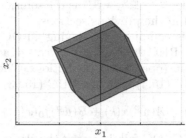

(a) Example 15.4. Solution to the CROC-OL problem. Polyhedral partition of the state space.

(b) Example 15.4. Solution to the CROC-CL problem. Polyhedral partition of the state space.

Figure 15.2 Example 15.4. Polyhedral partition of the state space corresponding to the explicit solution of CROC-OL and CROC-CL at time $t = 0$.

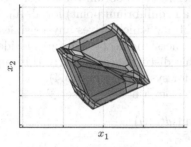
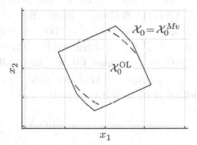

(a) Example 15.4. Solution to the CROC-Parametrized problem. Polyhedral partition of the state space.

(b) Example 15.4. Sets $\mathcal{X}_0^{\text{OL}}$ (dashed line), $\mathcal{X}_0$ (continuous line) and $\mathcal{X}_0^{Mv}$ (continuous line).

Figure 15.3 Example 15.4. (a) Polyhedral partition of the state space corresponding to the explicit solution of CROC-Parametrized at time $t = 0$. (b) Sets $\mathcal{X}_0^{\text{OL}}$, $\mathcal{X}_0$ and $\mathcal{X}_0^{Mv}$ corresponding to the controllers CROC-OL, CROC-CL and CROC-Parameterized, respectively.

CROC-Parameterized. The min-max state feedback control law $u^*(k) = f_k^{\text{par}}(x(0))$, $k = 1, \ldots, N - 1$ obtained by solving problem (15.71)–(15.73) using the approach described in Example 15.3. The resulting polyhedral partition for $k = 0$ consists of 136 regions and is depicted in Figure 15.3(a). Figure 15.3(b) shows the set of feasible initial states $\mathcal{X}_0^{\text{OL}}$, $\mathcal{X}_0$ and $\mathcal{X}_0^{Mv}$ corresponding to the controllers CROC-OL, CROC-CL and CROC-Parameterized, respectively.

15.7   Robust Receding Horizon Control

A robust receding horizon controller for system (15.1)–(15.4) which enforces the constraints (15.2) at each time t in spite of additive and parametric uncertainties can be obtained by setting

$$u(t) = f_0^*(x(t)), \tag{15.93}$$

where $f_0^*(x(0)) : \mathbb{R}^n \to \mathbb{R}^m$ is the solution to the CROC-OL or CROC-CL problems discussed in the previous sections.

If f_0 is computed by solving CROC-CL (15.14)–(15.17) (CROC-OL (15.11)), then the RHC law (15.93) is called a robust receding horizon controller with closed-loop (open-loop) predictions. The closed-loop system obtained by controlling (15.1)–(15.4) with the RHC (15.93) is

$$x(k+1) = A(w^p)x(k) + B(w^p)f_0(x(k)) + Ew^a = f_{cl}(x(k), w^p, w^a), \ k \geq 0 \quad (15.94)$$

If the CROC falls in one of the classes presented in Section 15.1.1 then from the theorems presented in this chapter we can immediately conclude that the robust RHC law (15.93) is piecewise affine and thus its on-line computation comprises a function evaluation.

As discussed in Chapter 12 convergence and persistent feasibility of the robust receding horizon controller are not guaranteed for nominal receding horizon controllers. In the robust RHC case it is desirable to obtain robust convergence to a set $\mathcal{O} \subseteq \mathcal{X}_f$ (rather than convergence to an equilibrium point) for all $\mathcal{X}_0$. In other words, the goal is to design a RHC control law which drives any feasible state in $\mathcal{X}_0$ into the set $\mathcal{O}$ for all admissible disturbances and keeps the states inside the set $\mathcal{O}$ for all future time and for all admissible disturbances. Clearly this is possible only if $\mathcal{O}$ is a robust control invariant set for system (15.1)–(15.4).

We define a distance of a point $x \in \mathbb{R}^n$ from a nonempty set $\mathcal{Y} \subset \mathbb{R}^n$ as:

$$d(x, \mathcal{Y}) = \inf_{y \in \mathcal{Y}} d(x, y) \quad (15.95)$$

The following theorem presents sufficient conditions for convergence and persistent feasibility in the robust case. It can be proven by using the arguments of Theorem 12.2. Slightly modified versions can be found with proof and further discussion in [247, p. 213–217, p. 618] and in [23, Section 7.6].

Theorem 15.9   *Consider system (15.1)–(15.4), the closed-loop RHC law (15.93) where $f_0(x)$ is obtained by solving the CROC-CL (15.14)–(15.16) with cost (15.20), (15.21) for $x(0) = x$ and the closed-loop system (15.94). Assume that*

(A0)   There exist constants $c_1, c_2, c_3, c_4 > 0$ such that

$$c_1 d(x, \mathcal{O}) \leq p(x) \leq c_2 d(x, \mathcal{O}) \ \ \forall x \in \mathcal{X}_0 \quad (15.96)$$

$$c_3 d(x, \mathcal{O}) \leq q(x, u) \leq c_4 d(x, \mathcal{O}) \ \ \forall (x, u) \in \mathcal{X}_0 \times \mathcal{U} \quad (15.97)$$

(A1)   The sets $\mathcal{X}, \mathcal{X}_f, \mathcal{U}, \mathcal{W}^a, \mathcal{W}^p$ are compact.

(A2) $\mathcal{X}_f$ and $\mathcal{O}$ are robust control invariants, $\mathcal{O} \subseteq \mathcal{X}_f \subseteq \mathcal{X}$.

(A3) $J^p(x) \leq 0 \ \forall x \in \mathcal{X}_f$ where

$$J^p(x) = \min_{u \in \mathcal{U}} \max_{w^a, w^p} p(x^+) - p(x) + q(x, u)$$

$$subj. \ to \ \begin{cases} w^a \in \mathcal{W}^a, \ w^p \in \mathcal{W}^p \\ x^+ = A(w^p)x + B(w^p)u + Ew^a \end{cases} \quad (15.98)$$

Then, for all $x \in \mathcal{X}_0$, $\lim_{k \to \infty} d(x(k), \mathcal{O}) = 0$.

Compare Theorem 15.9 with the nominal MPC stability results in Theorem 12.2.

- Assumption (A0) characterizes stage cost $q(x, u)$ and terminal cost $p(x)$ in both the nominal and the robust case. In the robust case q and p need to be zero in $\mathcal{O}$, positive and finitely determined outside $\mathcal{O}$ and continuous on the border of $\mathcal{O}$. Assumption (A0) of Theorem 15.9 guarantees these properties. Note that this cannot be obtained, for instance, with the cost (15.22) in (15.8).

- Assumption (A1) characterizes the initial sets in both the nominal and the robust case. Notice that the compactness of $\mathcal{W}^a$ and $\mathcal{W}^p$ is required to have a finite cost in the inner maximization problem. In unconstrained min-max problems often the stage cost is augmented with the term $-\rho^2 \|w\|$ to have a well defined maximization problem [247] [Section 3.3.3].

- Assumption (A2) characterizes the terminal set in both the nominal and the robust case. Recall that, as discussed earlier, the set $\mathcal{O}$ takes the role of the origin for the nominal case. The set $\mathcal{O}$ must also be a robust invariant.

- Assumption (A3) characterizes the terminal cost in both the nominal and the robust case. The minimization problem in the nominal cases is replaced by the min-max problem in the robust case. Robust control invariance of the set $\mathcal{X}_f$ in assumption (A2) of Theorem 15.9 implies that problem (15.98) in assumption (A3) is always feasible.

Remark 15.10   Robust control invariance of the set $\mathcal{O}$ in assumption (A2) of Theorem 15.9 is also implicitly guaranteed by assumption (A3) as shown next.

Because of assumption (A0), assumption (A3) in $\mathcal{O}$ becomes

$$\min_{u} \max_{w^a, w^p} \left(p(A(w^p)x + B(w^p)u + Ew^a) \right) \leq 0 \quad \forall x \in \mathcal{O}.$$

From assumption (A4), this can be verified only if it is equal to zero, i.e., if there exists a $u \in \mathcal{U}$ such that $A(w^p)x + B(w^p)u + Ew^a \in \mathcal{O}$ for all $w^a \in \mathcal{W}^a$ $w^p \in \mathcal{W}^p$ (since $\mathcal{O}$ is the only place where $p(x)$ can be zero). This implies the robust control invariance of $\mathcal{O}$.

15.8   Literature Review

An extensive treatment of robust invariant sets can be found in [57, 58, 51, 59]. The proof to Theorem 10.2 can be fond in [172, 100] For the derivation of the algorithms 10.4, 10.5 for computing robust invariant sets (and their finite termination) see [10, 49, 172, 124, 164].

Min-max robust constrained optimal control was originally proposed by Witsenhausen [289]. In the context of robust MPC, the problem was tackled by Campo and Morari [78], and further developed in [5] for SISO FIR plants. Kothare *et al.* [180] optimize robust performance for polytopic/multimodel and linear fractional uncertainty, Scokaert and Mayne [259] for additive disturbances, and Lee and Yu [189] for linear time-varying and time-invariant state space models depending on a vector of parameters $\theta \in \Theta$, where Θ is either an ellipsoid or a polyhedron. Other suboptimal CROC-CL strategies have been proposed in [180, 29, 181]. For stability and feasibility of the robust RHC (15.1), (15.93) we refer the reader to [43, 197, 204, 23].

The idea of the parametrization (15.69) appears in the work of Gartska & Wets in 1974 in the context of stochastic optimization [119]. Recently, it reappeared in robust optimization work by Guslitzer and Ben-Tal [136, 46], and in the context of robust MPC in the work of van Hessem & Bosgra, Löfberg and Goulart & Kerrigan [281, 194, 130].

Part V

Constrained Optimal Control
of Hybrid Systems

16

Models of Hybrid Systems

Hybrid systems describe the dynamical interaction between continuous and discrete signals in one common framework (see Figure 16.1). In this chapter we focus our attention on mathematical models of hybrid systems that are particularly suitable for solving finite time-constrained optimal control problems.

16.1 Models of Hybrid Systems

The mathematical model of a dynamical system is traditionally expressed through differential or difference equations, typically derived from physical laws governing the dynamics of the system under consideration. Consequently, most of the control theory and tools address models describing the evolution of real-valued signals according to smooth linear or nonlinear state transition functions, typically differential or difference equations. In many applications, however, the system to be controlled also contains discrete-valued signals satisfying Boolean relations, if-then-else conditions, on/off conditions, etc., that also involve the real-valued signals. An example would be an on/off alarm signal triggered by an analog variable exceeding a given threshold. *Hybrid systems* describe in a common framework the dynamics of real-valued variables, the dynamics of discrete variables, and their interaction.

In this chapter we will focus on discrete-time hybrid systems, which we will call *discrete hybrid automata* (DHA), whose continuous dynamics is described by linear difference equations and whose discrete dynamics is described by finite state machines, both synchronized by the same clock [276]. A particular case of DHA is the class of *piecewise affine* (PWA) systems [266]. Essentially, PWA systems are switched affine systems whose mode depends on the current location of the state vector, as depicted in Figure 16.2. PWA and DHA systems can be translated into a form, denoted as *mixed logical dynamical* (MLD) form, that is more suitable for solving optimization problems. In particular, complex finite time hybrid dynamical optimization problems can be recast into mixed-integer linear or quadratic programs as will be shown in Chapter 17.

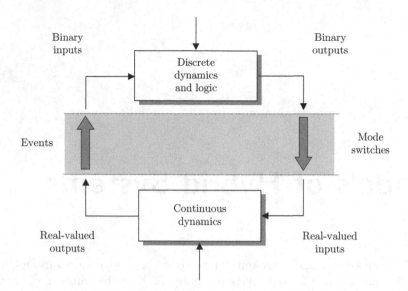

Figure 16.1 Hybrid systems. Logic-based discrete dynamics and continuous dynamics interact through events and mode switches.

In Section 16.7 we will introduce the tool HYSDEL (HYbrid Systems DEscription Language), a high level language for modeling and simulating DHA. Therefore, DHA will represent for us the starting point for modeling hybrid systems. We will show that DHA, PWA, and MLD systems are equivalent model classes, and in particular that DHA systems can be converted to an equivalent PWA or MLD form for solving optimal control problems.

After introducing PWA systems, we will go through the steps needed for modeling a system as a DHA. We will first detail the process of translating propositional logic involving Boolean variables and linear threshold events over continuous variables into mixed-integer linear inequalities, generalizing several results available in the literature, in order to get an equivalent MLD form of a DHA system. Finally, we will briefly present the tool HYSDEL that allows to describe the DHA in a textual form and to obtain equivalent MLD and PWA representations in MATLAB®.

16.2 Piecewise Affine Systems

PWA systems [266, 145] are defined by partitioning the space of states and inputs into polyhedral regions (cf. Figure 16.2) and associating with each region different affine state-update and output equations:

$$x(t+1) = A^{i(t)}x(t) + B^{i(t)}u(t) + f^{i(t)} \tag{16.1a}$$

$$y(t) = C^{i(t)}x(t) + D^{i(t)}u(t) + g^{i(t)} \tag{16.1b}$$

$$H^{i(t)}x(t) + J^{i(t)}u(t) \le K^{i(t)} \tag{16.1c}$$

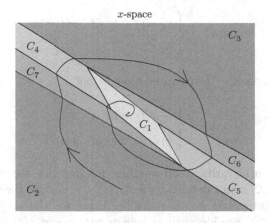

Figure 16.2 Piecewise affine (PWA) systems. Mode switches are triggered by linear threshold events. In each one of the shaded regions the affine mode dynamics is different.

where $x(t) \in \mathbb{R}^n$ is the state vector at time $t \in \mathbb{T}$ and $\mathbb{T} = \{0, 1, \ldots\}$ is the set of nonnegative integers, $u(t) \in \mathbb{R}^m$ is the input vector, $y(t) \in \mathbb{R}^p$ is the output vector, $i(t) \in \mathcal{I} = \{1, \ldots, s\}$ is the current *mode* of the system, the matrices $A^{i(t)}$, $B^{i(t)}$, $f^{i(t)}$, $C^{i(t)}$, $D^{i(t)}$, $g^{i(t)}$, $H^{i(t)}$, $J^{i(t)}$, $K^{i(t)}$ are constant and have suitable dimensions, and the inequalities in (16.1c) should be interpreted component-wise. Each linear inequality in (16.1c) defines a half-space in $\mathbb{R}^n$ and a corresponding hyperplane, that will be referred to as *guardline*. Each vector inequality (16.1c) defines a polyhedron $\mathcal{C}^i = \{[\begin{smallmatrix} x \\ u \end{smallmatrix}] \in \mathbb{R}^{n+m} : H^i x + J^i u \leq K^i\}$ in the state+input space $\mathbb{R}^{n+m}$ that will be referred to as *cell*, and the union of such polyhedral cells as *partition*. We assume that $\mathcal{C}^i$ are full-dimensional sets of $\mathbb{R}^{n+m}$, for all $i = 1, \ldots, s$.

A PWA system is called *well-posed* if it satisfies the following property [40]:

Definition 16.1   *Let P be a PWA system of the form (16.1) and let $\mathcal{C} = \cup_{i=1}^s \mathcal{C}^i \subseteq \mathbb{R}^{n+m}$ be the polyhedral partition associated with it. System P is called* well-posed *if for all pairs $(x(t), u(t)) \in \mathcal{C}$ there exists only one index $i(t)$ satisfying (16.1).*

Definition 16.1 implies that $x(t+1)$, $y(t)$ are single-valued functions of $x(t)$ and $u(t)$, and therefore that state and output trajectories are uniquely determined by the initial state and input trajectory. A relaxation of definition 16.1 is to let polyhedral cells $\mathcal{C}^i$ share one or more hyperplanes. In this case the index $i(t)$ is not uniquely defined, and therefore the PWA system is not well-posed. However, if the mappings $(x(t), u(t)) \to x(t+1)$ and $(x(t), u(t)) \to y(t)$ are continuous across the guardlines that are facets of two or more cells (and, therefore, they are continuous on their domain of definition), such mappings are still single valued.

16.2.1   Modeling Discontinuities

Discontinuous dynamical behaviors can be modeled by disconnecting the domain. For instance, the state-update equation

$$x(t+1) = \begin{cases} \frac{1}{2}x(t) + 1 & \text{if} \quad x(t) \le 0 \\ 0 & \text{if} \quad x(t) > 0 \end{cases} \qquad (16.2a)$$

is discontinuous across $x = 0$. It can be modeled as

$$x(t+1) = \begin{cases} \frac{1}{2}x(t) + 1 & \text{if} \quad x(t) \le 0 \\ 0 & \text{if} \quad x(t) \ge \epsilon \end{cases} \qquad (16.2b)$$

where $\epsilon > 0$ is an arbitrarily small number, for instance the machine precision. Clearly, system (16.2) is not defined for $0 < x(t) < \epsilon$, i.e., for the values of the state that cannot be represented in the machine. However, the trajectories produced by (16.2a) and (16.2b) are identical as long as $x(t) > \epsilon$ or $x(t) \le 0$, $\forall t \in \mathbb{N}$.

As remarked above, multiple definitions of the state-update and output functions over common boundaries of sets $\mathcal{C}^i$ is a technical issue that arises only when the PWA mapping is discontinuous. Rather than disconnecting the domain, another way of dealing with discontinuous PWA mappings is to allow strict inequalities in the definition of the polyhedral cells in (16.1), or by dealing with open polyhedra and boundaries separately as in [266]. We prefer to assume that in the definition of the PWA dynamics (16.1) the polyhedral cells $\mathcal{C}^{i(t)}$ are closed sets. As will be clear in the next chapter, the closed-polyhedra description is mainly motivated by the fact that numerical solvers cannot handle open sets.

Example 16.1 The following PWA system

$$\begin{cases} x(t+1) = 0.8 \begin{bmatrix} \cos \alpha(t) & -\sin \alpha(t) \\ \sin \alpha(t) & \cos \alpha(t) \end{bmatrix} x(t) + \begin{bmatrix} 0 \\ 1 \end{bmatrix} u(t) \\ y(t) = \begin{bmatrix} 0 & 1 \end{bmatrix} x(t) \\ \alpha(t) = \begin{cases} \frac{\pi}{3} & \text{if} \quad \begin{bmatrix} 1 & 0 \end{bmatrix} x(t) \ge 0 \\ -\frac{\pi}{3} & \text{if} \quad \begin{bmatrix} 1 & 0 \end{bmatrix} x(t) < 0 \end{cases} \end{cases} \qquad (16.3)$$

is discontinuous at $x = \begin{bmatrix} 0 \\ x_2 \end{bmatrix}$, $\forall x_2 \ne 0$. It can be described in form (16.1) as

$$\begin{cases} x(t+1) = \begin{cases} 0.4 \begin{bmatrix} 1 & -\sqrt{3} \\ \sqrt{3} & 1 \end{bmatrix} x(t) + \begin{bmatrix} 0 \\ 1 \end{bmatrix} u(t) & \text{if} \quad \begin{bmatrix} 1 & 0 \end{bmatrix} x(t) \ge 0 \\ 0.4 \begin{bmatrix} 1 & \sqrt{3} \\ -\sqrt{3} & 1 \end{bmatrix} x(t) + \begin{bmatrix} 0 \\ 1 \end{bmatrix} u(t) & \text{if} \quad \begin{bmatrix} 1 & 0 \end{bmatrix} x(t) \le -\epsilon \end{cases} \\ y(t) = \begin{bmatrix} 0 & 1 \end{bmatrix} x(t) \end{cases}$$

$$(16.4)$$

for all $x_1 \in (-\infty, -\epsilon] \cup [0, +\infty)$, $x_2 \in \mathbb{R}$, $u \in \mathbb{R}$, and $\epsilon > 0$.

Figure 16.3 shows the free response of the systems (open-loop simulation of the system for a constant input $u = 0$) starting from the initial condition $x(0) = [1 \ 0]$ with sampling time equal to 0.5s and $\epsilon = 10^{-6}$.

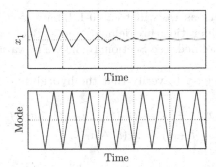

Figure 16.3 Example 16.1. Free response of state x_1 for $x(0) = [1\ 0]$.

In case the partition $\mathcal{C}$ does not cover the whole space $\mathbb{R}^{n+m}$, well-posedness does not imply that trajectories are *persistent*, i.e., that for all $t \in \mathbb{N}$ a successor state $x(t+1)$ and an output $y(t)$ are defined. A typical case of $\mathcal{C} \neq \mathbb{R}^{n+m}$ is when we are dealing with bounded inputs and bounded states $u_{\min} \leq u(t) \leq u_{\max}$, $x_{\min} \leq x(t) \leq x_{\max}$. By embedding such ranges in the inequalities (16.1c), the system becomes undefined outside the bounds, as no index i exists that satisfies any set of inequalities (16.1c).

As will be clearer in the next chapter, when model (16.1) is used in an optimal control formulation, any input sequence and initial state that are feasible for the related optimization problem automatically define unique trajectories over the whole optimal control horizon.

PWA systems can model a large number of physical processes, as they can model static nonlinearities through a piecewise affine approximation, or approximate nonlinear dynamics via multiple linearizations at different operating points. Moreover, tools exist for obtaining piecewise affine approximations automatically (see Section 16.8).

When the mode $i(t)$ is an exogenous variable, condition (16.1c) disappears and we refer to (16.1) as a *switched affine system* (SAS), see Section 16.3.1.

16.2.2   Binary States, Inputs and Outputs

When dealing with hybrid systems, quite often one encounters some signals that can only assume a binary value, namely either 0 or 1. In the most general form, let us assume that the state vector $x = \begin{bmatrix} x_c \\ x_\ell \end{bmatrix}$ where $x_c \in \mathbb{R}^{n_c}$ are the continuous states, $x_\ell \in \mathbb{R}^{n_\ell}$ are the binary states, and $n = n_c + n_\ell$. Similarly, let $y \in \mathbb{R}^{p_c} \times \{0,1\}^{p_\ell}$, $p = p_c + p_\ell$, $u \in \mathbb{R}^{m_c} \times \{0,1\}^{m_\ell}$, $m = m_c + m_\ell$. By defining a polyhedral partition $\{\mathcal{C}^i\}_{i=0}^{s-1}$ of the sets of state and input space $\mathbb{R}^{n+m}$, for any $x_\ell \in \{0,1\}$ and $u_\ell \in \{0,1\}$ a sufficient condition for the PWA system (16.1) to be well posed is that the rows and columns of matrices A^i, B^i, C^i, D^i corresponding to binary states and binary outputs are zero and that the corresponding rows of matrices f^i, g^i are either 0 or 1, i.e., that the binary state update and output equations are binary piecewise constant functions.

In the following sections we will treat 0-1 binary variables both as numbers (over which arithmetic operations are defined) and as Boolean variables (over which Boolean functions are defined, see Section 16.3.3). The variable type will be clear from the context.

As an example, it is easy to verify that the hybrid dynamical system

$$x_c(t+1) = 2x_c(t) + u_c(t) - 3u_\ell(t) \tag{16.5a}$$

$$x_\ell(t+1) = x_\ell(t) \wedge u_\ell(t) \tag{16.5b}$$

where "$\wedge$" represents the logic operator "and," can be represented in the PWA form

$$\begin{bmatrix} x_c \\ x_\ell \end{bmatrix}(t+1) = \begin{cases} \begin{bmatrix} 2x_c(t) + u_c(t) \\ 0 \end{bmatrix} & \text{if } x_\ell \leq \frac{1}{2}, u_\ell \leq \frac{1}{2} \\[2ex] \begin{bmatrix} 2x_c(t) + u_c(t) - 3 \\ 0 \end{bmatrix} & \text{if } x_\ell \leq \frac{1}{2}, u_\ell \geq \frac{1}{2} + \epsilon \\[2ex] \begin{bmatrix} 2x_c(t) + u_c(t) \\ 0 \end{bmatrix} & \text{if } x_\ell \geq \frac{1}{2} + \epsilon, u_\ell \leq \frac{1}{2} \\[2ex] \begin{bmatrix} 2x_c(t) + u_c(t) - 3 \\ 1 \end{bmatrix} & \text{if } x_\ell \geq \frac{1}{2} + \epsilon, u_\ell \geq \frac{1}{2} + \epsilon \end{cases} \tag{16.5c}$$

by associating $x_\ell = 0$ with $x_\ell \leq \frac{1}{2}$ and $x_\ell = 1$ with $x_\ell \geq \frac{1}{2} + \epsilon$ for any $0 < \epsilon \leq \frac{1}{2}$. Note that, by assuming $x_\ell(0) \in \{0,1\}$ and $u_\ell(t) \in \{0,1\}$ for all $t \in \mathbb{T}$, $x_\ell(t)$ will be in $\{0,1\}$ for all $t \in \mathbb{T}$.

Example 16.2   Consider the spring-mass-damper system

$$M\dot{x}_2 = u_1 - k(x_1) - b(u_2)x_2$$

where x_1 and $x_2 = \dot{x}_1$ denote the position and the velocity of the mass, respectively, and u_1 a continuous force input. The binary input u_2 switches the friction coefficient

$$b(u_2) = \begin{cases} b_1 \text{ if } u_2 = 1 \\ b_2 \text{ if } u_2 = 0. \end{cases}$$

The spring coefficient switches to a different value at position x_m

$$k(x_1) = \begin{cases} k_1 x_1 + d_1 \text{ if } x_1 \leq x_m \\ k_2 x_1 + d_2 \text{ if } x_1 > x_m. \end{cases}$$

Assume the system description is valid for $-5 \leq x_1, x_2 \leq 5$, and $-10 \leq u_1 \leq 10$.

The system has four modes, depending on the binary input u_2 and the position x_1. Assuming that the system parameters are $M = 1$, $b_1 = 1$, $b_2 = 50$, $k_1 = 1$, $k_2 = 3$,

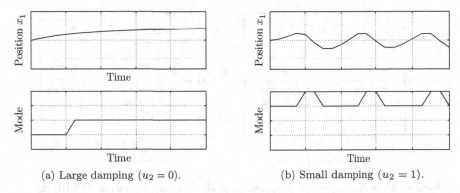

(a) Large damping ($u_2 = 0$).   (b) Small damping ($u_2 = 1$).

Figure 16.4 Example 16.2. Open-loop simulation of system (16.6) for $u_1 = 3$ and zero initial conditions.

$d_1 = 1$, $d_2 = 7.5$, $x_m = 1$, after discretizing the dynamics in each mode with a sampling time of 0.5 time units we obtain the following discrete-time PWA system

$x(t+1)$

$$
-\begin{cases}
\begin{array}{ll}
& \text{Mode 1} \\
\left[\begin{smallmatrix} 0.8956 & 0.0198 \\ -0.0198 & -0.0004 \end{smallmatrix}\right] x(t) + \left[\begin{smallmatrix} 0.1044 \\ 0.0198 \end{smallmatrix}\right] u_1(t) + \left[\begin{smallmatrix} -0.0096 \\ -0.0198 \end{smallmatrix}\right] & \text{if } x_1(t) \leq 1, u_2(t) \leq 0.5 \\
\\
& \text{Mode 2} \\
\left[\begin{smallmatrix} 0.8956 & 0.0195 \\ -0.0584 & -0.0012 \end{smallmatrix}\right] x(t) + \left[\begin{smallmatrix} 0.1044 \\ 0.0195 \end{smallmatrix}\right] u_1(t) + \left[\begin{smallmatrix} -0.0711 \\ -0.1459 \end{smallmatrix}\right] & \text{if } x_1(t) \geq 1 + \epsilon, u_2(t) \leq 0.5 \\
\\
& \text{Mode 3} \\
\left[\begin{smallmatrix} 0.8956 & 0.3773 \\ -0.3773 & 0.5182 \end{smallmatrix}\right] x(t) + \left[\begin{smallmatrix} 0.1044 \\ 0.3773 \end{smallmatrix}\right] u_1(t) + \left[\begin{smallmatrix} -0.1044 \\ -0.3773 \end{smallmatrix}\right] & \text{if } x_1(t) \leq 1, u_2(t) \geq 0.5 \\
\\
& \text{Mode 4} \\
\left[\begin{smallmatrix} 0.8956 & 0.3463 \\ -1.0389 & 0.3529 \end{smallmatrix}\right] x(t) + \left[\begin{smallmatrix} 0.1044 \\ 0.3463 \end{smallmatrix}\right] u_1(t) + \left[\begin{smallmatrix} -0.7519 \\ -2.5972 \end{smallmatrix}\right] & \text{if } x(t) \geq 1 + \epsilon, u_2(t) \geq 0.5
\end{array}
\end{cases}
$$
$$(16.6)$$

for $x_1(t) \in [-5, 1] \cup [1 + \epsilon, 5]$, $x_2(t) \in [-5, 5]$, $u_1(t) \in [-10, 10]$, and for any arbitrary small $\epsilon > 0$.

Figure 16.4 shows the open-loop simulation of the system for a constant continuous input $u_1 = 3$, starting from zero initial conditions and for $\epsilon = 10^{-6}$.

Example 16.3   Consider the following SISO system:

$$x_1(t+1) = ax_1(t) + bu(t). \tag{16.7}$$

A logic state $x_2 \in \{0, 1\}$ stores the information whether the state of system (16.7) has ever gone below a certain lower bound x_{lb} or not:

$$x_2(t+1) = x_2(t) \bigvee [x_1(t) \leq x_{lb}]. \tag{16.8}$$

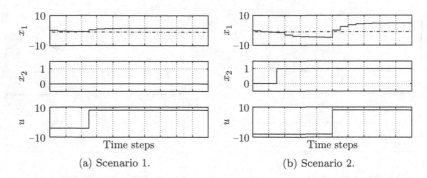

(a) Scenario 1.   (b) Scenario 2.

Figure 16.5 Example 16.3. Open-loop simulation of system (16.10) for different sequences of the input u.

Assume that the input coefficient is a function of the logic state:

$$b = \begin{cases} b_1 \text{ if } x_2 = 0 \\ b_2 \text{ if } x_2 = 1. \end{cases} \tag{16.9}$$

The system can be described by the PWA model:

$$x(t+1) = \begin{cases} \begin{bmatrix} a & 0 \\ 0 & 0 \end{bmatrix} x(t) + \begin{bmatrix} b_2 \\ 0 \end{bmatrix} u(t) + \begin{bmatrix} 0 \\ 1 \end{bmatrix} & \text{if} \quad \begin{bmatrix} 1 & 0 \end{bmatrix} x(t) \leq x_{lb} \\[12pt] \begin{bmatrix} a & 0 \\ 0 & 1 \end{bmatrix} x(t) + \begin{bmatrix} b_1 \\ 0 \end{bmatrix} u(t) & \text{if} \quad \begin{bmatrix} 1 & 0 \\ 0 & -1 \end{bmatrix} x(t) \geq \begin{bmatrix} x_{lb} + \epsilon \\ -0.5 \end{bmatrix} \\[12pt] \begin{bmatrix} a & 0 \\ 0 & 1 \end{bmatrix} x(t) + \begin{bmatrix} b_2 \\ 0 \end{bmatrix} u(t) & \text{if} \quad x(t) \geq \begin{bmatrix} x_{lb} + \epsilon \\ 0.5 \end{bmatrix} \end{cases} \tag{16.10}$$

for $u(t) \in \mathbb{R}$, $x_1(t) \in (-\infty, x_{lb}] \cup [x_{lb} + \epsilon, +\infty)$, $x_2 \in \{0, 1\}$, and for any $\epsilon > 0$.

Figure 16.5 shows two open-loop simulations of the system, for $a = 0.5$, $b_1 = 0.1$, $b_2 = 0.3$, $x_{lb} = -1$, $\epsilon = 10^{-6}$. The initial conditions is $x(0) = [1, \ 0]$. Note that when the continuous state $x_1(t)$ goes below $x_{lb} = -1$ at time t, then $x_\ell(t+1)$ switches to 1 and the input has a stronger effect on the states from time $t + 2$ on. Indeed, the steady state of x_1 is a function of the logic state x_2.

16.3 Discrete Hybrid Automata

As shown in Figure 16.6, a *discrete hybrid automaton* (DHA) is formed by generating the mode $i(t)$ of a switched affine system through a *mode selector* function that depends on (i) the discrete state of a *finite state machine*, (ii) *discrete events* generated by the continuous variables of the switched affine system exceeding given linear thresholds (the guardlines), (iii) exogenous discrete inputs [276]. We will detail each of the four blocks in the next sections.

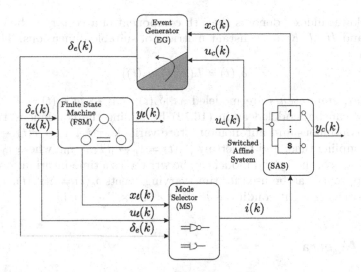

Figure 16.6 A discrete hybrid automaton (DHA) is the connection of a finite state machine (FSM) and a switched affine system (SAS), through a mode selector (MS) and an event generator (EG). The output signals are omitted for clarity.

16.3.1 Switched Affine System (SAS)

A switched affine system is a collection of affine systems:

$$x_c(t+1) = A^{i(t)}x_c(t) + B^{i(t)}u_c(t) + f^{i(t)} \tag{16.11a}$$

$$y_c(t) = C^{i(t)}x_c(t) + D^{i(t)}u_c(t) + g^{i(t)}, \tag{16.11b}$$

where $t \in \mathbb{T}$ is the time indicator, $x_c \in \mathbb{R}^{n_c}$ is the continuous state vector, $u_c \in \mathbb{R}^{m_c}$ is the exogenous continuous input vector, $y_c \in \mathbb{R}^{p_c}$ is the continuous output vector, $\{A^i, B^i, f^i, C^i, D^i, g^i\}_{i \in \mathcal{I}}$ is a collection of matrices of suitable dimensions, and the mode $i(t) \in \mathcal{I} = \{1, \dots, s\}$ is an input signal that determines the affine state update dynamics at time t. An SAS of the form (16.11) preserves the value of the state when a mode switch occurs, but it is possible to implement reset maps on an SAS as shown in [276].

16.3.2 Event Generator (EG)

An event generator is an object that generates a binary vector $\delta_e(t) \in \{0,1\}^{n_e}$ of *event conditions* according to the satisfaction of a linear (or affine) threshold condition. Let $h : \mathbb{R}^{n_c} \times \mathbb{R}^{n_c} \to \{0,1\}^{n_e}$ be a vector function defined as

$$h_i(x_c, u_c) = \begin{cases} 1 & \text{if} \quad H_i x_c + J_i u_c + K_i \leq 0 \\ 0 & \text{if} \quad H_i x_c + J_i u_c + K_i > 0 \end{cases}$$

where the lower index $_i$ denotes the i-th component of a vector or the i-th row of a matrix, and H, J, K are constant matrices of suitable dimensions. Then events are defined as

$$\delta_e(t) = h(x_c(t), u_c(t)). \tag{16.12}$$

In particular, *state events* are modeled as $[\delta_e(t) = 1] \leftrightarrow [a'x_c(t) \leq b]$. Note that *time events* can be modeled as in (16.12) by adding the continuous time as an additional continuous and autonomous state variable, $\tau(t+1) = \tau(t) + T_s$, where T_s is the sampling time, and by letting $[\delta_e(t) = 1] \leftrightarrow [tT_s \geq \tau_0]$, where τ_0 is a given time. By doing so, the hybrid model can be written as a time-invariant one. Clearly the same approach can be used for time-varying events $\delta_e(t) = h(x_c(t), u_c(t), t)$, by using time-varying event conditions $h : \mathbb{R}^{n_c} \times \mathbb{R}^m \times \mathbb{T} \to \{0, 1\}^{n_e}$.

16.3.3   Boolean Algebra

Before we deal in detail with the other blocks constituting the DHA and introduce further notation, we recall some basic definitions of Boolean algebra. A more comprehensive treatment of Boolean calculus can be found in digital circuit design texts, e.g., [86, 142]. For a rigorous exposition see, e.g., [208].

A variable δ is a Boolean variable if $\delta \in \{0, 1\}$, where "$\delta = 0$" means something is false, "$\delta = 1$" that it is true. A Boolean expression is obtained by combining Boolean variables through the logic operators $\neg$ (not), $\vee$ (or), $\wedge$ (and), $\leftarrow$ (implied by), $\rightarrow$ (implies), and $\leftrightarrow$ (iff). A Boolean function $f : \{0, 1\}^{n-1} \mapsto \{0, 1\}$ is used to define a Boolean variable δ_n as a logic function of other variables $\delta_1, \ldots, \delta_{n-1}$:

$$\delta_n = f(\delta_1, \delta_2, \ldots, \delta_{n-1}). \tag{16.13}$$

Given n Boolean variables $\delta_1, \ldots, \delta_n$, a Boolean formula F defines a relation

$$F(\delta_1, \ldots, \delta_n) \tag{16.14}$$

that must hold true. Every Boolean formula $F(\delta_1, \delta_2, \ldots, \delta_n)$ can be rewritten in the conjunctive normal form (CNF)

$$\text{(CNF)} \quad \bigwedge_{j=1}^{m} \left(\left(\bigvee_{i \in P_j} \delta_i \right) \vee \left(\bigvee_{i \in N_j} \sim \delta_i \right) \right) \tag{16.15}$$

$$N_j, P_j \subseteq \{1, \ldots, n\}, \ \forall j = 1, \ldots, m.$$

As mentioned in Section 16.2.2, often we will use the term binary variable and Boolean variable without distinction. An improper arithmetic operation over Boolean variables should be understood as an arithmetic operation over corresponding binary variables and, vice versa, an improper Boolean function of binary variables should be interpreted as the same function over the corresponding Boolean variables. Therefore, from now on we will call "binary" variables both 0-1 variables and Boolean variables.

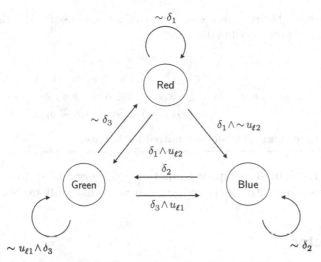

Figure 16.7 Example of finite state machine.

16.3.4 Finite State Machine (FSM)

A finite state machine (or automaton) (FSM) is a discrete dynamic process that evolves according to a Boolean state update function:

$$x_\ell(t+1) = f_\ell(x_\ell(t), u_\ell(t), \delta_e(t)), \tag{16.16a}$$

where $x_\ell \in \{0,1\}^{n_\ell}$ is the binary state, $u_\ell \in \{0,1\}^{m_\ell}$ is the exogenous binary input, $\delta_e(t)$ is the endogenous binary input coming from the EG, and $f_\ell : \{0,1\}^{n_\ell} \times \{0,1\}^{m_\ell} \times \{0,1\}^{n_e} \to \{0,1\}^{n_\ell}$ is a deterministic Boolean function. In this text we will only refer to synchronous finite state machines, where the transitions may happen only at sampling times. The adjective synchronous will be omitted for brevity.

An FSM can be conveniently represented using an oriented graph. An FSM may also have an associated binary output

$$y_\ell(t) = g_\ell(x_\ell(t), u_\ell(t), \delta_e(t)), \tag{16.16b}$$

where $y_\ell \in \{0,1\}^{p_\ell}$ and $g_\ell : \{0,1\}^{n_\ell} \times \{0,1\}^{m_\ell} \times \{0,1\}^{n_e} \mapsto \{0,1\}^{p_\ell}$.

Example 16.4   Figure 16.7 shows a finite state machine where $u_\ell = [u_{\ell 1}\ u_{\ell 2}]'$ is the input vector, and $\delta = [\delta_1 \ldots \delta_3]'$ is a vector of signals coming from the event generator. The Boolean state update function (also called *state transition function*) is:

$$x_\ell(t+1) = \begin{cases} \text{Red if } ((x_\ell(t) = \text{Green}) \wedge \sim \delta_3) \vee \\ \qquad ((x_\ell(t) = \text{Red}) \wedge \sim \delta_1), \\ \text{Green if } ((x_\ell(t) = \text{Red}) \wedge \delta_1 \wedge u_{\ell 2}) \vee \\ \qquad ((x_\ell(t) = \text{Blue}) \wedge \delta_2) \vee \\ \qquad ((x_\ell(t) = \text{Green}) \wedge \sim u_{\ell 1} \wedge \delta_3), \\ \text{Blue if } ((x_\ell(t) = \text{Red}) \wedge \delta_1 \wedge \sim u_{\ell 2}) \vee \\ \qquad ((x_\ell(t) = \text{Green}) \wedge (\delta_3 \wedge u_{\ell 1})) \vee \\ \qquad ((x_\ell(t) = \text{Blue}) \wedge \sim \delta_2). \end{cases} \tag{16.17}$$

By associating a binary vector $x_\ell = \begin{bmatrix} x_{\ell 1} \\ x_{\ell 2} \end{bmatrix}$ to each state (Red $= \begin{bmatrix} 0 \\ 0 \end{bmatrix}$, Green $= \begin{bmatrix} 0 \\ 1 \end{bmatrix}$, and Blue $= \begin{bmatrix} 1 \\ 0 \end{bmatrix}$), one can rewrite (16.17) as:

$$
\begin{aligned}
x_{\ell 1}(t+1) &= (\sim x_{\ell 1} \wedge \sim x_{\ell 2} \wedge \delta_1 \wedge \sim u_{\ell 2}) \vee \\
&\quad (x_{\ell 1} \wedge \sim \delta_2) \vee (x_{\ell 2} \wedge \delta_3 \wedge u_{\ell 1}) \\
x_{\ell 2}(t+1) &= (\sim x_{\ell 1} \wedge \sim x_{\ell 2} \wedge \delta_1 \wedge u_{\ell 2}) \vee \\
&\quad (x_{\ell 1} \wedge \delta_2) \vee (x_{\ell 2} \wedge \delta_3 \wedge \sim u_{\ell 1}),
\end{aligned}
$$

where the time index (t) has been omitted for brevity.

Since the Boolean state update function is deterministic, for each state the conditions associated with all the outgoing arcs are mutually exclusive.

16.3.5　Mode Selector

In a DHA, the dynamic mode $i(t) \in \mathcal{I} = \{1, \ldots, s\}$ of the SAS is a function of the binary state $x_\ell(t)$, the binary input $u_\ell(t)$, and the events $\delta_e(t)$. With a slight abuse of notation, let us indicate the mode $i(t)$ through its binary encoding, $i(t) \in \{0,1\}^{n_s}$ where $n_s = \lceil \log_2 s \rceil$, so that $i(t)$ can be treated as a vector of Boolean variables. Any discrete variable $\alpha \in \{\alpha_1, \ldots, \alpha_j\}$ admits a Boolean encoding $a \in \{0,1\}^{d(j)}$, where $d(j)$ is the number of bits used to represent $\alpha_1, \ldots, \alpha_j$. For example, $\alpha \in \{0,1,2\}$ may be encoded as $a \in \{0,1\}^2$ by associating $\begin{bmatrix} 0 \\ 0 \end{bmatrix} \to 0$, $\begin{bmatrix} 0 \\ 1 \end{bmatrix} \to 1$, $\begin{bmatrix} 1 \\ 0 \end{bmatrix} \to 2$.

Then, we define the *mode selector* by the Boolean function $f_{\mathrm{M}} : \{0,1\}^{n_\ell} \times \{0,1\}^{m_\ell} \times \{0,1\}^{n_e} \to \{0,1\}^{n_s}$. The output of this function

$$
i(t) = \mu(x_\ell(t), u_\ell(t), \delta_e(t)) \tag{16.18}
$$

is called the *active mode* of the DHA at time t. We say that a *mode switch* occurs at step t if $i(t) \neq i(t-1)$. Note that, in contrast to continuous-time hybrid models where switches can occur at any time, in our discrete-time setting, as mentioned earlier, a mode switch can only occur at sampling instants.

16.3.6　DHA Trajectories

For a given initial condition $\begin{bmatrix} x_c(0) \\ x_\ell(0) \end{bmatrix} \in \mathbb{R}^{n_c} \times \{0,1\}^{n_\ell}$, and inputs $\begin{bmatrix} u_c(t) \\ u_\ell(t) \end{bmatrix} \in \mathbb{R}^{m_c} \times \{0,1\}^{m_\ell}$, $t \in \mathbb{T}$, the state $x(t)$ of the system is computed for all $t \in \mathbb{T}$ by recursively iterating the set of equations:

$$
\begin{aligned}
\delta_e(t) &= h(x_c(t), u_c(t), t) & \text{(16.19a)} \\
i(t) &= \mu(x_\ell(t), u_\ell(t), \delta_e(t)) & \text{(16.19b)} \\
y_c(t) &= C^{i(t)} x_c(t) + D^{i(t)} u_c(t) + g^{i(t)} & \text{(16.19c)} \\
y_\ell(t) &= g_\ell(x_\ell(t), u_\ell(t), \delta_e(t)) & \text{(16.19d)} \\
x_c(t+1) &= A^{i(t)} x_c(t) + B^{i(t)} u_c(t) + f^{i(t)} & \text{(16.19e)} \\
x_\ell(t+1) &= f_\ell(x_\ell(t), u_\ell(t), \delta_e(t)). & \text{(16.19f)}
\end{aligned}
$$

A definition of well-posedness of DHA can be given similarly to Definition 16.1 by requiring that the successor states $x_c(t+1)$, $x_\ell(t+1)$ and the outputs $y_c(t)$, $y_\ell(t)$ are uniquely defined functions of $x_c(t)$, $x_\ell(t)$, $u_c(t)$, $u_\ell(t)$ defined by the DHA equations (16.19).

DHA can be considered as a subclass of *hybrid automata* (HA) [7]. The main difference is in the time model: DHA are based on discrete time, HA on continuous time. Moreover, DHA models do not allow instantaneous transitions, and are deterministic, contrary to HA where any enabled transition may occur in zero time. This has two consequences: (i) DHA do not admit live-locks (infinite switches in zero time); (ii) DHA do not admit Zeno behaviors (infinite switches in finite time). Finally, in DHA models, guards, reset maps and continuous dynamics are limited to linear (or affine) functions.

16.4 Logic and Mixed-Integer Inequalities

Despite the fact that DHA are rich in their expressiveness and are therefore quite suitable for modeling and simulating a wide class of hybrid dynamical systems, they are not directly suitable for solving optimal control problems because of their heterogeneous discrete and continuous nature. In this section we want to describe how DHA can be translated into different hybrid models that are more suitable for optimization. We highlight the main techniques of the translation process, by generalizing several results that appeared in the literature [126, 244, 287, 213, 83, 42, 81, 277, 286, 219].

16.4.1 Transformation of Boolean Relations

Boolean formulas can be equivalently represented as integer linear inequalities. For instance, $\delta_1 \vee \delta_2 = 1$ is equivalent to $\delta_1 + \delta_2 \geq 1$ [287]. Some equivalences are reported in Table 16.1. The results of the table can be generalized as follows.

Lemma 16.1 *For every Boolean formula $F(\delta_1, \delta_2, \ldots, \delta_n)$ there exists a polyhedral set P such that a set of binary values $\{\delta_1, \delta_2, \ldots, \delta_n\}$ satisfies the Boolean formula F if and only if $\delta = [\delta_1\ \delta_2\ \ldots\ \delta_n]' \in P$.*

Proof: Given a formula F, one way of constructing a polyhedron P is to rewrite F in the conjunctive normal form (16.15), and then simply define P as

$$P = \left\{ \delta \in \{0,1\}^n : \begin{array}{c} 1 \leq \displaystyle\sum_{i \in P_1} \delta_i + \sum_{i \in N_1}(1 - \delta_i) \\ \vdots \\ 1 \leq \displaystyle\sum_{i \in P_m} \delta_i + \sum_{i \in N_m}(1 - \delta_i) \end{array} \right\}. \tag{16.20}$$

∎

Table 16.1 Basic conversion of Boolean relations into mixed-integer inequalities. Relations involving the inverted literals $\sim \delta$ can be obtained by substituting $(1 - \delta)$ for δ in the corresponding inequalities. More conversions are reported in [210], or can be derived from (16.15)–(16.20)

Relation	Boolean	Linear constraints
AND	$\delta_1 \wedge \delta_2$	$\delta_1 = 1, \delta_2 = 1$ or $\delta_1 + \delta_2 \geq 2$
OR	$\delta_1 \vee X_2$	$\delta_1 + \delta_2 \geq 1$
NOT	$\sim \delta_1$	$\delta_1 = 0$
XOR	$\delta_1 \oplus \delta_2$	$\delta_1 + \delta_2 = 1$
IMPLY	$\delta_1 \rightarrow \delta_2$	$\delta_1 - \delta_2 \leq 0$
IFF	$\delta_1 \leftrightarrow \delta_2$	$\delta_1 - \delta_2 = 0$
ASSIGNMENT		$\delta_1 + (1 - \delta_3) \geq 1$
$\delta_3 = \delta_1 \wedge \delta_2$	$\delta_3 \leftrightarrow \delta_1 \wedge \delta_2$	$\delta_2 + (1 - \delta_3) \geq 1$
		$(1 - \delta_1) + (1 - \delta_2) + \delta_3 \geq 1$

The smallest polyhedron P associated with formula F has the following geometric interpretation: Assume to list all the 0-1 combinations of δ_i's satisfying F (namely, to generate the truth table of F), and think of each combination as an n-dimensional binary vector in $\mathbb{R}^n$, then P is the convex hull of such vectors [83, 153, 211]. For methods to compute convex hulls, we refer the reader to [111].

16.4.2   Translating DHA Components into Linear Mixed-Integer Relations

Events of the form (16.12) can be expressed equivalently as

$$h^i(x_c(t), u_c(t), t) \leq M^i(1 - \delta_e^i), \tag{16.21a}$$

$$h^i(x_c(t), u_c(t), t) > m^i \delta_e^i, \qquad i = 1, \ldots, n_e, \tag{16.21b}$$

where M^i, m^i are upper and lower bounds, respectively, on $h^i(x_c(t), u_c(t), t)$. As we have pointed out in Section 16.2.1, from a computational viewpoint it is convenient to avoid strict inequalities. As suggested in [287], we modify the strict inequality (16.21b) into

$$h^i(x_c(t), u_c(t), t) \geq \epsilon + (m^i - \epsilon)\delta_e^i \tag{16.21c}$$

where ϵ is a small positive scalar (e.g., the machine precision). Clearly, as for the case of discontinuous PWA discussed in Section 16.2.1, Equations (16.21) or (16.12) are equivalent to (16.21c) only for $h^i(x_c(t), u_c(t), t) \leq 0$ and $h^i(x_c(t), u_c(t), t) \geq \epsilon$.

Regarding switched affine dynamics, we first rewrite the state-update equation (16.11a) as the following combination of affine terms and *if-then-else* conditions:

$$z_1(t) = \begin{cases} A^1 x_c(t) + B^1 u_c(t) + f^1, & \text{if } (i(t) = 1), \\ 0, & \text{otherwise,} \end{cases} \tag{16.22a}$$

$$\vdots$$

$$z_s(t) = \begin{cases} A^s x_c(t) + B^s u_c(t) + f^s, & \text{if } (i(t) = s), \\ 0, & \text{otherwise,} \end{cases} \tag{16.22b}$$

$$x_c(t+1) = \sum_{i=1}^{s} z_i(t), \tag{16.22c}$$

where $z_i(t) \in \mathbb{R}^{n_c}, i = 1, \ldots, s$. The output Equation (16.11b) admits a similar transformation.

A generic if-then-else construct of the form

$$\text{IF } \delta \text{ THEN } z = a^{1'} x + b^{1'} u + f^1 \text{ ELSE } z = a^{2'} x + b^{2'} u + f^2, \tag{16.23}$$

where $\delta \in \{0,1\}$, $z \in \mathbb{R}$, $x \in \mathbb{R}^n$, $u \in \mathbb{R}^m$, and $a^1, b^1, f^1, a^2, b^2, f^2$ are constants of suitable dimensions, can be translated into [45]

$$(m_2 - M_1)\delta + z \le a^{2'} x + b^{2'} u + f^2, \tag{16.24a}$$

$$(m_1 - M_2)\delta - z \le -a^{2'} x - b^{2'} u - f^2, \tag{16.24b}$$

$$(m_1 - M_2)(1 - \delta) + z \le a^{1'} x + b^{1'} u + f^1, \tag{16.24c}$$

$$(m_2 - M_1)(1 - \delta) - z \le -a^{1'} x - b^{1'} u - f^1, \tag{16.24d}$$

where M_i, m_i are upper and lower bounds on $a^i x + b^i u + f^i$, $i = 1, 2$.

Finally, the mode selector function and binary state-update function of the automaton are Boolean functions that can be translated into integer linear inequalities as described in Section 16.4.1. The idea of transforming a well-posed FSM into a set of Boolean equalities was also presented in [229] where the authors performed model checking using (mixed) integer optimization on an equivalent set of integer inequalities.

16.5 **Mixed Logical Dynamical Systems**

Given a DHA representation of a hybrid process, by following the techniques described in the previous section for converting logical relations into inequalities we obtain an equivalent representation of the DHA as a *mixed logical dynamical* (MLD) system [42] described by the following relations:

$$x(t+1) = Ax(t) + B_1 u(t) + B_2 \delta(t) + B_3 z(t) + B_5, \tag{16.25a}$$

$$y(t) = Cx(t) + D_1 u(t) + D_2 \delta(t) + D_3 z(t) + D_5, \tag{16.25b}$$

$$E_2 \delta(t) + E_3 z(t) \le E_1 u(t) + E_4 x(t) + E_5, \tag{16.25c}$$

where $x \in \mathbb{R}^{n_c} \times \{0,1\}^{n_\ell}$ is a vector of continuous and binary states, $u \in \mathbb{R}^{m_c} \times \{0,1\}^{m_\ell}$ are the inputs, $y \in \mathbb{R}^{p_c} \times \{0,1\}^{p_\ell}$ the outputs, $\delta \in \{0,1\}^{r_\ell}$ are auxiliary binary variables, $z \in \mathbb{R}^{r_c}$ are auxiliary continuous variables which arise in the transformation (see Example 16.5), and $A, B_1, B_2, B_3, C, D_1, D_2, D_3, E_1, \ldots, E_5$ are matrices of suitable dimensions. Given the current state $x(t)$ and input $u(t)$, the time-evolution of (16.25) is determined by finding a feasible value for $\delta(t)$ and $z(t)$ satisfying (16.25c), and then by computing $x(t+1)$ and $y(t)$ from (16.25a)–(16.25b).

As MLD models consist of a collection of linear difference equations involving both real and binary variables and a set of linear inequality constraints, they are model representations of hybrid systems that can be easily used in optimization algorithms, as will be described in Chapter 17.

A definition of well-posedness of the MLD system (16.25) can be given similarly to Definition 16.1 by requiring that for all $x(t)$ and $u(t)$ within a given bounded set the pair of variables $\delta(t)$, $z(t)$ satisfying (16.25c) is unique, so that the successor state $x(t+1)$ and output $y(t)$ are also uniquely defined functions of $x(t)$, $u(t)$ through (16.25a)–(16.25b). Such a well-posedness assumption is usually guaranteed by the procedure described in Section 16.3.3 used to generate the linear inequalities (16.25c). A numerical test for well posedness is reported in [42, Appendix 1]. For a more general definition of well posedness of MLD systems see [42].

Note that the constraints (16.25c) allow one to specify additional linear constraints on continuous variables (e.g., constraints over physical variables of the system), and logical constraints over Boolean variables. The ability to include constraints, constraint prioritization, and heuristics adds to the expressiveness and generality of the MLD framework. Note also that despite the fact that the description (16.25) seems to be linear, clearly the nonlinearity is concentrated in the integrality constraints over binary variables.

Example 16.5   Consider the following simple switched linear system [42]

$$x(t+1) = \begin{cases} 0.8x(t) + u(t) & \text{if} \quad x(t) \geq 0 \\ -0.8x(t) + u(t) & \text{if} \quad x(t) < 0 \end{cases} \tag{16.26}$$

where $x(t) \in [-10, 10]$, and $u(t) \in [-1, 1]$. The condition $x(t) \geq 0$ can be associated to an event variable $\delta(t) \in \{0, 1\}$ defined as

$$[\delta(t) = 1] \;\leftrightarrow\; [x(t) \geq 0]. \tag{16.27}$$

By using the transformations (16.21a)–(16.21c), Equation (16.27) can be expressed by the inequalities

$$-m\delta(t) \leq x(t) - m \tag{16.28a}$$
$$-(M + \epsilon)\delta(t) \leq -x(t) - \epsilon \tag{16.28b}$$

where $M = -m = 10$, and ϵ is an arbitrarily small positive scalar. Then (16.26) can be rewritten as

$$x(t+1) = 1.6\delta(t)x(t) - 0.8x(t) + u(t). \tag{16.29}$$

By defining a new variable $z(t) = \delta(t)x(t)$ which, by (16.23)–(16.24) can be expressed as

$$z(t) \leq M\delta(t) \tag{16.30a}$$
$$z(t) \geq m\delta(t) \tag{16.30b}$$
$$z(t) \leq x(t) - m(1 - \delta(t)) \tag{16.30c}$$
$$z(t) \geq x(t) - M(1 - \delta(t)) \tag{16.30d}$$

the evolution of system (16.26) is ruled by the linear equation

$$x(t + 1) = 1.6z(t) - 0.8x(t) + u(t) \qquad (16.31)$$

subject to the linear constraints (16.28) and (16.30). Therefore, the MLD equivalent representation of (16.26) for $x \in [-10, -\epsilon] \cup [0, 10]$ and $u \in [-1, 1]$ is given by collecting Equations (16.31), (16.28) and (16.30).

16.6 Model Equivalence

In the previous chapters we have presented three different classes of discrete-time hybrid models: PWA systems, DHA, and MLD systems. For what we described in Section 16.3.3, under the assumption that the set of valid states and inputs is bounded, DHA systems can always be equivalently described as MLD systems. Also, a PWA system is a special case of a DHA whose threshold events and mode selector function are defined by the PWA partition (16.1c). Therefore, a PWA system with bounded partition $\mathcal{C}$ can always be described as an MLD system (an efficient way of modeling PWA systems in MLD form is reported in [42]). The converse result, namely that MLD systems (and therefore DHA) can be represented as PWA systems, is less obvious. For any choice of δ, model (16.25) represents an affine system defined over a polyhedral domain. Under the assumption of well posedness these domains do not overlap. This result was proved formally in [36, 30, 122].

Such equivalence results are of interest because DHA are most suitable in the modeling phase, but MLD systems are most suitable for solving open-loop finite time optimal control problems, and PWA systems are most suitable for solving finite time optimal control problems in state feedback form, as will be described in Chapter 17.

16.7 The HYSDEL Modeling Language

A modeling language was proposed in [276] to describe DHA models, called HYbrid System DEscription Language (HYSDEL). The HYSDEL description of a DHA is an abstract modeling step. The associated HYSDEL compiler then translates the description into several computational models, in particular into an MLD model using the technique presented in Section 16.4, and a PWA form using either the approach of [122] or the approach of [30]. HYSDEL can generate also a simulator that runs as a function in MATLAB®. Both the HYSDEL compiler and the Hybrid Toolbox can import and convert HYSDEL models.

In this section we illustrate the functionality of HYSDEL through a set of examples. For more examples and the detailed syntax we refer the interested reader to [149].

Example 16.6 Consider the DHA system:

$$\text{SAS: } x_c'(t) = \begin{cases} x_c(t) + u_c(t) - 1, & \text{if} \quad i(t) = 1, \\ 2x_c(t), & \text{if} \quad i(t) = 2, \\ 2, & \text{if} \quad i(t) = 3, \end{cases} \tag{16.32a}$$

$$\text{EG: } \begin{cases} \delta_e(t) = [x_c(t) \geq 0], \\ \delta_f(t) = [x_c(t) + u_c(t) - 1 \geq 0], \end{cases} \tag{16.32b}$$

$$\text{MS: } i(t) = \begin{cases} 1, & \text{if} \quad \begin{bmatrix} \delta_e(t) \\ \delta_f(t) \end{bmatrix} = \begin{bmatrix} 0 \\ 0 \end{bmatrix}, \\ 2, & \text{if} \quad \delta_e(t) = 1, \\ 3, & \text{if} \quad \begin{bmatrix} \delta_e(t) \\ \delta_f(t) \end{bmatrix} = \begin{bmatrix} 0 \\ 1 \end{bmatrix}. \end{cases} \tag{16.32c}$$

The corresponding HYSDEL list is reported in Table 16.2.

The HYSDEL list is composed of two parts. The first one, called INTERFACE, contains the declaration of all variables and parameters, so that it is possible to make the proper type checks. The second part, IMPLEMENTATION, is composed of specialized sections where the relations among the variables are described.

The HYSDEL section AUX contains the declaration of the auxiliary variables used in the model. The HYSDEL section AD allows one to define Boolean variables from continuous ones, and is based exactly on the semantics of the event generator (EG)

Table 16.2 Sample HYSDEL list of system (16.32).

```
SYSTEM sample {
INTERFACE {
  STATE {
    REAL xr [-10, 10]; }
  INPUT {
          REAL ur [-2, 2]; }
}
IMPLEMENTATION {
  AUX {
    REAL z1, z2, z3;
    BOOL de, df, d1, d2, d3; }
  AD {
    de = xr >= 0;
    df = xr + ur - 1 >= 0; }
  LOGIC {
    d1 = ~de & ~df;
    d2 = de;
    d3 = ~de & df; }
  DA {
    z1 = {IF d1 THEN xr + ur - 1 };
    z2 = {IF d2 THEN 2 * xr };
    z3 = {IF (~de & df)  THEN 2 }; }
  CONTINUOUS {
    xr = z1 + z2 + z3; }
}}
```

described earlier. The HYSDEL section DA defines continuous variables according to if-then-else conditions. This section models part of the switched affine system (SAS), namely the variables z_i defined in (16.22a)–(16.22b). The CONTINUOUS section describes the linear dynamics, expressed as difference equations. This section models (16.22c). The section LOGIC allows one to specify arbitrary functions of Boolean variables.

Example 16.7 Consider again the PWA system described in Example 16.1. Assume that $[-5, 5] \times [-5, 5]$ is the set of states $x(t)$ of interest and $u(t) \in [-1, 1]$. By using HYSDEL the PWA system (16.3) is described as in Table 16.3 and the equivalent MLD form is obtained

Table 16.3 HYSDEL model of the PWA system described in Example 16.1

```
/* 2x2 PWA system */

SYSTEM pwa {

INTERFACE {
    STATE { REAL x1 [-5,5];
            REAL x2 [-5,5];
        }
    INPUT { REAL u [-1,1];
        }
    OUTPUT{ REAL y;
        }
    PARAMETER {
        REAL alpha = 60*pi/180;
        REAL C = cos(alpha);
        REAL S = sin(alpha);
        REAL MLD_epsilon = 1e-6;    }
    }

IMPLEMENTATION {
        AUX { REAL z1,z2;
              BOOL sign; }
        AD  { sign = x1<=0; }

        DA  { z1 = {IF sign THEN 0.8*(C*x1+S*x2)
                    ELSE 0.8*(C*x1-S*x2) };
              z2 = {IF sign THEN 0.8*(-S*x1+C*x2)
                    ELSE 0.8*(S*x1+C*x2) };  }

        CONTINUOUS { x1 = z1;
                     x2 = z2+u; }

        OUTPUT { y = x2;  }
    }
}
```

$$x(t+1) = \begin{bmatrix} 1 & 0 \\ 0 & 1 \end{bmatrix} z(t) + \begin{bmatrix} 0 \\ 1 \end{bmatrix} u(t)$$
$$y(t) = \begin{bmatrix} 0 & 1 \end{bmatrix} x(t)$$

$$\begin{bmatrix} -5-\epsilon \\ 5 \\ c_1 \\ c_1 \\ -c_1 \\ -c_1 \\ c_1 \\ c_1 \\ -c_1 \\ -c_1 \end{bmatrix} \delta(t) + \begin{bmatrix} 0 & 0 \\ 0 & 0 \\ -1 & 0 \\ 1 & 0 \\ -1 & 0 \\ 1 & 0 \\ 0 & -1 \\ 0 & 1 \\ 0 & -1 \\ 0 & 1 \end{bmatrix} z(t) \le \begin{bmatrix} 1 & 0 \\ -1 & 0 \\ -0.4 & -c_2 \\ 0.4 & c_2 \\ -0.4 & c_2 \\ 0.4 & -c_2 \\ c_2 & -0.4 \\ -c_2 & 0.4 \\ -c_2 & -0.4 \\ c_2 & 0.4 \end{bmatrix} x(t) + \begin{bmatrix} -\epsilon \\ 5 \\ c_1 \\ c_1 \\ 0 \\ 0 \\ c_1 \\ c_1 \\ 0 \\ 0 \end{bmatrix}$$

where $c_1 = 4(\sqrt{3}+1)$, $c_2 = 0.4\sqrt{3}$, $\epsilon = 10^{-6}$. Note that in Table 16.3 the OUTPUT section allows to specify a linear map for the output vector y.

Example 16.8 Consider again the hybrid spring-mass-damper system described in Example 16.2. Assume that $[-5,5] \times [-5,5]$ is the set of states x and $[-10,10]$ the set of continuous inputs u_1 of interest. By using HYSDEL, system (16.6) is described as in Table 16.4 and an equivalent MLD model with 2 continuous states, 1 continuous input, 1 binary input, 9 auxiliary binary variables, 8 continuous variables, and 58 mixed-integer inequalities is obtained.

Example 16.9 Consider again the system with a logic state described in Example 16.3. The MLD model obtained by compiling the HYSDEL list of Table 16.5 is

$$x(t+1) = \begin{bmatrix} 0 & 0 \\ 0 & 1 \end{bmatrix} \delta(t) + \begin{bmatrix} 1 \\ 0 \end{bmatrix} z(t)$$

$$\begin{bmatrix} -9 & 0 \\ 11 & 0 \\ 0 & 0 \\ 0 & 0 \\ 0 & 0 \\ 0 & 0 \\ 0 & -1 \\ 1 & -1 \\ -1 & 1 \end{bmatrix} \delta(t) + \begin{bmatrix} 0 \\ 0 \\ -1 \\ 1 \\ -1 \\ 1 \\ 0 \\ 0 \\ 0 \end{bmatrix} z(t) \le \begin{bmatrix} 0 \\ 0 \\ -0.3 \\ 0.3 \\ -0.1 \\ 0.1 \\ 0 \\ 0 \\ 0 \end{bmatrix} u(t) + \begin{bmatrix} 1 & 0 \\ -1 & 0 \\ -0.5 & -14 \\ 0.5 & -14 \\ -0.5 & 14 \\ 0.5 & 14 \\ 0 & -1 \\ 0 & 0 \\ 0 & 1 \end{bmatrix} x(t) + \begin{bmatrix} 1 \\ 10 \\ 14 \\ 14 \\ 0 \\ 0 \\ 0 \\ 0 \\ 0 \end{bmatrix}$$

where we have assumed that $[-10,10]$ is the set of states x_1 and $[-10,10]$ the set of continuous inputs u of interest. In Table 16.5 the AUTOMATA section specifies the state transition equations of the finite state machine (FSM) as a collection of Boolean functions.

16.8 Literature Review

The lack of a general theory and of systematic design tools for systems having such a heterogeneous dynamical discrete and continuous nature led to a considerable interest in the study of *hybrid systems*. After the seminal work published in 1966 by Witsenhausen [288], who examined an optimal control problem for a class of hybrid-state continuous-time dynamical systems, there has been a renewed interest in the study of hybrid systems. The main reason for such an interest is probably the recent advent of technological innovations, in particular in the domain of embedded systems, where a logical/discrete decision device is "embedded" in a

Table 16.4 HYSDEL model of the spring-mass-damper system described in Example 16.2.
The A and B values are set according to equation (16.6).

```
/* Spring-Mass-Damper System
*/

SYSTEM springmass  {

INTERFACE { /* Description of variables and constants */

    STATE {
            REAL x1 [-5,5];
            REAL x2 [-5,5];
    }

    INPUT { REAL u1 [-10,10];
            BOOL u2;
        }


    PARAMETER {
            /* Spring breakpoint */
            REAL xm;

            /* Dynamic coefficients */
            REAL A111,A112,A121,A122,A211,A212,A221,A222;
            REAL A311,A312,A321,A322,A411,A412,A421,A422;
            REAL B111,B112,B121,B122,B211,B212,B221,B222;
            REAL B311,B312,B321,B322,B411,B412,B421,B422;
    }
}

IMPLEMENTATION {
    AUX {
            REAL zx11,zx12,zx21,zx22,zx31,zx32,zx41,zx42;
            BOOL region;
    }

    AD {    /* spring region */
            region = x1-xm <= 0;
    }

    DA {
            zx11 = { IF region  & u2  THEN A111*x1+A112*x2+B111*u1+B112};
            zx12 = { IF region  & u2  THEN A121*x1+A122*x2+B121*u1+B122};
            zx21 = { IF region  & ~u2 THEN A211*x1+A212*x2+B211*u1+B212};
            zx22 = { IF region  & ~u2 THEN A221*x1+A222*x2+B221*u1+B222};
            zx31 = { IF ~region & u2  THEN A311*x1+A312*x2+B311*u1+B312};
            zx32 = { IF ~region & u2  THEN A321*x1+A322*x2+B321*u1+B322};
            zx41 = { IF ~region & ~u2 THEN A411*x1+A412*x2+B411*u1+B412};
            zx42 = { IF ~region & ~u2 THEN A421*x1+A422*x2+B421*u1+B422};
    }

    CONTINUOUS {    x1=zx11+zx21+zx31+zx41;
                    x2=zx12+zx22+zx32+zx42;
    }
  }
}
```

Table 16.5 HYSDEL model of the system with logic state described in Example 16.3.

```
/* System with logic state */

SYSTEM SLS  {
INTERFACE { /* Description of variables and constants */
    STATE {
            REAL x1 [-10,10];
            BOOL x2;
    }
    INPUT { REAL u [-10,10];
        }
    PARAMETER {

            /* Lower Bound Point */
            REAL xlb = -1;

            /* Dynamic coefficients */
            REAL a = .5;
            REAL b1 =.1;
            REAL b2 =.3;
    }
}

IMPLEMENTATION {
    AUX {BOOL region;
        REAL zx1;
    }
    AD {  /* PWA Region */
            region = x1-xlb <= 0;
    }
    DA { zx1={IF x2 THEN a*x1+b2*u  ELSE a*x1+b1*u};
    }
    CONTINUOUS {     x1=zx1;
    }
    AUTOMATA { x2= x2 | region;
    }
  }
}
```

physical dynamical environment to change the behavior of the environment itself. Another reason is the availability of several software packages for simulation and numerical/symbolic computation that support the theoretical developments.

Several modelling frameworks for hybrid systems have appeared in the literature. We refer the interested reader to [8, 127] and the references therein. Each class is usually tailored to solve a particular problem, and many of them look largely dissimilar, at least at first sight. Two main categories of hybrid systems were successfully adopted for analysis and synthesis [66]: *hybrid control systems* [195, 196], where continuous dynamical systems and discrete/logic automata interact (see Figure 16.1), and *switched systems* [266, 67, 155, 292, 261], where the state space is partitioned into regions, each one being associated with different continuous dynamics (see Figure 16.2).

Today, there is a widespread agreement in defining hybrid systems as dynamical systems that switch among many operating modes, where each mode is governed by its own characteristic dynamical laws, and mode transitions are triggered by variables crossing specific thresholds (state events), by the lapse of certain time periods (time events), or by external inputs (input events) [8]. In the literature, systems whose mode only depends on external inputs are usually called *switched systems*, the others *switching systems*.

Complex systems organized in a hierarchial manner, where, for instance, discrete planning algorithms at the higher level interact with continuous control algorithms and processes at the lower level, are another example of hybrid systems. In these systems, the hierarchical organization helps to manage the complexity of the system, as higher levels in the hierarchy require less detailed models (also called *abstractions*) of the lower levels functions.

Hybrid systems arise in a large number of application areas and are attracting increasing attention in both academic theory-oriented circles as well as in industry, for instance in the automotive industry [16, 157, 62, 125, 74, 214]. Moreover, many physical phenomena admit a natural hybrid description, like circuits involving relays or diodes [28], biomolecular networks [6], and TCP/IP networks in [150].

In this book we work exclusively with dynamical systems formulated in discrete time. Therefore this chapter focuses on hybrid models formulated in discrete time. Though the effects of sampling can be neglected in most applications, some interesting mathematical phenomena occurring in hybrid systems, such as Zeno behaviors [158] do not exist in discrete time, as switches can only occur at sampling instants. On the other hand, most of these phenomena are usually a consequence of the continuous-time switching model, rather than the real behavior. Our main motivation for concentrating on discrete-time models is that optimal control problems are easier to formulate and to solve numerically than continuous-time formulations.

In the theory of hybrid systems, several problems were investigated in the last few years. Besides the issues of existence and computation of trajectories described in Section 16.1, several other issues were considered. These include: equivalence of hybrid models, stability and passivity analysis, reachability analysis and verification of safety properties, controller synthesis, observability analysis, state estimation and fault detection schemes, system identification, stochastic and event-driven dynamics. We will briefly review some of these results in the next paragraphs and provide pointers to some relevant literature references.

Equivalence of Linear Hybrid Systems

Under the condition that the MLD system is well-posed, the result showing that an MLD systems admits an equivalent PWA representation was proved in [36]. A slightly different and more general proof is reported in [30], where the author also provides efficient MLD to PWA translation algorithms. A different algorithm for obtaining a PWA representation of a DHA is reported in [122].

The fact that PWA systems are equivalent to interconnections of linear systems and finite automata was pointed out by Sontag [267]. In [148, 40] the authors proved the equivalence of discrete-time PWA/MLD systems with other classes of discrete-time hybrid systems (under some assumptions) such as linear complementarity (LC) systems [146, 279, 147], extended linear complementarity (ELC) systems [94], and max-min-plus-scaling (MMPS) systems [95].

Stability Analysis

Piecewise quadratic Lyapunov stability is becoming a standard in the stability analysis of hybrid systems [155, 96, 102, 233, 234]. It is a deductive way to prove the stability of an equilibrium point of a subclass of hybrid systems (piecewise affine systems). The computational burden is usually low, at the price of a convex relaxation of the problem, that leads to possibly conservative results. Such conservativeness can be reduced by constructing piecewise polynomial Lyapunov functions via semidefinite programming by means of the sum of squares (SOS) decomposition of multivariate polynomials [230]. SOS methods for analyzing stability of continuous-time hybrid and switched systems are described in [238]. For the general class of switched systems of the form $\dot{x} = f_i(x)$, $i = 1, \ldots, s$, an extension of the Lyapunov criterion based on multiple Lyapunov functions was introduced in [67]. The reader is also referred to the book by Liberzon [191].

The research on stability criteria for PWA systems has been motivated by the fact that the stability of each component subsystem is not sufficient to guarantee stability of a PWA system (and vice versa). Branicky [67], gives an example where stable subsystems are suitably combined to generate an unstable PWA system. Stable systems constructed from unstable ones have been reported in [278]. These examples point out that restrictions on the switching have to be imposed in order to guarantee that a PWA composition of stable components remains stable.

Passivity analysis of hybrid models has received little attention, except for the contributions of [77, 199, 295] and [236], in which notions of passivity for continuous-time hybrid systems are formulated, and of [31], where passivity and synthesis of passifying controllers for discrete-time PWA systems are investigated.

Reachability Analysis and Verification of Safety Properties

Although simulation allows to probe a model for a certain initial condition and input excitation, any analysis based on simulation is likely to miss the subtle phenomena that a model may generate, especially in the case of hybrid models. Reachability analysis (also referred to as "safety analysis" or "formal verification"), aims at detecting if a hybrid model will eventually reach unsafe state configurations or satisfy a temporal logic formula [7] for *all* possible initial conditions and input excitations within a prescribed set. Reachability analysis relies on a reach set computation algorithm, which is strongly related to the mathematical model of the system. In the case of MLD systems, for example, the reachability analysis problem over a finite time horizon (also referred to as *bounded model checking*) can be cast as a mixed-integer feasibility problem. Reachability analysis was also investigated via *bisimulation* ideas, namely by analyzing the properties of a simpler, abstracted system instead of those of the original hybrid dynamics [182].

Timed automata and hybrid automata have proved to be a successful modeling framework for formal verification and have been widely used in the literature. The starting point for both models is a finite state machine equipped with continuous dynamics. In the theory of *timed automata*, the dynamic part is the continuous-time flow $\dot{x} = 1$. Efficient computational tools complete the theory of timed automata and allow one to perform verification and scheduling of such models. Timed automata were extended to *linear hybrid automata* [7], where the dynamics is modeled by the differential inclusion $a \leq \dot{x} \leq b$. Specific tools allow one to verify such models against safety and liveness requirements. Linear hybrid automata were further extended to *hybrid automata* where the continuous dynamics is governed by differential equations. Tools exist to model and analyze those systems, either directly or by approximating the model with timed automata or linear hybrid automata (see the survey paper [265]).

Control

The majority of the control approaches for hybrid systems is based on optimal control ideas (see the survey [291]). For continuous-time hybrid systems, most authors either studied necessary conditions for a trajectory to be optimal [235, 271], or focused on the computation of optimal/suboptimal solutions by means of dynamic programming or the maximum principle [128, 245, 143, 144, 80, 292, 192, 261, 264, 266, 273, 70, 141, 196, 128, 69, 252, 71] The hybrid optimal control problem becomes less complex when the dynamics is expressed in discrete-time, as the main source of complexity becomes the combinatorial (yet finite) number of possible switching sequences. As will be shown in Chapter 17, optimal control problems can be solved for discrete-time hybrid systems using either the PWA or the MLD models described in this chapter. The solution to optimal control problems for discrete-time hybrid systems was first outlined by Sontag in [266]. In his plenary presentation [202] at the 2001 European Control Conference Mayne presented an intuitively appealing characterization of the state feedback solution to optimal control problems for linear hybrid systems with performance criteria based on quadratic and piecewise linear norms. The detailed exposition presented in the first part of Chapter 17 follows a similar line of argumentation and shows that the state feedback solution to the finite time optimal control problem is a time-varying piecewise affine feedback control law, possibly defined over nonconvex regions.

Model Predictive Control for discrete-time PWA systems and their stability and robustness properties have been studied in [68, 60, 32, 188, 185, 186]. Invariant sets computation for PWA systems has also been studied in [187, 4].

Observability and State Estimation

Observability of hybrid systems is a fundamental concept for understanding if a state observer can be designed for a hybrid system and how well it will perform.

In [36] the authors show that observability properties (as well as reachability properties) can be very complex and present a number of counterexamples that rule out obvious conjectures about inheriting observability/controllability properties from the composing linear subsystems. They also provide observability tests based on linear and mixed-integer linear programming.

State estimation is the reconstruction of the value of unmeasurable state variables based on output measurements. While state estimation is primarily required for output-feedback control, it is also important in problems of monitoring and fault detection [41, 17]. Observability properties of hybrid systems were directly exploited for designing convergent state estimation schemes for hybrid systems in [103].

Identification

Identification techniques for piecewise affine systems were recently developed [104, 255, 163, 39, 165, 282, 227], that allow one to derive models (or parts of models) from input/output data.

Extensions to Event-driven and Stochastic Dynamics

The discrete-time methodologies described in this chapter were employed in [73] to tackle event-based continuous-time hybrid systems with integral continuous dynamics, called *integral continuous-time hybrid automata* (icHA). The hybrid dynamics is translated into an equivalent MLD form, where continuous-time is an additional state variable and the index t counts events rather than time steps. Extensions of DHA to discrete-time stochastic hybrid dynamics were proposed in [33], where discrete-state transitions depend on both deterministic and stochastic events.

17

Optimal Control of Hybrid Systems

In this chapter we study the finite time, infinite time and receding horizon optimal control problem for the class of hybrid systems presented in the previous chapter. We establish the structure of the optimal control law and derive several different algorithms for its computation. For finite time problems with linear and quadratic objective functions we show that the time varying feedback law is piecewise affine.

17.1 Problem Formulation

Consider the PWA system (16.1) subject to hard input and state constraints

$$Ex(t) + Lu(t) \leq M \tag{17.1}$$

for $t \geq 0$, and rewrite its restriction over the set of states and inputs defined by (17.1) as

$$x(t+1) = A^i x(t) + B^i u(t) + f^i \quad \text{if} \quad \begin{bmatrix} x(t) \\ u(t) \end{bmatrix} \in \tilde{\mathcal{C}}^i \tag{17.2}$$

where $\{\tilde{\mathcal{C}}^i\}_{i=0}^{s-1}$ is the new polyhedral partition of the sets of state+input space $\mathbb{R}^{n+m}$ obtained by intersecting the sets $\mathcal{C}^i$ in (16.1c) with the polyhedron described by (17.1). In this chapter we will assume that the sets $\mathcal{C}^i$ are polytopes.

Define the cost function

$$J_0(x(0), U_0) = p(x_N) + \sum_{k=0}^{N-1} q(x_k, u_k) \tag{17.3}$$

where x_k denotes the state vector at time k obtained by starting from the state $x_0 = x(0)$ and applying to the system model

$$x_{k+1} = A^i x_k + B^i u_k + f^i \quad \text{if} \quad \begin{bmatrix} x_k \\ u_k \end{bmatrix} \in \tilde{\mathcal{C}}^i \tag{17.4}$$

the input sequence $u_0, \ldots, u_{k-1}$.

If the 1-norm or ∞-norm is used in the cost function (17.3), then we set $p(x_N) = \|Px_N\|_p$ and $q(x_k, u_k) = \|Qx_k\|_p + \|Ru_k\|_p$ with $p = 1$ or $p = \infty$ and P, Q, R full column rank matrices. Cost (17.3) is rewritten as

$$J_0(x(0), U_0) = \|Px_N\|_p + \sum_{k=0}^{N-1} \|Qx_k\|_p + \|Ru_k\|_p. \qquad (17.5)$$

If the squared Euclidian norm is used in the cost function (17.3), then we set $p(x_N) = x_N' P x_N$ and $q(x_k, u_k) = x_k' Q x_k + u_k' R u_k$ with $P \succeq 0$, $Q \succeq 0$ and $R \succ 0$. Cost (17.3) is rewritten as

$$J_0(x(0), U_0) = x_N' P x_N + \sum_{k=0}^{N-1} x_k' Q x_k + u_k' R u_k. \qquad (17.6)$$

Consider the constrained finite time optimal control problem (CFTOC)

$$J_0^*(x(0)) = \min_{U_0} J_0(x(0), U_0) \qquad (17.7a)$$

$$\text{subj. to} \begin{cases} x_{k+1} = A^i x_k + B^i u_k + f^i & \text{if } \begin{bmatrix} x_k \\ u_k \end{bmatrix} \in \tilde{\mathcal{C}}^i \\ x_N \in \mathcal{X}_f \\ x_0 = x(0) \end{cases} \qquad (17.7b)$$

where the column vector $U_0 = [u_0', \ldots, u_{N-1}']' \in \mathbb{R}^{m_c N} \times \{0,1\}^{m_\ell N}$, is the optimization vector, N is the time horizon and $\mathcal{X}_f$ is the terminal region.

In general, the optimal control problem (17.7) may not have a minimizer for some feasible $x(0)$. This is caused by discontinuity of the PWA system in the input space. We will assume, however, that a minimizer $U_0^*(x(0))$ exists for all feasible $x(0)$. Also the optimizer function U_0^* may not be uniquely defined if the optimal set of problem (17.7) is not a singleton for some $x(0)$. In this case U_0^* denotes one of the optimal solutions.

We will denote by $\mathcal{X}_k \subseteq \mathbb{R}^{n_c} \times \{0,1\}^{n_\ell}$ the set of states x_k that are feasible for (17.7):

$$\mathcal{X}_k = \left\{ x \in \mathbb{R}^{n_c} \times \{0,1\}^{n_\ell} \;\middle|\; \begin{array}{l} \exists u \in \mathbb{R}^{m_c} \times \{0,1\}^{m_\ell}, \\ \exists i \in \{1, \ldots, s\} \\ \begin{bmatrix} x \\ u \end{bmatrix} \in \tilde{\mathcal{C}}^i \text{ and} \\ A^i x + B^i u + f^i \in \mathcal{X}_{k+1} \end{array} \right\},$$

$$k = 0, \ldots, N-1 \qquad (17.8)$$

$$\mathcal{X}_N = \mathcal{X}_f.$$

The definition of $\mathcal{X}_i$ requires that for any initial state $x_i \in \mathcal{X}_i$ there exists a feasible sequence of inputs $U_i = [u_i', \ldots, u_{N-1}']$ which keeps the state evolution in the feasible set $\mathcal{X}$ at future time instants $k = i+1, \ldots, N-1$ and forces x_N into $\mathcal{X}_f$ at time N. The sets $\mathcal{X}_k$ for $i = 0, \ldots, N$ play an important role in the solution of the optimal control problem. They are independent of the cost function and of the algorithm used to compute the solution to problem (17.7). As in the case of linear systems (see Chapter 11.2) there are two ways to rigourously define and compute the sets $\mathcal{X}_i$: the *batch approach* and the *recursive approach*. In this chapter we will not discuss the details on the computation of $\mathcal{X}_i$. Also, we will not discuss invariant and reachable sets for hybrid systems. While the basic concepts are identical to those presented in Section 10.1 for linear systems, the discussion of efficient algorithms requires a careful treatment which goes beyond the scope of this book. The interested reader is referred to the work in [131, 121] for a detailed discussion on reachable and invariant sets for hybrid systems.

In the following we need to distinguish between optimal control based on the squared 2-norm and optimal control based on the 1-norm or ∞-norm. Note that the results of this chapter also hold when the number of switches is weighted in the cost function (17.3), if this is meaningful in a particular situation.

In this chapter we will make use of the following definition. Consider system (17.2) and recall that, in general, $x = \left[\begin{smallmatrix} x_c \\ x_\ell \end{smallmatrix}\right]$ where $x_c \in \mathbb{R}^{n_c}$ are the continuous states and $x_\ell \in \mathbb{R}^{n_\ell}$ are the binary states and $u \in \mathbb{R}^{m_c} \times \{0,1\}^{m_\ell}$ where $u_c \in \mathbb{R}^{m_c}$ are the continuous inputs and $u_\ell \in \mathbb{R}^{m_\ell}$ are the binary inputs (Section 16.2.2). We will make the following assumption.

Assumption 17.1   *For the discrete-time PWA system (17.2) the mapping $(x_c(t), u_c(t)) \mapsto x_c(t+1)$ is continuous.*

Assumption 17.1 requires that the PWA function that defines the update of the continuous states is continuous on the boundaries of contiguous polyhedral cells, and therefore allows one to work with the closure of the sets $\tilde{\mathcal{C}}^i$ without the need of introducing multivalued state update equations. With abuse of notation in the next sections $\tilde{\mathcal{C}}^i$ will always denote the closure of $\tilde{\mathcal{C}}^i$. Discontinuous PWA systems will be discussed in Section 17.7.

17.2   Properties of the State Feedback Solution, 2-Norm Case

Theorem 17.1   *Consider the optimal control problem (17.7) with cost (17.6) and let assumption 17.1 hold. Then, there exists a solution in the form of a PWA state feedback control law*

$$u_k^*(x(k)) = F_k^i x(k) + g_k^i \quad \text{if } x(k) \in \mathcal{R}_k^i, \tag{17.9}$$

where $\mathcal{R}_k^i$, $i = 1, \ldots, N_k$ is a partition of the set $\mathcal{X}_k$ of feasible states $x(k)$, and the closure $\bar{\mathcal{R}}_k^i$ of the sets $\mathcal{R}_k^i$ has the following form:

$$\bar{\mathcal{R}}_k^i = \{x : \ x(k)' L_k^i(j) x(k) + M_k^i(j)' x(k) \le N_k^i(j),$$
$$j = 1, \ldots, n_k^i\}, \ k = 0, \ldots, N-1, \tag{17.10}$$

and

$$x(k+1) = A^i x(k) + B^i u_k^*(x(k)) + f^i$$
$$\text{if } \left[\begin{smallmatrix} x(k) \\ u_k^*(x(k)) \end{smallmatrix}\right] \in \tilde{\mathcal{C}}^i, \ i = \{1, \ldots, s\}. \tag{17.11}$$

Proof: The piecewise linearity of the solution was first mentioned by Sontag in [266]. In [202] Mayne sketched a proof. In the following we will give the proof for $u_0^*(x(0))$, the same arguments can be repeated for $u_1^*(x(1)), \ldots, u_{N-1}^*(x(N-1))$.

- **Case 1.** *($m_l = n_l = 0$) no binary inputs and states*
  Depending on the initial state $x(0)$ and on the input sequence $U = [u_0', \ldots, u_k']$, the state x_k is either infeasible or it belongs to a certain polyhedron $\tilde{C}^i$, $k = 0, \ldots, N-1$. The number of all possible locations of the state sequence $x_0, \ldots, x_{N-1}$ is equal to s^N. Denote by $\{v_i\}_{i=1}^{s^N}$ the set of all possible switching sequences over the horizon N, and by v_i^k the k-th element of the sequence v_i, i.e., $v_i^k = j$ if $x_k \in \tilde{C}^j$.

Fix a certain v_i and constrain the state to switch according to the sequence v_i. Problem (17.3)–(17.7) becomes

$$J_{v_i}^*(x(0)) = \min_{\{U_0\}} J_0(U_0, x(0)) \tag{17.12a}$$

$$\text{subj. to} \quad \begin{cases} x_{k+1} = A^{v_i^k} x_k + B^{v_i^k} u_k + f^{v_i^k} \\ \begin{bmatrix} x_k \\ u_k \end{bmatrix} \in \tilde{C}^{v_i^k} \\ \quad\quad k = 0, \ldots, N-1 \\ x_N \in \mathcal{X}_f \\ x_0 = x(0) \end{cases} \tag{17.12b}$$

Problem (17.12) is equivalent to a finite time optimal control problem for a linear time-varying system with time-varying constraints and can be solved by using the approach described in Chapter 11. The first move u_0 of its solution is the PPWA feedback control law

$$u_0^i(x(0)) = \tilde{F}^{i,j} x(0) + \tilde{g}^{i,j}, \quad \forall x(0) \in \mathcal{T}^{i,j}, \quad j = 1, \ldots, N^{ri} \tag{17.13}$$

where $\mathcal{D}^i = \bigcup_{j=1}^{N^{ri}} \mathcal{T}^{i,j}$ is a polyhedral partition of the convex set $\mathcal{D}^i$ of feasible states $x(0)$ for problem (17.12). N^{ri} is the number of regions of the polyhedral partition of the solution and it is a function of the number of constraints in problem (17.12). The upper index i in (17.13) denotes that the input $u_0^i(x(0))$ is optimal when the switching sequence v_i is fixed.

The set $\mathcal{X}_0$ of all feasible states at time 0 is $\mathcal{X}_0 = \bigcup_{i=1}^{s^N} \mathcal{D}^i$ and in general it is not convex. Indeed, as some initial states can be feasible for different switching sequences, the sets $\mathcal{D}^i$, $i = 1, \ldots, s^N$, in general, can overlap. The solution $u_0^*(x(0))$ to the original problem (17.3)–(17.7) can be computed in the following way. For every polyhedron $\mathcal{T}^{i,j}$ in (17.13),

1. If $\mathcal{T}^{i,j} \cap \mathcal{T}^{l,m} = \emptyset$ for all $l \neq i$, $l = 1, \ldots, s^N$, and for all $m \neq j$, $m = 1, \ldots, N^{rl}$, then the switching sequence v_i is the only feasible one for all the states belonging to $\mathcal{T}^{i,j}$ and therefore the optimal solution is given by (17.13), i.e.,

$$u_0^*(x(0)) = \tilde{F}^{i,j} x(0) + \tilde{g}^{i,j}, \quad \forall x(0) \in \mathcal{T}^{i,j}. \tag{17.14}$$

2. If $\mathcal{T}^{i,j}$ intersects one or more polyhedra $\mathcal{T}^{l_1,m_1}, \mathcal{T}^{l_2,m_2}, \ldots$, the states belonging to the intersection are feasible for more than one switching

sequence v_i, v_{l_1}, v_{l_2}, ... and therefore the corresponding value functions $J^*_{v_i}$, $J^*_{v_{l_1}}$, $J^*_{v_{l_2}}$, ... in (17.12a) have to be compared in order to compute the optimal control law.

Consider the simple case when only two polyhedra overlap, i.e., $\mathcal{T}^{i,j} \cap \mathcal{T}^{l,m} = \mathcal{T}^{(i,j),(l,m)} \neq \emptyset$. We will refer to $\mathcal{T}^{(i,j),(l,m)}$ as a *double feasibility polyhedron*. For all states belonging to $\mathcal{T}^{(i,j),(l,m)}$ the optimal solution is:

$$
u^*_0(x(0)) = \begin{cases} \tilde{F}^{i,j} x(0) + \tilde{g}^{i,j}, & \forall x(0) \in \mathcal{T}^{(i,j),(l,m)}: \\ & J^*_{v_i}(x(0)) < J^*_{v_l}(x(0)) \\ \tilde{F}^{l,m} x(0) + \tilde{g}^{l,m}, & \forall x(0) \in \mathcal{T}^{(i,j),(l,m)}: \\ & J^*_{v_i}(x(0)) > J^*_{v_l}(x(0)) \\ \begin{cases} \tilde{F}^{i,j} x(0) + \tilde{g}^{i,j} \text{ or} \\ \tilde{F}^{l,m} x(0) + \tilde{g}^{l,m} \end{cases} & \forall x(0) \in \mathcal{T}^{(i,j),(l,m)}: \\ & J^*_{v_i}(x(0)) = J^*_{v_l}(x(0)) \end{cases} \tag{17.15}
$$

Because $J^*_{v_i}$ and $J^*_{v_l}$ are quadratic functions of $x(0)$ on $\mathcal{T}^{i,j}$ and $\mathcal{T}^{l,m}$ respectively, we find the expression (17.10) of the control law domain. The sets $\mathcal{T}^{i,j} \setminus \mathcal{T}^{l,m}$ and $\mathcal{T}^{l,m} \setminus \mathcal{T}^{i,j}$ are two *single feasibility P-collections* which can be partitioned into a set of *single feasibility polyhedra*, and thus be described through (17.10) with $L^i_k = 0$.

In order to conclude the proof, the general case of n intersecting polyhedra has to be discussed. We follow three main steps. *Step 1:* generate one polyhedron of nth-ple feasibility and $2^n - 2$ P-collections possibly empty and disconnected, of single, double, ..., $(n-1)^{th}$ ple feasibility. *Step 2:* the ith-ple feasibility P-collection is partitioned into several i^{th}-ple feasibility polyhedra. *Step 3:* any i^{th}-ple feasibility polyhedron with $i > 1$ is further partitioned into at most i subsets (17.10) where in each one of them a certain feasible value function is greater than all the others. The procedure is depicted in Figure 17.1 when $n = 3$.

- **Case 2.** *binary inputs*, $m_\ell \neq 0$
  The proof can be repeated in the presence of binary inputs, $m_\ell \neq 0$. In this case the switching sequences v_i are given by all combinations of region indices and *binary inputs*, i.e., $i = 1, \ldots, (s \cdot m_\ell)^N$. For each sequence v_i there is an associated optimal continuous component of the input as calculated in Case 1.

- **Case 3.** *binary states*, $n_l \neq 0$
  The proof can be repeated in the presence of binary states by a simple enumeration of all the possible n_ℓ^N discrete state evolutions. ∎

From the result of the theorem above one immediately concludes that the value function J^*_0 is piecewise quadratic:

$$
J^*_0(x(0)) = x(0)' H^i_1 x(0) + H^{i'}_2 x(0) + H^i_3 \quad \text{if } x(0) \in \mathcal{R}^i_0. \tag{17.16}
$$

The proof of Theorem 17.1 gives useful insights into the properties of the sets $\mathcal{R}^i_k$ in (17.10). We will summarize them next.

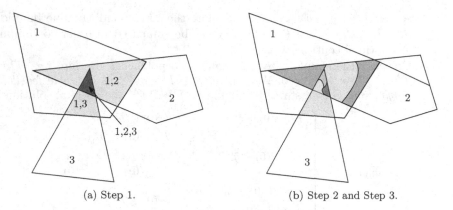

(a) Step 1.   (b) Step 2 and Step 3.

Figure 17.1 Graphical illustration of the main steps for the proof of Theorem 17.1 when 3 polyhedra intersect. Step 1: the three intersecting polyhedra are partitioned into: one polyhedron of triple feasibility (1,2,3), 2 polyhedra of double feasibility $(1, 2)$ and $(1, 3)$, 3 polyhedra of single feasibility $(1),(2),(3)$. The sets (1), (2) and $(1,2)$ are neither open nor closed polyhedra. Step 2: the single feasibility sets of (1), (2) and (3) are partitioned into six polyhedra. Step 3: value functions are compared inside the polyhedra of multiple feasibility.

Each set $\mathcal{R}_k^i$ has an associated multiplicity j which means that j switching sequences are feasible for problem (17.3)–(17.7) starting from a state $x(k) \in \mathcal{R}_k^i$. If $j = 1$, then $\mathcal{R}_k^i$ is a polyhedron. In general, if $j > 1$ the boundaries of $\mathcal{R}_k^i$ can be described either by an affine function or by a quadratic function. In the sequel boundaries which are described by quadratic functions but degenerate to hyperplanes or sets of hyperplanes will be considered affine boundaries.

Quadratic boundaries arise from the comparison of value functions associated with feasible switching sequences, thus a maximum of $j - 1$ quadratic boundaries can be present in a j-ple feasible set. The *affine* boundaries can be of three types.

- Type **a**: they are inherited from the original j-ple feasible P-collection. In this case across such boundaries the multiplicity of the feasibility changes.

- Type **b**: they are artificial cuts needed to describe the original j-ple feasible P-collection as a set of j-ple feasible polyhedra. Across type **b** boundaries the multiplicity of the feasibility does not change.

- Type **c**: they arise from the comparison of quadratic value functions which degenerate in an affine boundary.

In conclusion, we can state the following lemma

Lemma 17.1   *The value function J_k^**

1. *is a quadratic function of the states inside each $\mathcal{R}_k^i$*

2. *is continuous on quadratic and affine boundaries of type **b** and **c***

3. *may be discontinuous on affine boundaries of type **a**,*

and the optimizer u_k^*

1. *is an affine function of the states inside each* $\mathcal{R}_k^i$

2. *is continuous across and unique on affine boundaries of type* **b**

3. *is nonunique on quadratic boundaries, except possibly at isolated points*

4. *may be nonunique on affine boundaries of type* **c**

5. *may be discontinuous across affine boundaries of type* **a**

Based on Lemma 17.1 above one can highlight the only source of discontinuity of the value function: affine boundaries of *type a*. The following corollary gives a useful insight on the class of possible value functions.

Corollary 17.1 J_0^* *is a lower-semicontinuous PWQ function on* $\mathcal{X}_0$.

Proof: The proof follows from the result on the minimization of lower-semicontinuous point-to-set maps in [47]. Below we give a simple proof without introducing the notion of point-to-set maps.

Only points where a discontinuity occurs are relevant for the proof, i.e., states belonging to boundaries of type **a**. From assumption 17.1 it follows that the feasible switching sequences for a given state $x(0)$ are all the feasible switching sequences associated with any set $\mathcal{R}_0^j$ whose closure $\bar{\mathcal{R}}_0^j$ contains $x(0)$. Consider a state $x(0)$ belonging to boundaries of type **a** and the proof of Theorem 17.1. The only case of discontinuity can occur when *(i)* a j-ple feasible set $\mathcal{P}_1$ intersects an i-ple feasible set $\mathcal{P}_2$ with $i < j$, *(ii)* there exists a point $x(0) \in \mathcal{P}_1$, $\mathcal{P}_2$ and a neighborhood $\mathcal{N}(x(0))$ with $x, y \in \mathcal{N}(x(0))$, $x \in \mathcal{P}_1$, $x \notin \mathcal{P}_2$ and $y \in \mathcal{P}_2$, $y \notin \mathcal{P}_1$. The proof follows from the previous statements and the fact that $J_0^*(x(0))$ is the minimum of all $J_{v_i}^*(x(0))$ for all feasible switching sequences v_i. ∎

The result of Corollary 17.1 will be used extensively in the next sections. Even if value function and optimizer are discontinuous, one can work with the closure $\bar{\mathcal{R}}_k^j$ of the original sets $\mathcal{R}_k^j$ without explicitly considering their boundaries. In fact, if a given state $x(0)$ belongs to several regions $\bar{\mathcal{R}}_0^1, \ldots, \bar{\mathcal{R}}_0^p$, then the minimum value among the optimal values (17.16) associated with each region $\bar{\mathcal{R}}_0^1, \ldots, \bar{\mathcal{R}}_0^p$ allows us to identify the region of the set $\mathcal{R}_0^1, \ldots, \mathcal{R}_0^p$ containing $x(0)$.

Next we show some interesting properties of the optimal control law when we restrict our attention to smaller classes of PWA systems.

Corollary 17.2   *Assume that the PWA system (17.2) is continuous, and that $E = 0$ in (17.1) and $\mathcal{X}_f = \mathbb{R}^n$ in (17.7) (which means that there are no state constraints, i.e., $\tilde{P}$ is unbounded in the x-space). Then, the value function J_0^* in (17.7) is continuous.*

Proof: Problem (17.7) becomes a multiparametric program with only input constraints when the state at time k is expressed as a function of the state at time 0 and the input sequence $u_0, \ldots, u_{k-1}$, i.e., $x_k = f_{PWA}((\cdots(f_{PWA}(x_0, u_0), u_1), \ldots, u_{k-2}), u_{k-1})$. J_0 in (17.3) will be a continuous function of x_0 and

$u_0, \ldots, u_{N-1}$ since it is the composition of continuous functions. By assumptions the input constraints on $u_0, \ldots, u_{N-1}$ are convex and the resulting set is compact. The proof follows from the continuity of J and Theorem 5.2.   ∎

Note that $E = 0$ is a sufficient condition for ensuring that constraints (17.1) are convex in the optimization variables $u_0, \ldots, u_n$. In general, even for continuous PWA systems with state constraints it is difficult to find weak assumptions ensuring the continuity of the value function J_0^*. Ensuring the continuity of the optimal control law $u(k) = u_k^*(x(k))$ is even more difficult. A list of sufficient conditions for U_0^* to be continuous can be found in [106]. In general, they require the convexity (or a relaxed form of it) of the cost $J_0(U_0, x(0))$ in U_0 for each $x(0)$ and the convexity of the constraints in (17.7) in U_0 for each $x(0)$. Such conditions are clearly very restrictive since the cost and the constraints in problem (17.7) are a composition of quadratic and linear functions, respectively, with the piecewise affine dynamics of the system.

The next theorem provides a condition under which the solution $u_k^*(x(k))$ of the optimal control problem (17.3)–(17.7) is a PPWA state feedback control law.

Theorem 17.2   *Assume that the optimizer $U_0^*(x(0))$ of (17.3)–(17.7) is unique for all $x(0)$. Then the solution to the optimal control problem (17.3)–(17.7) is a PPWA state feedback control law of the form*

$$u_k^*(x(k)) = F_k^i x(k) + g_k^i \quad \text{if } x(k) \in \mathcal{P}_k^i \quad k = 0, \ldots, N-1, \tag{17.17}$$

where $\mathcal{P}_k^i$, $i = 1, \ldots, N_k^r$, is a polyhedral partition of the set $\mathcal{X}_k$ of feasible states $x(k)$.

Proof: In Lemma 17.1 we concluded that the value function $J_0^*(x(0))$ is continuous on quadratic type boundaries. By hypothesis, the optimizer $u_0^*(x(0))$ is unique. Theorem 17.1 implies that $\tilde{F}^{i,j} x(0) + \tilde{g}^{i,j} = \tilde{F}^{l,m} x(0) + \tilde{g}^{l,m}$, $\forall x(0)$ belonging to the quadratic boundary. This can occur only if the quadratic boundary degenerates to a single feasible point or to affine boundaries. The same arguments can be repeated for $u_k^*(x(k))$, $k = 1, \ldots, N-1$.   ∎

Remark 17.1   Theorem 17.2 relies on a rather strong uniqueness assumption. Sometimes, problem (17.3)–(17.7) can be modified in order to obtain uniqueness of the solution and use the result of Theorem 17.2 which excludes the existence of ellipsoidal sets. It is reasonable to believe that there are other conditions under which the state feedback solution is PPWA without claiming uniqueness.

Example 17.1   Consider the following simple system

$$\begin{cases} x(t+1) = \begin{cases} \begin{bmatrix} 0 & -1 \\ 1 & 0 \end{bmatrix} x(t) + \begin{bmatrix} 0.2 \\ 0 \end{bmatrix} u(t) + \begin{bmatrix} -0.2 \\ 0.2 \end{bmatrix} \\ \qquad \text{if } x(t) \in \mathcal{C}^1 = \{x : [0\ 1]x \geq 0\} \\[1em] \begin{bmatrix} 0 & -1 \\ 1 & 0 \end{bmatrix} x(t) + \begin{bmatrix} 0.2 \\ 0 \end{bmatrix} u(t) - \begin{bmatrix} 0.2 \\ -0.2 \end{bmatrix} \\ \qquad \text{if } x(t) \in \mathcal{C}^2 = \{x : [0\ 1]x < 0\} \end{cases} \\ x(t) \in [-0.5, -0.5] \times [0.5, 0.5] \\ u(t) \in [-1000, 1000] \end{cases} \tag{17.18}$$

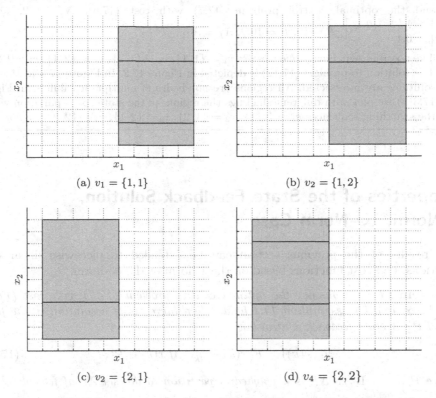

(a) $v_1 = \{1, 1\}$ (b) $v_2 = \{1, 2\}$

(c) $v_2 = \{2, 1\}$ (d) $v_4 = \{2, 2\}$

Figure 17.2 Example 17.1. Problem (17.12) with cost (17.6) is solved for different v_i, $i = 1, \ldots, 4$. Control law $u_0^*(x_0)$ is PPWA. State space partition is shown.

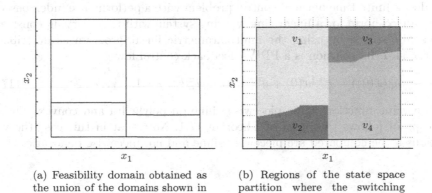

(a) Feasibility domain obtained as the union of the domains shown in Figure 17.2. All regions are polyhedra of multiple feasibility.

(b) Regions of the state space partition where the switching sequences $v_1 = \{1, 1\}$, $v_2 = \{1, 2\}$, $v_3 = \{2, 1\}$, $v_4 = \{2, 2\}$ are optimal.

Figure 17.3 Example 17.1. State space partition corresponding to the optimal control law.

and the optimal control problem (17.7) with cost (17.6), $N = 2$, $Q = \begin{bmatrix} 1 & -0.1 \\ -0.1 & 1 \end{bmatrix}$, $R = 10$, $P = 10Q$, $\mathcal{X}_f = \mathcal{X}$.

The possible switching sequences are $v_1 = \{1, 1\}$, $v_2 = \{1, 2\}$, $v_3 = \{2, 1\}$, $v_4 = \{2, 2\}$. The solution to problem (17.12) is depicted in Figure 17.2. In Figure 17.3(a) the four solutions are intersected. All regions are polyhedra of multiple feasibility. In Figure 17.3(b) we plot with different shadings the regions of the state space partition where the switching sequences $v_1 = \{1, 1\}$, $v_2 = \{1, 2\}$, $v_3 = \{2, 1\}$, $v_4 = \{2, 2\}$ are optimal.

17.3 Properties of the State Feedback Solution, 1-Norm, ∞-Norm Case

The results of the previous section can be extended to piecewise linear cost functions, i.e., cost functions based on the 1-norm or the ∞-norm.

Theorem 17.3 *Consider the optimal control problem (17.7) with cost (17.5), $p = 1$, ∞ and let assumption 17.1 hold. Then there exists a solution in the form of a PPWA state feedback control law*

$$u_k^*(x(k)) = F_k^i x(k) + g_k^i \quad \text{if } x(k) \in \mathcal{P}_k^i, \tag{17.19}$$

where $\mathcal{P}_k^i$, $i = 1, \dots, N_k^r$ is a polyhedral partition of the set $\mathcal{X}_k$ of feasible states $x(k)$.

Proof: The proof is similar to the proof of Theorem 17.1. Fix a certain switching sequence v_i, consider the problem (17.3)–(17.7) and constrain the state to switch according to the sequence v_i to obtain problem (17.12). Problem (17.12) can be viewed as a finite time optimal control problem with a performance index based on 1-norm or ∞-norm for a linear time varying system with time varying constraints and can be solved by using the multiparametric linear program as described in Chapter 11.4. Its solution is a PPWA feedback control law

$$u_0^i(x(0)) = \tilde{F}^{i,j} x(0) + \tilde{g}^{i,j}, \quad \forall x \in \mathcal{T}^{i,j}, \quad j = 1, \dots, N^{ri}, \tag{17.20}$$

and the value function $J_{v_i}^*$ is piecewise affine on polyhedra and convex. The rest of the proof follows the proof of Theorem 17.1. Note that in this case the value functions to be compared are piecewise affine and not piecewise quadratic. ∎

17.4 Computation of the Optimal Control Input via Mixed Integer Programming

In the previous section the properties enjoyed by the solution to hybrid optimal control problems were investigated. Despite the fact that the proofs are constructive (as shown in the figures), they are based on the enumeration of all the possible switching sequences of the hybrid system, the number of which grows exponentially with the time horizon. Although the computation is performed off-line (the on-line complexity is the one associated with the evaluation of the PWA control law (17.17)), more efficient methods than enumeration are desirable. Here we show that the MLD framework can be used to avoid the complete enumeration. In fact, when the model of the system is an MLD model and the performance index is quadratic, the optimization problem can be cast as a Mixed-Integer Quadratic Program (MIQP). Similarly, 1-norm and ∞-norm performance indices lead to Mixed-Integer Linear Programs (MILPs) problems. In the following we detail the translation of problem (17.7) with cost (17.5) or (17.6) into a mixed integer linear or quadratic program, respectively, for which efficient branch and bound algorithms exist.

Consider the equivalent MLD representation (16.25) of the PWA system (17.2). Problem (17.7) is rewritten as:

$$J_0^*(x(0)) = \min_{U_0} J_0(x(0), U_0) \qquad (17.21a)$$

$$\text{subj. to} \begin{cases} x_{k+1} = Ax_k + B_1 u_k + B_2 \delta_k + B_3 z_k \\ y_k = Cx_k + D_1 u_k + D_2 \delta_k + D_3 z_k + D_5 \\ E_2 \delta_k + E_3 z_k \le E_1 u_k + E_4 x_k + E_5 \\ x_N \in \mathcal{X}_f \\ x_0 = x(0) \end{cases} \qquad (17.21b)$$

Note that the cost function (17.21a) is of the form

$$J_0(x(0), U_0) = \|Px_N\|_p + \sum_{k=0}^{N-1} \|Qx_k\|_p + \|Ru_k\|_p + \|Q_\delta \delta_k\|_p + \|Q_z z_k\|_p \quad (17.22)$$

when $p = 1$ or $p = \infty$ or

$$J_0(x(0), U_0) = x_N' P x_N + \sum_{k=0}^{N-1} x_k' Q x_k + u_k' R u_k + \delta_k' Q_\delta \delta_k + z_k' Q_z z_k \qquad (17.23)$$

when p = 2.

The optimal control problem (17.21) with the cost (17.22) can be formulated as a *Mixed Integer Linear Program* (MILP). The optimal control problem (17.21) with the cost (17.23) can be formulated as a *Mixed Integer Quadratic Program* (MIQP). The compact form for both cases is

$$\min_{\varepsilon} \quad \varepsilon' H_1 \varepsilon + \varepsilon' H_2 x(0) + x(0)' H_3 x(0) + c_1' \varepsilon + c_2' x(0) + c$$

$$\text{subj. to} \quad G\varepsilon \le w + Sx(0) \qquad (17.24)$$

where H_1, H_2, H_3, c_1, c_2, G, w, S are matrices of suitable dimensions, $\varepsilon = [\varepsilon_c', \varepsilon_d']$ where ε_c, ε_d represent continuous and discrete variables, respectively and H_1, H_2, H_3, are null matrices if problem (17.24) is an MILP.

The translation of (17.21) with cost (17.23) into (17.24) is simply obtained by substituting the state update equation

$$x_k = A^k x_0 + \sum_{j=0}^{k-1} A^j (B_1 u_{k-1-j} + B_2 \delta_{k-1-j} + B_3 z_{k-1-j}) \tag{17.25}$$

The optimization vector ε in (17.24) is $\varepsilon = \{u_0, \ldots, u_{N-1}, \delta_0, \ldots, \delta_{N-1}, z_0, \ldots, z_{N-1}\}$.

The translation of (17.21) with cost (17.22) into (17.24) requires the introductions of slack variables as shown in Section 11.4 for the case of linear systems. In particular, for $p = \infty$, $Q_z = 0$ and $Q_\delta = 0$, the sum of the components of any vector $\{\varepsilon_0^u, \ldots, \varepsilon_{N-1}^u, \varepsilon_0^x, \ldots, \varepsilon_N^x\}$ that satisfies

$$\begin{aligned}
-\mathbf{1}_m \varepsilon_k^u &\leq Ru_k, \ k = 0, 1, \ldots, N-1 \\
-\mathbf{1}_m \varepsilon_k^u &\leq -Ru_k, \ k = 0, 1, \ldots, N-1 \\
-\mathbf{1}_n \varepsilon_k^x &\leq Qx_k, \ k = 0, 1, \ldots, N-1 \\
-\mathbf{1}_n \varepsilon_k^x &\leq -Qx_k, \ k = 0, 1, \ldots, N-1 \\
-\mathbf{1}_n \varepsilon_N^x &\leq Px_N, \\
-\mathbf{1}_n \varepsilon_N^x &\leq -Px_N,
\end{aligned} \tag{17.26}$$

represents an upper bound on $J_0^*(x(0))$, where $\mathbf{1}_k$ is a column vector of ones of length k, and where $x(k)$ is expressed as in (17.25). Similarly to what was shown in [78], it is easy to prove that the vector $\varepsilon = \{\varepsilon_0^u, \ldots, \varepsilon_{N-1}^u, \varepsilon_0^x, \ldots, \varepsilon_N^x, u(0), \ldots, u(N-1)\}$ that satisfies equations (17.26) and simultaneously minimizes

$$J(\varepsilon) = \varepsilon_0^u + \cdots + \varepsilon_{N-1}^u + \varepsilon_0^x + \cdots + \varepsilon_N^x \tag{17.27}$$

also solves the original problem, i.e., the same optimum $J_0^*(x(0))$ is achieved. Therefore, problem (17.21) with cost (17.22) can be reformulated as the following MILP problem

$$\begin{aligned}
\min_{\varepsilon} \quad & J(\varepsilon) \\
\text{subj. to} \ -\mathbf{1}_m \varepsilon_k^u &\leq \pm Ru_k, \ k = 0, 1, \ldots, N-1 \\
-\mathbf{1}_n \varepsilon_k^x &\leq \pm Q \left(A^k x_0 + \sum_{j=0}^{k-1} A^j (B_1 u_{k-1-j} + \right. \\
& \qquad \left. B_2 \delta_{k-1-j} + B_3 z_{k-1-j}) \right) \ k = 0, \ldots, N-1 \\
-\mathbf{1}_n \varepsilon_N^x &\leq \pm P \left(A^N x_0 + \sum_{j=0}^{N-1} A^j (B_1 u_{k-1-j} + \right. \\
& \qquad \left. B_2 \delta_{k-1-j} + B_3 z_{k-1-j}) \right) \\
x_{k+1} &= Ax_k + B_1 u_k + B_2 \delta_k + B_3 z_k, \ k \geq 0 \\
E_2 \delta_k + E_3 z_k &\leq E_1 u_k + E_4 x_k + E_5, \ k \geq 0 \\
x_N &\in \mathcal{X}_f \\
x_0 &= x(0)
\end{aligned} \tag{17.28}$$

where the variable $x(0)$ in (17.28) appears only in the constraints as a parameter vector.

Given a value of the initial state $x(0)$, the MIQP (17.24) or the MILP (17.28) can be solved to get the optimizer $\varepsilon^*(x(0))$ and therefore the optimal input $U_0^*(0)$. In the next Section 17.5 we will show how multiparametric programming can be also used to efficiently compute the piecewise affine state feedback optimal control law (17.9) or (17.19).

Example 17.2 Consider the problem of steering in three steps the simple piecewise affine system presented in Example 16.1 to a small region around the origin. The system state update equations are

$$
\begin{cases}
x(t+1) = \begin{cases}
0.4 \begin{bmatrix} 1 & -\sqrt{3} \\ \sqrt{3} & 1 \end{bmatrix} x(t) + \begin{bmatrix} 0 \\ 1 \end{bmatrix} u(t) & \text{if } \begin{bmatrix} 1 & 0 \end{bmatrix} x(t) \geq 0 \\[2em]
0.4 \begin{bmatrix} 1 & \sqrt{3} \\ -\sqrt{3} & 1 \end{bmatrix} x(t) + \begin{bmatrix} 0 \\ 1 \end{bmatrix} u(t) & \text{if } \begin{bmatrix} 1 & 0 \end{bmatrix} x(t) \leq -\epsilon
\end{cases} \\[3em]
y(t) = \begin{bmatrix} 0 & 1 \end{bmatrix} x(t)
\end{cases}
$$

(17.29)

subject to the constraints

$$
\begin{aligned}
x(t) &\in [-5, 5] \times [-5, 5] \\
u(t) &\subset [-1, 1].
\end{aligned}
$$

(17.30)

The MLD representation of system (17.29)–(17.30) was reported in Example 16.7.

The finite time constrained optimal control problem (17.7) with cost (17.5) with $p = \infty$, $N = 3$, $P = Q = \begin{bmatrix} 1 & 0 \\ 0 & 1 \end{bmatrix}$, $R = 1$, and $\mathcal{X}_f = [-0.01, 0.01] \times [-0.01, 0.01]$, can be solved by considering the optimal control problem (17.21) with cost (17.22) for the equivalent MLD representation and solving the associated MILP problem (17.24).

The resulting MILP has 27 variables ($|\varepsilon| = 27$) which are $x \in \mathbb{R}^2$, $\delta \in \{0, 1\}$, $z \in \mathbb{R}^2$, $u \in \mathbb{R}$, $y \in \mathbb{R}$ over 3 steps plus the real-valued slack variables ε_0^u, ε_1^u, ε_2^u and ε_1^x, ε_2^x, ε_3^x (note that ε_0^x is not needed). The number of mixed-integer equality constraints is 9 resulting from the 3 equality constraints of the MLD systems over 3 steps. The number of mixed-integer inequality constraints is 110.

The finite time constrained optimal control problem (17.7) with cost (17.6), $N = 3$, $P = Q = \begin{bmatrix} 1 & 0 \\ 0 & 1 \end{bmatrix}$, $R = 1$, and $\mathcal{X}_f = [-0.01\ 0.01] \times [-0.01\ 0.01]$, can be solved by considering the optimal control problem (17.21) with cost (17.23) for the equivalent MLD representation and solving the associated MIQP problem (17.24).

The resulting MIQP has 21 variables ($|\varepsilon| = 21$) which are $x \in \mathbb{R}^2$, $\delta \in \{0, 1\}$, $z \in \mathbb{R}^2$, $u \in \mathbb{R}$, $y \in \mathbb{R}$ over 3 steps. The number of mixed-integer equality constraints is 9 resulting from the 3 equality constraints of the MLD systems over 3 steps. The number of mixed-integer inequality constraints is 74.

17.4.1 Mixed-Integer Optimization Methods

With the exception of particular structures, mixed-integer programming problems involving 0-1 variables are classified as $\mathcal{N}P$-complete, which means that in the worst case, the solution time grows exponentially with the problem size.

Despite the combinatorial nature of these problems, excellent solvers for MILPs and MIQPs have become available over the last 10–15 years. As discussed in Chapter 2.4, the most obvious way to solve an MIQP (MILP) is to enumerate all the integer values and solve the corresponding QPs (LPs). By comparing the QPs (LPs) optimal costs one can derive the optimizer and the optimal cost of the MIQP (MILP). This is far from being efficient. The book [108] is a good reference for a long list of algorithmic approaches to efficiently solve mixed integer programming problems.

One simple idea called "Branch and Bound" has become the basis of many efficient solvers. We will explain the basic ideas behind Branch and Bound solvers with the help of a simple example.

Example 17.3 Illustrative Example for Branch and Bound Procedure   We refer to Figure 17.4. Assume that we are solving an MILP with three 0-1 integer variables $x_{d,1}, x_{d,2}, x_{d,3}$.

1. In a first step we relax the integer constraints, i.e., we allow the variables $x_{d,1}, x_{d,2}, x_{d,3}$ to vary continuously between 0 and 1. We find the minimum of the Relaxed LP to be 10, which must be a lower bound on the solution.

2. We fix one of the integer variables $x_{d,1}$ at 0 and 1, respectively, and leave the others relaxed. We solve the two resulting LPs and obtain 13 and 12, which are new improved lower bounds on the solutions along the two branches.

3. Heuristically we decide to continue on the branch with $x_{d,1} = 1$, because the lower bound is lower than with $x_{d,1} = 0$. We now fix one of the remaining integer variables $x_{d,2}$ at 0 and 1, respectively, and leave $x_{d,3}$ relaxed. We solve the two resulting LPs and obtain 16 and 14, which are new improved lower bounds on the solution along the two branches.

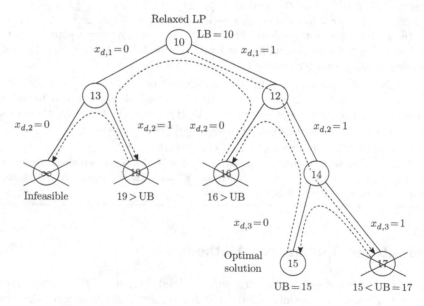

Figure 17.4 Illustration of Branch and Bound procedure.

4. Heuristically we decide again to continue on the branch with $x_{d,2} = 1$, because the lower bound is lower than with $x_{d,2} = 0$. We now fix the remaining integer variable $x_{d,3}$ at 0 and 1, respectively. We solve the two resulting LPs and obtain 15 and 17. We can conclude now that 15 is an upper bound on the solution and the integer choices leading to 17 are nonoptimal. We can also stop the further exploration of node 16 because this lower bound is higher than the established upper bound of 15.

5. We explore node 13 and find the integer choices either to be infeasible or to lead to a lower bound of 19 which is again higher than the established upper bound of 15. Thus no further explorations are necessary and 15 is confirmed to be the optimal solution.

More generally, the Branch and Bound algorithm for MILPs involves solving and generating a set of LP subproblems in accordance with a tree search, where the nodes of the tree correspond to LP subproblems. Branching generates child-nodes from parent-nodes according to branching rules, which can be based, for instance, on a-priori specified priorities on integer variables, or on the amount by which the integer constraints are violated. Nodes are labeled either as pending, if the corresponding LP problem has not yet been solved, or fathomed, if the node has already been fully explored. The algorithm stops when all nodes have been fathomed.

The success of the branch and bound algorithm relies on the fact that whole subtrees can be excluded from further exploration by fathoming the corresponding root nodes. This happens if the corresponding LP subproblem is either infeasible or an integer solution is obtained. In the second case, the corresponding value of the cost function serves as an upper bound on the optimal solution of the MILP problem, and is used to further fathom any other nodes having either a greater optimal value or lower bound.

17.5   State Feedback Solution via Batch Approach

Multiparametric programming [115, 101, 44, 63] has been used to compute the PWA form of the optimal state feedback control law $u^*(x(k))$ for linear system. By generalizing the results of the previous chapters to hybrid systems, the state vector $x(0)$, which appears in the objective function and in the linear part of the right-hand side of the constraints (17.24), can be considered as a vector of parameters. Then, for performance indices based on the ∞-norm or 1-norm, the optimization problem can be treated as a *multiparametric MILP* (mp-MILP), while for performance indices based on the 2-norm, the optimization problem can be treated as a *multiparametric MIQP* (mp-MIQP). Solving an mp-MILP (mp-MIQP) amounts to expressing the solution of the MILP (MIQP) (17.24) as a function of the parameters $x(0)$.

In Section 6.4.1 we have presented an algorithm for solving mp-MILP problems, while, to the authors' knowledge, there does not exist an efficient method for solving general mp-MIQPs. In Section 17.6 we will present an algorithm that efficiently solves the specific mp-MIQPs derived from optimal control problems for discrete-time hybrid systems.

17.6   State Feedback Solution via Recursive Approach

In this section we propose an efficient algorithm for computing the solution to the finite time optimal control problem for discrete-time linear hybrid systems. It is based on a dynamic programming recursion and a multiparametric linear or quadratic programming solver. The approach represents an alternative to the mixed-integer parametric approach presented earlier.

The PWA solution (17.9) will be computed proceeding backwards in time using two tools: a linear or quadratic multiparametric programming solver (depending on the cost function used) and a special technique to store the solution which will be illustrated in the next sections. The algorithm will be presented for optimal control based on a quadratic performance criterion. Its extension to optimal control based on linear performance criteria is straightforward.

17.6.1   Preliminaries and Basic Steps

Consider the PWA map ζ defined as

$$\zeta : \ x \in \mathcal{R}_i \mapsto F_i x + g_i \quad \text{for} \quad i = 1, \dots, N_{\mathcal{R}}, \tag{17.31}$$

where $\mathcal{R}_i$, $i = 1, \dots, N_{\mathcal{R}}$ are subsets of the x−space. Note that if there exist $l, m \in \{1, \dots, N_{\mathcal{R}}\}$ such that for $x \in \mathcal{R}_l \cap \mathcal{R}_m$, $F_l x + g_l \neq F_m x + g_m$ the map ζ (17.31) is not single valued.

Definition 17.1   *Given a PWA map (17.31) we define $f_{PWA}(x) = \zeta_o(x)$ as the* ordered region single-valued *function associated with (17.31) when*

$$\zeta_o(x) = F_j x + g_j \mid x \in \mathcal{R}_j \text{ and } \forall i < j : x \notin \mathcal{R}_i,$$
$$j \in \{1, \dots, N_{\mathcal{R}}\},$$

and write it in the following form

$$\zeta_o(x) = \left. \begin{matrix} F_1 x + g_1 & \text{if } x \in \mathcal{P}_1 \\ \vdots & \\ F_{N_{\mathcal{R}}} x + g_{N_{\mathcal{R}}} & \text{if } x \in \mathcal{P}_{N_{\mathcal{R}}}. \end{matrix} \right.$$

Note that given a PWA map (17.31) the corresponding *ordered region single-valued* function ζ_o changes if the order used to store the regions $\mathcal{R}_i$ and the corresponding affine gains change. For illustration purposes consider the example depicted in Figure 17.5, where $x \in \mathbb{R}$, $N_{\mathcal{R}} = 2$, $F_1 = 0$, $g_1 = 0$, $\mathcal{R}_1 = [-2, 1]$, $F_2 = 1$, $g_2 = 0$, $\mathcal{R}_2 = [0, 2]$.

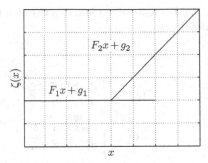

(a) Multivalued PWA map ζ.

$$\zeta_{12}(x) = \Big\downarrow \begin{array}{l} F_1 x + g_1 \text{ if } x \in \mathcal{R}_1 \\ F_2 x + g_2 \text{ if } x \in \mathcal{R}_2 \end{array} \qquad \zeta_{21}(x) = \Big\downarrow \begin{array}{l} F_2 x + g_2 \text{ if } x \in \mathcal{R}_2 \\ F_1 x + g_1 \text{ if } x \in \mathcal{R}_1 \end{array}$$

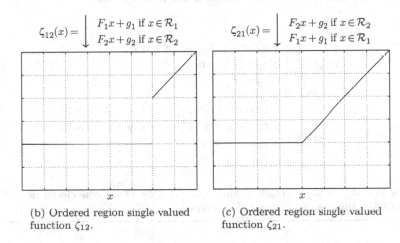

(b) Ordered region single valued   (c) Ordered region single valued
function ζ_{12}.   function ζ_{21}.

Figure 17.5 Illustration of the ordered region single valued function.

In the following, we assume that the sets $\mathcal{R}_i^k$ in the optimal solution (17.9) can overlap. When we refer to the PWA function $u_k^*(x(k))$ in (17.9) we will implicitly mean the ordered region single-valued function associated with the map (17.9).

Example 17.4   Let $J_1^* : \mathcal{P}_1 \to \mathbb{R}$ and $J_2^* : \mathcal{P}_2 \to \mathbb{R}$ be two quadratic functions, $J_1^*(x) = x' L_1 x + M_1' x + N_1$ and $J_2^*(x) = x' L_2 x + M_2' x + N_2$, where $\mathcal{P}_1$ and $\mathcal{P}_2$ are convex polyhedra and $J_i^*(x) = +\infty$ if $x \notin \mathcal{P}_i$, $i \in \{1, 2\}$. Let $u_1^* : \mathcal{P}_1 \to \mathbb{R}^m$, $u_2^* : \mathcal{P}_2 \to \mathbb{R}^m$ be vector functions. Let $\mathcal{P}_1 \cap \mathcal{P}_2 = \mathcal{P}_3 \neq \emptyset$ and define

$$J^*(x) = \min\{J_1^*(x), J_2^*(x)\} \tag{17.32}$$

$$u^*(x) = \begin{cases} u_1^*(x) & \text{if } J_1^*(x) \leq J_2^*(x) \\ u_2^*(x) & \text{if } J_1^*(x) \geq J_2^*(x) \end{cases} \tag{17.33}$$

where $u^*(x)$ can be a set valued function. Let $L_3 = L_2 - L_1$, $M_3 = M_2 - M_1$, $N_3 = N_2 - N_1$. Then, corresponding to the three following cases

(i) $J_1^*(x) \leq J_2^*(x) \; \forall x \in \mathcal{P}_3$

(ii) $J_1^*(x) \geq J_2^*(x) \; \forall x \in \mathcal{P}_3$

(iii) $\exists x_1, x_2 \in \mathcal{P}_3 | J_1^*(x_1) < J_2^*(x_1)$ and $J_1^*(x_2) > J_2^*(x_2)$

the expressions (17.32) and a real-valued function that can be extracted from (17.33) can be written equivalently as:

(i)

$$J^*(x) = \left\lfloor \begin{array}{ll} J_1^*(x) & \text{if } x \in \mathcal{P}_1 \\ J_2^*(x) & \text{if } x \in \mathcal{P}_2 \end{array} \right. \tag{17.34}$$

$$u^*(x) = \left\lfloor \begin{array}{ll} u_1^*(x) & \text{if } x \in \mathcal{P}_1 \\ u_2^*(x) & \text{if } x \in \mathcal{P}_2 \end{array} \right. \tag{17.35}$$

(ii) as in (17.34) and (17.35) by switching the indices 1 and 2

(iii)

$$J^*(x) = \left\lfloor \begin{array}{ll} \min\{J_1^*(x), J_2^*(x)\} & \text{if } x \in \mathcal{P}_3 \\ J_1^*(x) & \text{if } x \in \mathcal{P}_1 \\ J_2^*(x) & \text{if } x \in \mathcal{P}_2 \end{array} \right. \tag{17.36}$$

$$u^*(x) = \left\lfloor \begin{array}{ll} u_1^*(x) & \text{if } x \in \mathcal{P}_3 \bigcap \{x \mid x'L_3x + M_3'x + N_3 \geq 0\} \\ u_2^*(x) & \text{if } x \in \mathcal{P}_3 \bigcap \{x \mid x'L_3x + M_3'x + N_3 \leq 0\} \\ u_1^*(x) & \text{if } x \in \mathcal{P}_1 \\ u_2^*(x) & \text{if } x \in \mathcal{P}_2 \end{array} \right. \tag{17.37}$$

where (17.34), (17.35), (17.36), and (17.37) have to be considered as PWA and PPWQ functions in the *ordered region* sense.

Example 17.4 shows how to

- avoid the storage of the intersections of two polyhedra in Case (i) and (ii)

- avoid the storage of possibly nonconvex regions $\mathcal{P}_1 \setminus \mathcal{P}_3$ and $\mathcal{P}_2 \setminus \mathcal{P}_3$

- work with multiple quadratic functions instead of quadratic functions defined over nonconvex and nonpolyhedral regions.

The three points listed above will be the three basic ingredients for storing and simplifying the optimal control law (17.9). Next we will show how to compute it.

Remark 17.2   In Example 17.4 the description (17.36)–(17.37) of Case (iii) can always be used but the on-line evaluation of the control $u^*(x)$ is rather involved requiring a series of set membership and comparison evaluations. To assess if the simpler description of $u^*(x)$ of Case (i) or Case (ii) could be used instead, one needs to solve indefinite quadratic programs of the form

$$\begin{array}{ll} \min_x & x'L_3x + M_3'x + N_3 \\ \text{subj. to} & x \in \mathcal{P}_3 \end{array} \tag{17.38}$$

which are nontrivial, in general.

17.6.2 Multiparametric Programming with Multiple Quadratic Functions

Consider the multiparametric program

$$J^*(x) = \min_u \quad l(x,u) + q(f(x,u)) \tag{17.39}$$
$$\text{subj. to} \quad f(x,u) \in \mathcal{P},$$

where $\mathcal{P} \subseteq \mathbb{R}^n$ is a compact set, $f : \mathbb{R}^n \times \mathbb{R}^m \to \mathbb{R}^n$, $q : \mathcal{P} \to \mathbb{R}$, and $l : \mathbb{R}^n \times \mathbb{R}^m \to \mathbb{R}$ is a convex quadratic function of x and u. We aim at determining the region $\mathcal{X}$ of variables x such that the program (17.39) is feasible and the optimum $J^*(x)$ is finite, and at finding the expression $u^*(x)$ of (one of) the optimizer(s). We point out that the constraint $f(x,u) \in \mathcal{P}$ implies a constraint on u as a function of x since u can assume only values where $f(x,u)$ is defined.

Next we show how to solve several forms of problem (17.39).

Lemma 17.2 (one to one problem) *Problem (17.39) where f is linear, q is quadratic and strictly convex, and $\mathcal{P}$ is a polyhedron can be solved by one mp-QP.*

Proof: See Chapter 6.3.1.

Lemma 17.3 (one to one problem of multiplicity d) *Problem (17.39) where f is linear, q is a multiple quadratic function of multiplicity d and $\mathcal{P}$ is a polyhedron can be solved by d mp-QPs.*

Proof: The multiparametric program to be solved is

$$J^*(x) = \min_u \quad l(x,u) + \min\{q_1(f(x,u)), \dots, q_d(f(x,u))\} \tag{17.40}$$
$$\text{subj. to} \quad f(x,u) \in \mathcal{P}$$

and it is equivalent to

$$J^*(x) = \min \left\{ \begin{array}{l} \min_u l(x,u) + q_1(f(x,u)), \\ \text{subj. to } f(x,u) \in \mathcal{P}, \\ \quad \vdots \\ \min_u l(x,u) + q_d(f(x,u)) \\ \text{subj. to } f(x,u) \in \mathcal{P} \end{array} \right\}. \tag{17.41}$$

The ith subproblems in (17.41)

$$J_i^*(x) = \min_u l(x,u) + q_i(f(x,u)) \tag{17.42}$$
$$\text{subj. to } f(x,u) \in \mathcal{P} \tag{17.43}$$

is a *one to one problem* and therefore it is solvable by an mp-QP. Let the solution of the i-th mp-QPs be

$$u^i(x) = \tilde{F}^{i,j}x + \tilde{g}^{i,j}, \quad \forall x \in \mathcal{T}^{i,j}, \quad j = 1, \dots, N^{ri}, \tag{17.44}$$

where $\mathcal{T}^i = \bigcup_{j=1}^{N^{ri}} \mathcal{T}^{i,j}$ is a polyhedral partition of the convex set $\mathcal{T}^i$ of feasible x for the ith sub-problem and N^{ri} is the corresponding number of polyhedral regions.

The feasible set $\mathcal{X}$ satisfies $\mathcal{X} = \mathcal{T}^1 = \cdots = \mathcal{T}^d$ since the constraints of the d sub-problems are identical.

The solution $u^*(x)$ to the original problem (17.40) is obtained by comparing and storing the solution of d mp-QP subproblems (17.42)–(17.43) as explained in Example 17.4. Consider the case $d = 2$, and consider the intersection of the polyhedra $\mathcal{T}^{1,i}$ and $\mathcal{T}^{2,l}$ for $i = 1, \ldots, N^{r1}$, $l = 1, \ldots, N^{r2}$. For all $\mathcal{T}^{1,i} \cap \mathcal{T}^{2,l} = \mathcal{T}^{(1,i),(2,l)} \neq \emptyset$ the optimal solution is stored in an ordered way as described in Example 17.4, while paying attention to the fact that a region could be already stored. Moreover, when storing a new polyhedron with the corresponding value function and optimizer, the relative order of the regions already stored must not be changed. The result of this *Intersect and Compare* procedure is

$$u^*(x) = F^i x + g^i \text{ if } x \in \mathcal{R}^i,$$
$$\mathcal{R}^i = \{x : x' L^i(j)x + M^i(j)'x \le N^i(j), \ j = 1, \ldots, n^i\}, \tag{17.45}$$

where $\mathcal{R} = \bigcup_{j=1}^{N_\mathcal{R}} \mathcal{R}^j$ is a polyhedron and the value function

$$J^*(x) = \tilde{J}_j^*(x) \quad \text{if } x \in \mathcal{D}^j, j = 1, \ldots, N^\mathcal{D}, \tag{17.46}$$

where $\tilde{J}_j^*(x)$ are multiple quadratic functions defined over the convex polyhedra $\mathcal{D}^j$. The polyhedron $\mathcal{D}^j$ can contain several regions $\mathcal{R}^i$ or can coincide with one of them. Note that (17.45) and (17.46) have to be considered as PWA and PPWQ functions in the *ordered region* sense.

If $d > 2$ then the value function in (17.46) is intersected with the solution of the third mp-QP subproblem and the procedure is iterated by making sure not to change the relative order of the polyhedra and the corresponding gain of the solution constructed in the previous steps. The solution will still have the same form (17.45)–(17.46). ∎

Lemma 17.4 (one to r problem)   *Problem (17.39) where f is linear, q is a lower-semicontinuous PPWQ function defined over r polyhedral regions and strictly convex on each polyhedron, and $\mathcal{P}$ is a polyhedron, can be solved by r mp-QPs.*

Proof: Let $q(x) = q_i$ if $x \in \mathcal{P}_i$ the PWQ function where the closures $\bar{\mathcal{P}}_i$ of $\mathcal{P}_i$ are polyhedra and q_i strictly convex quadratic functions. The multiparametric program to solve is

$$J^*(x) = \min \left\{ \begin{array}{l} \min_u l(x,u) + q_1(f(x,u)), \\ \text{subj. to } f(x,u) \in \bar{\mathcal{P}}_1 \\ \qquad\quad f(x,u) \in \mathcal{P} \\ \vdots \\ \min_u l(x,u) + q_r(f(x,u))\} \\ \text{subj. to } f(x,u) \in \bar{\mathcal{P}}_r \\ \qquad\quad f(x,u) \in \mathcal{P} \end{array} \right\}. \tag{17.47}$$

The proof follows the lines of the proof of the previous theorem with the exception that the constraints of the i-th mp-QP subproblem differ from the j-th mp-QP subproblem, $i \neq j$.

The lower-semicontinuity assumption on $q(x)$ allows one to use the closure of the sets $\mathcal{P}_i$ in (17.47).The cost function in problem (17.39) is lower-semicontinuous since it is a composition of a lower-semicontinuous function and a continuous function. Then, since the domain is compact, problem (17.47) admits a minimum. Therefore for a given x, there exists one mp-QP in problem (17.47) which yields the optimal solution. The procedure based on solving mp-QPs and storing the results as in Example 17.4 will be the same as in Lemma 17.3 but the domain $\mathcal{R} = \bigcup_{j=1}^{N^{\mathcal{R}}} \mathcal{R}^j$ of the solution can be a P-collection. ∎

If f is PPWA defined over s regions then we have a *s to X problem* where X can belong to any of the problems listed above. In particular, we have an *s to r problem of multiplicity d* if f is PPWA and defined over s regions and q is a multiple PPWQ function of multiplicity d, defined over r polyhedral regions. The following lemma can be proven along the lines of the proofs given before.

Lemma 17.5   *Problem (17.39) where f is linear and q is a lower-semicontinuous PPWQ function of multiplicity d, defined over r polyhedral regions and strictly convex on each polyhedron, is a one to r problem of multiplicity d and can be solved by $r \cdot d$ mp-QPs.*
An s to r problem of multiplicity d can be decomposed into s one to r problems of multiplicity d. An s to one problem can be decomposed into s one to one problems.

17.6.3   Algorithmic Solution of the Bellman Equations

In the following we will substitute the PWA system Equation (17.2) with the shorter form

$$x(k+1) = \tilde{f}_{PWA}(x(k), u(k)) \tag{17.48}$$

where $\tilde{f}_{PWA} : \tilde{\mathcal{C}} \to \mathbb{R}^n$ and $\tilde{f}_{PWA}(x, u) = A^i x + B^i u + f^i$ if $\left[\begin{smallmatrix} x \\ u \end{smallmatrix}\right] \in \tilde{\mathcal{C}}^i$, $i = 1, \ldots, s$, and $\{\tilde{\mathcal{C}}^i\}$ is a polyhedral partition of $\tilde{\mathcal{C}}$.

Consider the dynamic programming formulation of the CFTOC problem (17.6)-(17.7),

$$J_j^*(x(j)) = \min_{u_j} \quad x_j' Q x_j + u_j' R u_j + J_{j+1}^*(\tilde{f}_{PWA}(x(j), u_j)) \tag{17.49}$$

$$\text{subj. to} \quad \tilde{f}_{PWA}(x(j), u_j) \in \mathcal{X}_{j+1} \tag{17.50}$$

for $j = N - 1, \ldots, 0$, with terminal conditions

$$\mathcal{X}_N = \mathcal{X}_f \tag{17.51}$$

$$J_N^*(x) = x' P x, \tag{17.52}$$

where $\mathcal{X}_j$ is the set of all states $x(j)$ for which problem (17.49)–(17.50) is feasible:

$$\mathcal{X}_j = \{x \in \mathbb{R}^n | \; \exists u, \; \tilde{f}_{PWA}(x, u) \in \mathcal{X}_{j+1}\}. \tag{17.53}$$

Assume for the moment that there are no binary inputs and binary states, $m_\ell = n_\ell = 0$. The Bellman Equations (17.49)–(17.52) can be solved backwards in time by using a multiparametric quadratic programming solver and the results of the previous section.

Consider the first step of the dynamic program (17.49)–(17.52)

$$J^*_{N-1}(x_{N-1}) = \min_{\{u_{N-1}\}} \; x'_{N-1} Q x_{N-1} + u'_{N-1} R u_{N-1} + J^*_N(\tilde{f}_{PWA}(x_{N-1}, u_{N-1}))$$

(17.54)

$$\text{subj. to} \quad \tilde{f}_{PWA}(x_{N-1}, u_{N-1}) \in \mathcal{X}_f.$$

(17.55)

The cost to go function $J^*_N(x)$ in (17.54) is quadratic, the terminal region $\mathcal{X}_f$ is a polyhedron and the constraints are piecewise affine. Problems (17.54)–(17.55) is an *s to one problem* that can be solved by solving s mp-QPs. From the second step $j = N - 2$ to the last one $j = 0$ the cost to go function $J^*_{j+1}(x)$ is a lower-semicontinuous PPWQ with a certain multiplicity d_{j+1}, the terminal region $\mathcal{X}_{j+1}$ is a P-collection and the constraints are piecewise affine. Therefore, problem (17.49)–(17.52) is an *s to N^r_{j+1} problem with multiplicity d_{j+1}* (where N^r_{j+1} is the number of polyhedra of the cost to go function J^*_{j+1}), that can be solved by solving $s N^r_{j+1} d_{j+1}$ mp-QPs (Lemma 17.5). The resulting optimal solution will have the form (17.9) considered in the ordered region sense.

In the presence of binary inputs the procedure can be repeated, with the difference that all the possible combinations of binary inputs must be enumerated. Therefore, a *one to one problem* becomes a *2^{m_ℓ} to one problem* and so on. In the presence of binary states the procedure can be repeated either by enumerating them all or by solving a dynamic programming algorithm at time step k from a relaxed state space to the set of binary states feasible at time $k + 1$.

Next we summarize the main steps of the dynamic programming algorithm discussed in this section. We use boldface characters to denote sets of polyhedra, i.e., $\mathbf{R} = \{\mathcal{R}_i\}_{i=1,\dots,|\mathbf{R}|}$, where $\mathcal{R}_i$ is a polyhedron and $|\mathbf{R}|$ is the cardinality of the set $\mathbf{R}$. Furthermore, when we say SOLVE an mp-QP we mean to compute and store the triplet $S_{k,i,j}$ of expressions for the value function, the optimizer, and the polyhedral partition of the feasible space.

In Algorithm 17.1, the structure $S_{k,i,j}$ stores the matrices defining quadratic function $J^*_{k,i,j,l}(\cdot)$, affine function $u^*_{k,i,j,l}(\cdot)$, and polyhedra $\mathcal{R}_{k,i,j,l}$, for all l:

$$S_{k,i,j} = \bigcup_l \left\{ \left(J^*_{k,i,j,l}(x), \; u^*_{k,i,j,l}(x), \; \mathcal{R}_{k,i,j,l} \right) \right\},$$

(17.56)

where the indices in (17.56) have the following meaning: k is the time step, i indexes the piece of the "cost-to-go" function that the DP algorithm is considering, j indexes the piece of the PWA dynamics the DP algorithm is considering, and l indexes the polyhedron in the mp-QP solution of the (k, i, j)th mp-QP problem.

The "KEEP only triplets" Step of Algorithm 17.1 aims at discarding regions $\mathcal{R}_{k,h}$ that are completely covered by some other regions that have lower cost. Obviously, if there are some parts of the region $\mathcal{R}_{k,h}$ that are not covered at all by other regions (first condition) we need to keep them. Note that comparing the cost functions is, in general, nonconvex optimization problem. One might consider solving the problem exactly, but since algorithm works even if some removable regions are kept, we usually formulate an LMI relaxation of the problem at hand.

The output of Algorithm 17.1 is the state feedback control law (17.9) considered in the ordered region sense. The online implementation of the control law requires simply the evaluation of the PWA controller (17.9) in the ordered region sense.

Algorithm 17.1

Input: CFTOC problem (17.3)–(17.7)
Output: Solution (17.9) in the ordered region sense

$\mathbf{R}_N \leftarrow \{\mathcal{X}_f\}$
$J_{N,1}^*(x) \leftarrow x'Px$
For $k = N - 1, \dots, 1$
    For $i = 1, \dots, |\mathbf{R}_{k+1}|$
        For $j = 1, \dots, s$
            $\mathcal{S}_{k,i,j} \leftarrow \{\}$
            Solve the mp-QP

$$\mathcal{S}_{k,i,j} \leftarrow \quad \min_{u_k} \quad x_k'Qx_k + u_k'Ru_k + J_{k+1,i}^*(A_jx_k + B_ju_k + f_j)$$

$$\text{subj. to} \quad \begin{cases} A_jx_k + B_ju_k + f_j \in \mathcal{R}_{k+1,i} \\ \left[\begin{smallmatrix} x_k \\ u_k \end{smallmatrix}\right] \in \tilde{\mathcal{C}}^j \end{cases}$$

        End
    End
    $\mathbf{R}_k \leftarrow \{\mathcal{R}_{k,i,j,l}\}_{i,j,l}$. Denote by $\mathcal{R}_{k,h}$ its elements, and by $J_{k,h}^*$ and $u_{k,h}^*(x)$ the associated costs and optimizers, with $h \in \{1, \dots, |\mathbf{R}_k|\}$
    Keep only triplets $(J_{k,h}^*(x),\ u_{k,h}^*(x),\ \mathcal{R}_{k,h})$ for which $\exists x \in \mathcal{R}_{k,h} : x \notin \mathcal{R}_{k,d}, \forall d \neq h$ OR $\exists x \in \mathcal{R}_{k,h} : J_{k,h}^*(x) < J_{k,d}^*(x), \forall d \neq h$
    Create multiplicity information and additional regions for an ordered region solution as explained in Example 17.4
End

17.6.4   Examples

Example 17.5   Consider the control problem of steering the piecewise affine system (17.29) to a small region around the origin. The constraints and cost function of Example 17.2 are used. The state feedback solution $u_0^*(x(0))$ was determined by using the approach presented in the previous section. When the infinity norm is used in the cost function, the feasible state space $\mathcal{X}_0$ at time 0 is divided into 84 polyhedral regions and it is depicted in Figure 17.6(a). When the squared Euclidean norm is used in the cost function, the feasible state space $\mathcal{X}_0$ at time 0 is divided into 92 polyhedral regions and is depicted in Figure 17.6(b).

Note that as explained in Section 17.6.2 for the case of the squared Euclidian norm, the optimal control law is stored in a special data structure where:

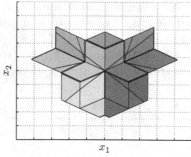
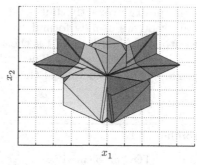

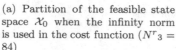

(a) Partition of the feasible state space $\mathcal{X}_0$ when the infinity norm is used in the cost function ($N^r{}_3 = 84$)

(b) Partition of the feasible state space $\mathcal{X}_0$ when the squared Euclidian norm is used in the cost function ($N^r{}_3 = 92$)

Figure 17.6 Example 17.5. State space control partition for $u_0^*(x(0))$

1. The ordering of the regions is important.

2. The polyhedra can overlap.

3. The polyhedra can have an associated value function of multiplicity $d > 1$. Thus, d quadratic functions have to be compared on-line in order to compute the optimal control action.

Example 17.6 Consider the hybrid spring-mass-damper system described in Example 16.2

$$x(t+1) = \begin{cases} \begin{bmatrix} 0.90 & 0.02 \\ -0.02 & -0.00 \end{bmatrix} x(t) + \begin{bmatrix} 0.10 \\ 0.02 \end{bmatrix} u_1(t) + \begin{bmatrix} -0.01 \\ -0.02 \end{bmatrix} & \text{if } x_1(t) \leq 1, u_2(t) \leq 0.5 \\[12pt] \begin{bmatrix} 0.90 & 0.02 \\ -0.06 & -0.00 \end{bmatrix} x(t) + \begin{bmatrix} 0.10 \\ 0.02 \end{bmatrix} u_1(t) + \begin{bmatrix} -0.07 \\ -0.15 \end{bmatrix} & \text{if } x_1(t) \geq 1+\epsilon, u_2(t) \leq 0.5 \\[12pt] \begin{bmatrix} 0.90 & 0.38 \\ -0.38 & 0.52 \end{bmatrix} x(t) + \begin{bmatrix} 0.10 \\ 0.38 \end{bmatrix} u_1(t) + \begin{bmatrix} -0.10 \\ -0.38 \end{bmatrix} & \text{if } x_1(t) \leq 1, u_2(t) \geq 0.5 \\[12pt] \begin{bmatrix} 0.90 & 0.35 \\ -1.04 & 0.35 \end{bmatrix} x(t) + \begin{bmatrix} 0.10 \\ 0.35 \end{bmatrix} u_1(t) + \begin{bmatrix} -0.75 \\ -2.60 \end{bmatrix} & \text{if } x(t) \geq 1+\epsilon, u_2(t) \geq 0.5 \end{cases}$$
$$(17.57)$$

subject to the constraints

$$\begin{aligned} x(t) &\in [-5,5] \times [-5,5] \\ u(t) &\in [-1,1]. \end{aligned} \qquad (17.58)$$

We solve the finite time constrained optimal control problem (17.7) with cost (17.6) with $N = 3$, $P = Q = \begin{bmatrix} 1 & 0 \\ 0 & 1 \end{bmatrix}$, $R = \begin{bmatrix} 0.2 & 0 \\ 0 & 1 \end{bmatrix}$. The state feedback solution was determined by using the approach presented in the previous section in two cases: without terminal constraint (Figure 17.7(a)) and with terminal constraint $\mathcal{X}_f = [-0.01 \ 0.01] \times [-0.01 \ 0.01]$ (Figure 17.7(b)).

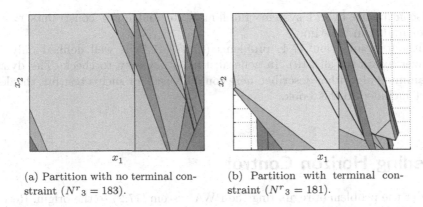

(a) Partition with no terminal constraint ($N^r_3 = 183$).

(b) Partition with terminal constraint ($N^r_3 = 181$).

Figure 17.7 Example 17.6. State space optimal control partition for $u_0^*(x(0))$.

17.7 Discontinuous PWA Systems

Without assumption 17.1 the optimal control problems (17.3)–(17.7) may be
feasible but may not admit an optimizer for some $x(0)$ (the problem in this case
would be to find an infimum rather than the minimum).

Under the assumption that the optimizer exists for all states $x(k)$, the approach
explained in the previous sections can be applied to discontinuous systems by
considering three elements. First, the PWA system (17.2) has to be defined
on each polyhedron of its domain *and all its lower dimensional facets*. Second,
dynamic programming has to be performed "from" and "to" any lower dimensional
facet of each polyhedron of the PWA domain. Finally, value functions are not
lower-semicontinuous, which implies that Lemma 17.4 cannot by used. Therefore,
when considering the closure of polyhedral domains in multiparametric programming (17.47), a postprocessing is necessary in order to remove multiparametric
optimal solutions which do not belong to the original set but only to its closure.
The tedious details of the dynamic programming algorithm for discontinuous PWA
systems are not included here but can be immediately extracted from the results
of the previous sections.

In practice, the approach just described for discontinuous PWA systems can
easily be numerically prohibitive. The simplest approach from a practical point of
view is to introduce gaps between the boundaries of any two polyhedra belonging
to the PWA domain (or, equivalently, to shrink by a quantity ε the size of every
polyhedron of the original PWA system). This way, one deals with PWA systems
defined over a disconnected union of closed polyhedra. By doing so, one can use
the approach discussed in this chapter for continuous PWA systems. However,
the optimal controller will not be defined at the points in the gaps. Also, the
computed solution might be arbitrarily different from the original solution to
problems (17.3)–(17.7) at any feasible point x. Despite this, if the magnitude ε
of the gaps is close to the machine precision and comparable to sensor/estimation
errors, such an approach can very appealing in practice. In some cases this approach
might be the only one that is computationally tractable for computing controllers

for discontinuous hybrid systems fulfilling state and input constraints that are implementable in real-time.

Without assumption 17.1, problems (17.3)–(17.7) is well defined only if an optimizer exists for all $x(0)$. In general, this is not easy to check. The dynamic programming algorithm described here could be used for such a test but the details are not included in this book.

17.8 Receding Horizon Control

Consider the problem of regulating the PWA system (17.2) to the origin. Receding Horizon Control (RHC) can be used to solve such a constrained regulation problem. The control algorithm is identical to the one outlined in Chapter 12 for linear systems. Assume that a full measurement of the state $x(t)$ is available at the current time t. Then, the finite time optimal control problem

$$J_t^*(x(t)) = \min_{U_{t \to t+N|t}} J_t(x(t), U_{t \to t+N|t}) \tag{17.59a}$$

$$\text{subj. to} \begin{cases} x_{t+k+1|t} = A^i x_{t+k|t} + B^i u_{t+k|t} + f^i & \text{if } \begin{bmatrix} x_{t+k|t} \\ u_{t+k|t} \end{bmatrix} \in \tilde{\mathcal{C}}^i \\ x_{t+N|t} \in \mathcal{X}_f \\ x_{t|t} = x(t) \end{cases} \tag{17.59b}$$

is solved at each time t, where $U_{t \to t+N|t} = \{u_{t|t}, \ldots, u_{t+N-1|t}\}$.

Let $U_t^* = \{u_{t|t}^*, \ldots, u_{t+N-1|t}^*\}$ be the optimal solution of (17.59) at time t. Then, the first sample of U_t^* is applied to system (17.2):

$$u(t) = u_{t|t}^*. \tag{17.60}$$

The optimization (17.59) is repeated at time $t+1$, based on the new state $x_{t+1|t+1} = x(t+1)$, yielding a *moving* or *receding horizon* control strategy.

Based on the results of previous sections the state feedback receding horizon controller (17.59)–(17.60) can be immediately obtained in two ways: (i) solve the MIQP/MILP (17.24) for $x_{t|t} = x(t)$ or (ii) by setting

$$u(t) = f_0^*(x(t)), \tag{17.61}$$

where $f_0^* : \mathbb{R}^n \to \mathbb{R}^{n_u}$ is the piecewise affine solution to the CFTOC (17.59) computed as explained in Section 17.6. The explicit form (17.61) has the advantage of being easier to implement, and provides insight on the type of action of the controller in different regions of the state space.

17.8.1 Stability and Feasibility Issues

As discussed in Chapter 12 the feasibility and stability of the receding horizon controller (17.59)–(17.60) is, in general, not guaranteed.

Theorem 12.2 in Chapter 12 can be immediately modified for the hybrid case: persistent feasibility and Lyapunov stability are guaranteed if the terminal constraint $\mathcal{X}_f$ is a control invariant set (assumption (A2) in Theorem 12.2) and the terminal cost $p(x_N)$ is a control Lyapunov function (assumption (A3) in Theorem 12.2). In the hybrid case the computation of control invariant sets and control Lyapunov functions is computationally more involved. As in the linear case, $\mathcal{X}_f = 0$ satisfies the aforementioned properties and it thus represents a very simple, yet restrictive, choice for guaranteeing persistent feasibility and Lyapunov stability.

17.8.2 Examples

Example 17.7 Consider the problem of regulating the piecewise affine system (17.29) to the origin. The constrained finite time optimal control problem (17.7) is solved with $N = 3$, $P = Q = \begin{bmatrix} 1 & 0 \\ 0 & 1 \end{bmatrix}$, $R = 1$, and $\mathcal{X}_f = [-0.01 \ 0.01] \times [-0.01 \ 0.01]$. Its state feedback solution (17.9) $u^*(x(0)) = f_0^*(x(0))$ at time 0 is implemented in a receding horizon fashion, i.e., $u(x(k)) = f_0^*(x(k))$.

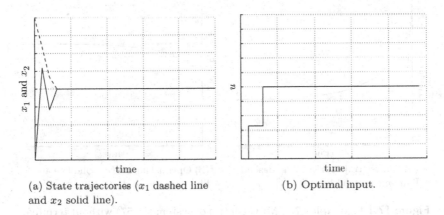

(a) State trajectories (x_1 dashed line and x_2 solid line).

(b) Optimal input.

Figure 17.8 Example 17.7. MPC control of system (17.29) when the infinity norm is used in the objective.

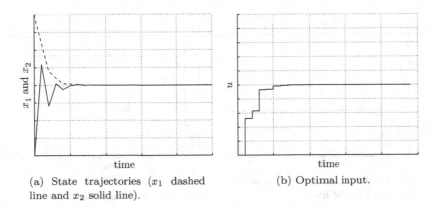

(a) State trajectories (x_1 dashed line and x_2 solid line).

(b) Optimal input.

Figure 17.9 Example 17.7. MPC control of system (17.29) when the squared Euclidian norm is used in the objective.

The state feedback control law with cost (17.5) with $p = \infty$ was computed in Example 17.5 and consists of 84 polyhedral regions. None of them has multiplicity higher than 1. Figure 17.8 shows the corresponding closed-loop trajectories starting from the initial state $x(0) = [-2\ 2]'$.

The state feedback control law with cost (17.6) was computed in Example 17.5 and consists of 92 polyhedral regions, some of which have multiplicity higher than 1. Figure 17.9 shows the corresponding closed-loop trajectories starting from the initial state $x(0) = [-2\ 2]'$.

Example 17.8   Consider the problem of regulating the hybrid spring-mass-damper system (17.57) described in Examples 16.2 and 17.6 to the origin. The finite time constrained optimal control problem (17.7) with cost (17.6) is solved with $N = 3$, $P = Q = \left[\begin{smallmatrix} 1 & 0 \\ 0 & 1 \end{smallmatrix}\right]$, $R = \left[\begin{smallmatrix} 0.2 & 0 \\ 0 & 1 \end{smallmatrix}\right]$. Its state feedback solution (17.9) $u^*(x(0)) = f_0^*(x(0))$ at time 0 is implemented in a receding horizon fashion, i.e., $u(x(k)) = f_0^*(x(k))$.

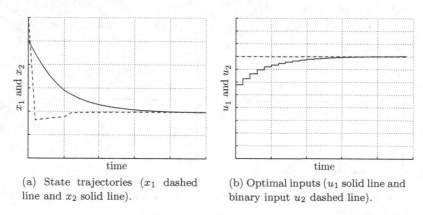

(a) State trajectories (x_1 dashed line and x_2 solid line).

(b) Optimal inputs (u_1 solid line and binary input u_2 dashed line).

Figure 17.10 Example 17.8. MPC control of system (17.57) without terminal constraint.

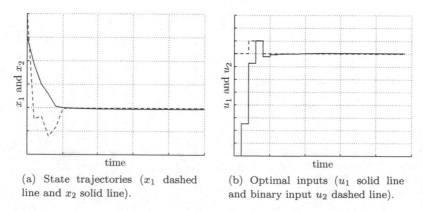

(a) State trajectories (x_1 dashed line and x_2 solid line).

(b) Optimal inputs (u_1 solid line and binary input u_2 dashed line).

Figure 17.11 Example 17.8. MPC control of system (17.57) with terminal constraint.

The state feedback solution was determined in Example 17.6 for the case of no terminal constraint (Figure 17.7(a)). Figure 17.10 depicts the corresponding closed-loop trajectories starting from the initial state $x(0) = [3\ 4]'$.

The state feedback solution was determined in Example 17.6 for terminal constraint $\mathcal{X}_f = [-0.01,\ 0.01] \times [-0.01,\ 0.01]$ (Figure 17.7(b)). Figure 17.11 depicts the corresponding closed-loop trajectories starting from the initial state $x(0) = [3\ 4]'$.

Comparing the two controllers, we find that the terminal constraint leads to a more aggressive behavior. In addition, in the case with terminal constraint we observe the coordination of the binary and continuous inputs. The small damping coefficient is switched on to increase the effect of the continuous input. The damping is increased again once the steady-state value is approached.

References

[1] J. Acevedo and E.N. Pistikopoulos. A multiparametric programming approach for linear process engineering problems under uncertainty. *Ind. Eng. Chem. Res.*, 36:717–728, 1997.

[2] I. Adler and R.D.C. Monteiro. A geometric view of parametric linear programming. *Algorithmica*, 8(2):161–176, 1992.

[3] A. Alessio and A. Bemporad. A survey on explicit model predictive control. In L. Magni, D. Raimondo, and F. Allgöwer, eds., *Nonlinear Model Predictive Control*, Vol. 384 of *Lecture Notes in Control and Information Sciences*, 345–369. Springer Verlag, 2009.

[4] A. Alessio, M. Lazar, A. Bemporad, and W.P.M.H. Heemels. Squaring the circle: An algorithm for obtaining polyhedral invariant sets from ellipsoidal ones. *Automatica*, 43(12):2096–2103, 2007.

[5] J.C. Allwright and G.C. Papavasiliou. On linear programming and robust model-predictive control using impulse-responses. *Systems & Control Letters*, 18:159–164, 1992.

[6] R. Alur, C. Belta, F. Ivančić, V. Kumar, M. Mintz, G.J. Pappas, H. Rubin, and J. Schug. Hybrid modeling and simulation of biomolecular networks. In M.D. Di Benedetto and A. Sangiovanni Vincentelli, eds., *Hybrid Systems: Computation and Control*, Vol. 2034 of *Lecture Notes in Computer Science*, 19–33. Springer Verlag, 2001.

[7] R. Alur, C. Courcoubetis, T.A. Henzinger, and P.H. Ho. Hybrid automata: An algorithmic approach to the specification and verification of hybrid systems. In R.L. Grossman, A. Nerode, A.P. Ravn, and H. Rischel, eds., *Hybrid Systems*, Vol. 736 of *Lecture Notes in Computer Science*, 209–229. Springer Verlag, 1993.

[8] P.J. Antsaklis. A brief introduction to the theory and applications of hybrid systems. *Proc. IEEE, Special Issue on Hybrid Systems: Theory and Applications*, 88(7): 879–886, July 2000.

[9] L. Armijo. Minimization of functions having Lipschitz continuous first partial derivatives. *Pacific J. of Mathematics*, 16(1):1–3, 1966.

[10] J.P. Aubin. *Viability theory*. Systems & Control: Foundations & Applications. Birkhäuser, 1991.

[11] D. Avis. A revised implementation of the reverse search vertex enumeration algorithm. In Gil Kalai and Günter M. Ziegler, eds., *Polytopes — Combinatorics and Computation*, pages 177–198. Birkhäuser Basel, Basel, 2000.

[12] D. Avis, D. Bremner, and R. Seidel. How good are convex hull algorithms? *Computational Geometry*, 7(5):265–301, 1997.

[13] T.A. Badgwell and K.R. Muske. Disturbance model design for linear model predictive control. In *Proc. American Control Conf.*, Vol. 2, 1621–1626, 2002.

[14] V.L. Bageshwar and F. Borrelli. On a property of a class of offset-free model predictive controllers. *IEEE Trans. Automat. Control*, 54(3):663–669, March 2009.

[15] E. Balas. Projection with a minimum system of inequalities. *Computational Optimization and Applications*, 10:189–193, 1998.

[16] A. Balluchi, L. Benvenuti, M. Di Benedetto, C. Pinello, and A. Sangiovanni-Vincentelli. Automotive engine control and hybrid systems: Challenges and opportunities. *Proc. IEEE*, 88(7):888–912, 2000.

[17] A. Balluchi, L. Benvenuti, M.D. Di Benedetto, and A. Sangiovanni-Vincentelli. Design of observers for hybrid systems. In C.J. Tomlin and M.R. Greenstreet, eds., *Hybrid Systems: Computation and Control*, Vol. 2289 of *Lecture Notes in Computer Science*, 76–89. Springer Verlag, Berlin Heidelberg New York, 2002.

[18] B. Bank, J. Guddat, D. Klatte, B. Kummer, and K. Tammer. *Non-Linear Parametric Optimization*. Akademie-Verlag, Berlin, 1982.

[19] M. Baotić. An efficient algorithm for multi-parametric quadratic programming. Technical Report AUT02-05, Automatic Control Laboratory, ETH Zurich, Switzerland, February 2002.

[20] M. Baotić. Gradient of the value function in parametric convex optimization problems. Technical report, Faculty of Electrical Engineering and Computing. University of Zagreb, Croatia, 2013. URL: https://arxiv.org/abs/1607.00366.

[21] M. Baotić, F. Borrelli, A. Bemporad, and M. Morari. Efficient on-line computation of constrained optimal control. *SIAM J. Control Optim.*, 5:2470–2489, September 2008.

[22] M. Baotić and F.D. Torrisi. Polycover. Technical Report AUT03-11, Automatic Control Laboratory, ETHZ, Switzerland, 2003.

[23] M. Barić. *Constrained Control – Computations, Performance and Robustness*. Dr. sc. thesis, Swiss Federal Institute of Technology (ETH), Zürich, Switzerland, October 2008.

[24] T. Basar and G.J. Olsder. *Dynamic Noncooperative Game Theory*. Classics in Applied Mathematics. SIAM, Philadelphia, 2nd edn., 1998.

[25] F. Bayat, T.A. Johansen, and A.A. Jalali. Using hash tables to manage the time-storage complexity in a point location problem: Application to explicit model predictive control. *Automatica*, 47(3):571–577, 2011.

[26] M.S. Bazaraa, J.J. Jarvis, and H.D. Sherali. *Linear Programming and Network Flows*. John Wiley & Sons, Inc., New York, 4th edn., December 2009.

[27] M.S. Bazaraa, H.D. Sherali, and C.M. Shetty. *Nonlinear Programming: Theory and Algorithms*. John Wiley & Sons, Inc., New York, 2nd edn., 1993.

[28] A.G. Beccuti, G. Papafotiou, R. Frasca, and M. Morari. Explicit hybrid model predictive control of the DC-DC boost converter. In Proc. *IEEE PESC*, Orlando, FL, USA, June 2007.

[29] A. Bemporad. Reducing conservativeness in predictive control of constrained systems with disturbances. In *Proc. 37th IEEE Conf. on Decision and Control*, pages 1384–1391, Tampa, FL, USA, 1998.

[30] A. Bemporad. Efficient conversion of mixed logical dynamical systems into an equivalent piecewise affine form. *IEEE Trans. Automat. Control*, 49(5):832–838, 2004.

[31] A. Bemporad, G. Bianchini, and F. Brogi. Passivity analysis and passification of discrete-time hybrid systems. *IEEE Trans. Automat. Control*, 54(4):1004–1009, 2008.

[32] A. Bemporad, F. Borrelli, and M. Morari. Piecewise linear robust model predictive control. In *Proc. European Control Conf.*, Porto, Portugal, October 2001.

[33] A. Bemporad and S. Di Cairano. Optimal control of discrete hybrid stochastic automata. In M. Morari and L. Thiele, eds., *Hybrid Systems: Computation and Control*, Vol. 3414 of *Lecture Notes in Computer Science*, pages 151–167. Springer Verlag, 2005.

[34] A. Bemporad, A. Casavola, and E. Mosca. Nonlinear control of constrained linear systems via predictive reference management. *IEEE Trans. Automat. Control*, 42(3):340–349, 1997.

[35] A. Bemporad, L. Chisci, and E. Mosca. On the stabilizing property of SIORHC. *Automatica*, 30(12):2013–2015, 1994.

[36] A. Bemporad, G. Ferrari-Trecate, and M. Morari. Observability and controllability of piecewise affine and hybrid systems. *IEEE Trans. Automat. Control*, 45(10):1864–1876, 2000.

[37] A. Bemporad and C. Filippi. An algorithm for approximate multiparametric convex programming. *Comput. Optim. Appl.*, 35(1):87–108, 2006.

[38] A. Bemporad, K. Fukuda, and F.D. Torrisi. Convexity recognition of the union of polyhedra. *Computational Geometry*, 18:141–154, 2001.

[39] A. Bemporad, A. Garulli, S. Paoletti, and A. Vicino. A bounded-error approach to piecewise affine system identification. *IEEE Trans. Automat. Control*, 50(10):1567–1580, October 2005.

[40] A. Bemporad, W.P.M.H. Heemels, and B. De Schutter. On hybrid systems and closed-loop MPC systems. *IEEE Trans. Automat. Control*, 47(5):863–869, May 2002.

[41] A. Bemporad, D. Mignone, and M. Morari. Moving horizon estimation for hybrid systems and fault detection. In *Proc. American Control Conf.*, pages 2471–2475, Chicago, IL, June 1999.

[42] A. Bemporad and M. Morari. Control of systems integrating logic, dynamics, and constraints. *Automatica*, 35(3):407–427, March 1999.

[43] A. Bemporad and M. Morari. Robust model predictive control: A survey. In A. Garulli, A. Tesi, and A. Vicino, eds. *Robustness in Identification and Control*, Vol. 245 of *Lecture Notes in Control and Information Sciences*, pages 207–226. Springer Verlag, 1999.

[44] A. Bemporad, M. Morari, V. Dua, and E.N. Pistikopoulos. The explicit linear quadratic regulator for constrained systems. *Automatica*, 38(1):3–20, January 2002.

[45] A. Bemporad, F.D. Torrisi, and M. Morari. Discrete-time hybrid modeling and verification of the batch evaporator process benchmark. *European J. Control*, 7(4):382–399, 2001.

[46] A. Ben-Tal, A. Goryashko, E. Guslitzer, and A. Nemirovski. Adjustable robust solutions of uncertain linear programs. *Math. Program.*, 99:351–376, 2004.

[47] C. Berge. *Topological Spaces*. Dover Publications, Mineola, NY, 1997.

[48] A.B. Berkelaar, K. Roos, and T. Terlaky. The optimal set and optimal partition approach to linear and quadratic programming. In T. Gal and H.J. Greenberg, eds. *Advances in Sensitivity Analysis and Parametric Programming*, Vol. 6 of *International Series in Operations Research & Management Science*, chapter 6. Kluwer Academic Publishers, 1997.

[49] D.P. Bertsekas. *Control of Uncertain Systems with a set–membership description of the uncertainty.* PhD thesis, MIT, 1971.

[50] D.P. Bertsekas. Infinite-time reachability of state-space regions by using feedback control. *IEEE Trans. Automat. Control*, 17:604–613, October 1972.

[51] D.P. Bertsekas. *Dynamic Programming and Optimal Control.* Athena Scientific, Belmont, Massachusetts, 1995.

[52] D.P. Bertsekas. *Nonlinear Programming.* Athena Scientific, Belmont, Massachusetts, 2nd edn., 1999.

[53] D.P. Bertsekas. *Dynamic Programming and Optimal Control*, Vol. II. Athena Scientific, Belmont, Massachusetts, 2nd edn., 2001.

[54] D.P. Bertsekas and I.B. Rhodes. On the minimax reachability of target sets and target tubes. *Automatica*, 7:233–247, 1971.

[55] L.T. Biegler and V.M. Zavala. Large-scale nonlinear programming using IPOPT: An integrating framework for enterprise-wide dynamic optimization. *Computers & Chemical Engineering*, 33(3):575–582, 2009.

[56] R.R. Bitmead, M. Gevers, and V. Wertz. *Adaptive Optimal Control: The Thinking Man's GPC.* International Series in Systems and Control Engineering. Prentice Hall, 1990.

[57] F. Blanchini. Ultimate boundedness control for uncertain discrete-time systems via set-induced Lyapunov functions. *IEEE Trans. Automat. Control*, 39(2):428–433, February 1994.

[58] F. Blanchini. Set invariance in control — a survey. *Automatica*, 35(11):1747–1768, November 1999.

[59] F. Blanchini and S. Miani. *Set-Theoretic Methods in Control.* Birkhäuser, 2009.

[60] F. Borrelli, M. Baotić, A. Bemporad, and M. Morari. Dynamic programming for constrained optimal control of discrete-time linear hybrid systems. *Automatica*, 41:1709–1721, October 2005.

[61] F. Borrelli, M. Baotić, J. Pekar, and G. Stewart. On the computation of linear model predictive control laws. *Automatica*, 46(6):1035–1041, 2010.

[62] F. Borrelli, A. Bemporad, M. Fodor, and D. Hrovat. An MPC/hybrid system approach to traction control. *IEEE Trans. Control Syst. Tech.*, 14(3):541–552, May 2006.

[63] F. Borrelli, A. Bemporad, and M. Morari. A geometric algorithm for multi-parametric linear programming. *J. Opt. Theory and Applications*, 118(3), September 2003.

[64] S. Boyd, N. Parikh, E. Chu, B. Peleato, and J. Eckstein. Distributed optimization and statistical learning via the alternating direction method of multipliers. *Foundations and Trends in Machine Learning*, 3(1):1–122, 2011.

[65] S. Boyd and L. Vandenberghe. *Convex Optimization.* Cambridge University Press, March 2004.

[66] M.S. Branicky. *Studies in hybrid systems: modeling, analysis, and control.* PhD thesis, LIDS-TH 2304, MIT, Cambridge, MA, 1995.

[67] M.S. Branicky. Multiple Lyapunov functions and other analysis tools for switched and hybrid systems. *IEEE Trans. Automat. Control*, 43(4):475–482, April 1998.

[68] M.S. Branicky, V.S. Borkar, and S.K. Mitter. A unified framework for hybrid control: model and optimal control theory. *IEEE Trans. Automat. Control*, 43(1):31–45, 1998.

[69] M.S. Branicky and S.K. Mitter. Algorithms for optimal hybrid control. In *Proc. 34th IEEE Conf. on Decision and Control*, New Orleans, LA, USA, December 1995.

[70] M.S. Branicky and G. Zhang. Solving hybrid control problems: Level sets and behavioral programming. In *Proc. American Control Conf.*, Chicago, Illinois, USA, June 2000.

[71] M. Buss, O. von Stryk, R. Bulirsch, and G. Schmidt. Towards hybrid optimal control. *AT – Automatisierungstechnik*, 48:448–459, 2000.

[72] R. Cagienard, P. Grieder, E.C. Kerrigan, and M. Morari. Move blocking strategies in receding horizon control. *J. of Proc. Control*, 17(6):563–570, 2007.

[73] S. Di Cairano, A. Bemporad, and J. Júlvez. Event-driven optimization-based control of hybrid systems with integral continuous-time dynamics. *Automatica*, 45(5): 1243–1251, 2009.

[74] S. Di Cairano, A. Bemporad, I. Kolmanovsky, and D. Hrovat. Model predictive control of magnetically actuated mass spring dampers for automotive applications. *Int. J. Control*, 80(11):1701–1716, 2007.

[75] F.M. Callier and C.A. Desoer. *Linear System Theory*. Springer Texts in Electrical Engineering. Springer-Verlag New York, 1991.

[76] E.F. Camacho and C. Bordons. *Model Predictive Control*. Advanced Textbooks in Control and Signal Processing. Springer, London, 1999.

[77] M.K. Camlibel, W.P.M.H. Heemels, and J.M. Schumacher. On linear passive complementarity systems. *European J. of Control*, 8(3):220–237, 2002.

[78] P.J. Campo and M. Morari. Robust model predictive control. In *Proc. American Control Conf.*, Vol. 2, pages 1021–1026, 1987.

[79] P.J. Campo and M. Morari. Model predictive optimal averaging level control. *AIChE J.*, 35(4):579–591, 1989.

[80] C.G. Cassandras, D.L. Pepyne, and Y.Wardi. Optimal control of a class of hybrid systems. *IEEE Trans. Automat. Control*, 46(3):3981–415, 2001.

[81] T.M. Cavalier, P.M. Pardalos, and A.L. Soyster. Modeling and integer programming techniques applied to propositional calculus. *Comput. Oper. Res.*, 17(6):561–570, 1990.

[82] S.N. Cernikov. Contraction of finite systems of linear inequalities (in Russian). *Doklady Akademiia Nauk SSSR*, 152(5):1075–1078, 1963. (English translation in Soviet Mathematics - Doklady, Vol. 4, No. 5 (1963), pp.1520–1524).

[83] V. Chandru and J.N. Hooker. *Optimization methods for logical inference*. Wiley-Interscience, 1999.

[84] H. Chen and F. Allgöwer. A quasi-infinite horizon nonlinear model predictive control scheme with guaranteed stability. *Automatica*, 304(10):1205–1218, 1998.

[85] D. Chmielewski and V. Manousiouthakis. On constrained infinite-time linear quadratic optimal control. *Systems & Control Letters*, 29(3):121–130, November 1996.

[86] D. Christiansen. *Electronics Engineers' Handbook*, 4th edn. IEEE Press/McGraw Hill, Inc., 1997.

[87] F.J. Christophersen. *Optimal Control and Analysis for Constrained Piecewise Affine Systems*. Dr. sc. thesis, ETH, Zürich, Switzerland, August 2006.

[88] F.J. Christophersen and M. Morari. Further Results on 'Infinity Norms as Lyapunov Functions for Linear Systems'. *IEEE Trans. Automat. Control*, 52(3):547–553, March 2007.

[89] D.W. Clarke, C. Mohtadi, and P.S. Tuffs. Generalized Predictive Control – part I. the basic algorithm. *Automatica*, 23(2):137–148, March 1987.

[90] C. R. Cutler. Personal communication to M. Morari, 2010.

[91] C.R. Cutler and B.C. Ramaker. Dynamic matrix control—a computer control algorithm. In *Proc. American Control Conf.*, Vol. WP5-B, San Francisco, CA, USA, 1980.

[92] G.B. Dantzig. *Linear Programming and Extensions*. Princeton University Press, Princeton, NJ, June 1963.

[93] G.B. Dantzig, J. Folkman, and N. Shapiro. On the continuity of the minimum set of a continuous function. *J. of Mathematical Analysis and Applications*, 17:519–548, 1967.

[94] B. De Schutter and B. De Moor. The extended linear complementarity problem and the modeling and analysis of hybrid systems. In P. Antsaklis, W. Kohn, M. Lemmon, A. Nerode, and S. Sastry, eds. *Hybrid Systems V*, Vol. 1567 of *Lecture Notes in Computer Science*, pages 70–85. Springer, 1999.

[95] B. De Schutter and T. van den Boom. Model predictive control for max-plus-linear discrete event systems. *Automatica*, 37(7):1049–1056, July 2001.

[96] R. DeCarlo, M. Branicky, S. Pettersson, and B. Lennartson. Perspectives and results on the stability and stabilizability of hybrid systems. *Proc. IEEE*, 88(7):1069–1082, 2000.

[97] O. Devolder, F. Glineur, and Y. Nesterov. First-order methods of smooth convex optimization with inexact oracle. *Mathematical Programming*, 146:37–75, 2014.

[98] A. Domahidi, A. Zgraggen, M.N. Zeilinger, M. Morari, and C.N. Jones. Efficient interior point methods for multistage problems arising in receding horizon control. In *Proc. 51st IEEE Conf. on Decision and Control*, 668–674, Maui, HI, USA, December 2012.

[99] J.A. De Doná. *Input Constrained Linear Control*. PhD thesis, Control Group, Department of Engineering, University of Cambridge, Cambridge, 2000.

[100] C.E.T. Dórea and J.C. Hennet. (a,b)-invariant polyhedral sets of linear discrete-time systems. *J. Opt. Theory and Applications*, 103(3):521–542, 1999.

[101] V. Dua and E.N. Pistikopoulos. An algorithm for the solution of multiparametric mixed integer linear programming problems. *Ann. Oper. Res.*, 99:123–139, 2000.

[102] G. Ferrari-Trecate, F.A. Cuzzola, D. Mignone, and M. Morari. Analysis of discrete-time piecewise affine and hybrid systems. *Automatica*, 38(12):2139–2146, 2002.

[103] G. Ferrari-Trecate, D. Mignone, and M. Morari. Moving horizon estimation for hybrid systems. *IEEE Trans. Automat. Control*, 47(10):1663–1676, 2002.

[104] G. Ferrari-Trecate, M. Muselli, D. Liberati, and M. Morari. A clustering technique for the identification of piecewise affine systems. *Automatica*, 39(2):205–217, February 2003.

[105] H.J. Ferreau, M. Diehl, and H.G. Bock. An online active set strategy to overcome the limitations of explicit MPC. *Int. J. Robust Nonlinear Control*, 18:816–830, 2008.

[106] A.V. Fiacco. Sensitivity analysis for nonlinear programming using penalty methods. *Math. Program.*, 10(3):287–311, 1976.

[107] A.V. Fiacco. *Introduction to sensitivity and stability analysis in nonlinear programming*. Academic Press, London, UK, 1983.

[108] C.A. Floudas. *Nonlinear and Mixed-Integer Optimization*. Oxford University Press, 1995.

[109] S. Fortune. Voronoi diagrams and Delaunay triangulations. In J.E. Goodman and J. O'Rourke, eds. *Handbook of Discrete and Computational Geometry*, pages 513–528. CRC Press, 2nd edn., 2004.

[110] G. Frison, H.H.B. Sørensen, B. Dammann, and J.B. Jørgensen. High-performance small-scale solvers for linear Model Predictive Control. In *Proc. European Control Conf.*, pages 128–133, June 2014.

[111] K. Fukuda. *cdd, cddplus and cddlib homepage*, 2016. Swiss Federal Institute of Technology, Zurich. URL: http://www.inf.ethz.ch/personal/fukudak/cdd_home/index.html.

[112] K. Fukuda. Polyhedral computation FAQ, 2004. URL: https://www.inf.ethz.ch/personal/fukudak/polyfaq/polyfaq.html.

[113] K. Fukuda, Th.M. Liebling, and C. Lütolf. Extended convex hull. *Computational Geometry*, 20:13–23, 2001.

[114] K. Fukuda and A. Prodon. Double description method revisited. In M. Deza, R. Euler, and I. Manoussakis, eds. *Combinatorics and Computer Science*, Vol. 1120 of *Lecture Notes in Computer Science*, pages 91–111. Springer-Verlag, 1996.

[115] T. Gal. *Postoptimal Analyses, Parametric Programming, and Related Topics*. de Gruyter, Berlin, 2nd edn., 1995.

[116] T. Gal and H.J. Greenberg (Eds.). *Advances in Sensitivity Analysis and Parametric Programming*, Vol. 6 of *International Series in Operations Research & Management Science*. Kluwer Academic Publishers, 1997.

[117] T. Gal and J. Nedoma. Multiparametric linear programming. *Management Science*, 18:406–442, 1972.

[118] C.E. García and A.M. Morshedi. Quadratic programming solution of dynamic matrix control (QDMC). *Chem. Eng. Communications*, 46:73–87, 1986.

[119] S.J. Gartska and R.J.B. Wets. On decision rules in stochastic programming. *Mathematical Programming*, 7:117–143, 1974.

[120] S.I. Gass and T.L. Saaty. The computational algorithm for the parametric objective function. *Naval Research Logistics Quarterly*, 2:39–45, 1955.

[121] T. Geyer. *Low Complexity Model Predictive Control in Power Electronics and Power Systems*. Dr. sc. thesis, ETH, Zürich, Switzerland, March 2005.

[122] T. Geyer, F.D. Torrisi, and M. Morari. Efficient Mode Enumeration of Compositional Hybrid Models. In A. Pnueli and O. Maler, eds. *Hybrid Systems: Computation and Control*, Vol. 2623 of *Lecture Notes in Computer Science*, pages 216–232. Springer Verlag, 2003.

[123] T. Geyer, F.D. Torrisi, and M. Morari. Optimal complexity reduction of polyhedral piecewise affine systems. *Automatica*, 44(7):1728–1740, July 2008.

[124] E.G. Gilbert and K.T. Tan. Linear systems with state and control constraints: the theory and applications of maximal output admissible sets. *IEEE Trans. Automat. Control*, 36(9):1008–1020, 1991.

[125] N. Giorgetti, A. Bemporad, H.E. Tseng, and D. Hrovat. Hybrid model predictive control application towards optimal semi-active suspension. *Int. J. Control*, 79(5):521–533, 2006.

[126] F. Glover. Improved linear integer programming formulations of nonlinear integer problems. *Management Science*, 22(4):455–460, 1975.

[127] R. Goebel, R.G. Sanfelice, and A.R. Teel. Hybrid dynamical systems. *IEEE Control Systems Magazine*, 29(2):28–93, 2009.

[128] K. Gokbayrak and C.G. Cassandras. A hierarchical decomposition method for optimal control of hybrid systems. In *Proc. 38th IEEE Conf. on Decision and Control*, pages 1816–1821, Phoenix, AZ, December 1999.

[129] G.C. Goodwin and K. S. Sin. *Adaptive Filtering, Prediction and Control*. Prentice-Hall, Englewood Cliffs, NJ, 1984.

[130] P.J. Goulart, E.C. Kerrigan, and J.M. Maciejowski. Optimization over state feedback policies for robust control with constraints. *Automatica*, 42(4):523–533, 2006.

[131] P. Grieder. *Efficient Computation of Feedback Controllers for Constrained Systems*. Dr. sc. thesis, ETH, Zürich, Switzerland, 2004.

[132] P. Grieder, F. Borrelli, F.D. Torrisi, and M. Morari. Computaion of the constrained infinite horizon linear quadratic regulator. Technical Report AUT02-09, Automatic Control Laboratory, ETH Zurich, Switzerland, July 2002.

[133] P. Grieder, F. Borrelli, F.D. Torrisi, and M. Morari. Computation of the constrained infinite time linear quadratic regulator. *Automatica*, 40(4):701–708, April 2004.

[134] P. Grieder, M. Kvasnica, M. Baotić, and M. Morari. Low complexity control of piecewise affine systems with stability guarantee. In *Proc. American Control Conf.*, pages 1196–1201, Boston, USA, June 2004.

[135] B. Grünbaum. *Convex Polytopes*. Springer Verlag, 2nd edn., 2003.

[136] E. Guslitzer. Uncertainty-immunized solutions in linear programming. Master's thesis, Technion (Israel Institute of Technology), Haifa, Israel, 2002.

[137] P.O. Gutman. Online use of a linear programming controller. In G. Ferrate and E.A. Puente, eds. *Software for Computer Control 1982. IFAC/IFIP Symposium*, pages 313–318. Pergamon, Oxford, 1983.

[138] P.O. Gutman. A linear programming regulator applied to hydroelectric reservoir level control. *Automatica*, 22(5):533–541, 1986.

[139] P.O. Gutman and M. Cwikel. Admissible sets and feedback control for discrete-time linear dynamical systems with bounded control and states. *IEEE Trans. Automat. Control*, 31(4):373–376, 1986.

[140] P.O. Gutman and M. Cwikel. An algorithm to find maximal state constraint sets for discrete-time linear dynamical systems with bounded control and states. *IEEE Trans. Automat. Control*, 32(3):251–254, 1987.

[141] A. Hassibi and S. Boyd. Quadratic stabilization and control of piecewise-linear systems. In *Proc. American Control Conf.*, Philadelphia, Pennsylvania, USA, June 1998.

[142] J.P. Hayes. *Introduction to Digital Logic Design*. Addison-Wesley Publishing Company, Inc., 1993.

[143] S. Hedlund and A. Rantzer. Optimal control of hybrid systems. In *Proc. 38th IEEE Conf. on Decision and Control*, pages 3972–3976, Phoenix, AZ, December 1999.

[144] S. Hedlund and A. Rantzer. Convex dynamic programming for hybrid systems. *IEEE Trans. Automat. Control*, 47(9):1536–1540, September 2002.

[145] W.P.M.H Heemels, B. de Schutter, and A. Bemporad. On the equivalence of classes of hybrid dynamical models. In *Proc. 40th IEEE Conf. on Decision and Control*, pages 364–369, Orlando, Florida, 2001.

[146] W.P.M.H. Heemels. *Linear complementarity systems: a study in hybrid dynamics*. PhD thesis, Dept. of Electrical Engineering, Eindhoven University of Technology, The Netherlands, 1999.

[147] W.P.M.H. Heemels, J.M. Schumacher, and S. Weiland. Linear complementarity systems. *SIAM J. Appl. Math.*, 60(4):1234–1269, 2000.

[148] W.P.M.H. Heemels, B. de Schutter, and A. Bemporad. Equivalence of hybrid dynamical models. *Automatica*, 37(7):1085–1091, July 2001.

[149] M. Herceg, M. Kvasnica, C.N. Jones, and M. Morari. Multi-Parametric Toolbox 3.0. In *Proc. European Control Conf.*, pages 502–510, Zürich, Switzerland, July 17–19 2013.

[150] J.P. Hespanha, S. Bohacek, K. Obraczka, and J. Lee. Hybrid modeling of TCP congestion control. In M.D. Di Benedetto and A. Sangiovanni Vincentelli, eds. *Hybrid Systems: Computation and Control*, Vol. 2034 of *Lecture Notes in Computer Science*, pages 291–304. Springer Verlag, 2001.

[151] J.-B. Hiriart-Urruty and C. Lemaréchal. *Fundamentals of Convex Analysis*. Springer, 2001.

[152] W.M. Hogan. Point-to-set maps in mathematical programming. *SIAM Rev.*, 15(3):591–603, July 1973.

[153] J.N. Hooker. *Logic-Based Methods for Optimization: Combining Optimization and Constraint Satisfaction*. Wiley, New York, 2000.

[154] A. Jadbabaie, Y. Jie, and J. Hauser. Stabilizing receding horizon control of nonlinear systems: a control Lyapunov function approach. In *Proc. American Control Conf.*, June 1999.

[155] M. Johannson and A. Rantzer. Computation of piece-wise quadratic Lyapunov functions for hybrid systems. *IEEE Trans. Automat. Control*, 43(4):555–559, 1998.

[156] T.A. Johansen and A. Grancharova. Approximate explicit constrained linear model predictive control via orthogonal search tree. *IEEE Trans. Automat. Control*, 48(5):810–815, May 2003.

[157] T.A. Johansen, J. Kalkkuhl, J. Lüdemann, and I. Petersen. Hybrid control strategies in ABS. In *Proc. American Control Conf.*, Arlington, VA, June 2001.

[158] K.H. Johansson, M. Egerstedt, J. Lygeros, and S. Sastry. On the regularization of Zeno hybrid automata. *Systems & Control Letters*, 38:141–150, 1999.

[159] C.N. Jones, M. Barić, and M. Morari. Multiparametric Linear Programming with Applications to Control. *European J. Control*, 13(2-3):152–170, March 2007.

[160] C.N. Jones, P. Grieder, and S. Raković. A Logarithmic-Time Solution to the Point Location Problem for Closed-Form Linear MPC. In *Proc. IFAC World Congress*, Prague, Czech Republic, July 2005.

[161] C.N. Jones, E.C. Kerrigan, and J.M. Maciejowski. Equality set projection: A new algorithm for the projection of polytopes in halfspace representation. Technical Report CUED Technical Report CUED/F-INFENG/TR.463, Department of Engineering, Cambridge University, UK, 2004.

[162] C.N. Jones and M. Morari. Polytopic approximation of explicit model predictive controllers. *IEEE Trans. Automat. Control*, 55(11):2542–2553, November 2010.

[163] P. Julian, M. Jordan, and A. Desages. Canonical piecewise-linear approximation of smooth functions. *IEEE Trans. Circuits and Systems — I: Fundamental Theory and Applications*, 45(5):567–571, May 1998.

[164] A.A. Julius and A.J. van der Schaft. The maximal controlled invariant set of switched linear systems. In *Proc. 41st IEEE Conf. on Decision and Control*, Las Vegas, NV, USA, December 2002.

[165] A. Juloski, S. Weiland, and M. Heemels. A Bayesian approach to identification of hybrid systems. In *Proc. 43rd IEEE Conf. on Decision and Control*, 2004.

[166] N.K. Karmarkar. A new polynomial-time algorithm for linear programming. *Combinatorica*, 4:373–395, 1984.

[167] S. Keerthi and E. Gilbert. Computation of minimum-time feedback control laws for discrete-time systems with state-control constraints. *IEEE Trans. Automat. Control*, 32(5):432–435, 1987.

[168] S.S. Keerthi and E.G. Gilbert. Optimal infinite-horizon feedback control laws for a general class of constrained discrete-time systems: stability and moving-horizon approximations. *J. Opt. Theory and Applications*, 57:265–293, 1988.

[169] S.S. Keerthi and K. Sridharan. Solution of parametrized linear inequalities by Fourier elimination and its applications. *J. Opt. Theory and Applications*, 65(1): 161–169, 1990.

[170] A. Kelman and F. Borrelli. Parallel nonlinear predictive control. In *Proc. 50th Annual Allerton Conf. on Communication, Control, and Computing (Allerton)*, pages 71–78, Oct 2012.

[171] A. Kelman, J. Kong, S. Vichik, K. Chiang, and F. Borrelli. BLOM: The berkeley library for optimization modeling. In *Proc. American Control Conf.*, pages 2900–2905, June 2014.

[172] E.C. Kerrigan. *Robust Constraint Satisfaction: Invariant Sets and Predictive Control*. PhD thesis, Department of Engineering, University of Cambridge, Cambridge, UK, 2000.

[173] E.C. Kerrigan and J.M. Maciejowski. Soft constraints and exact penalty functions in model predictive control. In *Proc. UKACC International Conf. (Control 2000)*, Cambridge, UK, September 2000.

[174] E.C. Kerrigan and J.M. Maciejowski. Designing model predictive controllers with prioritised constraints and objectives. In *Proc. IEEE International Symposium on Computer Aided Control System Design*, pages 33–38, 2002.

[175] E.C. Kerrigan and J.M. Maciejowski. On robust optimization and the optimal control of constrained linear systems with bounded state disturbances. In *Proc. European Control Conf.*, Cambridge, UK, September 2003.

[176] H.K. Khalil. *Nonlinear Systems*. Prentice Hall, 2nd edn, 1996.

[177] H. Kiendl, J. Adamy, and P. Stelzner. Vector norms as Lyapunov functions for linear systems. *IEEE Trans. Automat. Control*, 37(6):839–842, June 1992.

[178] D. Klatte and G. Thiere. Error bounds for solutions of linear equations and inequalities. *ZOR - Math. Methods Oper. Res.*, 41:191–214, 1995.

[179] I. Kolmanovsky and E.G. Gilbert. Theory and computation of disturbance invariant sets for discrete-time linear systems. *Math. Probl. Eng.*, 4:317–367, 1998.

[180] M.V. Kothare, V. Balakrishnan, and M. Morari. Robust constrained model predictive control using linear matrix inequalities. *Automatica*, 32(10):1361–1379, 1996.

[181] B. Kouvaritakis, J.A. Rossiter, and J. Schuurmans. Efficient robust predictive control. *IEEE Trans. Automat. Control*, 45(8):1545–1549, 2000.

[182] G. Lafferriere, G.J. Pappas, and S. Sastry. Reachability analysis of hybrid systems using bisimulations. In *Proc. 37th IEEE Conf. on Decision and Control*, pages 1623–1628, Tampa, FL, USA, 1998.

[183] J.P. LaSalle. Stability theory for difference equations. In Jack Hale, eds. *Studies in Ordinary Differential Equations*, Vol. 14 of MAA *Studies in Mathematics*, pages 1–31. Mathematical Assoc. of Amer., 1977.

[184] J.P. LaSalle. *The Stability and Control of Discrete Processes*, Vol. 62 of *Applied Mathematical Sciences*. Springer Verlag, New York, 1986.

[185] M. Lazar and W.P.M.H. Heemels. Predictive control of hybrid systems: Input-to-state stability results for sub-optimal solutions. *Automatica*, 45(1):180–185, 2009.

[186] M. Lazar, W.P.M.H. Heemels, and A.R. Teel. Lyapunov functions, stability and input-to-state stability subtleties for discrete-time discontinuous systems. *IEEE Trans. Automat. Control*, 54(10):2421–2425, 2009.

[187] M. Lazar, W.P.M.H. Heemels, S. Weiland, and A. Bemporad. Stabilizing model predictive control of hybrid systems. *IEEE Trans. Automat. Control*, 51(11): 1813–1818, 2006.

[188] M. Lazar, D. M. de la Peña, W.P.M.H. Heemels, and T. Alamo. On input-to-state stability of min-max nonlinear model predictive control. *Systems & Control Letters*, 57:39–48, 2008.

[189] J. H. Lee and Z. Yu. Worst-case formulations of model predictive control for systems with bounded parameters. *Automatica*, 33(5):763–781, 1997.

[190] F.L. Lewis and V.L. Syrmos. *Optimal Control.* John Wiley & Sons, Inc., New York, 1995.

[191] D. Liberzon. *Switching in Systems and Control.* Systems and Control: Foundations and Application. Birkhäuser, Boston, MA, June 2003.

[192] B. Lincoln and A. Rantzer. Optimizing linear system switching. In *Proc. 40th IEEE Conf. on Decision and Control*, pages 2063–2068, Orlando, FL, USA, 2001.

[193] Y.-C. Liu and C. B. Brosilow. Simulation of large scale dynamic systems—I. modular integration methods. *Computers & Chemical Engineering*, 11(3):241–253, 1987.

[194] J. Löfberg. *Minimax approaches to robust model predictive control.* PhD thesis, Linköping University, Sweden, April 2003.

[195] J. Lygeros, D.N. Godbole, and S. Sastry. A game theoretic approach to hybrid system design. In R. Alur and T. Henzinger, eds. *Hybrid Systems III*, Vol. 1066 of *Lecture Notes in Computer Science*, pages 1–12. Springer Verlag, 1996.

[196] J. Lygeros, C. Tomlin, and S. Sastry. Controllers for reachability specifications for hybrid systems. *Automatica*, 35(3):349–370, 1999.

[197] J.M. Maciejowski. *Predictive Control with Constraints.* Prentice Hall, 2002.

[198] U. Maeder, F. Borrelli, and M. Morari. Linear offset-free Model Predictive Control. *Automatica*, 45(10):2214–2222, 2009.

[199] S. Mahapatra. *Stability of Hybrid Haptic Systems.* PhD thesis, University of Illinois, Chicago, Illinois, 2003.

[200] O.L. Mangasarian and J.B. Rosen. Inequalities for stochastic nonlinear programming problems. *Operations Research*, 12(1):143–154, 1964.

[201] J. Mattingley and S. Boyd. CVXGEN: A code generator for embedded convex optimization. *Optimization and Engineering*, 13:1–27, 2012.

[202] D.Q. Mayne. Constrained Optimal Control. European Control Conference, Seminario de Vilar, Porto, Portugal, Plenary Lecture, September 2001.

[203] D.Q. Mayne. Control of constrained dynamic systems. *European J. of Control*, 7:87–99, 2001.

[204] D.Q. Mayne, J.B. Rawlings, C.V. Rao, and P.O.M. Scokaert. Constrained model predictive control: Stability and optimality. *Automatica*, 36(6):789–814, June 2000.

[205] T.A. Meadowcroft, G. Stephanopoulos, and C. Brosilow. The Modular Multivariable Controller: 1: Steady-state properties. *AIChE J.*, 38(8):1254–1278, 1992.

[206] S. Mehrotra. On the implementation of a primal-dual interior point method. *SIAM J. on Optimization*, 2(4):575–601, November 1992.

[207] S. Mehrotra and R.D.C. Monteiro. Parametric and range analysis for interior point methods. Technical report, Dept. of Systems and Industrial Engineering, University of Arizona, Tucson, USA, 1992.

[208] E. Mendelson. *Introduction to mathematical logic.* Van Nostrand, 1964.

[209] J.A. Mendez, B. Kouvaritakis, and J.A. Rossiter. State space approach to interpolation in MPC. *Int. J. Robust Nonlinear Control*, 10(1):27–38, January 2000.

[210] D. Mignone. *Control and Estimation of Hybrid Systems with Mathematical Optimization*. Dr. sc. thesis, ETH, Zürich, Switzerland, 2002.

[211] D. Mignone, A. Bemporad, and M. Morari. A framework for control, fault detection, state estimation and verification of hybrid systems. In *Proc. American Control Conf.*, pages 134–138, June 1999.

[212] R. Milman and E.J. Davison. A fast MPC algorithm using nonfeasible active set methods. *J. Opt. Theory and Applications*, 139(3):591–616, 2008.

[213] G. Mitra, C. Lucas, and S. Moody. Tool for reformulating logical forms into zero-one mixed integer programs. *European J. Oper. Res.*, 71:262–276, 1994.

[214] R. Möbus, M. Baotić, and M. Morari. Multi-objective adaptive cruise control. In O. Maler and A. Pnueli, eds. *Hybrid Systems: Computation and Control*, Vol. 2623 of *Lecture Notes in Computer Science*, pages 359–374. Springer Verlag, 2003.

[215] M. Morari and G. Stephanopoulos. Minimizing unobservability in inferential control schemes. *Int. J. Control*, 31:367–377, 1980.

[216] M. Morari and G. Stephanopoulos. Studies in the synthesis of control structures for chemical processes; Part III: Optimal selection of secondary measurements within the framework of state estimation in the presence of persistent unknown disturbances. *AIChE J.*, 26:247–260, 1980.

[217] K.G. Murty. *Linear Programming*. Wiley-Interscience Publication, 1983. 1st edn.

[218] K.R. Muske and T.A. Badgwell. Disturbance modeling for offset-free linear model predictive control. *J. of Proc. Control*, 12:617–632, 2002.

[219] G.L. Nemhauser and L.A. Wolsey. *Integer and Combinatorial Optimization*. Wiley, 1988.

[220] Y. Nesterov. *Introductory Lectures on Convex Optimization. A Basic Course.* Springer, 2004.

[221] Y. Nesterov. Smooth minimization of non-smooth functions. *Mathematical Programming*, 103(1):127–152, 2005.

[222] Y. Nesterov and A. Nemirovski. *Interior-Point Polynomial Algorithms in Convex Programming*. Society for Industrial Mathematics, 1994.

[223] J. Nocedal and S.J. Wright. *Numerical Optimization*. Springer, New York, 2nd edn., 2006.

[224] J.M. Ortega and W.C. Rheinboldt. *Iterative solution of nonlinear equations in several variables*, Vol. 30. Society for Industrial and Applied Mathematics, 1987.

[225] G. Pannocchia. Robust disturbance modeling for model predictive control with application to multivariable ill-conditioned processes. *J. of Proc. Control*, 13(8):693–701, 2003.

[226] G. Pannocchia and J.B. Rawlings. Disturbance models for offset-free model predictive control. *AIChE J.*, 49(2):426–437, 2003.

[227] S. Paoletti. *Identification of Piecewise Affine Models*. PhD thesis, Dept. Information Engineering, University of Siena, Italy, 2004.

[228] N. Parikh and S. Boyd. Proximal algorithms. *Foundations and Trends in optimization*, 1(3):123–231, 2013.

[229] T. Park and P.I. Barton. Implicit model checking of logic-based control systems. *AIChE J.*, 43(9):2246–2260, 1997.

[230] P.A. Parrilo. *Structured semidefinite programs and semialgebraic geometry methods in robustness and optimization.* Ph.D. thesis, California Institute of Technology, Pasadena, CA, USA, 2000.

[231] P. Patrinos and H. Sarimveis. Convex parametric piecewise quadratic optimization: Theory and Algorithms. *Automatica*, 47(8):1770–1777, 2011.

[232] D. M. De La Pena, A. Bemporad, and C. Filippi. Robust explicit MPC based on approximate multiparametric convex programming. *IEEE Trans. Automat. Control*, 51(8):1399–1403, 2006.

[233] S. Pettersson and B. Lennartson. Stability and robustness for hybrid systems. In *Proc. 35th IEEE Conf. on Decision and Control*, pages 1202–1207, Kobe, Japan, 1996.

[234] S. Pettersson and B. Lennartson. Exponential stability of hybrid systems using piecewise quadratic Lyapunov functions resulting in LMIs. In *Proc. IFAC World Congress*, pages 103–108, Beijing, China, July 1999.

[235] B. Piccoli. Necessary conditions for hybrid optimization. In *Proc. 38th IEEE Conf. on Decision and Control*, Phoenix, AZ, USA, December 1999.

[236] A. Pogromski, M. Jirstrand, and P. Spangeus. On stability and passivity of a class of hybrid systems. In *Proc. 37th IEEE Conf. on Decision and Control*, pages 3705–3710, Tampa, Florida, USA, 1998.

[237] A. Polanski. On infinity norms as Lyapunov functions for linear systems. *IEEE Trans. Automat. Control*, 40(7):1270–1274, July 1995.

[238] S. Prajna and A. Papachristodoulou. Analysis of switched and hybrid systems – beyond piecewise quadratic methods. In *Proc. American Control Conf.*, Vol. 4, pages 2779–2784, 2003.

[239] A.I. Propoi. Use of linear programming methods for synthesizing sampled–data automatic systems. *Autom. Remote Control*, 24(7):837–844, 1963.

[240] S.J. Qin and T.A. Badgwell. An overview of industrial model predictive control technology. In *Chemical Process Control - V*, Vol. 93, no. 316, pages 232–256. AIChE Symposium Series - American Institute of Chemical Engineers, 1997.

[241] S.J. Qin and T.A. Badgwell. A survey of industrial model predictive control technology. *Control Engineering Practice*, 11:733–764, 2003.

[242] S.V. Raković, P. Grieder, M. Kvasnica, D.Q. Mayne, and M. Morari. Computation of invariant sets for piecewise affine discrete time systems subject to bounded disturbances. In *Proc. 43rd IEEE Conf. on Decision and Control*, pages 1418–1423, December 2004.

[243] S.V. Rakovic, E.C. Kerrigan, and D.Q. Mayne. Reachability computations for constrained discrete-time systems with state- and input-dependent disturbances. In *Proc. 42nd IEEE Conf. on Decision and Control*, pages 3905–3910, December 2003.

[244] R. Raman and I.E. Grossmann. Relation between MILP modeling and logical inference for chemical process synthesis. *Computers & Chemical Engineering*, 15(2):73–84, 1991.

[245] A. Rantzer and M. Johansson. Piecewise linear quadratic optimal control. *IEEE Trans. Automat. Control*, 45(4):629–637, April 2000.

[246] C.V. Rao, S.J. Wright, and J.B. Rawlings. Application of Interior-Point methods to model predictive control. *J. Opt. Theory and Applications*, 99(3):723–757, December 1998.

[247] J.B. Rawlings and D.Q. Mayne. *Model Predictive Control: Theory and Design.* Nob Hill Publishing, 2009.

[248] J. Richalet, A. Rault, J.L. Testud, and J. Papon. Algorithmic control of industrial processes. In *Proc. 4th IFAC symposium on identification and system parameter estimation*, Vol. WP5-B, pages 1119–1167, 1976.

[249] J. Richalet, A. Rault, J.L. Testud, and J. Papon. Model predictive heuristic control-application to industrial processes. *Automatica*, 14:413–428, 1978.

[250] S. Richter. *Computational Complexity Certification of Gradient Methods for Real-Time Model Predictive Control*. Dr. sc. thesis, ETH, Zurich, Switzerland, November 2012.

[251] S. Richter, C.N. Jones, and M. Morari. Certification Aspects of the Fast Gradient Method for Solving the Dual of Parametric Convex Programs. *Math. Methods Oper. Res.*, 77(3):305–321, January 2013.

[252] P. Riedinger, F.Kratz, C. Iung, and C. Zanne. Linear quadratic optimization for hybrid systems. In *Proc. 38th IEEE Conf. on Decision and Control*, Phoenix, Arizona USA, December 1999.

[253] S.M. Robinson. Some continuity properties of polyhedral multifunctions. *Mathematical Programing Study*, 14:206–214, 1981.

[254] S.M. Robinson and R.H. Day. A sufficient condition for continuity of optimal sets in mathematical programming. *J. Math. Anal. Appl.*, 45:506–511, 1974.

[255] J. Roll, A. Bemporad, and L. Ljung. Identification of piecewise affine systems via mixed-integer programming. *Automatica*, 40(1):37–50, 2004.

[256] L.O. Santos, P.A.F.N.A. Afonso, J.A.A.M. Castro, N.M.C. Oliveira, and L.T. Biegler. On-line implementation of nonlinear MPC: an experimental case study. *Control Engineering Practice*, 9(8):847–857, 2001.

[257] M. Schechter. Polyhedral functions and multiparametric linear programming. *J. Opt. Theory and Applications*, 53(2):269–280, May 1987.

[258] C. Scherer and S. Weiland. Linear Matrix Inequalities in Control. Technical report, Center for Systems and Control, Delft University of Technology, The Netherlands, January 2005. Available from http://goo.gl/D82c0Y.

[259] P.O.M. Scokaert and D.Q. Mayne. Min-max feedback model predictive control for constrained linear systems. *IEEE Trans. Automat. Control*, 43(8):1136–1142, 1998.

[260] P.O.M. Scokaert and J.B. Rawlings. Constrained linear quadratic regulation. *IEEE Trans. Automat. Control*, 43(8):1163–1169, 1998.

[261] C. Seatzu, D. Corona, A. Giua, and A. Bemporad. Optimal control of continuous-time switched affine systems. *IEEE Trans. Automat. Control*, 51(5):726–741, 2006.

[262] M.M. Seron, J.A. DeDoná, and G.C. Goodwin. Global analytical model predictive control with input constraints. In *Proc. 39th IEEE Conf. on Decision and Control*, pages 154–159, 2000.

[263] J. Serra. *Image Analysis and Mathematical Morphology, Vol II: Theoretical Advances*. Academic Press, 1988.

[264] M.S. Shaikh and P.E. Caines. On the optimal control of hybrid systems: optimization of trajectories, switching times and location schedules. In *6th Int. Workshop on Hybrid Systems: Computation and Control*, Prague, The Czech Republic, 2003.

[265] B.I. Silva, O. Stursberg, B.H. Krogh, and S. Engell. An assessment of the current status of algorithmic approaches to the verification of hybrid systems. In *Proc. 40th IEEE Conf. on Decision and Control*, pages 2867–2874, Orlando, FL, December 2001.

[266] E.D. Sontag. Nonlinear regulation: The piecewise linear approach. *IEEE Trans. Automat. Control*, 26(2):346–358, April 1981.

[267] E.D. Sontag. Interconnected automata and linear systems: A theoretical framework in discrete-time. In R. Alur, T.A. Henzinger, and E.D. Sontag, eds. *Hybrid Systems III—Verification and Control*, number 1066 in Lecture Notes in Computer Science, pages 436–448. Springer Verlag, 1996.

[268] J. Spjotvold, E.C. Kerrigan, C.N. Jones, P. Tøndel, and T.A Johansen. On the facet-to-facet property of solutions to convex parametric quadratic programs. *Automatica*, 42(12):2209–2214, December 2006.

[269] R. Suard, J. Löfberg, P. Grieder, M. Kvasnica, and M. Morari. Efficient computation of controller partitions in multi-parametric programming. In *Proc. 43rd IEEE Conf. on Decision and Control*, pages 3643–3648, Bahamas, December 2004.

[270] S. Summers, C.N. Jones, J. Lygeros, and M. Morari. A multiresolution approximation method for fast explicit model predictive control. *IEEE Trans. Automat. Control*, 56(11):2530–2541, Nov 2011.

[271] H.J. Sussmann. A maximum principle for hybrid optimal control problems. In *Proc. 38th IEEE Conf. on Decision and Control*, Phoenix, Arizona USA, December 1999.

[272] M. Sznaier and M.J. Damborg. Suboptimal control of linear systems with state and control inequality constraints. In *Proc. 26th IEEE Conf. on Decision and Control*, Vol. 1, pages 761–762, 1987.

[273] C.J. Tomlin, J. Lygeros, and S.S. Sastry. A game theoretic approach to controller design for hybrid systems. *Proc. IEEE*, 88(7):949–970, July 2000.

[274] P. Tøndel, T.A. Johansen, and A. Bemporad. An algorithm for multi-parametric quadratic programming and explicit MPC solutions. In *Proc. 40th IEEE Conf. on Decision and Control*, December 2001.

[275] P. Tøndel, T.A. Johansen, and A. Bemporad. Evaluation of piecewise affine control via binary search tree. *Automatica*, 39(5):945–950, 2003.

[276] F.D. Torrisi and A. Bemporad. HYSDEL — A tool for generating computational hybrid models. *IEEE Trans. Control Syst. Tech.*, 12(2):235–249, March 2004.

[277] M.L. Tyler and M. Morari. Propositional logic in control and monitoring problems. *Automatica*, 35(4):565–582, 1999.

[278] V.I. Utkin. Variable structure systems with sliding modes. *IEEE Trans. Automat. Control*, 22(2):212–222, April 1977.

[279] A.J. van der Schaft and J.M. Schumacher. Complementarity modelling of hybrid systems. *IEEE Trans. Automat. Control*, 43:483–490, 1998.

[280] R.J. Vanderbei. *Linear Programming*, Vol. 196 of *International Series in Operations Research & Management Science*. Springer US, Boston, MA, 2014.

[281] D.H. van Hessem and O.H. Bosgra. A conic reformulation of model predictive control including bounded and stochastic disturbances under state and input constraints. In *Proc. 41st IEEE Conf. on Decision and Control*, pages 4643–4648, Las Vegas, NV, USA, 2002.

[282] R. Vidal, S. Soatto, Y. Ma, and S. Sastry. An algebraic geometric approach to the identification of a class of linear hybrid systems. In *Proc. 42nd IEEE Conf. on Decision and Control*, pages 167–172, Maui, Hawaii, 2003.

[283] D.W. Walkup and R.J.-B. Wets. A Lipschitzian characterizations of convex polyhedra. *Proc. American Mathematical Society*, 20:167–173, 1969.

[284] Y. Wang and S. Boyd. Fast model predictive control using online optimization. *IEEE Trans. Control Syst. Tech.*, 18(2):267 –278, March 2010.

[285] J. Warren, S. Schaefer, A. Hirani, and M. Desbrun. Barycentric coordinates for convex sets. *Adv. Comput. Math.*, 27(3):319–338, 2007.

[286] H.P. Williams. Logical problems and integer programming. *Bulletin of the Institute of Mathematics and Its Applications*, 13:18–20, 1977.

[287] H.P. Williams. *Model Building in Mathematical Programming*. John Wiley & Sons, 3rd edn., 1993.

[288] H. Witsenhausen. A class of hybrid-state continuous-time dynamic systems. *IEEE Trans. Automat. Control*, 11(2):161–167, 1966.

[289] H.S. Witsenhausen. A min-max control problem for sampled linear systems. *IEEE Trans. Automat. Control*, 13(1):5–21, 1968.

[290] S.J. Wright. *Primal-Dual Interior Point Methods*. SIAM, 1997.

[291] X. Xu and P.J. Antsaklis. Results and perspectives on computational methods for optimal control of switched systems. In O. Maler and A. Pnueli, eds. *Hybrid Systems: Computation and Control*, Vol. 2623 of *Lecture Notes in Computer Science*, pages 540–555. Springer Verlag, 2003.

[292] X. Xu and P.J. Antsaklis. Optimal control of switched systems based on parameterization of the switching instants. *IEEE Trans. Automat. Control*, 49(1):2–16, 2004.

[293] L.A. Zadeh and L.H. Whalen. On optimal control and linear programming. *IRE Trans. Automat. Control*, 7:45–46, 1962.

[294] E. Zafiriou and M. Morari. A general controller synthesis methodology based on the IMC structure and the H_2-, H_∞- and μ-optimal control theories. *Computers & Chemical Engineering*, 12(7):757–765, 1988.

[295] M. Zefran, F. Bullo, and M. Stein. A notion of passivity for hybrid systems. In *Proc. 40th IEEE Conf. on Decision and Control*, pages 768–773, Orlando, FL, 2001.

[296] G.M. Ziegler. *Lectures on Polytopes*. Graduate Texts in Mathematics. Springer Verlag, 1994.

Index

Printed in the United States
by Bookmasters

Printed in the United States
By Bookmasters